To Reach the High Frontier

TO
REACH
THE HIGH
FRONTIER

A History of U.S. Launch Vehicles

Edited by Roger D. Launius and Dennis R. Jenkins

THE UNIVERSITY PRESS OF KENTUCKY

Publication of this volume was made possible in part by
a grant from the National Endowment for the Humanities.

Editorial and Sales Offices: The University Press of Kentucky
663 South Limestone Street, Lexington, Kentucky 40508-4008
www.kentuckypress.com

11 10 09 08 07   5 4 3 2

Library of Congress Cataloging-in-Publication Data

Launius, Roger D.
To reach the high frontier : a history of U.S. launch vehicles /
Roger D. Launius and Dennis R. Jenkins.
p. cm.
Includes bibliographical references and index.
ISBN-10: 0-8131-2245-7 (cloth alk. paper)
ISBN-13: 978-0-8131-2245-8 (cloth alk. paper)
1. Launch vehicles (Astronautics)—United States—History. 2.
Rocketry—United States—History. I. Jenkins, Dennis R. II. Title.
TL785.8.L3 L385 2002
629.47'5—dc21
2002011433

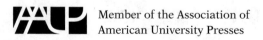

For J.W.

Contents

Introduction: Episodes in the Evolution of Launch
Vehicle Technology
Roger D. Launius   1

1. Rocketry and the Origins of Space Flight
   Ray A. Williamson and Roger D. Launius   33

2. Stage-and-a-Half: The Atlas Launch Vehicle
   Dennis R. Jenkins   70

3. Delta: The Ultimate Thor
   Kevin S. Forsyth   103

4. Titan: Some Heavy Lifting Required
   Roger D. Launius   147

5. History and Development of U.S. Small Launch
   Vehicles
   *Matt Bille, Pat Johnson, Robyn Kane, and
   Erika R. Lishock*   186

6. Minuteman and the Development of Solid Rocket Launch Technology
   J.D. Hunley   229

7. The Biggest of Them All: Reconsidering the Saturn V
   Ray A. Williamson   301

8. Taming Liquid Hydrogen: The Centaur Saga
   Virginia P. Dawson   334

9. Broken in Midstride: Space Shuttle as a Launch Vehicle
   Dennis R. Jenkins   357

10. Eclipsed by Tragedy: The Fated Mating of the Shuttle and Centaur
    Mark D. Bowles   415

11. The Quest for Reusability
    Andrew J. Butrica   443

12. Epilogue: "To the Very Limit of Our Ability": Reflections on Forty Years of Military-Civil Partnership in Space Launch
    David N. Spires and Rick W. Sturdevant   470

    Notes on Contributors   502

    Index   505

Introduction

Episodes in the Evolution of Launch Vehicle Technology

Roger D. Launius

Access. No single word better describes the primary concern of everyone interested in the exploration and development of space. Every participant in space activities—civil, military, or commercial—needs affordable, reliable, frequent, and flexible access to space. Comparisons with Earth-based activities are illustrative of the problems encountered with the current situation in space access.

Terrestrial shipping services deliver on a fairly predictable schedule, function in a variety of climates, often book shipments on short notice, and hardly ever destroy the cargo. The same cannot be said for space launch services. Delivery is frequently held up by technical glitches and adverse weather, flights need to be booked two to four years in advance, and reliability rates for large launchers range from about 82 percent to 99 (shuttle) percent. Clearly, this situation must improve.

After more than four decades of effort, access to space remains one of the most difficult challenges for the spacefaring industry in the United States. While the issue is international, this book is oriented toward assessing the development of American launchers, and with due recognition to the launch vehicles of other nations, this work does not discuss their development. Of course, everyone recognizes that because of the extraordinary difficulties involved, space transport services should not be measured by terrestrial standards. But if the grand plans of space visionaries and entrepreneurs are to be carried out, there is a real need to move beyond currently available technologies. Unfortunately, the high cost associated with space launch from 1950 to 2000 has demonstrated the slowest rate of improvement of all space technologies. Everyone in space activities shares a responsibility for addressing this critical technical problem. The overwhelming influence that space access has on all aspects of civil, commercial, and military space efforts indicate that it should enjoy a top priority.[1]

Of course, a key element in the spacefaring vision long held in the United States is the belief that inexpensive, reliable, safe, and easy space flight is attainable. Indeed, from virtually the beginning of the twentieth century, those interested in the human exploration of space have viewed as central to that endeavor the development of vehicles of flight that travel easily to and from Earth orbit. The more technically minded recognized that once humans had achieved Earth orbit about 200 miles up, the vast majority of the atmosphere and the gravity well had been conquered and persons were now about halfway to anywhere they might want to go.[2]

It seems appropriate to break down the development of launch vehicles into six major time periods, each characterized by specific technological challenges, political and economic environments, design priorities, mission objectives, and legacies and lessons (fig. 1). The discussion that follows traces this evolutionary process and offers observations about key issues affecting the process of technological innovation to the present.

Cold War Origins

The primary U.S. space launch capabilities were created largely because of the challenge of an exceptionally desperate cold war rivalry with the Soviet Union. Accordingly, the orbiting of scientific satellites into space and the development and deployment of ballistic missiles and space-based intelligence-gathering capabilities were all critical to ensuring the national security of the United States (fig. 2).[3] For the first eight years of the space age, the U.S. space effort operated and evolved in response to consistently focused governmental policy with the highest national priority. Then, the focus was on test and evaluation of intercontinental ballistic missiles (ICBMs) and on developing these systems into space launch vehicles, primarily to support the reconnaissance mission operated from Vandenberg Air Force Base (AFB) on the West Coast, and on the robotic space flight mission operated from Cape Canaveral on the East Coast. The initial development, test, and evaluation of ballistic missiles also drove the development of extensive tracking, telemetry collection, and precise photographic capabilities.

During this earliest period the United States began developing the principal launchers—Atlas, Thor/Delta, and Titan—that are still in use. It is hard to believe in the year 2002, but the United States still relies on the descendants of these three ballistic missiles for the bulk of its space access requirements. Even though the three families of space boosters, each with numerous variants, have enjoyed incremental improvement since first flight, there seems no way to escape their beginnings in technology (dating back to the early 1950s) and their primary task of launching nuclear warheads.

Image 0-1: Robert H. Goddard was the father of liquid rocketry in the United States. Shown here is Goddard at the time of his 16 March 1926 launch of the first liquid-fueled rocket from a farm at Amherst, Massachusetts. This historic event served as the starting point for a range of activities in rocketry thereafter. (NASA Photo)

Figure 1

Era	Political/ Economic Environment	Mission Objectives	Technological Challenges	Design Parameters	Lessons and Legacies
Cold War Origins (1948–1956)	Cold War begins; Soviets viewed as tecnologically aggressive; crash rocket programs emerge; U.S. economy stable	ICBMs; spy satellites; scientific satellites	Everything; is flight in space even possible?; long-range rockets; satellites; guidance and control	Big, dumb boosters; thrust enhancement; launch reliability poor	Program management concept; large-scale approach; funding no obstacle
Height of Cold War (1957–1965)	Cold War heightens; funding increases; economy thrives in 1960s; head-to-head space race; sense that Soviet Union leads world	ICBM operations-ICBMs used for piloted flight; reduction of costs; long-duration missions; lunar exploration	Improved reliability and schedule; concentration on miniaturization; sustained long-duration operations; Apollo	Studies of reusability; greater reliability; increased time for operations	Validation of program management concept; large-scale funding investment
Cold War confrontation wanes (1966–1972)	Cold War moderates; sense that U.S. ahead in space race; Apollo continues; U.S. economy expands; Vietnam sours American public on government activities	Little impetus for new military launchers; early warning system developed	Improved reliability and schedule; sustained long-duration operations; Apollo; continued miniaturization	Incremental reduction of cost and reliability of launch schedules	Continued validation of program management concept; large-scale funding investment

Era	Political/Economic Environment	Mission Objectives	Technological Challenges	Design Parameters	Lessons and Legacies
Visions of routine access to space (1972–1985)	Space as U.S. province; Americans voice little concern for space issues; U.S. economy stagflates; little concern for Soviets	Emphasis on multiple payloads with launcher; make space transportation like aviation; cost reduction major emphasis	Reusability; SSTO; reduction of cost of access; continued miniaturization	Entirely new generation of launchers, older systems abandoned	Replacement of program management concept in favor of lead center approach; attempted standardization of all U.S. payloads for flight on shuttle; attempted abandonment of all ELVs
Assuring access to space (1986–1989)	*Challenger* accident discredits NASA; ELVs reemerge; commercial markets emerge; U.S. deficit rages; Soviet Union collapses; Strategic Defense Initiative; loss of market to Ariane	Emphasis on multiple payloads with launchers; New ELV concepts; new military missions	Reusability; SSTO reemerges; reduction of cost of access; continued miniaturization	ELVs make incremental improvements and seek commercial payloads	Space access to be like aviation; space flight must pay; NASA less powerful, no longer can dictate policy
Commercial space access begins (1990–2002)	Space viewed largely as marketplace; reevaluation of future military concerns about use of space	Emphasis on multiple payloads with launcher	Reusability; SSTO; reduction of cost of access; continued miniaturization; nanotechnology	Develop EELVs, SSTOs	Commercial payoffs difficult; space access not self-supporting

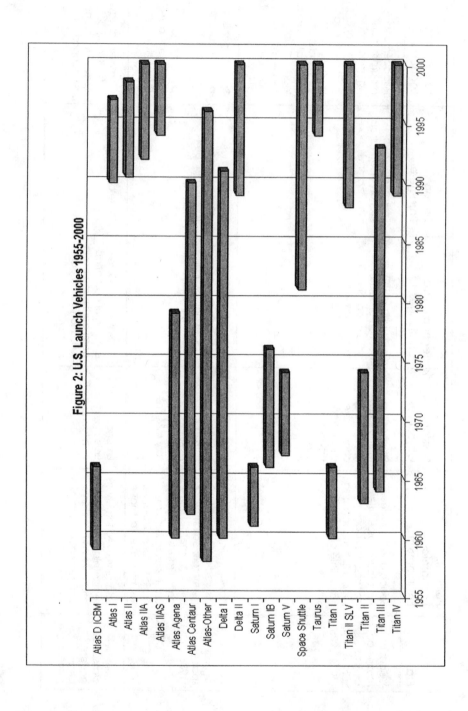

Figure 2: U.S. Launch Vehicles 1955-2000

The first-generation ICBM, the Atlas, was flight tested beginning 11 June 1955 and made operational in 1959. A second ballistic missile, the Thor intermediate-range ballistic missile (IRBM), also dates from the 1950s and, along with its evolved form named Delta, became an early workhorse in America's fleet of launchers. The Titan ICBM quickly followed the Atlas and Thor into service with the U.S. Air Force in 1959 and remained on alert until the end of the cold war at the end of the 1980s.[4]

Since these space launch vehicles began existence as national defense assets, they reflected both the benefits and liabilities of those origins. For example, the national defense requirements prompted the builders to emphasize schedule and operational reliability over launch costs. Consequently, these vehicles were exceptionally costly both to develop and operate.

Indeed, the Eisenhower administration poured enormous resources into the development of these first-generation space access vehicles. As a measure of government investment, through fiscal year 1957 the government spent $11.8 billion on military missile development (in fiscal year 1957 dollars). "The cost of continuing these programs from FY 1957 through FY 1963," Eisenhower was told, "would amount to approximately $36.1 billion, for a grand total of $47 billion."[5] In 2000 dollars, for comparison, this would have represented an investment of more than $230 billion. An investment of even 10 percent of that amount today would make possible an enormous advance of launch vehicle technology.

The Height of Cold War

The period between 1957 and 1965 might best be viewed as the height of the cold war era and the age of the great race for space. Engaged in a broad contest over the ideologies and allegiances of the nonaligned nations of the world, space exploration was one major area contested. The Soviets gained the upper hand in this competition on 4 October 1957 when they launched *Sputnik I,* the first artificial satellite to orbit Earth, as part of a larger scientific effort associated with the International Geophysical Year.[6] The Soviets did not relinquish their apparent lead in the space race until the mid-1960s. Meantime, the United States seemed incapable of conducting space operations effectively, notably during the failed launch of a Vanguard satellite on national television on 6 December 1957.

At the same time, visible successes in space during the late 1950s and early 1960s fostered a hubris within the Soviet Union not seen since the end of World War II and never to be experienced again in that empire's history. It represented a high-water mark of success, and Nikita Khrushchev's leadership exploited it to the fullest. Thereafter, with high priorities given the effort by the Soviet leadership, the communist state's rocketeers led the way in one visibly stunning success after another, as shown in figure 3.

In those first eight years of the space age, it looked as if the Soviet Union did everything right in space flight, and the United States publicly appeared at best a weakling without the kind of capabilities that the command economy of the "workers' state" in the Soviet Union had been able to muster. The result was that the United States mobilized to "catch up" to the apparent might of its cold war rival. As surely as the several crises in Berlin—the blockade and airlift, the wall—and the other flash points of competition, Sputnik served to fuel the antagonism and steel the resolve of both sides in the cold war.[7]

Without question, notwithstanding genuine accomplishments in space by the United States, during those first years the Soviet Union held the edge—at least in the public eye. And it was their rocket technology that allowed that to be the case. The large ICBM built by Sergei Korolev, the R-7, enabled the access to space so necessary to the long list of firsts piled up by the Soviet Union between 1957 and 1965.[8]

The United States worked hard to catch up to the Soviets in launcher technology during that period, and they did so by building on the ballistic missile accomplishments of the 1950s. As an example, combined with the

---

Figure 3
Soviet Firsts in Space, 1957-1965

The first living thing in orbit, the dog Layka launched on Sputnik II on 3 November 1957.

The first human-made object to escape Earth's gravity and to be placed in orbit around the Sun, Luna 1 in January 1959.

The first clear images of the Moon's surface in September 1959 from Luna 2.

The first pictures of the far side of the Moon, October 1959, taken by Luna 3.

The first return of living creatures from orbital flight, two dogs sent into space in August 1960 aboard Sputnik 5.

The first human in space, cosmonaut Yuri Gagarin, who flew a one-orbit mission aboard the spacecraft Vostok 1 on 12 April 1961.

The first day-long human space flight mission, August 1961, made by Vostok 2 with cosmonaut Gherman Titov aboard.

The first long duration space flight, cosmonaut Andrian Nicolayev spent four days in space aboard Vostok 3, August 1962.

The first woman in space, Cosmonaut Valentina Tereshkova, flew forty-eight orbits aboard Vostok 6 in June 1963.

The first multiperson mission into space, Voskhod 1 carrying cosmonauts Komarov, Yegorov, and Feoktistov in October 1964.

The first spacewalk or extravehicular activity, March 1965, by Alexei Leonov during the Voskhod 2 mission.

Image 0-2: This R-7 launch vehicle, developed by the Soviet Union in the latter 1950s made possible the orbiting of the first artificial satellite, *Sputnik I*, on 4 October 1957, ushering in the space age. (NASA Photo)

Agena upper stage, the Atlas propelled some of the earliest space probes, such as Mariner and Ranger, to the Moon and Mars. Another variant, the Atlas-Centaur, first flew in 1962 and underwent incremental improvements thereafter. It launched many of the National Aeronautics and Space Administration's (NASA) space probes to the planets and also launched numerous military and applications satellites.

For the first U.S. effort to orbit humans, Project Mercury, NASA also employed a modified Atlas rocket, at least for the later flights. But this decision was not without controversy. There were technical difficulties to be overcome in mating it to the Mercury capsule to be sure, but the biggest complication was a debate among NASA engineers over its propriety for human missions.[9]

Most of the problems were resolved by the first successful orbital flight of an unoccupied Mercury-Atlas combination in September 1961. On 29 November the final test flight took place, this time with the chimpanzee Enos occupying the capsule for a two-orbit ride before being successfully recovered in an ocean landing. Not until 20 February 1962, however, could NASA get ready for an orbital flight with an astronaut. On that date John Glenn became the first American to circle the Earth, making three orbits in his *Friendship 7* Mercury spacecraft. The flight had difficulties, and Glenn flew parts of the last two orbits manually because of an autopilot failure and left his normally jettisoned retrorocket pack attached to his capsule during reentry because of a suspected loose heat shield.[10]

A second ballistic missile, the Thor, also became a workhorse in America's fleet of launchers, both in its own right and later modified as the Delta. From 1960 to 1982 Thor and Delta underwent incremental improvements resulting in thirty-four separate configurations. The Titan ICBM, like Atlas and Delta, also underwent successive improvement during this era and was used to launch successive scientific, military, and a few commercial payloads.

The United States also undertook development of an entirely new space launch vehicle, the Saturn. NASA had inherited the effort to develop the Saturn family of boosters in 1960, when it acquired the Army Ballistic Missile Agency under Wernher von Braun.[11] By that time von Braun's engineers were hard at work on the first-generation Saturn launch vehicle, a cluster of eight Redstone boosters around a Jupiter fuel tank. Fueled by a combination of liquid oxygen (LOX) and RP-1 (a version of kerosene), the eight H-1 engines in the first stage of the Saturn I could generate a total of 1,296,000 lbf (pounds force, i.e., thrust).[12] This group also worked on a second stage that used a revolutionary fuel mixture of LOX and liquid hydrogen that could generate a greater thrust-to-weight ratio. The fuel choice made this second stage a difficult development effort because the mixture was highly volatile and could not be readily handled. But the S-IV stage could produce an additional 90,000 lbf using six RL-10 engines. The Saturn I was solely a research and development vehicle that would lead toward the accomplishment of Apollo, making ten flights between October 1961 and July 1965. The first four flights tested the first stage, but beginning with the fifth launch the second stage was active and these missions were used to place scientific payloads and Apollo test capsules into orbit.[13] The next step in Saturn development came with the maturation of the Saturn IB, an upgraded version of earlier vehicle, and the mighty Saturn V Moon rocket developed in the latter half of the 1960s.

Cold War Confrontation Wanes

About 1965 the cold war began to wane as a powerful motivator behind space activities. From the point where America began flying the Gemini spacecraft, and especially with the flights of Apollo (1968–72), it became obvious that the United States led the world in rocket technology. Accordingly, the space race began to wane during that era, as the nation was consumed with issues other than crises with the Soviet Union, especially the war in Vietnam.

Gemini had been conceived as a means of bridging the gap between the technological base required to land successfully on the Moon and that already in existence. NASA closed most of the gap by experimenting and training on the ground, but some issues required experience in space. Three major areas immediately arose where this was the case. The first was the ability in space to locate, maneuver toward, and rendezvous and dock with another spacecraft. The second was closely related: the ability of astronauts to work outside a spacecraft. The third involved the collection of more sophisticated physiological data about the human response to extended space flight.[14]

To gain experience in these areas before Apollo could be readied for flight, NASA devised Project Gemini. The two-person capsule was to be launched by the newly developed Titan II, another ballistic missile developed for the Air Force. The Titan II proved difficult; it had longitudinal oscillations called the "pogo effect" because it resembled the behavior of a child on a pogo stick. Overcoming this problem required engineering imagination and long hours of overtime to stabilize fuel flow and maintain vehicle control. The fuel cells leaked and had to be redesigned, and an Agena target vehicle used for docking suffered costly delays. All of these difficulties shot an estimated $350 million program to over $1 billion. The overruns were successfully justified by the space agency, however, as necessary in meeting the Apollo landing commitment.[15]

By the end of 1963 most of the difficulties with the Titan II had been resolved, albeit at great expense, and the program was ready for flight. Following two unoccupied orbital test flights, the first operational mission took place on 23 March 1965. Mercury astronaut Virgil I. "Gus" Grissom commanded the mission, with John W. Young, a naval aviator chosen as an astronaut in 1962, accompanying him. The next mission, flown in June 1965, stayed aloft for four days, and astronaut Edward H. White II performed the first U.S. extravehicular activity (EVA), or spacewalk.[16] Eight more missions followed through November 1966.[17]

If Gemini failed to convince anyone that the space race had been won by the United States, Apollo would. Using the Saturn IB, with more powerful engines generating 1,640,000 lbf from the first stage, the two-stage combination could place 62,000-pound payloads into Earth orbit. The first flight on 26 February 1966 tested the capability of the booster and the Apollo capsule in a suborbital flight. Two more flights followed in quick succession. The first astronaut-occupied flight (Apollo 7) of the Saturn IB took place between 11 and 22 October 1968, when Walter Schirra, Donn F. Eisele, and R. Walter Cunningham made 163 orbits testing Apollo equipment. Four other flights using the Saturn IB would be made after the Moon landings as part of the ASTP and Skylab programs.[18]

The final launch vehicle of this family, the Saturn V, represented the culmination of those earlier booster development programs. Standing 363 feet tall, with three stages, this was the vehicle that could take astronauts to the Moon and return them safely to Earth. The first stage generated 7,500,000 lbf from five massive Rocketdyne F-1 engines developed for the system. These engines were some of the most significant engineering accomplishments of the program, requiring the development of new alloys and different construction techniques to withstand the extreme heat and shock of firing. The thunderous sound of the first static test of this stage, taking place at Huntsville, Alabama, on 16 April 1965, brought home to many that the Kennedy goal was within the technological grasp. For others, it signaled the magic of technological effort; one engineer even characterized rocket engine technology as a "black art" without rational principles.

The second stage presented enormous challenges to NASA engineers and very nearly caused the lunar landing goal to be missed. Consisting of five engines burning LOX and liquid hydrogen, this stage could deliver 1,000,000 lbf. It was always behind schedule and required constant attention and additional funding to ensure completion by the deadline for a lunar landing. In contrast, both the first and third stages of this Saturn vehicle development program moved forward relatively smoothly. (The third stage was essentially identical to the second stage of the Saturn IB and had few developmental complications.)[19]

But even as the Apollo program achieved success, the NASA Administrator terminated the Saturn V production line. With no large-scale space exploration programs envisioned beyond Apollo, there were no new missions requiring the booster's power. NASA's original order was for fifteen Saturn V rockets. As early as 1968, the agency faced the issue of whether it would need more Saturn Vs and decided to wait before ordering the long-lead-time components involved. In struggling in 1971 and 1972 to obtain approval to develop a new space transportation system, the Space Shuttle, NASA officials reluctantly decided that they had no choice but to give up hopes of preserving the two remaining Saturn V boosters for future use and of maintaining production capabilities for additional vehicles. NASA thereby took the action to end the Saturn family's production.[20]

At the same time, the American government sponsored continued maturation of the Atlas, Titan, and Delta launchers, but most of the payloads continued to support official operations. Movement beyond these first-generation launchers has remained a dream in the opening of space to wider operations. Like the earlier experience with propeller-driven aircraft, the United States has sponsored incremental improvement of launchers for the last forty years

Image 0–3: The mighty Saturn V launch vehicle rocketed American astronauts to the Moon in the latter 1960s and early 1970s. Here the launch of an unmanned Saturn V rocket takes place in October 1967. (NASA Photo no. 67PC-0397)

without making a major breakthrough in technology. Accordingly, America today has a very efficient and mature expendable launch vehicle (ELV) launch capability that is still unable to overcome the limitations of the first-generation ICBMs on which it is based. After four decades of effort, access to space remains a difficult challenge.

Visions of Routine Access to Space

In 1972, with the completion of the Apollo program, President Richard M. Nixon acquiesced in the decision to build the Space Shuttle, a mostly reusable launch vehicle that sought to achieve the long-held vision of routine, reliable, low-cost U.S. access space. This represented a major shift in the national priority for access to space: no longer would the United States be relying on ELVs based on original ICBM designs for most of its space access requirements. Instead, the new national priority between 1972 and 1985 was to develop and operate the partially reusable Space Shuttle as the primary U.S. means of placing national security, scientific, applications, civil, and commercial satellites into space.[21]

In 1972 NASA promoted a reusable space shuttle as a means of reducing the cost to orbit from one thousand to one hundred dollars per pound. To conduct an aggressive space exploration effort, NASA officials declared in 1972, "efficient transportation to and from the earth is required." Some NASA officials even compared the older method of using ELVs like the Saturn V to operating a railroad and throwing away the locomotive and boxcars with every trip. The shuttle, they claimed, would provide the United States with low-cost, routine access to space.[22]

At that time some space observers calculated that a Titan IIIC could cost as much as $24 million to procure and launch, while each Saturn IB cost $55 million. Carrying 23,000 pounds to low-Earth orbit (LEO), the Titan IIIC delivered its payload at a cost per pound of about $1,000 then. The Saturn IB cost about $1,500 per pound to deliver its 37,000-pound payload. It was these launch costs that NASA officials sought to reduce by a much-heralded factor of ten.[23]

The Space Shuttle, therefore, became an attempt to provide "low-cost access [to space] by reusable chemical and nuclear rocket transportation systems."[24] George M. Low, NASA's Deputy Administrator, voiced the redefinition of this approach to the NASA leadership on 27 January 1970: "I think there is really only one objective for the Space Shuttle program, and that is `to provide a low-cost, economical space transportation system.' To meet this objective, one has to concentrate both on low development costs and on low operational costs."[25] "Low cost, economical" space transportation became NASA's criterion for the program, and it was an effort to deal with a real-time problem of public perception about space flight at the time: that it was too expensive.[26]

NASA had originally intended to achieve cost-effectiveness on the shuttle through economies of scale, as late as 1984 estimating that they could fly as many as twenty-four missions per year. This has proven an unattainable goal. Instead, NASA might have cut operational costs by investing more money in cost-saving technologies at the beginning of the program. Dale D. Myers, who served as NASA Deputy Administrator in the post-Challenger era, suggested that reductions in the cost of flight operations might have been achieved "had the design team concentrated on operations as strongly as they concentrated on development."[27]

While the effort to achieve "low cost, economical" access to space was an appropriate goal for NASA, it eventually proved an embarrassment to the space program. So far, in spite of high hopes, the shuttle has provided neither low cost nor routine access to space. The Space Shuttle—second to the Saturn V in both capability and cost—launches some 53,000 pounds of payload into orbit at a cost per launch of about $445 million. It is a high-end user, and the cost per flight is so astronomical that only the government can afford it. In addition, by January 1986, there had been only twenty-four shuttle flights, although in the 1970s NASA had projected more flights than that for every year. Although the system is mostly reusable, its complexity, coupled with the ever-present rigors of flying in an aerospace environment, means that the turn-around time between flights requires several months instead of several days.

Since neither the cost per launch nor the flight schedule has met expectations, many criticized NASA for failing to meet the promises made in gaining approval of the Shuttle program. In some respects, therefore, a consensus emerged in the last decade of the twentieth century that the shuttle has been both a triumph and a tragedy. It remains an engagingly ambitious program that operates an exceptionally sophisticated vehicle, one that no other nation on Earth could have built at the time. As such it has been an enormously successful program. At the same time, the shuttle is essentially a continuation of space spectaculars, à la Apollo, and its much-touted capabilities remain unrealized. It made far fewer flights and conducted far fewer scientific experiments than NASA publicly predicted.[28]

While the Space Shuttle represented an enormously significant, if ultimately unsuccessful, attempt to lower the cost to orbit, the other major launchers of the United States—the Atlas, Titan, and Delta—worked to achieve greater economy through the use of mature technologies incrementally improved and efficient operations honed to a fine edge over time. Unfortunately, this approach has also failed to significantly lower the costs of payloads to orbit. As shown in figure 4, no current launch vehicle is able to achieve orbit consistently at less than three thousand dollars per pound. Clearly this is unaccept-

Image 0–4: A Space Shuttle on its Crawler-Transporter after leaving the Vehicle Assembly Building (VAB) in September 1990. (NASA Photo no. 90PC-1324)

able for the opening of significant space operations. Commercial human flight in space—spaceplanes with the capability to move passengers to and from Earth orbit and around the globe—is often invoked as the ultimate goal but is commercially infeasible at the cost per pound shown here. For example, an individual and baggage totaling 220 pounds would have, at best, a ticket price of $733,260.

Under the national strategy of relying on the shuttle, the Department of Defense (DoD) and NASA were to launch the remainder of their ELVs and

Figure 4 The High Cost of Launch[29] (Fiscal Year 1993 dollars)			
Launch Vehicle	**Pounds to LEO**	**Cost per Launch**	**Cost per Pound**
Titan II	2,000–4,000	$40–$45 million	$10,000–$22,500
Delta II	5,000–11,000	$45–$50 million	$4,090–$10,000
Atlas II	12,000–18,000	$60–$70 million	$3,333–$5,833
Saturn V	253,000	$448 million	$1,771
Titan IV	30,000–50,000	$170–$220 million	$3,400–$7,333
Space Shuttle	53,000–56,000	$445 million	$8,036–$8,490

then to fly only on the Space Shuttle. This represented a profound shift in the strategy of space access for the United States. In addition to the presumed shutting down of the production line for these other launch vehicles, the launching and processing facilities for ELVs atrophied. Instead, the focus headed toward developing facilities and infrastructure to support only the Space Shuttle and ballistic missiles.[30]

From 1972 to 1985, NASA continued to conduct launches of communications satellites on behalf of U.S. commercial and foreign customers, as well as those of foreign scientific satellites. As U.S. civil and military launch rates declined, launches of communications satellites steadily increased, rising to a high of nine launches in 1982. During this entire thirteen-year period, commercial launches accounted for 22 percent of all U.S. launches. When the shuttle first flew in 1981, the Reagan administration moved quickly to declare it operational and to empower NASA to manifest commercial payloads for it. The shuttle first deployed commercial satellites in 1982, and by 1985 NASA had launched eleven commercial communications satellites on four shuttle flights and only three on Atlas-Centaur flights.[31]

While the U.S. government had mandated the phasing out of ELVs for its launches, manufacturers of these vehicles, as well as some users, proposed continuing production and competing directly with the Space Shuttle and newly operational European Ariane launch vehicle. However, the government's strategy to keep the shuttle price low and Europe's support to keep the Ariane price even lower remained major impediments to their commercial success. Despite their hesitation, in February 1984, President Ronald Reagan signed Executive Order 12465 on "Commercial Expendable Launch Vehicle Activities," and Congress passed the Commercial Space Launch Act (CSLA) of 1984 establishing a licensing and regulatory regime for nongovernment launch activities within the Department of Transportation. This established the fundamental framework for the current law.[32]

The CSLA of 1984 recognized that U.S. ELVs would no longer be needed for government use in light of the national policy to rely on the shuttle as the primary means of U.S. access to space. This law established the foundation and mechanisms necessary for U.S. companies to obtain use of or even ownership of these ELV-related facilities that were to become "excess or otherwise not needed for public use," as well as launch base and range support services from the Air Force and NASA that were similarly no longer "needed for public use."[33]

This was never a perfect situation, for during the early Reagan years the shuttle was shouldering the responsibility for all government launches and many commercial ones. It was, sadly, ill-equipped to satisfy these demands. Even with the best of intentions and with attractive payload pricing policies, the Space Shuttle remained a compromise vehicle. The desire for the shuttle to be all things to all people—research and development aerospace vehicle, operational space truck, commercial carrier, scientific platform—ensured that it would fully satisfy none of these singular and mutually exclusive missions.[34]

These inherently competing goals, coupled with the reality of primitive reusable launch vehicle (RLV) technology, led to disappointment and disillusionment with the Space Shuttle. By 1985 NASA had learned a great deal about the limits of its own abilities as the shuttle failed to deliver on its early promises, many of those promises of NASA's own making. Even so, NASA insisted on maintaining the shuttle as the preeminent launcher for the United States, a position that became less tenable with every year of its operation.[35]

Assuring Access to Space

The loss of Challenger on 28 January 1986 changed everything. By mid-1986 virtually all U.S. space launch systems had experienced launch failures. In addition to Challenger, a Titan 34D in May 1986 damaged both Titan launch pads at Vandenberg Air Force Base and Delta and Atlas failures at Cape Canaveral called into question the possibility of the United States having *any* access to space, much less assured access. This series of failures led to serious concerns regarding the reliability and resilience of U.S. national access to space, which in turn led to another important shift in national policy for the future of space access.[36]

The Challenger accident reinvigorated the debate over the use of the Space Shuttle to launch all U.S. satellites. In August 1986, President Reagan announced that the shuttle would no longer carry commercial satellites, a policy formalized in December 1986 in National Security Decision Directive 254, "United States Space Launch Strategy." A total of forty-four commercial and foreign payloads that had been manifested on the Space Shuttle were forced to find new ELV launchers.[37]

For the next three years the U.S. government worked to reinvigorate the American ELV production lines and to redesign and modify satellites to be launched on ELVs instead of the shuttle. The shift back to ELVs required additional government funding to fix the problems that had resulted from years of planning to retire these systems. As shown in figure 5, the United States practically ceased commercial launch activities for several years, conducting just three commercial satellite launches (one just prior to the Challenger flight) for only 6 percent of U.S. space launches from 1986 to 1989.[38]

During this period, however, two actions were initiated that enabled the emergence of a U.S. commercial launch industry. First, the DoD committed to purchasing a large number of ELVs as part of a strategy to maintain access to space using a mixed fleet of both the Space Shuttle and ELVs. This reopened the dormant U.S. ELV production lines at government expense and helped provide economies of scale necessary to enable U.S. companies to effectively compete against Ariane. Second, in 1988, Congress amended the CSLA to establish new insurance requirements whose effect was to limit liability for U.S. companies in case their launches caused damage to government property or third parties. The revised CSLA also established protections against government preemption of commercial launches on government ranges.[39]

As a result, the first U.S. commercial space launch took place in 1989, nearly five years after the CSLA was passed. Beginning in 1989, U.S. launches of commercial satellites were conducted by commercial launch companies (in most cases, the same companies providing launch services for DoD and NASA payloads as government contractors), not the U.S. government.

The development of revised ELVs followed quickly. For instance, the most recent version of the Atlas, the IIAS, began flying in 1993 and could send 19,000 pounds into LEO at the "bargain" price of $105 million per flight, about $5,500 per pound in current dollars. The three-stage Delta II entered operational service in 1989 and could place 3,190–4,060 pounds into orbit depending on configuration. Its cost is about $45–50 million per launch, about $12,000 per pound. The Titan IV, the only current operational version, is the largest and most powerful U.S. expendable launch vehicle in use. Capable of placing 39,000 pounds in Earth orbit, it costs more than $240 million per flight.[40]

Even so, the costs for space access remain exorbitant. The most modest space launchers placing relatively small satellites of less than 4,000 pounds into orbit, for example, still average some $25 million per flight, or about $10,000 per pound depending on the launch system. The mighty Saturn V Moon rocket, the most powerful launch system ever developed, could place into orbit a massive payload of 262,000 pounds, but to do so cost an enormous $113.1 million per launch ($455 million in 2000 dollars). And those are just basic launch costs to orbit; they do not include the cost of satellite development, indemnification, boost to optimum orbit, ground support and transportation, operations, and the like.[41]

Space travel started out and remains an exceptionally costly enterprise; the best expendable launch vehicles cost something approaching ten thousand dollars per pound from Earth to orbit. No wonder that it has been the province of the government, a few high-end communications satellite companies, and other unique users, despite attempts to encourage further development of launch technology.[42]

Commercial Space Access Begins

An important policy shift occurred within the government space launch management structure in the early 1990s. From the beginning, responsibilities for ELV acquisition, development, and operations, as well as the operation, maintenance, improvement, and modernization of the launch bases and ranges, resided in the acquisition and development arm of Air Force Systems Com-

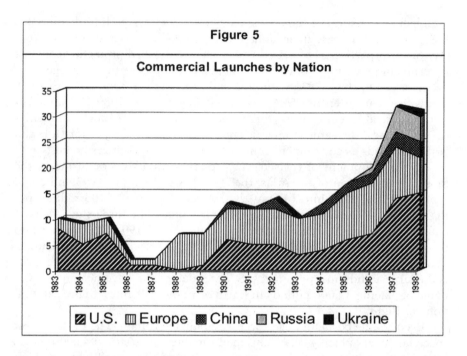

mand (AFSC). In 1990 AFSC ceased to exist as an independent entity, and the Air Force transferred access to space responsibilities from it to the operational arm under Air Force Space Command.[43] Just as the Air Force shifted its space launch focus from "development" to "operations," the U.S. commercial space launch industry entered a period of growth and expansion. Between 1990 to 1994, commercial launch activities climbed back to their pre-Challenger level of around 20 percent of U.S. space launches conducted. For the next few years, U.S. commercial launch providers engaged in intense competition with Arianespace—the European consortium that builds the Ariane Launcher for the European Space Agency—for leadership of the commercial launch industry. As the backlog of commercial payloads that had been delayed by the launch failures was flown out, an average of more than twelve launches per year were conducted from 1990 to 1994 (see fig. 5).

Recognizing the critical importance of space transportation to the U.S. national security, civil, and commercial space sectors, the Clinton administration issued the National Space Transportation Policy in 1994. A key feature of the policy was that it established a clear division of responsibilities: DoD would oversee the ELV fleet, and NASA was given primary responsibility for RLV technology development and demonstration. Both agencies were directed to involve the U.S. commercial space sector, including state governments, as partners, participants, and investors in these programs. This policy formed the genesis for DoD's Evolved Expendable Launch Vehicle (EELV) program and for NASA's X- series of RLV technology demonstrators.[44]

The EELV program emphasized the development of advanced Atlas and Delta launchers and the phaseout of the Titan rocket. In several versions, Atlas and Delta were to continue to be a major U.S. launch vehicle for the foreseeable future. Beginning in 1995 the United States government begin to phase the Titan out of the launch inventory. It had no commercial appeal because of its price tag per flight, and its principal user, the DoD, was finding fewer payloads for it with every year. In the fiscal year 1995 DoD appropriation, Congress included language that "terminates the Titan IV program after completion of the current contract," which provided for a total of forty launches (with one spare) through 2003. This legislation called the Titan IV "excessively expensive" and foreswore the continuation of this launcher on grounds of cost. The Titan launcher will be extinct before 2005, replaced by a new heavy-lift derivative of the Delta IV Evolved Expendable Launch Vehicle.[45]

Additionally, NASA charted a new course by placing greater responsibility for space flight operations with its private sector contractor, United Space Alliance (USA). Since 1995 USA may have increased performance by almost a third, cut ground processing time nearly in half, and reduced operating costs by more than a third when adjusted for inflation. Or so it seems. A shift in shuttle operations into the International Space Station era makes direct comparisons with previous operations difficult. NASA also invested about $100 million per year in Space Shuttle improvements to address safety and obsolescence, while USA has invested millions to improve shuttle operations.[46] At the same time, NASA invested heavily in the development of RLV technology with the X-33, X-34, and X-37 programs—all of them in partnership with industry—but by 2001 they had either been canceled or stretched out to an indefinite future.[47]

Driven by expanding market demand in the mid-1990s, the private sector also began making substantial investments in the development of new launch vehicles. For example, by the end of the century, no fewer than nine start-up companies had emerged with the objective of building their own launch vehicle with little or no government involvement.[48] Sadly, most have subsequently failed and been dissolved. Partnerships with other aerospace organizations from other nations also began during the last decade of the twentieth century. Following the end of the cold war, U.S. aerospace companies formed a number of joint ventures with Russian and Ukrainian launch companies to provide launch services on former Soviet launch vehicles. Prominent among these new relationships is International Launch Services (ILS), a joint venture between Lockheed Martin and Russia's Khrunichev and Energia to market the Proton launch vehicle. Another is Sea Launch, a joint venture between Boeing, Ukraine's Yuzhnoye, Russia's Energia, and Norway's Kvaerner to launch a Zenit rocket from a launch platform in the middle of the Pacific Ocean. In addition, various Russian technologies have been added to later U.S. launch vehicles—both Atlas V and Delta IV are using engines that originated behind the former iron curtain.[49]

The Value of RLVs versus ELVs

Many aerospace engineers believe that the long-term solutions to the world's launch needs are a series of completely reusable launch vehicles. A debate has raged between those who believe RLVs are the only—or at least the best—answer and those who emphasize the continuing place of expendable launch vehicles in future space access operations. RLV advocates have been convincing in their argument that the only course leading to "efficient transportation to and from the earth" would be RLVs and have made the case repeatedly since the late 1960s.[50] Their model for a prosperous future in space is the airline industry, with its thousands of flights per year and its exceptionally safe and reliable operations. Since the advent of the Space Shuttle, NASA has been committed to advancing this model with shuttle follow-on efforts since the 1980s. Few remember that the Space Shuttle promised similar advancements—in fact, each of the shuttle development proposals originally included an airline partner.

One especially important effort for a next-generation RLV emerged during the Reagan administration when senior government officials began to talk about the "Orient Express," a hybrid air- and spaceplane that would enable ordinary people to travel between New York City and Tokyo in about one hour. Such a concept was quite simple in theory although enormously complex in reality. It required developing a passenger spaceplane with the capability to fly from an ordinary runway like a conventional jet. Flying supersonic, it would reach an altitude of about 45,000 feet when the pilot would start scramjet engines, a more efficient, faster engine that has the potential to reach hypersonic speeds in the Mach 6 realm. These take the vehicle to the edge of space for a flight to the opposite side of the globe, from whence the process is reversed and the vehicle lands like a conventional airplane. It never would reach orbit, but it would still fly in space, and the result is the same as orbital flight for passengers but for less time. It would even be possible, RLV supporters insisted, to build such a spaceplane that could reach orbit. It had been investigated before—during the 1950s and again in the 1960s—each time as an expensive nonstarter.[51]

One of the most significant efforts to develop this reusable spaceplane was the National AeroSpace Plane (NASP), a joint NASA–U.S. Air Force technology demonstrator begun during the Reagan administration. Touted as a single-stage-to-orbit (SSTO) fully reusable vehicle using air-breathing engines and wings, NASP never progressed to flight stage. It finally died a merciful death, trapped as it was in bureaucratic politics and seemingly endless technological difficulty, in 1994.[52]

NASA began its own RLV program after the demise of NASP, and the agency's leadership expressed high hopes for the X-33, a small suborbital vehicle that would demonstrate the technologies required for an operational SSTO launcher. This is the first of a projected set of four stages that would lead to a routine spacefaring capability. The X-33 project, undertaken in partnership with Lockheed Martin, had an ambitious timetable to fly by 2001. But what would happen after its tests were completed was unclear. Even assuming complete success in meeting its R&D objectives, the time and money necessary to build, test, and certify a full-scale operational follow-on VentureStar™ remained problematic. Who would pay for such an operational vehicle also remained a mystery, especially since the private sector had become less enamored with the joint project over the years and had eased itself away from the venture after the failure of such high-visibility projects as Iridium. Continuing technical difficulties finally resulted in NASA withdrawing from the program in April 2001, although Lockheed Martin is hoping that the Air Force will continue to fund the single demonstrator vehicle as part of its new Bush administration space initiatives.[53]

There is also an understanding that the technical hurdles have proven more daunting than anticipated, as was the case thirty years ago with the Space Shuttle and more recently with the NASP. Any SSTO, and X-33 held true to this pattern, would require breakthroughs in a number of technologies, particularly in propulsion and materials. And when designers begin work on the full-scale SSTO, they may find that available technologies limit payload size so severely that the new vehicle provides little or no cost savings compared to old launchers. If this becomes the case, then everyone must understand that NASA will receive the same barbs from critics as were seen with the shuttle. They condemned NASA for "selling" the Space Shuttle program as a practical and cost-effective means of routine access to space and then failing to deliver on that promise.[54]

This is not to say that SSTO could never work. It has always been NASA's job to take risks and push the technological envelope. But while the goal may be the development of a launch system that is significantly cheaper, more reliable, and more flexible than what is presently available, it is possible to envision a future system that cannot meet those objectives. This is all the more true in a situation where breakthrough, revolutionary technologies do not emerge.[55]

Then there is an alternative position that suggests that the most appropriate approach to space access is through the use of throwaway "big, dumb boosters" that are inexpensive to manufacture and operate. While reusable rockets may seem to be an attractive, cost-saving alternative to expendables

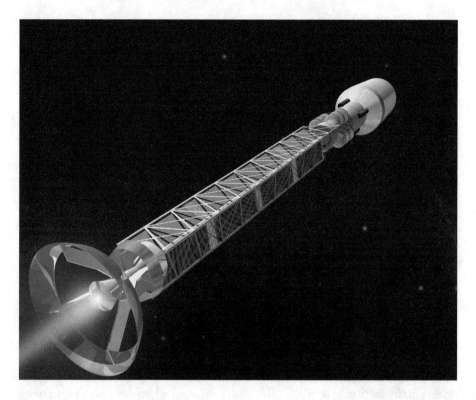

Image 0–5: This is an artist's rendition of what one future space access system may look like, an antimatter propulsion system. Matter-antimatter annihilation offers the highest possible physical energy density of any known reaction substance. It is about ten billion times more powerful than chemical energy such as hydrogen and oxygen combustion. Antimatter would be the perfect rocket fuel, but the problem is that the basic component of antimatter, antiprotons, does not exist in nature and has to be manufactured. Antimatter development is ongoing and making some strides, but production of a propulsion system is far into the future. (NASA Photo no. 9906272)

because they allow repeated use of critical components such as rocket motors and structural elements, ELV advocates claim that actually they offer a false promise of savings. This is because all RLV savings are predicated on maximizing usage of a small number of vehicles over a very long period of time for all types of space launch requirements. Accordingly, cost savings are realized only when an RLV flies many times over many years. That goal is unattainable, they claim, because it assumes that there will be no (or very few) accidents in the reusable fleet throughout its life span.[56]

The reality, ELV advocates warn, is that the probability of all RLV components operating without catastrophic failure throughout the lifetime of the vehicle cannot be assumed to be 100 percent. Indeed, the launch reliability rate of even relatively "simple" ELVs—those without upper stages or space-craft propulsion modules and with significant operational experience—peaks at 98 percent with the Delta II, and that took thirty years of operations to achieve. To be sure, most ELVs achieve a reliability rate of 90–92 percent, again only after a maturing of the system has taken place. The Space Shuttle, a partially reusable system, has attained a launch reliability rate of over 99 percent, but only through extensive and costly redundant systems and safety checks. In the case of a new RLV, or a new ELV for that matter, a higher failure rate has to be assumed because of a lack of experience with the system. More-over, RLV use doubles the time of exposure of the vehicle to failure because it must also be recovered and be reusable after refurbishment. To counter this challenge, more and better reliability has to be built into the system, and this exponentially increases both R&D and operational costs.[57]

Designing for one use only, those arguing for ELV development suggest, simplifies the system enormously. One use of a rocket motor, guidance system, and the like means that the item only needs to function correctly one time. Acceptance of an operational reliability of 90 percent or even less would further reduce the costs incurred in designing and developing a new ELV. Indeed, many experts believe that reliability rates cannot be advanced more than another 1.5 percent above the 90 percent mark without enormous effort—effort that would be strikingly cost-inefficient.[58]

Some provocatively suggest that new ELVs should be designed with the types of payloads to be carried clearly in mind, accepting the risk inherent in a space launch environment where 90 percent reliability would be the norm. For expensive and one-of-a-kind scientific and military satellites, as well as expensive commercial spacecraft, spacecraft with a reliability rate and a higher price tag could be acceptable. But for most payloads, especially logistics sup-plies and the like going to the International Space Station, reliabilities as low as 80 percent might be acceptable. And, it goes without saying that for human space flights NASA's longstanding goal of 99.9 percent operational reliability is not too high a goal to seek. The debate continues and will not end until a truly outstanding launch vehicle emerges that achieves what has been pro-posed and thereby quells critics.

An Overview of This Volume

The essays that follow are attempts to capture the history of individual launch vehicles, both their development and their operation, from inception to the present. As such, they represent the only systematic attempt in nearly forty years to capture the history of the major launch vehicle programs of the United States. In 1964, just as the space age was beginning, Eugene M. Emme edited a collection of essays of a similar nature. Each of those was written by a prac-titioner and told in anecdotal form the story of the development of individual rockets from the early twentieth century to the early 1960s.[59] As important as Emme's *History of Rocket Technology* truly was—and readers shall see it cited in virtually every chapter—it represented only a small first step in the effort to record the history of important space access technology. Compendiums of statistical data about launchers also exist, but few seek to take a historical perspective on the subject and serve more as resources for current technical data useful to engineers rather than studies of how America arrived at the current state of launch capability.[60] This book seeks to build on these earlier efforts to present a rounded and generally comprehensive history of space access for the United States.

Any collection of historically oriented essays raise an immediate and valid criticism: Why are some subjects included while others, of seemingly equal importance, are omitted? There can never be a satisfactory answer to this critique. Our own criteria for choosing particular subjects and omitting others rested on three interrelated factors. One, the subjects had to be of interest and importance to the history of space access in and of themselves. There also had to have been a reasonable chance of furthering the study of rocketry and space access in the United States through profiling the subject—no attempt is made to discuss launch vehicles from Russia, Europe, or other nations. Two, although there has not been much work published on the overall history of space access, there seemed no reason to duplicate serious recent historical scholarship. Three, the subjects included were constrained somewhat by whether or not a scholar was willing to take on the subject. In some instances individuals were asked to write essays but were forced to decline for a variety of issues. We very much wanted to include an essay on the Agena upper stage, but for a variety of reasons that essay could not be completed.

The twelve chapters that follow relate directly to the development of space launch technology from the 1950s to the present. Each explores the origins and development of individual launchers. Many of the launchers emerged as ballistic missiles from the nation's national security arena and have been modified and updated ever since. A major theme in these essays is the process of evolution from nascent idea through first conceptions and test flights to operational status. They include discussions of not only technical components and systems but also of the formation of the invisible infrastructure and its relationship to the vehicles. Taken altogether, this collection presents a tentative exploration of broad themes in space access.

Conclusion

Since the beginning of space flight more than forty years ago, those who seek to travel in space have been, in essence, between a rocket and a hard place. The enormous release of energy made possible through the development of chemical rocket technology allowed the first generation of launch vehicles to free humanity and its robots from the constraints of Earth's gravity. It allowed the still exceptionally limited exploitation of space technology for all manner of activities important on Earth—communications, weather, Global Positioning System, and a host of other remote sensing satellites—to such an extent that many individuals in the United States today cannot conceive of a world in which they did not exist. This same chemical rocket technology made possible human flight into space, albeit for a very limited number of exceptional people, and the visiting of robotic probes from this planet to our neighbors in the solar system.

These have been enormously significant, and overwhelmingly positive, developments. They have also been enormously expensive, despite sustained efforts to reduce the cost of space flight. One development is to use rocket propulsion and, with new materials and clever engineering, to make a launcher that is not only recoverable but also robust. An other is to use air-breathing launchers, and thus to employ the potentially large mass fractions that air breathing theoretically promises in building a robust launcher. There are other options still. Most launch vehicle efforts throughout the history of the space age, unfortunately, have committed a fair measure of self-deception and wishful thinking. A large, ambitious program is created, hyped, and then fails as a result of unrealistic management, especially with regard to technical risk. These typically have blurred the line, which should be bright, between revolutionary, high-risk, high-payoff R&D efforts and evolutionary, low-risk, marginal-payoff efforts to improve operational systems. Efforts to break the bonds of this deception may well lead to remarkable new directions in future launcher development efforts. Only once that happens will we be able to escape the nether world "between a rocket and a hard place."

Acknowledgments

Whenever historians take on a project of historical analysis such as this, they stand squarely on the shoulders of earlier investigators and incur a good many intellectual debts.

The editors and authors would like to acknowledge the assistance of several individuals who aided in the preparation of this collection of essays. It was only through the assistance of several key people that we have been able to assemble the essays and put together this volume. For their many contributions in completing this project we wish especially to thank Jane Odom, Colin Fries, and Mark Kahn, who helped track down information and correct inconsistencies; Stephen J. Garber, who offered valuable advice; Nadine Andreassen and Louise Alstork, who helped with proofreading and compilation; the staffs of the NASA Headquarters Library and the Scientific and Technical Information Program, who provided assistance in locating materials; and archivists at various presidential libraries and the National Archives and Records Administration, who aided with research efforts. In addition to these individuals, we wish to acknowledge the following scholars who aided in a variety of ways: George W. Bradley, David Brandt, Andrew J. Butrica, Tom D. Crouch, Dwayne A. Day, Andrew Dunar, Robert H. Ferrell,G. Michael Green, Lori B. Garver, Michael H. Gorn, Charles J. Gross, John F. Guilmartin Jr., Barton C. Hacker, R. Cargill Hall, Richard P. Hallion, Roger Handberg, T.A. Heppenheimer, Francis T. Hoban, David A. Hounshell, Perry D. Jamieson, Stephen B. Johnson, W.D. Kay, Richard H. Kohn, Sylvia K. Kraemer, John Krige, Alan M. Ladwig, W. Henry Lambright, John M. Logsdon, John Lonnquest, John L. Loos, Beth McCormick, Howard E. McCurdy, Jonathan C. McDowell, George E. Mueller, Valerie Neal, Allan A. Needell, Michael J. Neufeld, Frederick I. Ordway III, Dominick Pisano, Anthony M. Springer, David Stumpf, Glen E. Swanson, and Stephen P. Waring. All of these people would disagree with some of the conclusions offered here, but such is both the boon and the bane of historical inquiry. I also wish to thank the authors of the individual articles for their patience and helpfulness.

Notes

1. More than fifty space access studies have reached this conclusion over the last forty years. See Roger D. Launius and Howard E. McCurdy, *Imagining Space: Achievements, Predictions, Possibilities, 1950–2050* (San Francisco: Chronicle Books, 2001), chap. 4; U.S. Congress, Office of Technology Assessment, *Launch Options for the Future: Special Report* (Washington, D.C.: Government Printing Office, 1984); Vice President's Space Policy Advisory Board, "The Future of U.S. Space Launch Capability," Task Group Report, November 1992, NASA Historical Reference Collection, NASA History Office, Washington, D.C. (hereafter cited as NASA Historical Reference Collection); NASA Office of Space Systems Development, *Access to Space Study: Summary Report* (Washington, D.C.: NASA Report, 1994).

2. G. Harry Stine, *Halfway to Anywhere: Achieving America's Destiny in Space* (New York: M. Evans, 1996).

3. The centrality of space technology to cold war foreign policy objectives is amply demonstrated in Walter A. McDougall's Pulitzer Prize–winning book, *The Heavens and the Earth: A Political History of the Space Age* (New York: Basic Books, 1995).

4. Steven J. Isakowitz, Joseph P. Hopkins Jr., and Joshua B. Hopkins, *International Reference Guide to Space Launch Systems*, 3d ed. (Reston, Va.: American Institute for Aeronautics and Astronautics, 1999), passim; David K. Stumpf, *Titan II: A History of a Cold War Missile Program* (Fayetteville: University of Arkansas Press, 2000); John L. Chapman, *Atlas: The Story of a Missile* (New York: Harper and Brothers, 1960); Jacob Neufeld, *The Development of Ballistic Missiles in the United States Air Force, 1945–1960* (Washington, D.C.: Office of Air Force History, 1990).

5. S. Everett Gleason, "Discussion at the 329th Meeting of the National Security Council, Wednesday, July 3, 1957," 5 July 1957, p. 2, NSC Records, DDE Presidential Papers, Dwight D. Eisenhower Library, Abilene, Kans.

6. See Roger D. Launius, John M. Logsdon, and Robert W. Smith, eds., *Reconsidering Sputnik: Forty Years Since the Soviet Satellite* (Amsterdam: Harwood Academic Publishers, 2000).

7. The standard account of Soviet space endeavors is Asif A. Siddiqi, *Challenge to Apollo: The Soviet Union and the Space Race, 1945–1974*, NASA SP-2000-4408 (Washington, D.C.: NASA, 2000).

8. See James J. Harford, *Korolev: How One Man Masterminded the Soviet Drive to Beat America to the Moon* (New York: John Wiley & Sons, 1997).

9. Wernher von Braun, "The Redstone, Jupiter, and Juno," in Eugene M. Emme, ed., *The History of Rocket Technology: Essays on Research, Development, and Utility* (Detroit: Wayne State University Press, 1964), pp. 107–22.

10. Loyd S. Swenson, James M. Grimwood, and Charles C. Alexander, *This New Ocean: A History of Project Mercury,* NASA SP-4201 (Washington, D.C.: NASA, 1966), pp. 422–36.

11. U.S. Senate Committee on Aeronautical and Space Sciences, NASA Authorization Subcommittee, *Transfer of Von Braun Team to NASA,* 86th Cong., 2d sess. (Washington, D.C.: Government Printing Office, 1960); Robert M. Rosholt, *An Administrative History of NASA, 1958–1963,* NASA SP-4101 (Washington, D.C.: NASA, 1966), pp. 46–47, 117–20.

12. Peter Always, *Rockets of the World* (Ann Arbor, Mich.: Saturn Publications, 1992), passim.

13. Roger E. Bilstein, *Stages to Saturn: A Technological History of the Apollo/Saturn,* NASA SP-4206 (Washington, D.C.: NASA, 1980), pp. 155–258; Linda Neumann Ezell, *NASA Historical Data Book,* vol. 2, NASA SP-4012 (Washington, D.C.: NASA, 1986), pp. 54–61.

14. Barton C. Hacker, "The Idea of Rendezvous: From Space Station to Orbital Operations, in Space-Travel Thought, 1895–1951," *Technology and Culture* 15 (July 1974): 373–88; Barton C. Hacker, "The Genesis of Project Apollo: The Idea of Rendezvous, 1929–1961," in *Actes 10: Historic des techniques* (Paris: Congress of the History of Science, 1971), pp. 41–46; Barton C. Hacker and James M. Grimwood, *On the Shoulders of Titans: A History of Project Gemini,* NASA SP-4203 (Washington, D.C.: NASA, 1977), pp. 1–26.

15. James M. Grimwood and Ivan D. Ertal, "Project Gemini," *Southwestern Historical Quarterly* 81 (January 1968): 393–418; James M. Grimwood, Barton C. Hacker, and Peter J. Vorzimmer, *Project Gemini Technology and Operations,* NASA SP-4002 (Washington, D.C.: NASA, 1969); Robert N. Lindley, "Discussing Gemini: A`Flight' Interview with Robert Lindley of McDonnell," *Flight International,* 24 March 1966, pp. 488–89.

16. Reginald M. Machell, ed., *Summary of Gemini Extravehicular Activity,* NASA SP-149 (Washington, D.C.: NASA, 1968).

17. *Gemini Summary Conference,* NASA SP-138 (Washington, D.C.: NASA, 1967); Ezell, *NASA Historical Data Book* 2:149–70.

18. Ezell, *NASA Historical Data Book* 2:58–59.

19. Roger E. Bilstein, "From the S-IV to the S-IVB: The Evolution of a Rocket Stage for Space Exploration," *Journal of the British Interplanetary Society* 32 (December 1979): 452–58; Richard P. Hallion, "The Development of American Launch Vehicles since 1945," in Paul A. Hanle and Von Del Chamberlain," eds., *Space Science Comes of Age: Perspectives in the History of the Space Sciences* (Washington, D.C.: Smithsonian Institution Press, 1981), pp. 126–32.

20. James E. Webb, Administrator, memorandum to Dr. George Mueller, Associate Administrator for Manned Space Flight, "Termination of the Contract for Procurement of Long Lead Time Items for Vehicles 516 and 517," 1 August 1968; W. R. Lucas, Deputy Director, Technical, memorandum to Philip E. Culbertson, NASA Headquarters, "Long Term Storage and Launch of a Saturn V Vehicle in the Mid-1980's," 24 May 1972; Dale D. Myers, Associate Administrator for Manned Space Flight, memorandum to James C. Fletcher, NASA Administrator, "Saturn V Production Capability," 3 August 1972; all in NASA Historical Reference Collection.

21. T.A. Heppenheimer, *The Space Shuttle Decision: NASA Search for a Reusable Space Vehicle,* NASA SP-4220 (Washington, D.C.: NASA, 1999); Dennis R. Jenkins, Space Shuttle: *The History of the Space Transportation System—The First 100 Missions* (Cape Canaveral, Fla.: Specialty Press, 2001).

22. Roger D. Launius, "NASA and the Decision to Build the Space Shuttle, 1969–72,"

Historian 57 (Autumn 1994): 17–34; John M. Logsdon, "The Decision to Develop the Space Shuttle," *Space Policy* 2 (May 1986): 103–19; John M. Logsdon, "The Space Shuttle Decision: Technology and Political Choice," *Journal of Contemporary Business* 7 (1978): 13–30.

23. NASA, "Space Shuttle Economics Simplified," 1992, copy in author's possession; House Committee on Science and Technology, Subcommittee on Space Science and Applications, *Operational Cost Estimates: Space Shuttle*, 94th Cong., 2d sess., 1976.

24. NASA, *The Post-Apollo Space Program: A Report for the Space Task Group* (Washington, D.C.: NASA, September 1969), p. 6.

25. George M. Low to Dale D. Myers, "Space Shuttle Objectives," 27 January 1970, NASA Historical Reference Collection.

26. In January 1970 Thomas O. Paine, Richard Nixon's appointee as the NASA Administrator, described a somber meeting with the president in which Nixon told him that both public opinion polls and political advisors indicated that the mood of the country suggested hard cuts in the space and defense programs. Memorandum by Thomas O. Paine, "Meeting with the President, January 22, 1970," 22 January 1970, NASA Historical Reference Collection; Caspar W. Weinberger interview by John M. Logsdon, 23 August 1977, NASA Historical Reference Collection.

27. Dale D. Myers, "The Shuttle: A Balancing of Design and Politics," in Francis T. Hoban, ed., *Issues in NASA Program and Project Management* (Springfield, Va.: National Technical Information Service, 1992), p. 43.

28. John M. Logsdon, "The Space Shuttle Program: A Policy Failure," Science 232 (30 May 1986): 1099–1105; Roger D. Launius, *NASA: A History of the U.S. Civil Space Program* (Malabar, Fla: Krieger, 1994), pp. 114–15.

29. Launch costs vary with circumstances—and circumstances definitely vary. Estimates similar to those on this chart, however, are used extensively within the space community to compare the four main U.S. expendable launchers. When a payload cannot be accommodated on Atlas II, it is an enormous jump in cost to put it on Titan IV. Of course, the Saturn V is the most economical of any of these vehicles, but its use would only be economical if one needed to launch an enormous payload. There are none approaching the 250,000-pound capability that the Saturn V had save the vehicles that flew to the Moon or a space station launched whole into orbit. See, John T. Correll, "Fogbound in Space," *Air Force Magazine*, January 1994.

30. Jack Scarborough, "The Privatization of Expendable Launch Vehicles: Reconciliation of Conflicting Policy Objectives," *Policy Studies Review* 10 (1991): 12–30.

31. Isakowitz, Hopkins, and Hopkins, *International Reference Guide*, passim.

32. "Commercial Space Launch Act of 1984, Public Law 98–575," in John M. Logsdon, gen. ed., *Exploring the Unknown: Selected Documents in the History of the U.S. Civil Space Program*, vol. 4, *Accessing Space*, NASA SP-4407 (Washington, D.C.: NASA, 1999), pp. 431–40.

33. Ibid.; Space Launch Policy Working Group, "Report on Commercialization of U.S. Expendable Launch Vehicles," 13 April 1983, p. 3, NASA Historical Reference Collection; W.D. Kay, "Space Policy Redefined: The Reagan Administration and the Commercialization of Space," *Business and Economic History* 27 (Fall 1998): 237–47; W. Henry Lambright and Dianne Rahm, "Ronald Reagan and Space Policy," *Policy Studies Journal* 17 (1989): 515–28.

34. Few individuals have yet spoken to the competing priorities that the shuttle was asked to fulfill. It seems more true as time passes, however, that the one-size-fits-all approach to technological challenges that the shuttle was asked to solve was unfair to the launch vehicle, the people who made it fly, and the organization that built and launched it. This would not be the first time in American history when such had taken place. The Air Force had been forced in the 1960s to accept a combination fighter and bomber, the F/B-111, when it recommended against it. That airplane proved a disaster from start to finish. The individuals operating the Space Shuttle soldiered on as best they could to fulfill all

expectations, but the task was essentially impossible. See Michael F. Brown, *Flying Blind: The Politics of the U.S. Strategic Bomber Program* (Ithaca, N.Y.: Cornell University Press, 1992); David S. Sorenson, *The Politics of Strategic Aircraft Modernization* (Westport, Conn.: Praeger, 1995).

35. David J. Whalen, "NASA, USAF, and the Provisioning of Commercial Launch Services to Geosynchronous Orbit," paper presented at "Developing U.S. Launch Capability: The Role of Civil-Military Cooperation" symposium, 5 November 1999, Washington, D.C.

36. *Aeronautics and Space Report of the President, 1986, NASA Annual Report* (Washington, D.C.: NASA, 1987), Appendix B.

37. "NSDD-254," in Logsdon, gen. ed., *Exploring the Unknown* 4:382–85.

38. John M. Logsdon and Craig Reed, "Commercializing Space Transportation," in ibid. 4:405–22.

39. "Commercial Space Launch Act Amendments of 1988," in ibid. 4:458–65.

40. Isakowitz, Hopkins, and Hopkins, *International Reference Guide*, passim; Stumpf, *Titan II;* Chapman, *Atlas;* Neufeld, *The Development of Ballistic Missiles.*

41. Isakowitz, Hopkins, and Hopkins, *International Reference Guide*, passim.

42. Howard E. McCurdy, "The Cost of Space Flight," *Space Policy* 10 (November 1994): 277–89.

43. Rick W. Sturdevant, "The United States Air Force Organizes for Space: The Operational Quest, 1943–1993," in Roger D. Launius, ed., *Organizing for the Use of Space: Historical Perspectives on a Persistent Issue*, AAS History Series, vol. 18 (San Diego: Univelt, 1995), pp. 63–86.

44. "National Space Transportation Policy," 5 August 1994, in Logsdon, gen. ed., *Exploring the Unknown* 4:626–31.

45. U.S. House Of Representatives, "Department of Defense Appropriations Bill, 1995," Committee Report, 103d Cong., 2d sess. (PL 103-562), 27 June 1994.

46. NASA Office of Space Systems Development, "Access to Space Study—Summary Report," January 1994, and DoD, "Space Launch Modernization Plan—Executive Summary," May 1994, both in Logsdon, gen. ed., *Exploring the Unknown* 4:584–626; NASA News Release 95–205, "NASA to Pursue Non-Competivice Shuttle Contract with U.S. Alliance," 7 November 1995, NASA Historical Reference Collection; Craig Covault, "United Space Alliance Leads Shuttle Operations Revolution," *Aviation Week & Space Technology*, 16 June 1997, pp. 204–5, 210.

47. Andrew J. Butrica, "The Commercial Launch Industry, Reusable Space Vehicles, and Technological Change," *Business and Economic History* 27 (Fall 1998): 212–21.

48. These include, among others, Kistler Aerospace Corporation, Rotary Rocket Company, Pioneer Rocketplane, Kelly Space & Technology, Inc., Beal Aerospace, Advent Launch Services, Platform International's, and Space Access.

49. Paul Proctor, "Sea Launch Venture Eyes Mid-1998 First Flight," *Aviation Week & Space Technology*, 29 July 1996, pp. 56–57; John E. Draim, "The U.S. Navy's Hydra Project, and Other Floating Launch Programs," IAA-97-IAA.2.2.02, paper delivered at 48th International Astronautical Congress, 6–10 October 1997, Turin, Italy.

50. This was the argument made to obtain approval for the Space Shuttle. See NASA, *Post-Apollo Space Program*, pp. 1, 6.

51. Fred Hiatt, "Space Plane Soars on Reagan's Support," *Washington Post*, 6 February 1986, p. A4; Roger Handberg and Joan Johnson-Freese, "If Darkness Falls: The Consequences of a United States No-Go on a Hypersonic Vehicle," *Space Flight* 32 (April 1990): 128–31; Linda R. Cohen, Susan A. Edelman, and Roger G. Noll, "The National Aerospace Plane: An American Technological Long Shot, Japanese Style," *American Economic Review* 81 (1991): 50–53; Roger Handberg and Joan Johnson-Freese, "Pursuing the Hypersonic Option Now More than Ever," *Space Commerce* 1 (1991): 167–74.

52. Roger Handberg and Joan Johnson-Freese, "NASP as an American Orphan: Bureaucratic Politics and the Development of Hypersonic Flight," *Spaceflight* 33 (April 1991): 134–37; Larry E. Schweikart, "Hypersonic Hopes: Planning for NASP," *Air Power History* 41 (Spring 1994): 36–48; Larry E. Schweikart, "Managing a Revolutionary Technology, American Style: The National Aerospace Plane," *Essays in Business and Economic History* 12 (1994): 118–32; Larry E. Schweikart, "Command Innovation: Lessons from the National Aerospace Plane Program," in Roger D. Launius, ed., *Innovation and the Development of Flight* (College Station: Texas A&M University Press, 1999), pp. 299–323.

53. Frank Sietzen, "VentureStar™ Will Need Public Funding," *SpaceDaily Express*, 16 February 1998, NASA Historical Reference Collection.

54. Greg Easterbrook, "The Case Against NASA," *New Republic*, 8 July 1991, pp. 18–24; Alex Roland, "Priorities in Space for the USA," *Space Policy* 3 (May 1987): 104–14; Alex Roland, "The Shuttle's Uncertain Future," *Final Frontier*, April 1988, pp. 24–27.

55. James A. Vedda, "Long-term Visions for U.S. Space Policy," background paper prepared for the Subcommittee on National Security, International Affairs, and Criminal Justice of the House Committee on Government Reform and Oversight, May 1997.

56. Barbara A. Luxenberg, "Space Shuttle Issue Brief #IB73091," Library of Congress Congressional Research Service Major Issues System, 7 July 1981, NASA Historical Reference Collection; *Economic Analysis of New Space Transportation Systems: Executive Summary* (Princeton, N.J.: Mathematica, 1971); General Accounting Office, *Analysis of Cost Estimates for the Space Shuttle and Two Alternate Programs* (Washington, D.C.: General Accounting Office, 1973); William G. Holder and William D. Siuru Jr., "Some Thoughts on Reusable Launch Vehicles," *Air University Review* 22 (November–December 1970): 51–58; U.S. Congress, Office of Technology Assessment, *Reducing Launch Operations Costs: New Technologies and Practices* (Washington, D.C.: GPO, 1988).

57. Stephen A. Book, "Inventory Requirements for Reusable Launch Vehicles," paper presented at Space Technology & Applications International Forum (STAIF-99), copy in possession of author.

58. B. Peter Leonard and William A. Kisko, "Predicting Launch Vehicle Failure," *Aerospace America*, September 1989, pp. 36–38, 46; Robert G. Bramscher, "A Survey of Launch Vehicle Failures," *Spaceflight* 22 (November–December 1980): 51–58.

59. Eugene M. Emme, ed., *The History of Rocket Technology: Essays on Research, Development, and Utility* (Detroit: Wayne State University Press, 1964).

60. Isakowitz, Hopkins, and Hopkins, *International Reference Guide*, is the most significant of these works.

– 1 –

Rocketry and the Origins of Space Flight

Ray A Williamson and Roger D. Launius

Introduction

Curiosity about the universe and other worlds has been one of the few constants in the history of humankind. Prior to the twentieth century, however, there was little opportunity to explore the universe except in fiction and through astronomical observations. These early explorations led to the compilation of a body of knowledge that inspired and in some respects informed the efforts of certain scientists and engineers who began to think about applying rocket technology to the challenge of space flight in the early part of the twentieth century. These individuals were essentially the first space-flight pioneers, translating centuries of dreams into a reality that matched in some measure the expectations of the public that watched and the governments that supported their efforts. During the period between 1926, when Robert H. Goddard launched his first rocket, and 1957, when the first orbital spacecraft was launched, a dedicated group of rocketeers made the space age a reality, although by the beginning of World War II much of the technology was being developed by government organizations as potential weapons.

Progenitors of the Space Age

While the technology of rocketry was moving forward on other fronts, some individuals began to see its use for space travel. There were three great pioneering figures in this category—collectively, they were the progenitors of the modern space age. The earliest was the Russian theoretician Konstantin Eduardovich Tsiolkovsky. An obscure schoolteacher in a remote part of tsarist

Russia in 1898, he submitted for publication to the Russian journal *Nauchnoye Obozreniye (Science Review)* a work based upon years of calculations that laid out many of the principles of modern space flight. His article was not published until 1903, but it opened the door to future writings on the subject. In it Tsiolkovsky described in depth the use of rockets for launching orbital space ships. Tsiolkovsky continued to theorize on the subject of space flight until his death, describing in great detail both methods of flight and the technical requirements of space stations. Significantly, he never had the resources—nor perhaps the inclination—to experiment with rockets himself. His theoretical work, however, influenced later rocketeers both in his native land and abroad and served as the foundation of the Soviet space program.[1]

A second rocketry pioneer was Hermann Oberth, by birth a Transyl-vanian but by nationality a German. Oberth began studying the nature of space flight about the time of World War I and published his classic study *Die Rakete zu den Planetenraümen (Rockets in Planetary Space)* in 1923. It was a thorough discussion of almost every phase of rocket travel. He posited that a rocket could travel in the void of space and that it could move faster than the velocity of its own exhaust gases. He noted that with the proper velocity a rocket could launch a payload into orbit around the earth, and to accomplish this goal he reviewed several propellant mixtures to increase speed. He also designed a rocket that he believed had the capability to reach the upper atmosphere by using a combination of alcohol and hydrogen as fuel. Oberth also discussed other aspects of rocketry and the prospects of space travel. He became the father of German rocketry.[2] Among his protégés was Wernher von Braun, the senior member of the rocket team that built NASA's Saturn launch vehicle for the trip to the Moon in the 1960s.[3]

Finally, the American Robert H. Goddard pioneered the use of rockets for space flight.[4] Motivated by reading science fiction as a boy, Goddard became excited by the possibility of exploring space. In 1901 he wrote "The Navigation of Space," a short paper that argued that movement could take place by firing several cannon, "arranged like a 'nest' of beakers" in a single direction. He tried unsuccessfully to publish this article in *Popular Science News*.[5] At his high school oration in 1904 he summarized his future life's work: "It is difficult to say what is impossible, for the dream of yesterday is the hope of today and the reality of tomorrow."[6] In 1907 he wrote another paper on the possibility of using radioactive materials to propel a rocket through interplanetary space. He sent this article to several magazines, and all rejected it.[7] Still not dissuaded, as a young physics graduate student he worked on rocket propulsion and actually received two patents in 1914. One was the first for a rocket using solid and liquid propellants and the other for a multistage rocket.[8]

After a stint with the military in World War I, where he worked on solid rocket technology for use in combat, Goddard became a professor of physics at Clark College (later University) in Worcester, Massachusetts. There he turned his attention to liquid rocket propulsion, theorizing that liquid oxygen and liquid hydrogen were the best propellants but learning that oxygen and gasoline were less volatile and therefore more practical. To support his investigations, Goddard applied to the Smithsonian Institution for assistance in 1916 and received a five-thousand-dollar grant from its Hodgkins Fund.[9] His research was ultimately published by the Smithsonian as the classic study *A Method of Reaching Extreme Altitudes* in 1919. In it Goddard argued from a firm theoretical base that rockets could be used to explore the upper atmosphere. Moreover, he suggested that with a velocity of 17,500 miles per hour (mph), without air resistance, an object could escape Earth's gravity and head into infinity, or to other celestial bodies.[10] This became known as Earth's "escape velocity."

It also became a great joke for those who believed space flight either impossible or impractical. Some ridiculed Goddard's ideas in the popular press, much to the consternation of the already-shy Goddard. Soon after the appearance of his publication, he commented that he had been "interviewed a number of times, and on each occasion have been as uncommunicative as possible."[11] The *New York Times* was especially harsh in its criticisms, referring to him as a dreamer whose ideas had no scientific validity. It also compared his theories to those advanced by novelist Jules Verne, indicating that such musing is "pardonable enough in him as a romancer, but its like is not so easily explained when made by a savant who isn't writing a novel of adventure." The *Times* questioned both Goddard's credentials as a scientist and the Smithsonian's rationale for funding his research and publishing his results.[12]

The negative press Goddard received prompted him to be even more secretive and reclusive. It did not, however, stop his work, and he eventually registered 214 patents on various components of rockets. He concentrated on the design of a liquid-fueled rocket, the first such development, and the attendant fuel pumps, motors, and control components. On 16 March 1926 near Auburn, Massachusetts, Goddard launched his first successful rocket, a liquid oxygen and gasoline vehicle that rose 184 feet in 2.5 seconds.[13] This event heralded the modern age of rocketry. He continued to experiment with rockets and fuels for the next several years. A spectacular launch took place on 17 July 1929, when he flew the first instrumented payload—an aneroid barometer, a thermometer, and a camera—to record the readings. The launch failed; after rising about 40 feet the rocket turned and struck the ground 171 feet away. It caused such a fire that neighbors complained to the state fire marshal and Goddard was enjoined from making further tests in Massachusetts.[14]

This experience, as well as his personal shyness, led him to seek a more remote setting to conduct his experiments. His ability to shroud his research in mystery was greatly enhanced by Charles A. Lindbergh, fresh from his transatlantic solo flight, who helped Goddard obtain a series of grants from the Guggenheim Fund fostering aeronautical activities. This enabled him to pur-

Image 1-1: Robert H. Goddard is shown in 1930 working on one of his liquid-fueled rockets. (NASA Photo no. G-32-04)

chase a large tract of desolate land near Roswell, New Mexico, and to set up an independent laboratory to conduct rocket experiments far away from anyone else. Between 1930 and 1941 Goddard carried out more ambitious tests of rocket components in the relative isolation of New Mexico, much of which he summarized in a 1936 study, *Liquid-Propellant Rocket Development*. The culmination of this effort was a successful launch of a rocket to an altitude of 9,000 feet in 1941.[15] In late 1941 Goddard entered naval service and spent the duration of World War II developing a jet-assisted takeoff (JATO) rocket to shorten the distance required for heavy aircraft launches. Some of this work led to the development of the throttleable Curtis-Wright XLR25–CW-1 rocket engine that later powered the Bell X-2. Goddard did not live to see this; he died in Annapolis, Maryland, on 10 August 1945.[16]

Goddard accomplished much, but because of his secrecy few people knew about his achievements during his lifetime. These included the following pioneering activities:

> Theorizing on the possibilities of jet-powered aircraft, rocket-borne mail and express, passenger travel in space, nuclear-powered rockets, and journeys to the Moon and other planets (1904–45).
>
> First mathematical exploration of the practicality of using rockets to reach high altitudes and achieve escape velocity (1912).
>
> First patent on the idea of multistage rockets (1914).
>
> First experimental proof that a rocket could provide thrust in a vacuum (1915).
>
> The basic idea of antitank missiles, developed and demonstrated during work for the U.S. Army in World War I. This was the prototype for the bazooka infantry weapon (1918).
>
> First publication in the United States of the basic mathematical theory underlying rocket propulsion and space flight (1919).
>
> First development of a rocket motor burning liquid propellants (1920–26).
>
> First development of self-cooling rocket motors, variable-thrust rocket motors, practical rocket landing devices, pumps suitable for liquid fuels, and associated components (1920–41).
>
> First design, construction, and launch of a successful liquid-fueled rocket (1926).
>
> First development of gyro-stabilization equipment for rockets (1932).
>
> First use of deflector vanes in the blast of the rocket motor as a method of stabilizing and guiding rockets (1932).[17]

The U.S. government's recognition of Goddard's work came in 1960, when the Department of Defense (DoD) and the NASA awarded his estate $1 million for the use of his patents.[18]

Parallel Developments

Concomitant with Goddard's research into liquid-fuel rockets, and perhaps more immediately significant because the results were more widely disseminated, were activities in several other quarters. By the late 1930s, experimenters in Germany, Russia, and the United States had successfully flown liquid-fueled rockets of various types and capacities. Many experimenters belonged to rocket societies, which assisted the progress of rocket development both by developing new technological approaches and by creating broad interest in rocketry.[19]

The largest and most significant of these ventures was the German organization Verein für Raumschiffahrt (Society for Spaceship Travel, or VfR). Although space-flight aficionados and technicians had organized at other times and in other places, the VfR under the able leadership of Berlin engineer Max Valier emerged soon after its founding on 5 July 1927 as the leading space-travel group. Specifically organized to raise money to test Oberth's rocketry ideas, it was successful in building a base of support in Germany, publishing a magazine and scholarly studies, and constructing and launching small rockets. One of VfR's strengths from the beginning, however, was its ability to publicize both its activities and the dream of space flight.[20]

The VfR made good on some of those dreams on 21 February 1931, when it launched the liquid oxygen–methane liquid-fuel rocket HW-1 near Dessau to an altitude of approximately 2,000 feet. The organization's public relations arm went into high gear after this mission and emphasized the launch's importance as the first successful European liquid-fuel rocket flight.[21] Wernher von Braun, then a neophyte learning the principles of rocketry, was both enthralled with this flight and impressed with the publicity it engendered. Later, he became the quintessential and movingly eloquent advocate for the dream of space flight and a leading architect of its technical development. He began developing both skills while working with the VfR.[22]

There were other national rocketry societies that sprang up during this same period, each contributing to the base of technical knowledge and the popular conception of space flight. The American Interplanetary Society (AIS) was one of the more powerful of these institutions. Organized in 1930, within two years the AIS had begun a program of rocket experimentation. On 12 November 1931 the AIS tested its first static test of a liquid oxygen–gasoline rocket. It actually launched a rocket on 14 May 1932, attaining an altitude of only 250 feet. But its second and last launch on 9 September 1934 went over 1,300 feet into the atmosphere. Because of the great cost and risk to people involved, after this launch the group concentrated throughout the rest of the 1930s on static firings of engines and published results of its research, the culmination of which proved significant for later experimentation in rocketry. Almost concomitant with its withdrawal from rocket experimentation, and out of a desire to improve the image of the organization, the AIS changed its name to the American Rocket Society (ARS).[23]

In December 1941, just as the United States was entering World War II, four members of the ARS formed Reaction Motors, Inc., the first U.S. firm to build liquid-fuel rockets. Using ideas on cooling originally learned from reading one of Eugen Sänger's[24] papers, the Reaction Motors team developed a regeneratively cooled rocket engine[25] that circulated liquid oxygen (LOX)[26] in a cooling jacket around the engine.[27] In 1947 the Army used this engine in the

Bell X-1, the first aircraft to purposely exceed the speed of sound.[28] In the USSR, several groups emerged to study rocketry, the most important of which was the Moscow Group for the Study of Reactive Motion (MosGIRD), led by Sergei P. Korolev, who until his death in 1966 led the Soviet rocket program.[29]

Restructuring of the AIS may also have been prompted in part by the organization of the British Interplanetary Society (BIS) on 13 October 1933 at Liverpool, England. Oriented more toward theoretical studies than rocket experimentation, in the 1930s the BIS became a haven for writers and other intellectuals interested in the idea of space flight. By September 1939, at the beginning of World War II, the BIS numbered about one hundred members, including several Germans. The BIS periodical, the *Journal of the British Interplanetary Society*, began publication in January 1934, and it quickly became a persistent and powerful voice on behalf of space exploration. The BIS did not undertake field work with rockets (although several members did conduct some crude experiments with potential solid propellants), but in 1938–39 its members designed a lunar landing vehicle which later influenced the Lunar Module used in Project Apollo during the 1960s.[30]

While both the individual and societal precursors of space flight struggled along as best they could, beginning in 1936 the Guggenheim Aeronautical Laboratory at the California Institute of Technology (GALCIT) in Pasadena, California, began to pursue its own rocket research program.[31] Frank J. Malina, a young Caltech Ph.D. student at the time, persuaded GALCIT to adopt a research agenda for the design of a high-altitude sounding rocket and enthusiastically began experimentation. Using some of the ideas from the research of Eugen Sänger in Austria and Goddard in New Mexico, Malina and a design team—composed of, among others, Hsue Shen Tsien, a Chinese national who was later deported and became the architect of the intercontinental ballistic missile (ICBM) and space launcher programs for the People's Republic of China—began work. Nobel Prize–winning physicist Robert A. Millikin, president of Caltech, was especially supportive of this work because he wanted to use sounding rockets for cosmic ray research. Millikin and Malina tried to persuade Goddard to join this research project. Malina even visited him at his complex near Roswell in August 1936—but Goddard refused.[32] In a letter revealing much of Goddard's secretiveness, in September 1936 he wrote disparagingly of Malina to Robert Millikin. Goddard commented that he had tried to help Malina with some of his questions, but "I naturally cannot turn over the results of many years of investigation, still incomplete, for use as a student's thesis."[33]

Beginning in late 1936 Malina and his colleagues started the static testing of rocket engines in the canyons above the Rose Bowl, with mixed results. It was not until 28 November 1936, for example, that the motor ran at all, and then only for 15 seconds. A series of tests thereafter brought incremental improvements; a year later Malina and an associate had learned enough to distill the results into the first scholarly paper on rocketry to come out of GALCIT. The test results showed that with proper fuels and motor efficiency a rocket could be constructed with the capability to ascend as high as 1,000 miles.[34]

Because of this research GALCIT's rocketry team obtained funding from outside sources, among them Gen. Henry H. (Hap) Arnold, soon to become the U.S. Army Air Corps Chief of Staff; he visited GALCIT in the spring of 1938 and was enthusiastic about the work on rockets he saw Malina and co-workers doing. That fall he arranged for additional funding from the National Academy of Sciences to proceed with the project, with the specific goal of research on the possibilities of rocket-assisted takeoff for aircraft. The committee that approved this funding did so with some concern that it might be money poorly spent. Finally, Jerome Hunsaker, head of the Guggenheim Aeronautics Department of the Massachusetts Institute of Technology, told the committee that he would be glad to have Theodore von Kármán (1881–1963), director of GALCIT, "take the Buck Rogers job."[35] GALCIT accepted the task, and beginning in 1939 Malina and his rocket team began working on what became the JATO project. Although Malina always expressed misgivings about working on weaponry, and after World War II accepted employment with the United Nations so he could help prevent such occurrences from taking place again, the difficult political climate in 1939 prompted him to support the development of U.S. military capability as a deterrent to fascism. As a result, Malina and GALCIT engaged throughout the war years in rocketry research for military purposes.[36]

The Basics of Chemical Rocket Technology

Space access has rested firmly on the shoulders of chemical rockets for boosting payloads into Earth orbit and beyond. From the first experiments by Robert H. Goddard in the 1920s through the pathbreaking V-2 missile of World War II and the mighty Saturn V Moon rocket, to the most sophisticated spacecraft ever built, the Space Shuttle, the basic principles have not changed.[37] However, chemical rockets are notoriously inefficient and costly to operate. In future generations, space flight must move beyond this technology to embrace another approach to reaching space.

With the dawn of the twentieth century, the technical developments that led to rocket propulsion made possible the beginning of the space age. The rocket is a reaction device, based on Sir Isaac Newton's third law of motion: For every action there is an equal and opposite reaction. The action of a rocket is created by the pressure of gas inside a rocket engine that must escape through a nozzle. As the gas escapes at ever-higher velocities, it produces thrust. The reaction, therefore, propels the rocket upward. It will continue to accelerate until all of the propellant in the rocket engine is gone.[38] Unfortunately, this is an enormously simple explanation of an enormously complex technology. The story of space access in the twentieth century has been one of systematically applying this principle to technological systems and incrementally advancing the capabilities of those systems.

The unbroken path in rocketry from Goddard to the Space Shuttle rests on one simple mathematical equation. The product of the propellant mass flow (m) and its exhaust velocity (C) equals the thrust (T) generated by a rocket: T = mC. The higher the exhaust velocity, the more thrust generated per propellant mass flow. Rocketeers rate engines using two core measures. The first of these is the thrust of the rocket (rated in pounds-force, or lbf), that which gives a rocket the kick necessary to overcome gravity and reach orbit. It is analogous to the horsepower rating used in the automotive industry. It is, in essence, the heavy lifting of space flight and requires achieving an escape velocity of 17,500 mph. The heavier the vehicle, the more difficult the task of attaining this speed.

Accordingly, most launch vehicles are characterized by thrust generated and by the payload mass delivered to low-Earth orbit (LEO) and to geosynchronous orbit. Each of the three Space Shuttle Main Engines, as an example, has an average thrust of 470,000 lbf in a vacuum. Combined[39] with the two Solid Rocket Boosters, each producing 3,100,000 lbf, the Space Shuttle can deliver a payload of up to 55,000 pounds to LEO. Depending on exact altitude and on the inclination of the orbit over the earth's surface, payload delivery capabilities always rise or fall somewhat. Many aerospace professionals make a living calculating to multiple decimal points these critical parameters for every space launch.[40]

The second measurement used by rocketeers to rate engines is known as specific impulse (I_{sp})—the number of seconds a rocket engine can produce one pound of thrust from one pound of propellant. The I_{sp}, in effect, measures the efficiency of the rocket engine. If thrust is analogous to horsepower in an automobile, I_{sp} is analogous to miles per gallon of gas. I_{sp} is usually expressed in terms of seconds; for instance, an engine might have an I_{sp} of 300 seconds or 300 s. The best chemical rocket engines, the ones that power all of the major launch vehicles, are limited theoretically to an I_{sp} of about 500 s. As an example, the Space Shuttle Main Engine I_{sp} generates about 455 s, and it represents the current state of the art in launch technology.

All launch systems, however, are dependent on the types of chemicals used as propellant in generating their I_{sp}. Hydrazine, which powers many small rockets, has a value of about 200 s. A combination of kerosene and oxygen— a fuel used by Wernher von Braun when developing the V-2 ballistic missile during World War II—has an I_{sp} of about 350 s, while a liquid hydrogen– liquid oxygen mixture used on the shuttle can generate an I_{sp} of 450 s. Depending on the propellant system, the complexity of the engine, the efficiency of the nozzle, and a host of other attributes, engineers may be able to tinker around the edges of these chemical boundaries to deliver a slightly higher I_{sp} for any given engine.

A key issue wrestled with by rocketeers throughout the space age has been how best to maximize the efficiency of these chemical propulsion systems. The major aerospace corporations have honed the launch capabilities—especially the workhorse Atlas, Delta, and Titan vehicles—to a high degree and have achieved a level of maturity for their systems that they are loathe to abandon in favor of an entirely different approach to space access. But the I_{sp} delivered by chemical rockets is actually quite inefficient and certainly insufficient for long-duration interplanetary space flight. Because of chemical propulsion's limitations, robotic spacecraft (the only ones that have been sent) must use the slingshot effect of gravity-assist trajectories to explore the outer planets. This necessitates limited launch windows when the planets are properly lined up, and even then the flights take years to complete. Clearly, human missions to Mars are difficult with chemical rockets because of the time of flight and the long exposure to dangerous levels of radiation that the crew would face in deep space. Human flight is most assuredly impractical for missions beyond Mars for this same reason.

Developing the Vengeance Weapon 2 (V-2)

Modern rocketry is a legacy of World War II and its aftermath, the cold war. During World War II, France, Germany, Japan, the United Kingdom, and the United States attempted to build rockets in support of the war effort. Of these, only Germany seriously attempted to develop large rockets. Beginning in 1932, within about a decade a team of scientists and engineers led by Capt. Walter R. Dornberger designed, built, and tested the V-2 rocket. Starting in September 1944, the German army used the V-2 as an early ballistic missile to terrorize Allied military troops and civilian populations.

German rocketeers quickly went to work designing and testing a workable liquid-fuel engine. By December 1934 a team under Wernher von Braun had succeeded in building a motor powered by liquid oxygen and alcohol, which it used to send two small, gyroscopically controlled rockets about 6,500 feet high.[41] The team designated this design Aggregat-2, or A-2. The team's success attracted the interest of the German military, which wished to use rocket engines to assist propeller-driven aircraft at takeoff and to power aircraft and missiles. Out of this interest came a joint Army–Air Force establishment centered at Peenemünde, an island in the Baltic Sea.

By 1936 the experimenters had arrived at the basic design of the A-4, the vehicle that a few years later became the V-2 missile. Further design and testing produced an engine capable of generating the remarkable (for the time) thrust of 59,500 lbf[42] for 68 seconds. This engine, which was regeneratively

Image 1-2: The German V-2, developed by Wernher von Braun and his rocket team at Peenëmunde, was the first large ballistic missile. Shown here is a test launch in Germany during 1944. (NASA photo)

cooled, operated at a chamber pressure of 750 pounds per square inch. Kerosene and liquid oxygen were fed to the combustion chamber at rates of 50 gallons or more per second by steam-driven centrifugal pumps.[43] The A-4 stood nearly 50 feet high and was 5.4 feet in diameter. Fully loaded with fuel and a payload of 2,310 pounds, it weighed 28,229 pounds, and was capable of flying up to 3,500 miles per hour. The A-4 had a range of 190 miles and could reach 60 miles altitude. After the first two test flights ended in failure, the A-4 was successfully flown on 3 October 1942. Twenty-three months and some 65,000 technical alterations later, the A-4 became the operational Vengeance Weapon 2, the name given to the missile by Hitler. By early 1945, when Allied troops first entered the country, the German army had fired 3,225 warhead-carrying V-2 rockets, most of them toward London and Antwerp.

On 2 May 1945, von Braun, Dornberger, and 116 other rocket specialists surrendered to American officials in the Austrian Tyrol town of Reutte, just south of Bavaria.[44] Several months later, they were taken to the United States, along with about one hundred V-2 rockets, many rocket components, and truckloads of scientific documents. This "rocket team" formed one of the foundations of U.S. progress in missiles and rocket development for several decades to come.

Post–World War II American Rocket Research and Development

GALCIT, operated under the sponsorship of the U.S. Army, was renamed the Jet Propulsion Laboratory[45] (JPL) in 1944. Concentrating on rocketry during the 1940s and 1950s, it developed the most sophisticated understanding on the technology of space launch native to the United States. Among other rocket technologies, JPL developed slow-burning rocket propellant and storable liquid propellants that proved extremely useful after World War II.Even as these activities were taking place, in 1943 GALCIT engineers concluded in a report to the U.S. Army Air Forces that "the development of a long-range rocket projectile is within engineering feasibility" and asked for funding to bring it to a reality.[46] With some investment financing from the Army, JPL conducted research on engines and other components. Then on 16 January 1945, Malina sent to the Army's Ordnance Section a proposal for a liquid-fuel sounding rocket that would be able to launch a 25-pound payload to an altitude of 100,000 feet. What emerged from these recommendations was a decision to develop the WAC Corporal, first flown on 11 October 1945; the WAC Corporal became a significant launch vehicle in postwar rocket research, attaining an altitude of 45 miles.[47] An engine fueled by storable hypergolic fuels—red fuming nitric acid and aniline—powered this rocket.[48]

Intensive U.S. launch vehicle research and development essentially began with the testing of captured German V-2s on American soil following World

Image 1-3: Following World War II, American military leaders captured several V-2s, as well as Wernher von Braun and several of his colleagues, and began research using the rockets at the White Sands Proving Grounds in New Mexico. This photograph was taken during preparations for a launch on 10 May 1946. (NASA Photo)

War II. Nowhere can the close bonds between the development of weapon-carrying missiles and Earth-to-orbit launch vehicles be seen more clearly than in the use of these missiles to jump-start U.S. rocket development. The United States employed them not only to catch up to the conquered Germans in missile development but also to push the boundaries of space flight for scientific purposes. The V-2 technologies served as foundations for the development of U.S. sounding rockets and provided a vehicle for the first U.S. space science efforts, under the guidance of James van Allen, who directed the government's Upper Atmosphere Rocket Research Panel.[49] The investigation of V-2 technologies by the government and U.S. industry strengthened bonds that had begun in World War II. New firms were formed and old ones strengthened and enhanced by the partnership.

The Army set up launch facilities at White Sands, New Mexico, and hired the General Electric Company (GE) to carry out a long series of tests with the V-2s. The Army and GE test flew sixty-seven V-2s between 1946 and 1951, most of them at White Sands. Under Project Hermes, as the test program was called, GE and the Army also developed several different missiles. These included a series of launchers called Bumper, which used the WAC[50] Corporal as a second stage on top of a modified V-2.[51] Although the weight and propellant advantages of using several rocket stages, in which progressively smaller rockets took over after the previous stage had expended its propellant and fallen back to Earth, was well known, it had not been tried in a large rocket. Earlier experimenters had faced technical difficulties in both igniting an upper stage in flight and separating the two while controlling the upper stage, and indeed in the lack of a reliable upper-stage rocket. The WAC Corporal had proved sufficiently reliable as a sounding rocket. Testing of the Bumper was undertaken in part to reach high altitudes and in part to test the various techniques needed to control the ignition, separation, and control of a second stage. On 24 February 1949, one of these two-stage rockets reached into outer space at an altitude of 244 miles—a record that stood for several years. Bumper 8, the last of the series, was the first rocket to be launched from Cape Canaveral, Florida, on 24 July 1950. During these tests, the Army and GE experimented with developing a tactical missile using radio-inertial guidance.[52]

More successful than the WAC Corporal was the Aerobee, a scaled-up version of that earlier rocket developed by JPL, which could launch at a very economical cost a sizable payload to an altitude of 130 miles. The reliable little booster enjoyed a long career from its first instrumented firing on 27 November 1947 until the 17 January 1985 launch of the 1,037th and last Aerobee. Additionally, the Naval Research Laboratory (NRL) was involved in sounding rocket research, non-orbital instrument launches, using the Viking launch vehicle built by the Glenn L. Martin Company. *Viking 1* was launched from White Sands on 3 May 1949, while the twelfth and last Viking took off on 4 February 1955. The program produced significant scientific information about the upper atmosphere and took impressive high-altitude photographs of Earth. Most important, Viking pioneered the use of a gimbaled engine to control flight and paved the way for later orbiting scientific satellites.[53]

By making copies of the V-2 engines beginning in 1949, North American

Image 1-4: Bumper 8, shown here during launch on 24 July 1950. The V-2/Wac Corporal combination was the first missile launched at Cape Canaveral, Florida. (NASA Photo)

Aviation, Inc., was able to gain valuable experience in rocket motor design and construction that the company soon used to good effect in developing larger and more powerful rocket engines. North American was able by March 1950 to build and conduct successful tests on a LOX–alcohol engine that generated 75,000 lbf (the Experimental Liquid Rocket 43, or XLR43). By January 1956, Rocketdyne, North American's newly named rocket division,[54] had produced a version (the XLR83) containing three firing chambers that generated a then-astounding 415,000 lbf, burning LOX and kerosene.[55] Rocketdyne's engine was originally destined for incorporation into the experimental Navaho cruise missile, a development program begun by the U.S. Army Air Forces in 1946.[56] In July 1957, in a budget-cutting measure, the Army canceled the Navaho. However, the effort that had gone into developing the XLR83 resulted in a powerful engine that, in various modifications, served as the basis for many of America's future missiles and space launch vehicles.[57] For example, the lessons learned in building the Navaho engine were later put to good use for the very large F-1 engine, which powered the first stage of the Saturn V. The Navaho program produced a number of other technical advances, including the development of chemical milling for reducing structural weight while retaining strength, and the use of a titanium skin. It also developed an inertial guidance device that used the first transistorized launch vehicle computer.[58]

The cold war tensions of the 1950s, the development of nuclear weapons, and the Korean War spawned several additional missile-building programs.[59] Among them was the Redstone rocket, which originated in the Hermes C project. In July 1950, the Army Chief of Ordnance asked the Ordnance Guided Missile Center at Redstone Arsenal in Huntsville, Alabama, to study the feasibility of building a missile with a range of 500 miles. Wernher von Braun's team of scientists and engineers, which the Army had just moved from Texas to the Redstone Arsenal, was given the task. The team decided to use the XLR43, the engine from the Navaho, and an inertial guidance system using a stabilized platform and accelerometers, because they were "simple, reliable, accurate—and available."[60] It also employed many other features taken from the V-2. The pressures of the Korean War soon resulted in a redirection of the Hermes program to the development of a single-stage, surface-to-surface ballistic missile having only a 200-mile range but with high mobility, allowing field deployment. Christened the Redstone, the new missile first flew successfully on 20 August 1953 on a test flight of 8,000 yards. Between 1953 and 1958, the Arsenal fired thirty-seven Redstone test vehicles.[61] It was the first large ballistic missile developed in the United States and the first U.S. missile to use an inertial guidance system.

While the Redstone was under development, the Army and the Navy began a joint project to build an intermediate-range ballistic missile (IRBM) that could be launched at sea as well as on land. The Jupiter missile, as it was called, was developed in two versions, both using the Redstone as a basis. Jupiter A was an IRBM designed to carry a warhead. North Atlantic Treaty Organization (NATO) forces in Europe deployed it until 1963, after the Cuban Missile Crisis. Jupiter C, with the official name Jupiter Composite Re-entry Test Vehicle, was a vehicle primarily designed to test reentry technology. Before the United States could build and successfully operate a ballistic missile, it had to solve the difficult problem of reentry into the atmosphere. Opinions differed on how best to protect the nose cone of a nuclear warhead reentering the atmosphere from overheating and destroying the warhead before it reached its target. In 1953, H. Julian Allen, a scientist with the National Advisory Committee for Aeronautics (NACA), had postulated that a blunt rather than a sharp nose would more readily survive reentry. The Jupiter C nose cone was not only blunt but was coated with a fiberglass material that ablated, or burned off, as the surface of the nose cone heated up, thereby keeping the contents of the nose cone cool.[62]

JPL supplied the second and third upper stages for the Jupiter C. On 8 August 1957, the launch team used a Jupiter C to fire a warhead 600 miles high and 1,200 miles down range, where it was recovered from the Atlantic by U.S. Navy teams. The reentry nose cone on this flight was the first object crafted by humans to be recovered from space.[63] The success of this test, along with further refinements, later led to the incorporation of this technology into the design of the Mercury, Gemini, and Apollo capsules, and made possible the return of astronauts from space. As discussed below, this launch, or perhaps even one earlier in the test series, might have been able to launch an initial U.S. satellite, months before *Sputnik I.* History might then have been rather different.

Vanguard, Juno, and the First American Satellite

The ultimate goal of many of the early rocket researchers was to reach orbit. In the early 1950s, sounding rocket and balloon research on the upper atmosphere and growing interest in geophysics and radio propagation led to serious interest among scientists in launching a scientific research satellite. In 1954, meetings of the International Scientific Radio Union and the International Union of Geodesy and Geophysics passed resolutions calling for the launch of a scientific satellite during the International Geophysical Year (IGY), which had been set for 1957–58, when scientists expected peak sunspot activity. The United States and the Soviet Union in 1955 both announced their intentions to orbit a satellite sometime during the IGY.[64]

A committee within the Department of Defense picked the launch vehicle for this satellite from among three proposals: the Atlas ICBM, which was still in the development stage; the Jupiter, using several upper stages; and an unnamed vehicle that would use the Viking as a first stage, the Aerobee as a second stage, and a new solid-fuel third stage. The Viking and the Aerobee were liquid-fueled sounding rockets with proven launch records. The Viking/Aerobee combination, which had been proposed by NRL, had an advantage, since it used available sounding rockets and thus would not compete with the development of the higher-priority Atlas ICBM program. Further, the Viking rocket used gimbaled engines for control and had some growth potential. The resulting program, which was managed by NRL, was called Project Vanguard. The first and second stages used storable nitric acid and dimethylhydrazine as fuel. On 9 September 1955, the DoD authorized the Navy to proceed with Project Vanguard. As John P. Hagen, director of Project Vanguard, has noted, "The letter from the Secretary of Defense stated clearly that what was needed was a satellite (i.e., one) during the IGY which was in no way to interfere with the on-going military missile programs."[65]

The role of the aerospace industry in the construction of launch vehicles was an important, though not entirely easy, one. For example, NRL contracted with the Martin Company, the developer of Viking, to build the first stage of the Vanguard rocket and to oversee the vehicle's assembly. During the negotiations, NRL and Martin had protracted discussions about which organization should have responsibility for overall systems design and engineering. Despite strong arguments to the contrary from Martin, NRL maintained systems responsibility. As would become very evident ten years later in the development of Saturn V, such a division of labor sometimes led to friction between the contractor and the government office overseeing launch vehicle development.[66]

Project Vanguard selected Cape Canaveral, Florida, where there was already the beginnings of a missile test range, for its launch site, and established a worldwide tracking network using NRL's Minitrack system to maintain control over the launcher after it left the Cape. The Minitrack system also served to collect data from the orbiting satellite.

The debate over which of these satellite-launching proposals best served the nation's interests involved a good measure of interservice rivalry as well as rivalry among rocket teams. As Wernher von Braun later wrote, with a detachment that understates the strong feelings prevailing among the engineers at the Army Ballistic Missile Agency (ABMA), "While Project Vanguard was the approved U.S. satellite program, we at Huntsville knew that our rocket technology was fully capable of satellite application and could quickly be implemented." It is indeed likely that the von Braun team could have launched a simple satellite in 1957 (barring a launch failure), but it was prohibited from doing so after a 1956 Washington and White House review of that option. Von Braun's team got its chance only after the first attempt at launching a satellite with a Vanguard launch vehicle resulted, on 6 December 1957, in an embarrassing launch pad explosion.[67]

The scale of the embarrassment came about, in part, from President

Eisenhower's decision to announce the attempted liftoff well in advance. In October 1957, shortly after the surprise launch of *Sputnik*, Eisenhower was briefed on the situation and told of a planned Vanguard test launch in December. The test was to be the first launch of all three stages and the first launch of the second stage. The Vanguard team and the nation got their first taste of the political sensitivity of the space program when on 9 October 1957, President Eisenhower in a news conference announced that "the satellite project was assigned to the Naval Research Laboratory as Project Vanguard. . . . The first of these test vehicles is planned to be launched in December of this year." This put the launch team in the unenviable, and untenable, position of attempting in public the launch of an untried rocket—the three-stage Vanguard had never been tested as a unit. On 6 December 1957, the Vanguard Project team and the United States watched in dismay as the engines of the first stage ignited, then exploded in a fiery exhibition, while the world looked on over their shoulder. The press had a field day with the incident: "Vanguard was Kaputnik, Stayputnik, or Flopnik, and Americans swilled the Sputnik Cocktail: two parts vodka, one part sour grapes."[68]

It was the first and last failure of the first stage in the Vanguard program, but it set a tone that carried through the early days of the U.S. space program. Not only had the Soviets been first into space, but the United States was not even a near second. The Vanguard failure heightened the perception that U.S. engineers were space bunglers and stiffened U.S. resolve to best the Soviets. As some U.S. policy makers (but never President Eisenhower) saw it, winning the space race would demonstrate to the world, and to the nation, the superiority of the U.S. political and economic system. But first, rocket engineers had to launch a satellite.

A month before the Vanguard failure, after receiving White House permission to proceed with an alternative to Vanguard, the DoD had ordered the Army team at Redstone Arsenal to prepare its Jupiter launch vehicle for a satellite launch. The Army team quickly made itself ready. Adding an additional upper stage to the Jupiter C gave the vehicle the ability to reach orbit with a small satellite. When the order came to ABMA to attempt a satellite launch, the Jupiter C with the fourth upper stage became the Juno I, which on 31 January 1958 lifted the first U.S. satellite, *Explorer I*, into space. Because Juno I's lift capacity was limited, *Explorer I* weighed only eighteen pounds, but it carried instruments that made possible the discovery of one of Earth's natural radiation belts, now known as the Van Allen belts.[69]

The Vanguard rocket, too, finally achieved success on 17 March 1958, when it orbited the Vanguard I satellite. Although Vanguard was quickly superseded by other, more powerful rockets, its components, especially its Aerobee second stage and its solid-fuel third stage, had important roles in the later success of the Scout and Delta launchers.

In these early days of the U.S. space program, relatively small modifica-

Image 1-5: The launch of *Explorer I* took place on 31 January 1958 from Cape Canaveral, Florida. This was the first orbital satellite launched by the United States, atop a Jupiter-C ballistic missile built by Wernher von Braun's rocket team in Huntsville, Alabama. (NASA Photo no. SC 526122)

tions to the launch vehicles that were already available enabled designers to create launchers for ever more demanding projects. For example, by the end of April 1961, Juno II, which derived directly from the Jupiter IRBM, and was essentially a larger version of the Redstone, carried the deep space probes *Pioneer 3* and *Pioneer 4* toward the moon and *Explorers VII, VIII,* and *XI* into Earth orbit to return data about the physical characteristics of near-Earth space.

Missile Development and Modern Warfare

Until the early 1950s, missile designers had focused on the eventual development of large ICBMs produced to carry the massive nuclear warheads that the United States had developed immediately after World War II. The U.S.

strategic doctrine of the period depended on large bombers to carry nuclear warheads over the Soviet Union should hostilities between the two superpowers reach the flash point, and only a few dreamers expected ICBMs to gain ascendancy much before the middle of the 1960s. By 1953, however, scientists discovered how to make a relatively lightweight thermonuclear weapon, and U.S. officials realized that the Soviet Union had made considerable progress in developing a long-range missile. These events led to a reevaluation of the U.S. approach to ICBM development. In 1954, the Air Force Strategic Missiles Evaluation Committee (the "Teapot Committee"), chaired by mathematician John von Neumann, urged the development of a relatively small ICBM, capable of launching the newly developed weapons toward the Soviet Union. It also recommended creation of a special development group with sufficient funding and authority to proceed with dispatch. The Air Force created the Western Development Division, which later became the Ballistic Missile Division (BMD). The Space Technology Laboratories of the Ramo-Wooldridge Corporation, the precursor to TRW, provided systems engineering and technical direction for the BMD.[70]

During the mid-1950s, the Air Force started work on two major ICBM systems, both of which still play a major role in U.S. space transportation efforts—Atlas and Titan. It also started work on the Thor IRBM, which employed related technologies. These projects came to fruition in the late 1950s, adding to U.S. strength in the missile race and, soon after, to its ability to place satellites into orbit.[71]

The Atlas ICBM, a project that had originally started in 1945 as a classified Air Force effort (Project MX-774) and had died in 1947, was reborn in 1951 as a five-engine missile generating a takeoff thrust of 650,000 pounds. In 1954, it was redesigned and reduced to using three engines based on those originally developed for the Navaho. The Atlas incorporated several new design features, but one of the most important was the introduction of a pressurized stainless steel propellant tanks, designed to carry the structural burden. This innovation, introduced by Convair engineer Karel J. Bossart, reduced the need for stiffeners and made the Atlas much lighter for a given thrust than earlier designs.

The "steel balloon," as it was sometimes called, employed engineering techniques that ran counter to the conservative engineering approach used by Wernher von Braun and his "Rocket Team" at Huntsville, Alabama. Von Braun, according to Bossart, needlessly designed his boosters like "bridges," to withstand any possible shock. For his part, von Braun thought the Atlas was too flimsy to hold up during launch. The reservations began to melt away, however, when Bossart's team pressurized one of the boosters and dared one of von Braun's engineers to knock a hole in it with a sledgehammer. The blow left the booster unharmed, but the recoil from the hammer nearly clubbed the engineer. Bossart's team also introduced gimbaled thrust nozzles and a warhead that separated from the missile after burnout. The first successful flight of the Atlas occurred on 17 December 1957, after a series of both major and minor development problems. Its first use as a space launcher occurred on 18 December 1958, when one of the Atlas development vehicles launched the 142-pound SCORE communications payload. The entire Atlas rocket, minus the booster engines discarded during ascent, reached orbit with the SCORE payload attached—it was the largest manmade object in orbit at the time.[72]

Out of technical conservatism, and a desire to reduce the risk of depending on single industrial sources for the Atlas, the Air Force contracted with other firms to develop alternative approaches for the major systems. After assuring themselves that the Atlas design was on a sound track, in April 1955 Air Force officials approved the incorporation of several of the alternative subsystems, which involved more sophisticated technology development, into an alternate Titan missile, which was to be built by the Martin Corporation. Unlike the Atlas, the Titan missile had a monocoque airframe that absorbed much of the stress of flight, and a more sophisticated guidance system. It also had a different first-stage engine, built by Aerojet,[73] which used LOX/kerosene instead of LOX/alcohol. The Titan was also a true two-stage missile designed to be launched from a hardened, underground launch silo. The Titan I missile, guided by a combination radio-inertial guidance mechanism, had its first full test in February 1959 and was declared operational in 1962.

During the early 1950s, many Air Force officers had become convinced that the United States needed an intermediate-range ballistic missile, and in January 1955, the Department of Defense Scientific Advisory Board recommended that the Air Force proceed. However, the Army, which was developing the Jupiter, objected, as did the Navy, which also wanted its own program. The Joint Chiefs of Staff compromised by recommending to Secretary of Defense Charles Wilson in November 1955 that the Air Force develop the Thor, while the Army and the Navy worked jointly on the Jupiter. The Western Development Division got the Thor assignment a month later.[74]

Thor was undertaken as a high-risk program having the express goal of achieving flight within the shortest possible time. Using an engine originally developed for the Atlas, the Thor had its first complete launch pad test in January 1957 and a full-range flight test in September of that year. By 16 December 1958, the Strategic Air Command successfully launched a Thor from Vandenberg Air Force Base, California. The test marked the passage from development to initial military readiness.[75] On 28 February 1959, a Thor missile combined with an Agena second stage launched the first Air Force satellite, Discoverer I.[76] Under NASA's control, the Thor, using an Able upper stage and numerous detailed modifications, later evolved into the highly successful Delta launch vehicle, one of the standard vehicles used to launch NASA's scientific payloads and commercial communication satellites. The Delta evolved from the original model capable of placing a few hundred pounds into LEO to one (Delta II 7925) that by the mid-1990s was capable of launching payloads weighing 3,965 pounds to geostationary transfer orbit.

The V-2, Redstone, Jupiter, and Atlas missiles were all propelled by LOX/alcohol or LOX/kerosene. Liquid oxygen, has the serious drawback that it requires cooling and special handling, and therefore cannot be stored for long periods and must be loaded immediately prior to launch. Hence, liquid oxygen is ultimately unsuitable for use in military missiles, where launching quickly could be critical. In the late 1950s, missile designers spent considerable effort to develop storable liquid propellants and solid fuels.

The desire to operate from a "hardened" launch site, below ground and solidly encased in concrete, and to be ready to launch with only a few minutes notice, also acted to speed up development of storable hypergolic propellants. The Air Force embarked on the development of the Titan II missile, which later became a modest-capacity launch vehicle.[77] It used a mixture of unsymmetrical dimethylhydrazine (UDMH) and hydrazine, oxidized by nitrogen tetroxide. The two-stage Titan II represented a major leap in technology development over the Titan I. It not only used storable propellants but also had all-inertial guidance. NASA chose the Titan II to launch the Gemini spacecraft into orbit.[78]

The Atlas, Thor, and Titan launchers were all developed according to the management technique called "concurrency," in which all major systems and subsystems were developed in parallel. This technique called for the planning and construction of industrial production facilities and operational bases even before initial flight testing began. It put great pressure on the development team to oversee each step of development very closely. It also meant that (1) both the authority and responsibility for decisions had to be located in the same agency, (2) program managers had to have a high degree of technical competence, and (3) "funding and programming decisions outside of the authority of the program director had to be both timely and firm."[79]

All three of these programs achieved their objectives relatively quickly. As Robert Perry noted in the *History of Rocket Technology,* "The *management* of technology became the pacing element in the Air Force ballistic missile program. Moreover—as had not been true of any earlier missile program—technology involved not merely the creation of a single high-performance engine and related components in a single airframe, but the development of a family of compatible engines, guidance subsystems, test and launch site facilities, airframes, and a multitude of associated devices."[80]

Missile development also led to one other major technology advance that is now a common element of modern launch vehicles—the creation of solid-propellant rocket motors. Solid propellants are composed of an oxidizer such as ammonium perchlorate, a fuel such as aluminum powder, and an organic binder to create a mixture capable of being cast in a rocket motor casing. When ignited, the mixture continues to burn without benefit of an external source of oxygen. The advantages of using solid rocket fuel for a military missile are enormous. Rockets loaded with solid propellants can be built and stored for long periods, and they can be moved around readily. The Jet Propulsion Laboratory developed solid-fuel JATO rockets during World War II, and its small solid rocket motors were later used as upper stages in the Jupiter C.[81] However, the difficulties of mixing and casting solid propellant in motors large enough to carry a nuclear weapon, and the absence of a satisfactory igniter, had prevented its use in missiles. Ammonium perchlorate, the oxidizer of choice, is hard to handle in large quantities and difficult to mix evenly with an organic binder. In addition, Air Force scientists and engineers needed to develop methods for controlling the fuel's burn, its rate of thrust and its direction, as well as ways of constructing high-strength, lightweight engine cases.

By October 1957, the Air Force had made substantial progress toward building rocket motors large enough to propel a nuclear weapon but had no solid-fuel missile development project in place. Although solid-fuel missiles had been considered for tactical deployment, they had not reached a level of reliability and thrust sufficient to serve as ICBMs. However, the perceived crisis of responding to *Sputnik I*, coupled with the technical progress made in the 1950s, injected a new urgency into U.S. plans for developing a solid-fuel ICBM.[82] Studies developed the concept for Weapon System Q, a three-stage, solid-fueled ICBM, which would be deployed in large quantity in hardened missile pads. In September 1957, this was named Minuteman. By the end of 1957, the Ballistic Missile Division recommended that the Air Force begin a program to develop Minuteman, and in June 1958, Secretary of Defense Neil H. McElroy approved the request. Although the Air Force did not complete selection of the contractors for the rocket's stages and other major systems until July 1958, the first flight test took place only two and a half years later on 1 February 1961. It was highly successful. Developers used the same concurrency process that had worked well for the development of Atlas, Thor, and Titan. Of these four missiles, only the Minuteman has not enjoyed widespread success as a working Earth-orbit launch vehicle, although there have been moves in this direction as Minutemen have become excess to security requirements as a result of arms limitation agreements.[83]

In March 1956, the Navy had also gained permission to start its own missile program, which eventually led to the Polaris missile, launched from a submarine below the surface of the ocean. Like the Minuteman, the Polaris depended on a solid rocket motor for propulsion for much the same reasons that Air Force officials were drawn to it for the Minuteman—solid rocket motors can be fired nearly immediately and can be stored for long periods without degrading. They are also much, much easier to handle than liquid motors, making them especially attractive for launching from submarines. The Lockheed Aircraft Corporation was the prime contractor for Polaris. It conducted the first successful test of an inertially guided Polaris missile on

7 January 1960 from Cape Canaveral. On 20 July of that same year, the nuclear submarine *George Washington* conducted the first undersea firing of a Polaris.[84]

The experience gained in manufacturing solid rocket motors for the Minuteman and the Navy's Polaris[85] missile programs enabled NASA to develop the solid-fueled Scout small launch vehicle, which was first completed and launched in July 1960.

In the late 1950s, while developing rocket motors for missiles, rocket engineers also began to work on ever-larger solid rocket motors in hopes of creating a space booster capable of placing moderate-sized payloads into orbit. Rockets based solely on solid propellants require motors capable of generating several million pounds of thrust for durations of 100 seconds or more. Rocket designers faced the major problem of achieving a sustained, even burn, rather than igniting the entire mass of propellant at once. Among other things, this involved developing the means to disperse an oxygen-rich compound, commonly ammonium perchlorate, uniformly in an organic binder that would provide the fuel. It also involved building high-strength, lightweight engine cases. After considerable testing, they finally mastered the technique of casting solid propellant in large sizes and with internal shapes capable of sustaining an even burn rate. Nevertheless, it was clear that the enormous sizes (diameter and length) needed to develop millions of pounds of thrust would create difficult construction and transportation problems. However, if the rocket motors could be built in segments and bolted together on the launch pad, they would be much easier to construct and to transport to the launch site.

Starting in 1957 with funding from the Air Force, Aerojet General Corporation, which had manufactured JATO units during World War II, demonstrated that the concept was feasible by first cutting a 20-inch-diameter Regulus II booster rocket into three pieces, filling the pieces with propellant, reattaching them, and firing the segmented rocket motor. Following a successful test in early 1959, Aerojet attempted the same procedure with a 65-inch-diameter Minuteman rocket motor, which also fired successfully. On 17 February 1960 Aerojet successfully test-fired a three-segment, 100-inch-diameter rocket motor more than 400 inches long that produced an average 534,000 pounds of thrust for nearly 90 seconds.[86] The test program concluded in October 1962 after achieving partial success with tests of two longer 100-inch-diameter motors.[87] These developments demonstrated that reliable, segmented solid-fuel rockets could be built and fired in ground tests. Such experience enabled the Air Force and NASA to develop the large, segmented solid rocket boosters that were later used to power both the Titan III and IV launchers and the Space Shuttle.

During the early development of the Saturn liquid-fueled booster, proponents of solid rocket motors suggested their use in that program. NASA had explored the potential of solid rockets and decided that, while advantageous for some tasks, such as launching scientific payloads, they had not yet reached the level of development that would make them suitable for launching people into space. In the immediate aftermath of President Kennedy's May 1961 announcement that the United States would send people to the moon, a joint NASA–Department of Defense team examined the possible use of solid-fueled rockets in accomplishing that mission; however, NASA decided to stand by its earlier position. Hence, NASA carried out relatively little development work on solid rocket motors until they were under consideration for the Space Shuttle.

Launching People

Launching people into orbit introduced another set of considerations into booster design and manufacture. Although the armed services and NASA were concerned about launch vehicle reliability because of the costs involved in replacing an expensive payload, they had little concern about safety beyond the obvious issues of possible launch pad and range damage. Once the many tons of steel, aluminum, and propellants were on their way to space, loss of the vehicle primarily meant extra costs and loss of the payload and research results. However, the loss of human life was another matter, the costs of which could not be reckoned in dollars alone. The creation of Project Mercury, a highly visible U.S. human space-flight program, led to the need to reduce the risks of space flight, not only to protect the astronauts but also to protect the space program itself from cancellation. Astronauts were not merely test pilots; they were highly visible manifestations of U.S. technological and political accomplishments and soon became American icons. NASA began to institute different procedures for designing, building, and launching the rockets destined to carry humans. Because the Redstone had previously demonstrated relatively high reliability and flight stability, NASA requested eight Redstone launchers for the suborbital portion of Project Mercury. These boosters were modified to allow additional propellant to increase their lift capacity and to add an abort-sensing system to increase their safety.

"Man-rating" the Redstone also meant additional verifications of the reliability of launcher hardware and launch software and extensive testing for electronic and mechanical compatibility with the Mercury spacecraft payload. After an initial launch test to assure that all the systems and subsystems performed together, the first flight with a live passenger occurred on 31 January 1961, when the second Mercury/Redstone mission (MR-2) carried the chimpanzee Ham briefly into space and back on a parabolic trajectory. However, the Redstone boosted the Mercury capsule to a greater height than planned, and thus the capsule landed much farther down range than had been planned. The cause of the booster malfunction was quickly identified and remedied, but von Braun and his associates insisted on an additional test flight before committing an astronaut to a mission atop the Redstone. That additional flight took place on 23 March 1961 and was totally successful. If it had not been inserted into the Mercury schedule, the March flight could well have carried an astronaut, and it would have been an American, not the Soviet cosmonaut Yuri Gagarin, who would have been first into space (though not into orbit).[88]

On 5 May 1961, Lt. Cmdr. Alan B. Shepard Jr. of the Navy did become the first American human in space aboard *Freedom 7*. His flight was followed by the second and last crewed Mercury-Redstone flight on 21 July, which carried Capt. Virgil I. "Gus" Grissom of the Air Force into space and back. A more powerful rocket would be needed to place an astronaut into orbit.

For the orbital launch of the Mercury capsule, NASA officials decided at the start of the program to use the Atlas launcher, which was capable of carrying about 3,000 pounds into a 150-by-100-mile elliptical orbit. Using the Atlas to carry people required upgrading of the launcher to increase its safety margins; there was concern that there had been frequent failures during the use of the Atlas for unmanned space launches. In all, NASA procured ten specially built Atlas D launchers[89] from the Air Force for the task, of which four carried astronauts into orbit. NASA successfully completed the first orbital Mercury flight ten months after the Russian cosmonaut Yuri Gagarin first circled the globe. On 20 February 1962, astronaut John Glenn orbited the earth three times in *Friendship 7*, landing in the Atlantic Ocean southeast of Bermuda.

The Titan II became the second and last modified ICBM to be used for launching humans to orbit; it was employed in launching all ten spacecraft in the two-man Gemini program. The extra payload capacity of the Titan II compared to the Atlas made it possible to launch a heavier capsule, large enough to accommodate two individuals. Gemini was designed to develop the astronauts' skills in orbital rendezvous and docking as a precursor to the Apollo lunar program. It was also used to extend NASA's experience with space flight to a duration long enough to reach the moon and return and to test extravehicular activities.[90]

During the early 1960s, as military and national security payloads quickly grew in weight, it became clear that the Air Force would need a booster larger than the Titan II to lift its planned payloads to orbit. Hence, it modified the Titan II by adding an additional stage and solid "strap-on" booster rockets and designated the new rocket the Titan III. The first Titan IIIA, carrying a third "Transtage," successfully flew on 1 September 1964. Shortly thereafter, the Air Force used an Agena upper stage to create the Titan IIIB, capable of carrying 7,500 pounds into low-Earth orbit. Because still greater lift was needed to launch the Air Force's largest satellites, the Air Force added segmented solid rocket motors to create the Titan IIIC. It employed two strap-on boosters made up of five 10-foot-diameter segments that extended 86 feet in height. The boosters were developed and manufactured by United Technology Center, using techniques it developed in the late 1950s.[91] The first test flight took place in June 1965. The Titan IIIC was capable of lifting 28,000 pounds into low-Earth orbit. For even more massive loads, the Air Force contracted with Martin Marietta to build the Titan IIID and Titan IIIE, both of which used the solid rocket boosters from the Titan IIIC but had more powerful upper stages. The rocket combination with the greatest lift capacity was the IIIE, which employed a cryogenic[92] upper stage called the Centaur, first designed for use on an Atlas rocket. In the 1970s, NASA used the Titan IIIE with a Centaur upper stage to launch the two Mars Viking landers. The two successful flights are particularly notable for occurring within only three weeks of each other. *Viking 1* was launched on 20 August 1975, followed on 9 September by *Viking 2*.[93] They were launched from the Air Force–maintained Titan launch pads at Cape Canaveral, Florida.

The Centaur upper stage was developed by Convair with funding from the Advanced Research Projects Agency and the Air Force and successfully test flown on an Atlas rocket on 27 November 1963. The Centaur, variants of which are still in use, employs two Pratt & Whitney RL-10 engines and was the first liquid oxygen–liquid hydrogen (LOX-LH2) engine to demonstrate the capability to restart in space.[94] The Atlas-Centaur rocket has launched spacecraft to Mercury, Venus, Mars, Jupiter, and Saturn, and many communications satellites to geosynchronous orbit. Development of the Centaur provided the Air Force and NASA with significant experience with the problems encountered in using liquid hydrogen for propulsion, which assisted in the later development of the cryogenic engines used in the Saturn program.

Nuclear Propulsion

One of the more interesting aspects of rocket development was the partnership between NASA and the Atomic Energy Commission (AEC) in developing nuclear rocket engines. Seen strictly from the standpoint of available power for rocket thrust, a nuclear rocket generating heat from fission is much more efficient than chemical propulsion, allowing much higher thrust. Nuclear rockets have been of particular interest for interplanetary space flight, because they could markedly shorten trips to the planets. However, they also present formidable engineering and safety challenges. The notion of using atomic energy as a fuel source was briefly explored by Tsiolkovsky, Goddard, and others, but these early theoreticians and experimenters were daunted by the problem of controlling the enormous potential for explosive, rather than controlled, releases of energy. It was not until after controlled nuclear fission had been achieved in 1942 and after World War II that the technology began to receive serious attention in rocketry.

North American Aviation completed the first detailed (classified) study of the issue in 1947. It concluded that a nuclear rocket would be feasible if some serious technical hurdles could be overcome. Beginning in the late 1940s, the AEC also experimented with the use of nuclear power in aircraft, which generally contributed to the government's technical expertise in nuclear propulsion. R.W. Bussard, who worked on the nuclear aircraft program at the Oak Ridge National Laboratory, became interested in the challenge of nuclear rocketry and published an important report in 1953 that influenced the Air Force in its decision to start up a nuclear rocket program.[95] His work convinced officials that nuclear rockets might be feasible alternatives to chemical propulsion for ballistic missiles.[96]

Both Los Alamos and Lawrence Livermore National Laboratories established small programs to investigate nuclear propulsion technologies in detail. In November 1955, the Air Force and the AEC formally started Project Rover, with the goal of harnessing the enormous power of nuclear fission for space flight. Some Air Force officials felt that nuclear power would be of utility in powering ICBMs. Livermore was directed to focus on nuclear ramjets under Project Pluto, leaving Los Alamos to develop a nuclear reactor for a rocket engine. In 1957, program officials had chosen the area at the Nevada Test Site called Jackass Flats to conduct engine tests. Los Alamos developed the Kiwi experimental nuclear reactor,[97] testing several versions at Jackass Flats between 1959 and 1964. These tests demonstrated the use of carbide coatings to prevent hydrogen erosion of the graphite and established numerous crucial details about reactor design and control. Testing of the first version, KIWI-A, established the technical feasibility of creating a nuclear rocket. Nevertheless, it soon became clear that solid rocket propulsion was of much greater utility for ballistic missiles than nuclear engines. Among other things, nuclear bomb engineers had managed to create nuclear warheads of much-reduced mass, thereby relaxing the lift requirements for missiles.

Soon after the creation of NASA, the Eisenhower administration transferred the Air Force's responsibility for nuclear rocketry to NASA, which with the AEC created the NASA-AEC Space Nuclear Propulsion Office (SNPO) in August 1960. During his 25 May 1961 speech titled "Urgent National Needs," President Kennedy urged a speedup of the Rover nuclear rocket program, proposing a threefold increasing in funding.[98]

Soon after, the SNPO began the Nuclear Engine for Rocket Vehicle Application (NERVA) program, with the eventual goal of flight testing the NERVA engine on a Saturn rocket. Aerojet-General and Westinghouse Electric Corporation were awarded a contract to develop the NERVA engine, which was to be derived from the KIWI-B test engine then undergoing tests at Jackass Flats. In a program called Reactor-in-Flight Tests (RIFT), NASA planned to use a flight-rated version of the NERVA engine to power the third stage of a Saturn V.[99] A few NASA officials contemplated that it might serve as a second or third stage on the even larger Nova vehicle that some NASA engineers had been arguing for. In spring 1962, NASA selected Lockheed Missiles and Space Company as the prime contractor for the nuclear stage. As planned, the RIFT test vehicle was to consist of Saturn S-IC and S-II stages topped by the Saturn-N nuclear stage. In its lunar flight configuration, it would launch a crewed space-craft and lunar lander. After the first two stages carried the spacecraft beyond Earth's atmosphere, the nuclear engine would be started to carry the crew to the Moon.

By the end of 1963, the nuclear rocket effort was already in decline as NASA focused on making the Apollo program a success using more conventional rocket engines. Budget reductions forced NASA and AEC to terminate the RIFT project. They converted the NERVA Project to a technology effort using ground tests of nuclear engines and components. Between May 1964 and March 1969, the NERVA Project tested thirteen reactors, essentially completing the technology phase. The KIWI series was followed by a 5,000-megawatt reactor named Phoebus, designed to achieve higher temperatures and longer operating times at lower specific weights. NASA planned to use a flight-rated version of Phoebus for space travel.

In 1968, the project initiated work on a 75,000-lbf flight-rated engine having a specific impulse of 850 seconds,[100] but the program was nearing its end. As work on the proposed Space Shuttle increased, program officials even proposed that the shuttle would transport a NERVA engine into orbit for testing. Yet that effort fell on deaf ears, in part because the Nixon administration and Congress continued to reduce NASA's budget, reducing the need for propulsion to support interplanetary travel, but also because of mounting opposition to nuclear power. In 1972, Project Rover was terminated.

The nuclear rocket program had been quite ambitious, and it showed the technical feasibility of nuclear propulsion. As Dr. Glen T. Seaborg, Chairman of the Atomic Energy Commission, stated in 1970,

> Lest you get the impression that the development of such a nuclear rocket is as simple as its principle sounds, let me point out what is involved in it. What we must do is build a flyable reactor, little larger than an office desk, that will produce the 1500 megawatt power level of Hoover Dam and achieve this power in a matter of minutes from a cold start. During every minute of its operation, high-speed pumps must force nearly three tons of hydrogen, which has been stored in liquid form at 420 degrees F below zero, past the reactor's white-hot fuel elements, which reach a temperature of 4,000 degrees F. And this entire system must be capable of operating for hours and of being turned off and restarted with great reliability.[101]

Although the nuclear program had been relatively successful from a technical standpoint, and it still had many proponents within NASA and the Atomic Energy Commission, it could not survive the funding competition with programs that carried less technical risk, especially given the diminished prospects for interplanetary flight involving large payloads. Nuclear propulsion interested mission planners once again in the late 1980s and early 1990s after President Bush announced on 20 July 1989 a plan to send humans to Mars and back by 2019. However, that effort, which became known as the Space Exploration Initiative (SEI), was very short lived. Congressional proponents of NASA's other programs became worried that such a public effort, requiring many billions of dollars of investment, would use up funds planned for other NASA programs, including the long-planned International Space Station.

Conclusion

The last forty-plus years of the twentieth century offered both serious challenges and enormous potential for the development of new launch vehicles that helped to make possible America's entry into space. After four decades of effort, access to space remains a difficult challenge in the year 2000. This technical challenge—particularly the high costs associated with space launch from 1950 to 2000, the long lead times necessary for scheduling flights, and the poor reliability of rockets—has demonstrated the slowest rate of improvement of all space technologies. All space professionals have shared a responsibility for addressing these critical technical problems. The overwhelming influence that space access has on all aspects of civil, commercial, and military space efforts indicate that it should enjoy a top priority for the twenty-first century.[102]

Notes

1. Konstantin E. Tsiolkovsky, *Aerodynamics*, NASA TT F-236 (Washington, D.C.: NASA, 1965); Konstantin E. Tsiolkovsky, *Reactive Flying Machines*, NASA TT F-237 (Washington, D.C.: NASA, 1965); Konstantin E. Tsiolkovsky, *Works on Rocket Technology*, NASA TT F-243 (Washington, D.C.: NASA, 1965); Arkady Kosmodemyansky, *Konstantin Tsiolkovsky* (Moscow: Nauka, 1985). (NASA TT F-236 and F-237 are a two-volume complication of the English-language translations of Tsiolkovsky's work under the common title *Collected Works of K.E. Tsiolkovsky*.)

2. Hermann Oberth, *Ways to Spaceflight*, NASA TT F-622 (Washington, D.C.: NASA, 1972); Hermann Oberth, *Rockets into Planetary Space*, NASA TT F-9227 (Washington, D.C.: NASA, 1972); H.B. Walters, *Hermann Oberth: Father of Space Travel* (New York: Macmillan, 1962).

3. Interestingly, in 1969 Oberth attended the launch in the United States of Apollo 11 at the request of von Braun, who was then directing a major component of the lunar project.

4. The standard biography of Goddard is Milton Lehman, *This High Man* (New York: Farrar, Straus, 1963), although it is outdated and deserving of replacement.

5. Robert H. Goddard, "Material for an Autobiography," in Esther C. Goddard, ed., and G. Edward Pendray, assoc. ed., *The Papers of Robert H. Goddard* (New York: McGraw-Hill, 1970), 1:10.

6. Robert H. Goddard, "Of Taking Things for Granted," in ibid. 1:63–66.

7. Robert H. Goddard, "On the Possibility of Navigating Interplanetary Space," in ibid. 1:81–87; *Scientific American* to R.H. Goddard, 9 October 1907, in ibid. 1:87; William W. Payne to R.H. Goddard, 15 January 1908, in ibid. 1:88.

8. Goddard, "Material for an Autobiography," pp. 19–20; R.H. Goddard to Josephus Daniels, 25 July 1914, in Goddard and Pendray, eds., *Papers of Robert H. Goddard* 1:126–27.

9. C.D. Walcott to R.H. Goddard, 5 January 1917, in Goddard and Pendray, eds., *Papers of Robert H. Goddard* 1:190; Frederick C. Durant III, "Robert H. Goddard and the Smithsonian Institution," in Frederick C. Durant III and George S. James, eds., *First Steps Toward Space: Proceedings of the First and Second History Symposia of the International Academy of Astronautics* (Washington, D.C.: Smithsonian Institution Press, 1974), pp. 57–69.

10. Robert H. Goddard, *A Method of Reaching Extreme Altitudes* (Washington, D.C.: Smithsonian Miscellaneous Collections, vol. 71, no. 2, 1919), p. 54.

11. Robert H. Goddard, "Statement by R.H. Goddard for Newspapers," 18 January 1920, in Goddard and Pendray, eds., *Papers of Robert H. Goddard*, 1:409–10.

12. "Topics of the Times," *New York Times*, 18 January 1920, p. 12.

13. Robert H. Goddard, "R.H. Goddard's Diary," 16–17 March 1926, in Goddard and Pendray, eds., *Papers of Robert H. Goddard* 2:580–81; Lehman, *This High Man*, pp. 140–44.

14. Lehman, *This High Man*, pp. 156–62.

15. Ibid., pp. 161–312; Wernher von Braun, Frederick I. Ordway III, and Dave Dooling, *History of Rocketry and Space Travel* (New York: Thomas Y. Crowell, 1986), pp. 46–53. Many of these experiments were summarized in Robert H. Goddard, *Liquid-Propellant Rocket Development* (Washington, D.C.: Smithsonian Miscellaneous Collections, vol. 95, no. 3, 1936).

16. Frank H. Winter, *Rockets into Space* (Cambridge: Harvard University Press, 1990), pp. 33–34.

17. G. Edward Pendray, "Pioneer Rocket Development in the United States," in Eugene M. Emme, ed., *The History of Rocket Technology: Essays on Research, Development, and Utility* (Detroit: Wayne State University Press, 1964), pp. 19–23.

18. See the extensive documentation on this settlement in the "Goddard Patent Infringement" Folders, Biographical Collection, NASA Historical Reference Collection, NASA History Office, Washington, D.C. (hereafter cited as NASA Historical Reference Collection).

19. See, for example, a book by a captain in the Austrian army, Hermann Noordung (nom de plume of Herman Potôcnik), *The Problem of Space Travel: The Rocket Motor*, NASA SP-4026 (Washington, D.C.: NASA, 1995). This book examines many technical aspects of space travel, including space stations. It was originally published in Berlin in 1929 (*Problem der Befahrung des Weltraums* [Berlin: Schmidt, 1929]). For a discussion of the origins of many of the ideas regarding space travel, see John M. Logsdon, gen. ed., with Linda J. Lear, Jannelle Warren-Findley, Ray A. Williamson, and Dwayne A. Day, *Exploring the Unknown: Selected Documents in the History of the U.S. Civil Space Program*, vol. 1, *Organizing for Exploration*, NASA SP-4407 (Washington, D.C.: NASA, 1995), chap. 1.

20. The standard work on the rocket societies is Frank H. Winter, *Prelude to the Space Age: The Rocket Societies, 1924–1940* (Washington, D.C.: Smithsonian Institution Press, 1983). A briefer discussion is available in Winter, *Rockets into Space*, pp. 34–42.

21. Winter, *Rockets into Space*, p. 37.

22. Wernher von Braun, "German Rocketry," in Arthur C. Clarke, ed., *The Coming of the Space Age* (New York: Meredith Press, 1967), pp. 33–55. Von Braun's public relations skills were exceptional throughout his career. Evidence of this can be found in the more than 8 linear feet of materials by von Braun held in the Biographical Files of the NASA Historical Reference Collection.

23. Winter, *Prelude to the Space Age*, pp. 73–85; Eugene M. Emme, ed., *Aeronautics and Astronautics: An American Chronology of Science and Technology in the Exploration of Space, 1915–1960* (Washington, D.C.: National Aeronautics and Space Administration, 1961), p. 31.

24. Eugen Sänger was an Austrian scientist whose ideas about reusable spacecraft were commemorated in a 1980s German design for a two-stage launch system that carries his name. See E. Sänger, *Raketenflugtechnik* (1933). English version, *Rocket Flight Engineering*, NASA TT F-223 (Washington, D.C.: NASA, 1965).

25. Wernher von Braun, Frederick I. Ordway III, and Dave Dooling, *Space Travel: An*

Update of History of Rocketry and Space Travel, rev. 3d ed. (New York: Harper and Row, 1975), p. 82.

26. For years liquid oxygen was abbreviated LOX; however, during the 1980s it became popular to use its more correct abbreviation, LO2, in most technical works.

27. The advantage of regenerative cooling is that the propellant, while cooling the combustion chamber, is also preheated to make it more efficient in the burning cycle.

28. Known as the 6000C4, this engine was capable of generating 6,000 lbf. The Bell X-1 flew on nineteen contractor demonstration flights and fifty-nine Air Force test flights. The basic engine was also used in many of the later rocket-powered research airplanes, including early flights of the North American X-15, and all of the powered lifting bodies.

29. See James A. Harford, *Korolev: How One Man Masterminded the Soviet Drive to Beat America to the Moon* (New York: John Wiley & Sons, 1997).

30. Winter, *Prelude to the Space Age*, pp. 87–97; British Interplanetary Society, *The British Interplanetary Society: Origin and History* (London: British Interplanetary Society Diamond Jubilee Handbook, 1993), pp. 6, 17; H.E. Ross, "The British Interplanetary Society's Astronautical Studies, 1937–1939," in Durant and James, eds., *First Steps Toward Space*, pp. 209–16; H.E. Ross, "The British Interplanetary Society Spaceship," *Journal of the British Interplanetary Society* 5, no. 1 (January 1939): 4–9.

31. The history of organization has been well explored in Clayton R. Koppes, *JPL and the American Space Program: A History of the Jet Propulsion Laboratory* (New Haven, Conn.: Yale University Press, 1982).

32. Malina reflected that while Goddard was pleasant during a visit to his Roswell, New Mexico, facility in August 1936, he left two specific impressions. First, he was bitter toward the press and therefore exceptionally secretive about his work. Second, "he felt that rockets were his private preserve, so that others working on them took on the aspect of intruders." Malina recalled that Goddard showed him no technical details and declined to participate in the GALCIT program. See Frank J. Malina, "On the GALCIT Rocket Research Project, 1936–1938," in Durant and James, eds., *First Steps Toward Space*, pp. 113–27, quote from p. 117.

33. R.H. Goddard to Robert A. Millikin, 1 September 1936, in Goddard and Pendray, eds., *Papers of Robert H. Goddard* 3:1012–13.

34. Frank J. Malina and Apollo M.O. Smith, "Flight Analysis of the Sounding Rocket," *Journal of Aeronautical Sciences* 5 (1938): 199–202; Frank J. Malina, "The Jet Propulsion Laboratory: Its Origins and First Decade of Work," *Spaceflight* 6 (1964): 216–23; Frank J. Malina, "The Rocket Pioneers: Memoirs of the Infant Days of Rocketry at Caltech," *Engineering and Science* 31 (February 1968): 9–13, 30–32.

35. Theodore von Kármán with Lee Edson, *The Wind and Beyond: Theodore von Kármán, Pioneer in Aviation and Pathfinder in Space* (Boston: Little, Brown, 1967), p. 243.

36. Frank J. Malina to parents, 24 October 1938, in "Rocket Research and Development: Excerpts from Letters Written Home by Frank J. Malina between 1936 and 1946," pp. 22–23, Frank J. Malina Folder, Biographical Files, NASA Historical Reference Collection; von Kármán with Edson, *Wind and Beyond*, p. 244; Frank J. Malina, "The U.S. Army Air Corps Jet Propulsion Research Project, GALCIT Project No. 1, 1939–1946: A Memoir," in R. Cargill Hall, ed., *Essays on the History of Rocketry and Astronautics: Proceedings of the Third through the Sixth History Symposia of the International Academy of Astronautics* (San Diego: Univelt, 1986), pp. 154–60.

37. Basic histories of rocketry include David Baker, *The Rocket: The History and Development of Rocket and Missile Technology* (New York: Crown Books, 1978); Winter, *Rockets into Space;* von Braun, Ordway, and Dooling, *History of Rocketry and Space Travel*; Emme, ed., *History of Rocket Technology.*

38. This section is based on *Rockets: A Teacher's Guide with Activities in Science, Mathematics, and Technology* (Washington, D.C.: NASA Office of Human Resources and Educa-

tion, 1996); Robert Zubrin, *Entering Space: Creating a Spacefaring Civilization* (New York: Jeremy P. Tarcher/Putnam, 1999), pp. 35–38; Steven J. Isakowitz, Joseph P. Hopkins Jr., and Joshua B. Hopkins, *International Reference Guide to Space Launch Systems, Third Edition* (Reston, Va.: American Institute for Aeronautics and Astronautics, 1999), passim.

39. For a total of 8,030,000 lbf at liftoff.

40. On Space Shuttle early history, see T.A. Heppenheimer, *The Space Shuttle Decision: NASA's Quest for a Reusable Launch Vehicle,* NASA SP-4221 (Washington, D.C.: NASA, 1999) and Dennis R. Jenkins, *Space Shuttle: The History of the National Space Transportation System—The First 100 Missions* (Cape Canaveral, Fla.: Specialty Press, 2001). Specifically on the main engine, see T.A. Heppenheimer, "27,000 Second in Hell," *Air & Space/Smithsonian.*

41. These were named Max and Moritz after the Katzenjammer Kids of the popular comic strip of the day.

42. Compare the 6,000 lbf of the Reaction Motors engine used in the Bell X-1A decade later.

43. In searching for a manufacturer of pumps with the right specifications, von Braun made the interesting discovery that his needs could be satisfied by pumps very similar in pressure, rate, and size to those used by firefighters.

44. Von Braun and Dornberger feared being captured by the Russians and calculated that they would have a better chance of pursuing their rocket research on acceptable terms in the United States than in the Soviet Union. Hence, in February, after seeing the way the war was going, they led most of the upper echelon of German rocket scientists south to Bavaria in order to meet the Americans and avoid being captured by the Russians. See Frederick I. Ordway III and Mitchell Sharpe, *The Rocket Team* (New York: Crown, 1982), pp. 254–75.

45. The Jet Propulsion Laboratory was transferred to NASA on 3 December 1958. It is operated by the California Institute of Technology under contract to NASA.

46. Theodore von Kármán, "Memorandum on the Possibilities of Long-Range Rocket Projectiles," 20 November 1943, Frank J. Malina Folder, Biographical Files, NASA Historical Reference Collection.

47. Frank J. Malina, "America's First Long-Range-Missile and Space Exploration Program: The ORDCIT Project of the Jet Propulsion Laboratory, 1943–1946, a Memoir," in Hall, ed., *Essays on the History of Rocketry and Astronautics,* pp. 339–83; William R. Corliss, *NASA Sounding Rockets, 1958–1968: A Historical Summary,* NASA SP-4401 (Washington, D.C.: NASA, 1971), pp. 17–18.

48. Frank J. Malina, "Is the Sky the Limit?" *Army Ordnance,* July–August 1946.

49. The panel had representatives from the U.S. Army Signal Corps, the Johns Hopkins University Applied Physics Laboratory, the U.S. Army Air Forces, the Naval Research Laboratory, Princeton, Harvard, the California Institute of Technology, and the University of Michigan. See John P. Hagen, "Viking and Vanguard," in Emme, ed., *History of Rocket Technology,* p. 123.

50. Most sources attribute WAC to meaning "without attitude control"—denoting the simplicity of the vehicle.

51. Using a WAC Corporal as a second stage was suggested by engineer Frank Malina, who had a major role in developing the WAC Corporal. See Malina, "Is the Sky the Limit?" p. 45.

52. Radio-inertial guidance is a form of guidance in which the launch vehicle or missile is tracked by radar and commands are issued by radio to change attitude as the flight progresses. It is a technique that was used on the Titan launch vehicles until recently.

53. David H. DeVorkin, *Science with a Vengeance: How the Military Created the U.S. Space Sciences After World War II* (New York: Springer-Verlag, 1992), pp. 167–92.

54. Rocketdyne was formed as a separate division of North American Aviation, Inc., in 1955.

55. Compare the 393,800 lbf from the Space Shuttle Main Engines (at 104 percent thrust at sea level).

56. The version of this engine actually destined for the Navaho generated 120,000 pounds of thrust.

57. Julius H. Braun, "Development of the JUPITER Propulsion System," IAA-91-673, 42d Congress of the International Astronautical Federation, Montreal, Canada, October 1991.

58. Dale D. Myers, "The Navaho Cruise Missile: A Burst of Technology," IAA-91-679, 42d Congress of the International Astronautical Federation, Montreal, Canada, October 1991.

59. Technologies from sounding rockets, for example, were incorporated into medium-range missiles. Jacob Neufeld, *The Development of Ballistic Missiles in the United States Air Force, 1945–1960* (Washington, D.C.: Office of Air Force History, 1990).

60. Wernher von Braun, "The Redstone, Jupiter, and Juno," in Emme, ed., *History of Rocket Technology*, p. 109.

61. Twenty-five of these were essentially Jupiter A missiles.

62. Von Braun, "Redstone, Jupiter, and Juno," pp. 113–14; Philip Nash, *The Other Missiles of October: Eisenhower, Kennedy, and the Jupiters, 1957–1963* (Chapel Hill: University of North Carolina Press, 1997).

63. The von Braun team did not accept gracefully the 1955 decision to assign the satellite launch mission to the Naval Research Laboratory team and the Vanguard rocket. Throughout 1956, it kept pushing for a reconsideration of this decision and permission to attempt a satellite launch sometime in 1957. After review within the Pentagon, this suggestion was rejected, but still there was some sense that the Army team would try to launch a satellite without top-level permission. Thus, for this launch the upper stage was loaded with sand in order to prevent it from orbiting Earth.

64. A good account of the IGY satellite projects can be found in Rip Bulkeley, *The Sputniks Crisis and Early United States Space Policy* (Bloomington: Indiana University Press, 1989), pp. 89–122.

65. Kenneth A. Osgood, "Before Sputnik: National Security and the Formation of U.S. Outer Space Policy," in Roger D. Launius, John M. Logsdon, and Robert W. Smith, eds., *Reconsidering Sputnik: Forty Years Since the Soviet Satellite* (New York: Harwood Academic Publishers, 2000), chap. 7.

66. John Lisa Bromberg, *NASA and the Space Industry* (Baltimore: Johns Hopkins University Press, 1999), pp. 19–21.

67. Constance McL. Green and Milton Lomask, *Vanguard* (Washington, D.C.: Smithsonian Institution Press, 1970), pp. 6–39; R. Cargill Hall, "Origins and Early Development of the Vanguard and Explorer Satellite Programs," *Airpower Historian* 9 (October 1964): 102–8.

68. Walter A. McDougall, . . . *the Heavens and the Earth: A Political History of the Space Age* (New York: Basic Books, 1985), p. 154.

69. James C. Hagerty, Memorandum on Telephone Calls with Brigadier General Andrew J. Goodpaster, 31 January 1958 and 1 February 1958, Dwight D. Eisenhower Library, Abilene, Kans.

70. See Edmund Beard, *Developing the ICBM: A Study in Bureaucratic Politics* (New York: Columbia University Press, 1976).

71. This story is told in John L. Chapman, *Atlas: The Story of a Missile* (New York: Harper and Brothers, 1960); Micharl H. Armacost, *The Politics of Weapons Innovation: The Thor-Jupiter Controversy* (New York: Columbia University Press, 1969); James Baar and William E. Howard, *Combat Missileman* (New York: Harcourt, Brace, and World, 1961).

72. Richard E. Martin, *The Atlas and Centaur "Steel Balloon" Tanks: A Legacy of Karel Bossart* (San Diego: General Dynamics, 1989); Robert L. Perry, "The Atlas, Thor, Titan, and Minuteman," in Emme, ed., *History of Rocket Technology*, pp. 143–55; John L. Sloop, *Liquid*

Hydrogen as a Propulsion Fuel, 1945–1959, NASA SP-4404 (Washington, D.C.: NASA, 1978), pp. 173–77.

73. In keeping with its desire to maintain more than one supplier for critical launch technology, the Air Force chose Aerojet to build the engine for the Titan I. Aerojet used the same rocket motor technology used in the Atlas missile and originally developed for the Navaho missile. The Martin Company built the structure.

74. Neufeld, *The Development of Ballistic Missiles,* pp. 146–47.

75. Perry, "Atlas, Thor, Titan, and Minuteman," p. 151.

76. This was the first of many satellites in the Corona series of spy satellites. See Dwayne A. Day, John M. Logsdon, and Brian Latell, eds., *Eye in the Sky: The Story of the Corona Spy Satellites* (Washington, D.C.: Smithsonian Institution Press, 1998).

77. The Titan II was decommissioned as an ICBM between 1982 and 1987. Fourteen were refurbished as launchers and during the 1980s and 1990s have been used to launch a variety of automated payloads, including the NOAA series of polar-orbiting launch vehicles.

78. See especially Barton C. Hacker and James M. Grimwood, *On the Shoulders of Titans: A History of Project Gemini,* NASA SP-4203 (Washington, D.C.: NASA, 1970).

79. Perry, "Atlas, Thor, Titan, and Minuteman," p. 148.

80. Ibid., p. 150. See also O. J. Ritland, "Concurrency," *Air University Quarterly Review* 12 (Winter–Spring 1960–61): 32–41.

81. Jupiter C used eleven solid-fuel Baby Sargent rockets for its second stage, six of them for stage 3. To place Explorer I in orbit in 1958, ABMA employed a single Baby Sargent rocket as a fourth stage.

82. Neufeld, *The Development of Ballistic Missiles,* p. 227.

83. See Roy Neal, *Ace in the Hole: The Story of the Minuteman Missile* (Garden City, N.Y.: Doubleday, 1962).

84. Von Braun, Ordway, and Dooling, *Space Travel,* pp. 130–32; James Baar and William E. Howard, *Polaris!* (New York: Harcourt, Brace, 1960); Harvey M. Sapolsky, *The Polaris System Development: Bureaucratic and Programmatic Success in Government* (Cambridge, Mass.: MIT Press, 1972); Graham Spinardi, *From Polaris to Trident: The Development of U.S. Fleet Ballistic Missile Technology* (New York: Cambridge University Press, 1994).

85. Wyndham D. Miles, "The Polaris," in Emme, ed., *History of Rocket Technology,* pp. 162–75.

86. Karl Klager, "Segmented Rocket Demonstration: Historical Development Prior to Their Use as Space Boosters," IAA-91-687, 42d Congress of the International Astronautical Federation, 5–11 October 1991, Montreal, Canada.

87. The test failures were related to malfunctions of the motors' nozzle assembly, not the segment joints.

88. Loyd S. Swenson, James M. Grimwood, and Charles C. Alexander, *This New Ocean: A History of Project Mercury,* NASA SP-4201 (Washington, D.C.: NASA, 1966), pp. 167–222.

89. Designated LV-3B but more popularly called Mercury-Atlas D.

90. Hacker and Grimwood, *On the Shoulders of Titans.*

91. Winter, *Rockets into Space,* p. 92.

92. The term "cryogenic" refers to the low temperatures required to create and store substances such as liquid oxygen and liquid hydrogen.

93. Edward C. Ezell and Linda N. Ezell, *On Mars: Exploration of the Red Planet 1958–1978,* NASA SP-4212 (Washington, D.C.: NASA, 1984), pp. 325–26.

94. This test took place on 26 October 1966. The capability to start in the near vacuum of space was extremely important to the success of the Apollo program. See John L. Sloop, "Technological Innovation for Success: Liquid Hydrogen Propulsion," in Frederick C. Durant, ed., *Between Sputnik and Shuttle: New Perspectives on American Astronautics* (San Diego: Univelt, 1985), pp. 225–39.

95. R.W. Bussard, "Nuclear Energy for Rocket Propulsion," Oak Ridge National Laboratory, ORNL CF-53-6-6, 2 July 1953. This was published in *Reactor Science and Technology,* December 1953, pp. 79–170. This secret publication was declassified on 4 November 1960.

96. An historical summary of the early research in nuclear rockets appears in Robert W. Bussard, "Nuclear Rocketry—The First Bright Hope," *Astronautics,* December 1962, pp. 32–35.

97. The name Kiwi for the reactor derives from the name of the flightless bird native to New Zealand.

98. In his 25 May 1961 speech, President Kennedy announced that "an additional $23 million, together with $7 million already available, will accelerate development of the ROVER nuclear rocket. This gives promise of some day providing a means for even more exciting and ambitious exploration of space, perhaps beyond the moon, perhaps to the very ends of the solar system itself." Logsdon, gen. ed., *Exploring the Unknown* 1:III-12.

99. W. Scott Fellows, "RIFT," *Astronautics,* December 1963, pp. 38–47.

100. Compare, for example, the Space Shuttle Main Engine I_{SP} of 450 seconds.

101. Glenn T. Seaborg, "A Nuclear Space Odyssey." Remarks to the Commonwealth Club of California, San Francisco, 24 July 1970. U.S. Atomic Energy Commission release number S-27–70.

102. More than one thousand space access studies have reached this conclusion over the last forty years. See Roger D. Launius and Howard E. McCurdy, *Space Exploration in the Twenty-first Century: NASA and Beyond* (San Francisco: Chronicle Books, 2001), chap. 4; U.S. Congress, Office of Technology Assessment, *Launch Options for the Future: Special Report* (Washington, D.C.: Government Printing Office, 1984); Vice President's Space Policy Advisory Board, "The Future of U.S. Space Launch Capability," Task Group Report, November 1992, NASA Historical Reference Collection; NASA Office of Space Systems Development, *Access to Space Study: Summary Report* (Washington, D.C.: NASA Report, 1994).

– 2 –

Stage-and-a-Half

The Atlas Launch Vehicle

Dennis R. Jenkins

Introduction

All of the three major expendable U.S. launch vehicles during the last half of
the twentieth century can be traced directly back to the earliest ballistic missiles developed by the United States Air Force. In two cases the launch vehicles bore only slight resemblance to the original missile—the Delta was a
highly developed two-stage version of the original Thor intermediate-range
ballistic missile (IRBM) and the Titan III core was loosely based on the long-
range Titan intercontinental ballistic missile (ICBM). But the third major expendable launch vehicle, Atlas, remained much truer to its roots—until very
recently the basic Atlas booster looked remarkably similar to the ICBM that
originally spawned it. This chapter presents a short history of Atlas.[2]

Mutual distrust between the United States and the Soviet Union led to the
cold war that dominated the period from early 1950 until the 1990s. Both
sides engaged in an unbridled arms race in which any development by one
side quickly elicited an equivalent response by the other. The United States
had demonstrated the atomic bomb just prior to the end of World War II; the
Soviets exploded their version in 1949. It was the destructive power of nuclear
weapons that led strategists to consider the possibilities of intercontinental
missiles—their accuracy was insufficient for use with conventional munitions.
Three years later, the United States exploded the first thermonuclear device;
the Soviets followed in 1953. The so-called hydrogen bombs made the inaccu-
racies inherent in early strategic missiles even less important—missing by a
mile or so had little effect on the destructive power of the weapons. By the end
of the 1950s, both sides had demonstrated the ability to launch long-range bal-

listic missiles, and during the 1960s both sides began to deploy them in ever-increasing numbers.[3]

The First American Missiles

After the creation of a separate Air Force in 1947, the Army had continued rocket development, operating on the same assumption behind the German army's research in the 1930s—that rocketry was basically an extension of artillery. In June 1950, U.S. Army Ordnance moved its team of 130 German rocket scientists and engineers from Fort Bliss at El Paso to the Army's Redstone Arsenal at Huntsville, Alabama, along with some 800 military and General Electric employees. Headed by Wernher von Braun, who later became chief of the Guided Missile Development Division at Redstone Arsenal, the Army group began design studies on a liquid-fueled battlefield missile called the Hermes C1, a modified V-2. Soon the Huntsville engineers changed the design of the Hermes, which had been planned for a 500-mile range, to a 200-mile rocket capable of high mobility for field deployment. The Rocketdyne Division of North American Aviation modified the Navaho booster engine for the new weapon, and in 1952 the Army bombardment rocket was officially named Redstone.[4]

Always the favorite of the von Braun group working for the Army, the Redstone was a direct descendant of the V-2. The Redstone's liquid-fueled engine burned alcohol and liquid oxygen and produced about 75,000 lbf. Nearly 70 feet long and slightly under 6 feet in diameter, the battlefield missile had a speed at burnout, the point of propellant exhaustion, of 3,800 miles per hour. For guidance it utilized an all-inertial system featuring a gyroscopically stabilized platform, computers, a programmed flight path taped into the rocket before launch, and the activation of the steering mechanism by signals in flight. For control during powered ascent the Redstone depended on tail fins with movable rudders and refractory carbon vanes mounted in the rocket exhaust. The prime contract for the manufacture of Redstone test rockets went to the Chrysler Corporation. In August 1953 a Redstone fabricated at the Huntsville arsenal made a partially successful maiden flight of only 8,000 yards from the military's missile range at Cape Canaveral, Florida. During the next five years, thirty-seven Redstones were fired to test structure, engine performance, guidance and control, tracking, and telemetry.[5]

Developing a Concept

The developments that eventually led to the Atlas had actually begun on 31 October 1945 with a U.S. Army Air Technical Services Command request for proposals for a long-range missile system. Not sure of which route ultimately would be successful, the U.S. Army Air Forces specified both "A" winged cruise missiles and "B" ballistic missiles. Four weapons systems designations were eventually assigned: MX-770 to North American Aviation, MX-771 to Martin, MX-774 to Consolidated-Vultee,[6] and MX-775 to Northrop. Eventually these would lead to the development of the Navaho, Matador/Mace, Atlas, and Snark, respectively.

On 10 January 1946, Consolidated-Vultee, under the leadership of Belgian-born Karel J. "Charlie" Bossart, submitted proposals for studies of two 7,000-mile missiles: a subsonic turbojet-powered winged cruise missile and a rocket-powered ballistic missile. Bossart proposed several innovations for the ballistic missile, including the use of pressure-stabilized single-wall construction[7] for the propellant tanks that made up much of the airframe. The skin on these tanks was so thin that they could not support their own weight unless they were pressurized, but the construction technique promised to greatly reduce the empty weight of the airframe, thereby maximizing the missile's payload capacity.[8] Bossart and the engine manufacturer, Reaction Motors, Inc., proposed a technique basically new to American rocketry although patented by Goddard and tried on some German V-2s: controlling the rocket by swiveling the engines and hence varying the exit angle of the exhaust gases, using hydraulic actuators responding to commands from the autopilot and gyroscope. This technique was the precursor of the gimbaled engine method employed to control the Atlas and other later missiles.

On 19 April 1946, Consolidated-Vultee received a $1,893,000 contract for further definition of both missile types, but specifically including the fabrication and testing of ten MX-774B ballistic missiles to verify Bossart's innovative concepts. Ground testing of components designed for the MX-774B missiles began in San Diego in early 1947.

In June 1947, the Air Force directed Consolidated-Vultee to concentrate its efforts on the ballistic missile—Northrop and Martin had received contracts for development of the subsonic jet-powered designs. These early cruise missiles were largely technological and operational dead ends, but the ballistic missiles would become the largest and most important defense project of the cold war and eventually led to successful space launch vehicles. The ballistic missiles and their associated nuclear warheads also became the single most expensive project ever undertaken by the United States. But it would be five years before the crash effort would begin.

MX-774B

The RTV-A-2 HIROC (*high*-altitude *roc*ket) was initially going to be a slightly improved version of the German V-2, but the vehicle that emerged was a great deal more advanced. Its most significant innovation was its "balloon" construction in which the pressure-stabilized propellant tanks were an integral part of the missile structure. Carrying weight savings one step further than the normal monocoque construction used on aircraft, the walls of the propellant tanks were so thin that they could not support themselves without being pressurized with either propellants (operationally) or inert gases (for storage). This would become a trademark of all Atlas missiles until the Atlas V.[9]

The HIROC used four Reaction Motors liquid rocket engines that burned alcohol and liquid oxygen, generally similar to the XLR-11 motors used on the Bell X-1 and early North American X-15 rocket planes. Each of the motors could be gimbaled under control from a gyroscopic guidance system. This was a radical departure from the "vanes in exhaust" control system used on the V-2, although Robert Goddard had experimented with a similar system nine years earlier. The large fins at the base of the HIROC were not actually required for stability but provided an extra measure of emotional security for the designers. The third major advance brought by the HIROC was a warhead that could be separated from the vehicle; on the V-2[10] the vehicle was firmly attached to the warhead. The missile was 31 feet long, 2.5 feet in diameter, and had a fin span of approximately 6 feet. It weighed 1,200 pounds when empty.

In 1947, the Truman administration and the equally economy-minded Republican Eightieth Congress confronted the Air Force with the choice of having funds slashed for its intercontinental manned bombers and interceptors or cutting back on some of its advanced weapons designs. These budgetary constraints forced the cancellation of MX-774 in July 1947, just before the first HIROC flight vehicle was completed. However, some appropriations remained unspent on the contract, and Consolidated-Vultee had sufficient funds remaining to complete three test vehicles and to launch them at the White Sands Proving Grounds in New Mexico. A stand that had been used to launch captured V-2s was hurriedly converted to handle the new missile, and a static engine firing was conducted on 26 May 1948. Several more static tests followed.

The first launch was on 14 July 1948, and the vehicle performed flawlessly for the first 12.6 seconds of flight, then tumbled out of control. Most of the wreckage was recovered, and subsequent evaluations found that a vibration had caused a liquid-oxygen valve to close, starving all four engines. The next two flights performed better, lasting 48 and 52 seconds, but neither was categorized as a success. Nevertheless, the experience provided Consolidated-Vultee a great deal of information that would prove valuable a couple of years later when the design of Atlas began in earnest.[11] Because of Air Force budget constraints, further work at Consolidated-Vultee was reduced to low-priority development activity using company funds.

Atlas ICBM

As part of the National Security Act of 1947, the United States Air Force became a separate branch of the military in September 1947 and began searching for new weapon concepts to differentiate it from the Army and Navy. As part of these efforts, the Air Force asked the newly formed Project RAND[12] to evaluate several concepts, including the ballistic missile. RAND was enthusiastic about the possibilities, but the continuing tight fiscal environment did not allow the Air Force to pursue the concept further.

Defense spending in the immediate postwar period was relatively modest, based largely on the confidence the United States had in the atomic bomb. This began to change when the Soviet Union exploded their first atomic weapon in 1949. The outbreak of the Korean conflict in 1950 brought a drastic increase in defense spending and signaled the beginning of the forty-year cold war. The Air Force recognized the potential of an intercontinental missile to deliver nuclear weapons and began investigating what design offered the most promise. For the most part, engineers were still more comfortable with the more familiar cruise missile concept, but it was obvious that a V-2-type weapon offered many advantages—if technology and reliability problems could be overcome.

More studies were initiated by Project RAND on 1 October 1950, and there was sufficient confidence that an intercontinental missile could be built that on 16 January 1951 Consolidated-Vultee was awarded a contract for MX-1593. The new missile was to deliver an 8,000-pound thermonuclear warhead to a distance of 5,750 miles. Originally both a winged missile (boost-glide concept) and a ballistic rocket were carried as design options, but on 1 September 1951 the winged missile was dropped and all efforts concentrated on the ballistic version—both Consolidated-Vultee and the Air Force had concluded that the ballistic approach would be the best approach.

Convair soon reached an agreement with North American (Rocketdyne) to develop a modified version of the Navaho booster engine that used liquid oxygen and kerosene to power the new missile. A major propulsion problem during the early 1950s centered on the reliability of the rocket engine ignition systems. This seemed to be particularly true of the few attempts to "stage" a rocket in flight, where, as often as not, the second stage refused to ignite properly. To overcome this, Consolidated-Vultee developed the "stage-and-a-half" concept.[13] In Bossart's design, the missile would be powered by five engines: four "booster" engines and a single "sustainer" engine, all of which would be ignited on the ground prior to liftoff to allow confirmation that they were functioning properly before releasing the missile for flight. The four booster engines, along with their associated aerodynamic fairings and part of the structure, would be jettisoned a couple of minutes after liftoff, leaving the sustainer engine to provide the final thrust for the missile. All five engines used propellants from a single set of pressure-stabilized tanks.[14]

At this point the missile was 160 feet tall and 12 feet in diameter; it was an ambitious undertaking—and one that the Air Staff was not completely comfortable with. The Air Staff wanted to pursue a more conservative approach:

develop major subsystems, such as the engines, airframe, guidance system, and reentry vehicle, first, then proceed into full-scale development of the weapons system itself. Convair also wanted an opportunity to further refine some of the new technologies being applied to the Atlas[15] through an incremental flight-test program and proposed a series of test missiles that were subsequently designated X-11 and X-12. The X-11 would be used to test the overall airframe but would use only the single sustainer engine; the X-12 would be integrate more components including the four booster engines. In the meantime, work proceeded slowly.

During 1952, primarily in response to Army and Navy attempts to wrest control of the ICBM program away from the Air Force, the Air Staff finally approved a development program that would result in an operational missile in approximately ten years at a cost of some $378 million. The pace of work picked up.[16]

Several events combined to accelerate the development and testing of the Atlas as the first intercontinental ballistic missile (ICBM) developed by the United States. The U.S. test of a thermonuclear device in November 1952 was followed by the test of a Soviet hydrogen bomb in August 1953. Most intelligence experts also believed that the Soviets were far ahead of the United States in missile development and rocket technology. Finally, Trevor Gardner, as Special Assistant for Research and Development for the Secretary of the Air Force, selected ballistic missiles as a critical technology to be afforded a very high priority.

In 1953 the Air Force gave the Atlas program its highest available development priority (1A). Perhaps more important, the Atomic Energy Commission informed Consolidated-Vultee of several breakthroughs that permitted the development of much smaller nuclear warheads; the weight of the warhead was reduced from the original 8,000 pounds to only 3,000 pounds. This allowed Consolidated-Vultee to reduce the size of the new missile—although significantly smaller, the new missile was still a large vehicle, measuring over 110 feet tall and 12 feet in diameter. The gross liftoff weight was estimated at 440,000 pounds.[17]

The history of Atlas took a major turn in February 1954, when the Air Force Strategic Missiles Evaluation Committee (the "Teapot Committee") report was issued. Chaired by renowned mathematician John von Neumann, the committee recommended a "radical reorganization" of the entire program that would result in an operational missile in six years. The Air Force accepted the Teapot Committee recommendations, and on 14 May 1954 it accelerated the Atlas program to the "maximum extent that technology would permit." The Ramo-Wooldridge Corporation (now TRW) became responsible for overall systems engineering and technical direction for the entire ICBM program, while Convair remained the primary contractor responsible for de-

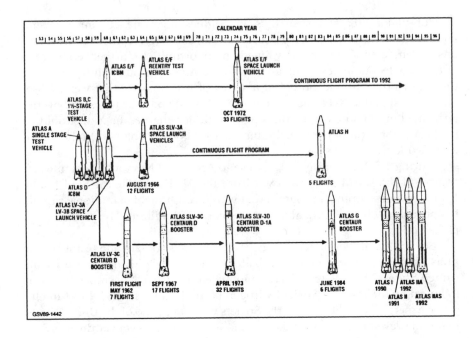

Image 2-1: The Atlas family of launch vehicles. (NASA Photo)

tailed design and manufacturing of the flight vehicle itself.[18] On 16 December 1954 the Air Force publicly announced for the first time that it was developing the Atlas ICBM. Nevertheless, concerns over the viability of the Atlas and its unique pressure-stabilized tanks subsequently caused the Air Force to initiate the development of the Titan ICBM by Martin as a backup capability.

Further reductions in the weight of the thermonuclear payload to 1,500 pounds allowed the Atlas to become even smaller. The diameter was reduced from 12 feet to 10 feet, and the gross liftoff weight went from 440,000 pounds to 240,000 pounds. The new design required only three engines—two jettisonable boosters and a single sustainer. The "balloon" pressure-stabilized structure that had been tested on the MX-774 project was still being used, only this time it would use stainless steel instead of the aluminum alloy used earlier. The tanks themselves constituted approximately 80 percent of the missile's mass and were manufactured from sheets of stainless steel ranging between 0.01 and 0.04 inch thick. The empty tanks needed to be pressurized with nitrogen gas at 5 pounds per square inch (psi) to maintain their shape until they were filled with propellants. The missile was designed to deliver a thermonuclear warhead over a range of 6,000–9,000 miles with a CEP (circular error probable)[19] of not more than 10 miles.

Atlas A

On 24 January 1951, Consolidated-Vultee was awarded Air Force contract AF33(038)-19956 for the development and manufacture of eight ground-test and eight flight-capable Atlas Series A[20] single-stage ballistic missiles. These were intended as test vehicles to demonstrate the technology required to proceed with the development of the XB-65[21] Atlas ICBM. Subsequently, the missiles were redesignated XB-65A, and then XSM-65A. Inside Convair the Atlas was designated as Model 7.[22] On 14 January 1955 the Air Force authorized the full-scale development of Weapon System WS107A-1,[23] with the missile itself designated B-65.[24]

Convair constructed a new facility in Kearny Mesa, California (outside San Diego) specifically for the Atlas engineering and manufacturing program. On 6 March 1955, faced with intelligence reports of Russian progress on their ICBM, the Atlas received the highest national development priority (1DX), a rating even higher than the Air Force 1A priority it had previously carried. The project became the largest and most complex production, testing, and construction program ever undertaken.[25]

Convair's plan again called for incremental testing of the overall concept. Propulsion system and component tests began in June 1956, and the initial captive- and flight-test missiles were completed later the same year. The Atlas was too large to be airlifted by the transports of the day, so the missiles were transported to Cape Canaveral aboard a special trailer constructed by the Goodyear Rubber Company. The trailer was 64 feet long and 14 feet wide and required two "back seat drivers" to maneuver around corners. The first non-flight-rated Atlas A (vehicle 1A) was completed in early August 1956 and trucked to a site in Sycamore Canyon, California, for static ground tests. After these were completed, the missile left San Diego on 1 October 1956 destined for the test range at Cape Canaveral, Florida. In an effort to maintain security and avoid low bridges, most of the trip was conducted on back roads around major cities. The missile completed its 2,622-mile journey to the Cape in nine days—soon the ICBM would be able to cover a similar distance in about 18 minutes.

The first flight-rated.Atlas A[26] (4A) arrived at the Cape in December 1956 and was launched on 11 June 1957. The first two launches were not particularly successful, but this was expected for a program pushing technology as far as Atlas was.[27] There had been a lot of skepticism about the pressure-stabilized propellant tanks, but these performed flawlessly. The same could not be said for everything else. The first two flights suffered base heating from the rocket exhaust, a burn-through of the base heat shield, and the burning or damaging of various hydraulic lines. Without hydraulic power, the booster engines went hard-over, causing the missiles to loop-the-loop before they could be destroyed. The missiles did all of this, however, without structural failure, validating the pressure-stabilized structure concept and convincing many doubters—the tests were a great deal more successful than they appeared.

A new urgency developed after the Soviet Union orbited *Sputnik I,* and on 5 October 1957 the Secretary of Defense authorized a greatly accelerated ICBM program with the goal of having the first four operational Atlas (and four operational Titan) squadrons in place by the end of 1962. Considering neither missile had made a successful test flight at that point, it was an ambitious schedule. Fortunately, the third Atlas A test flight proceeded better, and the missile flew 600 miles downrange powered by its two booster engines only. Although Atlas still had difficult times ahead of it, Convair had proven the concept would work, given enough time and money.[28]

The Atlas A vehicles were 75 feet 10 inches long and, like all subsequent Atlas missiles, had a diameter of 10 feet. The airframe weighed 12,490 pounds empty and had an all-up weight of 80,000 pounds. Unlike later Atlas versions, the warhead section was not separable and contained flight-test instrumentation. The two Rocketdyne XLR43–NA-3 booster engines provided 105,000 lbf each for 135 seconds using liquid oxygen and RP-1 (kerosene). The sustainer engine intended for production missiles was not fitted to the Atlas A.

Atlas B

The Atlas B was similar to the Atlas A, but it incorporated the planned sustainer engine to prove the stage-and-a-half concept. A total of thirteen vehicles were manufactured for flight and ground tests. All of the basic subsystems were tested in this series, including the MA-1 propulsion system, the Mod 1 radio guidance system, and the Mark 2 heat-sink reentry vehicle. The Atlas B series demonstrated booster staging and reentry vehicle separation and attained the specified range of 6,500 miles with vehicle 12B.[29]

The Atlas B[30] vehicles were 75 feet 10 inches long and contained provisions to separate the warhead section at the end of powered flight. The airframe weighed 18,333 pounds empty and had an all-up weight of 240,000 pounds. The MA-1 propulsion system used two Rocketdyne XLR43–NA-3 booster engines providing 105,000 lbf each for 100 seconds using liquid oxygen and RP-1, and a single 120,000-lbf XLR43–NA-5 sustainer engine that continued burning for up to 220 seconds. Vernier engines provided flight control after the gimbaling sustainer was shut down.[31]

Atlas-SCORE

President Eisenhower had considered plans for an Atlas satellite mission as early as 1956 but had rejected it because of potential interference with the development of the ICBM. After the launch of *Sputnik 3,* however, the president reconsidered and authorized the mission—provided a full-range demonstration flight of the ICBM was flown first. An Atlas B (vehicle 10B) was selected to carry the 142-pound SCORE (Signal Communications by Orbiting Relay

Equipment) payload provided by the U.S. Army's Signal Corps. The entire Atlas rocket, minus the booster engines discarded during ascent, would be placed into orbit with the SCORE payload attached. It would be the largest manmade object in orbit at that time.

After the Atlas 12B test on 28 November 1958 managed to fly the full 6,500 miles required for the ICBM, plans were made to launch SCORE. Less than three weeks later the missile was on the launch pad, and even most of the launch team was unaware of the true mission. On 18 December 1958, vehicle 10B made the first orbital delivery by an Atlas. The satellite was used to broadcast a previously recorded Christmas message from Eisenhower—later, voice and teletype messages were uplinked to the satellite, recorded, then rebroadcast. The satellite continued to transmit for thirteen days and remained in orbit attached to its Atlas booster for thirty-eight days.[32]

Because of the performance limitations of a stripped down Atlas B, the SCORE package had been limited to less than 150 pounds. Calculations indicated that the Atlas could loft this payload into an orbit with a perigee of about 100 miles and an apogee of between 500 and 800 miles. With such a relatively high orbit, real-time communication relay tests could be performed between ground stations as far as 1,000 miles from the satellite.

For the project, four mobile ground stations each consisting of appropriately equipped Army type V-51 vans and a quad-helix tracking antenna mounted on a searchlight base were located at Fort MacArthur, California; Fort Huachuca, Arizona; Fort Sam Houston, Texas; and Fort Stewart, Georgia. They were all connected to a control center at the Signal Corps Laboratory at Fort Monmouth, New Jersey, by telephone lines and HF radio.

The SCORE payload itself consisted of a redundant pair of battery-powered, vacuum tube–based UHF communication packages with a nominal design life of twenty-one days. The payload was housed in a cylindrical anodized metal housing at the nose of the missile designed to keep the temperature between 40 and 120 degrees F. The payload remained attached to the Atlas on-orbit, yielding an impressive total in-orbit mass of 8,750 pounds—slightly more than the mass of the Soviet *Sputnik 2* with its spent booster attached.[33]

Atlas C

Only six Atlas C test vehicles were built and flown, and this model was never deployed operationally as an ICBM or used for space launches. This missile emphasized weight reduction and improved accuracy and served as semiproduction prototypes for the ICBMs that would follow. The same 330,000-lbf MA-1 propulsion system used on the Atlas B was carried over intact. The 82.5-foot-long airframes used thinner skin on the propellant tanks and carried improved Mod 2 and Mod 3 radio guidance systems which used radars manufactured by General Electric coupled with Burroughs computers. Most of the Atlas C missions aimed for the Atlantic Missile Range Impact

Locator System area near Ascension Island in the South Atlantic. The next-to-the-last Atlas C flight was the first time that a reentry vehicle was recovered.[34]

Operational ICBMs

Atlas D

The Atlas D was the first operational version of the Atlas ICBM, initially designated SM-65D. It also was used as a launch vehicle for later Project Mercury flights (listed separately) and a wide variety of unmanned payloads. The Atlas D development flights were aimed at verifying the operational configuration of the missile and its radio-inertial guidance system. Although off to a disappointing start with three successive failures, these tests eventually confirmed that the missile was ready for operational deployment, and that the accuracy of the GE/Burroughs guidance system was within the revised 1-mile requirement. The Mark 3 ablative reentry vehicle was shown to be operationally sound, and several Atlas D flights demonstrated the ability to exceed a range of 9,000 miles.[35]

The General Electric Mark 3 reentry vehicle carried a single 1.44-megaton W49 thermonuclear warhead. Initially, the Atlas D ICBMs were built with MA-2 propulsion systems that generated 368,000 lbf; later units were reportedly equipped with MA-3 systems generating 392,000 lbf. Both systems consisted of two LR89–NA-3 boosters and a single LR105–NA-5 sustainer, differing only in details. At least seven of the Atlas D development test flights were used to prove components for the upcoming Atlas E/F all-inertial guidance system.[36]

On 16 November 1956 Cooke AFB (later renamed Vandenberg), California, was selected as the first operational ICBM site, and the first Atlas D squadron was activated on 1 July 1957—it was turned over to the Strategic Air Command on 15 January 1959. On the same day, as an emergency measure, the Air Force placed three development Atlas D missiles on alert on open vertical launch pads at the site. Completely exposed to the elements, the three missiles were serviced by a gantry crane, and at least one missile was on alert at all times. The initial operational capability was originally scheduled for 1 August 1957, but continuing development problems delayed this until 1 September 1959—even at that time it is unlikely the missiles were truly operational, although warheads were available and the missiles were on alert. The first Atlas pads were above-ground structures that look much like the space launch pads used today and were followed by "semihard," above-ground "coffin" launchers; hardened silos would come later.

Originally, the location of the Atlas launch sites was determined strictly by the missile's range—they had to be within 5,000 miles of their targets in the Soviet Union. Later, other factors that influenced the placement of the sites included that they be inland, out of range of Soviet submarine-launched missiles, close to support facilities, and, as a cost savings measure, built on government property wherever possible.

The first full Atlas D squadron became operational in 1960, deployed at so-called soft sites that could only withstand overpressures of 5 psi. The missiles were stored horizontally within 103-by-133-foot launch and service buildings built of reinforced concrete, and the missile bays were equipped with retractable roofs. To launch the missile, the roof was pulled back and the missile was raised to the vertical position, fueled, and fired—a time-consuming and cumbersome process.

An individual Atlas D launch site consisted of a launch and service building, a launch operations building, guidance operations building, generating plant, and communications facilities. The launch operations buildings were two-story structures built of reinforced concrete that housed the launch operations crew and were equipped with entrance tunnels, blast-proof doors, and escape tunnels. The guidance operations buildings housed the computers, which were used to send course corrections to the missile in flight.

The first Atlas D squadron at F.E. Warren AFB, Wyoming, used six launchers grouped together, controlled by two launch operations buildings and clustered around a central guidance control facility. This was the 3x2 configuration—two launch complexes of three missiles each constituted a squadron. At the two later Atlas D sites the missiles were based in a 3x3 configuration—three launchers and one combined guidance control/launch facility constituted a launch complex, and three complexes comprised a squadron. To reduce the risk that one powerful nuclear warhead could destroy multiple launch sites, the launch complexes were spread 20 to 30 miles apart.

During a typical Atlas D ICBM launch sequence, the two boosters and single sustainer were ignited on the ground, and two small vernier engines mounted above the sustainer came to life 2.5 seconds after liftoff. After leaving the launch pad the missile accelerated rapidly, gradually nosing over in a gentle arc towards the target. The booster engines burned for 140 seconds, and after receiving a signal from the controlling ground station, the booster engines and their turbopumps were jettisoned. The sustainer engine burned for another 130 seconds—final course and velocity corrections were made by the vernier engines. At the apogee of its elliptical flight path the missile reached an altitude of approximately 760 miles and a speed of 16,000 mph. The elapsed time to reach a target 6,500 miles away was roughly 43 minutes.

After they were retired, the Atlas D missiles were used extensively as targets for the Nike Zeus (Safeguard) antiballistic missile (ABM) system based at Kwajelein atoll in the Pacific, and others were used to support the Advanced Ballistic Reentry System (ABRES) research being conducted for the Minuteman warhead. A few were used to launch small satellites, but in general they were not converted to space launch vehicles.

Atlas E

The Atlas E, the initial fully operational version of the Atlas ICBM was deployed from 1961 to 1965. The Atlas E differed from the Atlas D primarily by using an American Bosch Arma all-inertial guidance system instead of the hybrid radio-inertial system. The all-inertial system did not require the ground-based computers used by the Atlas D system—the use of ground-based computers to signal course corrections had been considered susceptible to enemy jamming. An AVCO-manufactured Mark 4 ablative reentry vehicle carrying a single 4-megaton W38 warhead replaced the earlier Mark 3/W49 combination.[37] Since the missiles were no longer tied to a central guidance control facility, the launchers could be dispersed more widely. Thus, the three Atlas E squadrons were based in a 1x9 configuration—nine independent launch sites and a single launch operations building comprised a missile squadron.

The Atlas Es were based in semihard or coffin facilities that protected the missile against overpressures up to 25 psi. In this arrangement the missile, its support facilities, and the launch operations building were housed in reinforced concrete structures that were buried underground; only the roofs protruded above ground level. The missile launch and service building was a 105-by-100-foot structure with a central bay where the missile was stored horizontally. To launch a missile, the heavy roof was retracted, the missile raised to the vertical launch position, fueled, and then fired. The launch operations building was approximately 150 feet from the missile launch facility, and the two were connected by an underground passageway. The launch operations building contained the launch control facilities, crew's living quarters, and power plant. The Atlas E launch sites were spaced approximately 20 miles apart.

After retirement, the ICBMs were stored at Norton AFB, California; beginning in the late 1970s the missiles were refurbished, and in 1980 they began to be used as launch vehicles. Surprisingly, given the spotty success record they had experienced as ICBMs, they performed nearly flawlessly as space launch vehicles.

Atlas F

The final operational version of the Atlas ICBM was generally similar to the Atlas E but used an improved version of the all-inertial guidance system and was optimized to be stored vertically in "hard" silos. They were deployed as missiles from 1962 to 1965. With the exception of a pair of massive 45-ton doors, the silos, 174 feet deep and 52 feet in diameter, were completely underground. The walls of the silo were built of heavily reinforced concrete. Within the silo the missile and its support system were supported by a steel framework called the crib, which hung from the walls of the silo on four sets of huge springs.

Image 2-2: The Atlas ICBM basing concept with an underground silo. (USAF Photo)

Adjacent to the silo, and also buried underground, was the heavily rein-forced concrete launch control center that contained the launch control equip-ment plus living quarters for the crew. The control center was connected to the silo by a cylindrical tunnel 50 feet long and 8 feet in diameter that pro-vided access to the silo and served as a conduit for the launch control cabling.

In the firing sequence the missile was fueled, lifted by an elevator to the top of the silo, and then launched. Although the silo sites were by far the most difficult and costly sites to build, they offered protection from overpressures of up to 100 psi.

The Air Force deployed six Atlas F squadrons, one each at Schilling AFB, Kansas; Lincoln AFB, Nebraska; Altus AFB, Oklahoma; Dyess AFB, Texas; Walker AFB, New Mexico; and Plattsburgh AFB, New York (the only ICBMs ever based east of the Mississippi). Each squadron included twelve launch sites (1x12), and distances between the sites ranged from 20 to 30 miles.

After they were retired, a large number of Atlas F missiles were used to support the Advanced Ballistic Reentry System (ABRES) research being conducted for the Minuteman warhead. Beginning in 1968, the remaining missiles were refurbished and began to be used as space launch vehicles. A variety of upper stages were fitted to provide the final kick for satellites to obtain their correct orbits.

Retirement

The operational buildup and subsequent deactivation happened quickly—the development of smaller nuclear warheads and reliable solid rocket motors for the Minuteman ICBM rapidly made the first generation of Atlas and Titan ICBMs obsolete. In the end, only thirteen Atlas squadrons would be deployed, with the last being inactivated on 25 June 1965, only six years after the first had been activated.

When the missiles were removed from each of the operational bases, most were taken to the San Bernardino Air Materiel Area at Norton AFB, California, where they were preserved and stored awaiting other uses. Unfortunately, about twenty missiles were declared surplus and destroyed instead of being stored.

Fifty years later, it is hard to imagine the magnitude of the ICBM program. At the end of 1955 there were only 56 contractors actively working on the Atlas program—two years later there were more than 150. The total cost of Atlas, Titan, and Minuteman, in 1970 dollars, is estimated at $17 billion—compare this to a total outlay of only $2 billion for piloted aircraft during the same time period.[38]

Space Launchers

When the development of Atlas began, thermonuclear weapons were very large—this soon changed as the national laboratories managed to significantly reduce their size and weight. However, size is a relative concept, and the Atlas was still developed with a fairly substantial "throw weight." Once production of the ICBM version was well underway, it was only natural that the Atlas would be called upon to again perform orbital missions.

Initially, standard Atlas D ICBMs were modified on the assembly line as needed for space launch purposes and designated LV-3. The addition of an Agena upper stage created an LV-3A, while the use of a Centaur upper stage resulted in an LV-3C. The LV-3B designation was used for the special Atlas variants produced for NASA's Mercury program. Unfortunately, the individual tailoring that was required to convert Atlas ICBMs to specific space launch missions was complex and cumbersome and significantly raised the cost of the boosters.

To overcome this, in 1962 the Air Force awarded Convair a contract to

Figure 1
Operational Atlas ICBM Squadrons

Squadron	Support Base	Number & Type of Missiles	Basing Mode	Turned over to SAC	Inactivated
576 SMS	Vandenberg AFB, California	6 Atlas D 1 Atlas E 2 Atlas F	Vertical above ground (3 x 2) Horizontal above ground (1 x 1) Silo (1 x 2)	15 January 1959	2 April 1966
564 SMS	F.E. Warren AFB, Wyoming	6 Atlas D	Horizontal above ground (3 x 2)	30 August 1959	1 September 1964
565 SMS	F.E. Warren AFB, Wyoming	9 Atlas D	Horizontal above ground (3 x 3)	4 March 1961	1 December 1964
566 SMS[1] (549 SMS)	Offutt AFB, Nebraska	9 Atlas D	Horizontal above ground (3 x 3)	30 March 1961	15 December 1964
567 SMS	Fairchild AFB, Washington	9 Atlas E	Horizontal below ground (1 x 9)	28 September 1961	25 June 1965
548 SMS	Forbes AFB, Kansas	9 Atlas E	Horizontal below ground (1 x 9)	10 October 1961	25 March 1965
549 SMS[1] (566 SMS)	F.E. Warren AFB, Wyoming	9 Atlas E	Horizontal below ground (1 x 9)	20 November 1961	25 March 1965
550 SMS	Schilling AFB, Kansas	12 Atlas F	Silo (1 x 12)	7 September 1962	25 June 1965
551 SMS	Lincoln AFB, Nebraska	12 Atlas F	Silo (1 x 12)	15 September 1962	25 June 1965
577 SMS	Altus AFB, Oklahoma	12 Atlas F	Silo (1 x 12)	9 October 1962	25 March 1965
578 SMS	Dyess AFB, Texas	12 Atlas F	Silo (1 x 12)	4 November 1962	25 March 1965
579 SMS	Walker AFB, New Mexico	12 Atlas F	Silo (1 x 12)	30 November 1962	25 March 1965
556 SMS	Plattsburgh AFB, New York	12 Atlas F	Silo (1 x 12)	7 December 1962	25 March 1965

Source: Jacob Neufeld, *Ballistic Missiles in the United States Air Force: 1945–1960*, Washington, D.C.: Office of Air Force History, 1990), pp. 234–35.
[1] On 1 July 1961, the Atlas D squadron at offutt and the Atlas E squadron at Warren exchanged designators.

develop a standardized launch vehicle (SLV-3) based on the Atlas. The SLV-3 was essentially the same as the earlier LV-3A and generally used an Agena upper stage. The SLV-3A introduced propellant tanks that had been stretched by 9.75 feet and also incorporated a more powerful MA-5 propulsion system rated at 431,300 lbf. The SLV-3A used an improved Agena upper stage, while the similar SLV-3C used an improved Centaur. The overall configuration of the vehicle did not change, and they were still considered "stage-and-a-half" boosters—all three engines were ignited prior to liftoff, and the booster engines and their associated structure were jettisoned approximately 2 minutes into flight. The sustainer engine continued thrusting until fuel depletion (SLV-3A) or until the guidance section commanded it to stop (SLV-3C).

The SLV-3D was physically similar to the SLV-3C, including the Centaur upper stage. The primary difference was that the SLV-3D used the Centaur guidance system and autopilot during ascent—earlier Atlas boosters had used their own guidance systems, derived from the ICBM program, during the boost phase.

Although the standardized launch vehicle program greatly simplified the process and reduced costs, the ICBM-based boosters would make a return. When Atlas (and Titan I) was rather abruptly retired by the Air Force in preference to the solid-fueled Minuteman in 1965, many of the early ICBMs were placed into storage at Norton AFB, California. It was only a matter of time before they would be recognized as a valuable asset and find their way into the booster business. During January 1967 General Dynamics was authorized to begin converting Atlas D and E missiles into launch vehicles. A string of propulsion system failures on the reactivated ex-ICBMs led the Air Force to order the Canoga Overhaul Program (COP) to perform a complete overhaul of the MA-3 engines on all remaining Atlas D, E, and Fs.[39]

Mercury–Atlas D

Technically designated LV-3B, the Mercury–Atlas D was a modified version of the standard Atlas D ICBM used to carry Project Mercury capsules into orbit. Perhaps the most significant modification was a system that could sense problems aboard the Atlas and trigger the escape system that would pull the Mercury capsule free of the booster. Because the Mercury-Atlas configuration was about 20 feet taller than the ICBM, the rate gyro package had to be relocated higher in the airframe. A fiberglass shield was installed to cover the top dome of the liquid oxygen tank to protect it when the capsule's posigrade rockets fired at separation. Initially, the Mercury-Atlas boosters were constructed of a slightly thinner aluminum skin because they used a slower start sequence in the engines, but engineers soon found that the longer—but lighter—spacecraft exerted unexpected dynamic loads on the conical tank section just under the capsule. The answer was to add thicker skin in this area, negating most of the earlier weight savings.[40]

Image 2-3: The Mercury-Atlas launch vehicle that lofted astronaut Gordon Cooper into orbit on 15 May 1963. The 123-foot-tall Pad 14 gantry moves back during a systems check several days before the mission. The gantry moved away from the Atlas about one hour from lift-off on actual flight, and the umbilical cable to the tower at left was dropped a few minutes before lift-off. (NASA Photo no. 63-MA9-135)

It should be noted that these missiles were specially configured at the factory for their piloted missions and were not modified ICBMs per se. The boosters were optimized for launches to orbits inclined 28 degrees (due east from Cape Canaveral) and were capable of putting 3,000 pounds into a 115-mile orbit. Like the Atlas D, the booster used two 120,000-lbf LR89–NA-5 booster engines and a single 120,000-lbf LR105–NA-5 sustainer engine.[41]

Atlas-Able

The Atlas-Able was essentially an Atlas D ICBM with an Able 5 second stage and an Altair 1 third stage that was specifically configured to boost Pioneer satellites to the Moon.

Atlas-Vega

The Atlas-Vega would have consisted of an Atlas D booster with a new LOX/RP-1 upper stage using the General Electric 405 engine. NASA initiated the development effort to support deep space and planetary missions before the Atlas-Centaur was available. Work had already begun when NASA discovered that the Air Force had a similar launch vehicle in development for the MIDAS satellite programs. Atlas-Vega was accordingly canceled, and NASA adopted the Atlas-Agena vehicle.

LV-3A Atlas–Agena A

The LV-3A Atlas–Agena A was an Atlas D ICBM fitted with an Agena A second stage. The Agena was a storable propellant stage originally called "Hustler" and was based on a Bell Aircraft engine being developed for the canceled rocket-propelled nuclear warhead pod for the Convair B-58 Hustler bomber.

LV-3A Atlas–Agena B

Used extensively to launch early SAMOS surveillance and MIDAS early warning satellites, as well as the original Ranger probes to the Moon and Mariner probes to Venus, the LV-3A Atlas–Agena B was an Atlas D ICBM fitted with an improved Agena B second stage with increased propellant capacity.

LV-3A Atlas–Agena D

The LV-3A Atlas–Agena D was a further improved and lightened Agena upper stage, along with small third and fourth stages. It launched early KH-7 GAMBIT spy satellites plus the two Mariner probes to Mars. (The proposed Atlas–Agena C used an increased-diameter upper stage that was never built.)

LV-3C Atlas-Centaur

The LV-3C Atlas-Centaur was the first version of Atlas with a Centaur upper stage. The Centaur engine contract had been awarded on 1 October 1958, and by November 1960 the Centaur tracking network had been installed and was

Image 2-4: Preparations for the launch of an Atlas-Agena on 11 November 1966. (NASA Photo no. KSC-66-21891)

being tested. Development of production Centaurs began in August 1961 with an expected operational date of late 1961. However, on 28 September 1961 the first mission programmed for Centaur (Mariner) was moved to Agena D because of delays in the Centaur test program. The RL-10 engine for Centaur was finally flight qualified on 19 November 1961. Later flights of these test vehicles were used for the early Surveyor missions and tests.

SLV-3 Atlas

The SLV-3 Atlas was a standardized Atlas D launch vehicle with no (or a small solid) upper stage. Three of these vehicles were used to launch the PRIME SV-5D (X-23A) experimental reentry vehicles.

SLV-3 Atlas–Agena B

The SLV-3 Atlas–Agena B was the same as the LV-3A Atlas–Agena B, except that it used a standardized Atlas launch vehicle (instead of an ICBM) and a slightly improved Agena B upper stage.

SLV-3 Atlas–Agena D

The SLV-3 Atlas–Agena D was a standardized Atlas booster with an Agena D upper stage.

SLV-3A Atlas–Agena D

Used to launch classified government payloads into orbit, the SLV-3A Atlas–Agena D was an uprated standardized Atlas booster with an Agena D upper stage.

SLV-3C Atlas-Centaur

The SLV-3C Atlas-Centaur was an uprated standardized Atlas booster with Centaur upper stage.

SLV-3D Atlas-Centaur

The SLV-3D Atlas-Centaur was a fully developed version of Atlas with Centaur upper stage.

Atlas G and Atlas H

The last of the standardized launch vehicles (SLV-3D) led to two other vehicles. The Atlas G used stretched propellant tanks that added 4.25 feet to the length[42] of an Atlas D, and it incorporated a slightly improved MA-5 engine that provided 7,500 lbf more thrust. The Atlas H used the original, nonstretched propellant tanks and did not use a Centaur. This presented a challenge for its designers since the Atlas itself no longer had a guidance package. The solution was found by using surplus all-inertial guidance packages originally purchased for the Atlas E/F ICBMs. A solid kick motor allowed up to 4,400 pounds

Image 2-5: Atlas-Centaur 47 lifted from Pad A at Complex 36 on 4 May 1979 with a geosychronous communications satellite, FLTSATCOM, aboard. The launch marked the fiftieth mission for the Atlas-Centaur combination. (NASA Photo no. 79-HC-199)

to be delivered into polar low-Earth orbit. The development of further Atlas variants essentially stopped during the early 1980s as the Space Shuttle was designated the sole National Space Transportation System. Production of Atlas boosters slowed to a trickle in preparation of the shuttle taking over all U.S. launches.

Recoverable Atlas

Interestingly, Convair investigated ideas that would have allowed the reuse of the basic Atlas booster for more than a single mission. During 1957 Convair began participating in the Air Force SR-89774 study of reusable space boosters that lasted through 1965. To allow the reuse, a 51-foot span swept wing complete with end-plate vertical stabilizers and four General Electric CL-610 turbojet engines was grafted onto the Atlas booster. Similar studies were conducted by Boeing (Saturn) and Martin Marietta (Titan). All the studies concluded that a reusable booster could potentially lower the cost of access to space, but most of the effort got sidetracked into concepts that resembled space shuttles, and no reusable Atlas vehicles (or Titan or Saturns) were ever built.[43]

Atlas I

The Atlas I[44] program was begun in 1987 to fulfill commercial expendable launch vehicle requirements following the Space Shuttle *Challenger* accident, when it was decided to no longer use the shuttle to launch routine payloads. The Atlas I launch vehicle was derived from the Atlas G and included the same basic vehicle components (Atlas booster and Centaur upper stage). Significant improvements in the guidance and control system were made with an emphasis on replacing analog flight control components with digital units interconnected with a Mil-Std-1553B digital data bus. The Atlas I was equipped with a further improved 474,000-lbf MA-5A propulsion system and was capable of using both 11-foot- and 14-foot-diameter payload fairings.

Originally, eighteen Atlas I vehicles were planned for manufacture, but an Air Force contract for the development of the improved Atlas II Medium Launch Vehicle II (MLV-II) caused General Dynamics (which had absorbed Convair) to rescope Atlas I production commitments to eleven vehicles and convert the remaining commitments to the Atlas II/IIA/IIAS. The average launch price was just over $75 million, with the vehicle contributing approximately $70.3 million of that cost.[45]

Atlas II

The Atlas II booster was developed to support the U.S. Air Force Medium Launch Vehicle II program. The Atlas II booster was 8.75 feet longer than an Atlas I and included a further improved 490,000-lbf Rocketdyne MA-5A engine. The vernier engines used by all previous Atlas boosters were replaced with a hydrazine roll-control system, and the Centaur stage was stretched 3 feet, allowing a longer burn time to accommodate heavier payloads.

Atlas IIA

Used for both Air Force and commercial customers, the Atlas IIA was a further developed version of the basic Atlas II.

Atlas IIAS

The Atlas IIAS was essentially an Atlas IIA with four strap-on Castor IVA solid rocket boosters, each 37 feet long and 40 inches in diameter, that provided an average thrust of 112,000 lbf. The solid rockets fired two at a time—the first pair was ignited at liftoff and burned for 54 seconds while the second pair was ignited in flight after vehicle loading constraints were satisfied. Both pairs were jettisoned shortly after their respective burnouts. The average launch price was $105 million in fiscal year 1994 dollars.

Atlas III

To provide a substantial increase in capability, in 1995 Lockheed Martin began the development of an Atlas variant using a Russian engine in place of the MA-3/5 booster-sustainer group used on all previous models. With Atlas III,[46] the stage-and-a-half Atlas booster design is eliminated—the first stage still uses the unique Atlas pressure-stabilized tanks but incorporates a new RD-Amross[47] RD-180 engine in place of the booster/sustainer group used on all previous models. The RD-180 is a two-chamber throttleable engine that uses liquid oxygen and kerosene propellants to provide a total sea-level rated thrust of 585,000 lbf. The original rationale for selecting the RD-180 was its low cost and quick availability; unfortunately, neither ended up being totally realized. The Air Force was worried about being "held hostage" by a foreign engine supplier and insisted that a domestic engine company be qualified as a source of the RD-180. Pratt & Whitney was selected to manufacture the engine, but the efforts required to accomplish this, together with technology transfer concerns within Congress, have greatly increased the cost of the program. Lockheed Martin invested approximately $200 million in the development of the Atlas III vehicle, and Pratt & Whitney invested another $100 million in the RD-180. Pratt & Whitney has committed to acquire 101 engines from RD-Amross at a cost of just over $1 billion, the largest commercial U.S.-Russian aerospace project to date.[48]

The use of the RD-180, plus some producibility improvements in the booster structure, allowed the elimination of over fifteen thousand individual parts, greatly simplifying the assembly of the booster and also increasing its reliability. The use of the continuous throttling capability of the RD-180 allows the launch vehicle trajectory to be optimized for each mission—even after launch—thereby providing the end user maximum flexibility in tailoring

Figure 2
Comparison of Atlas Launch Vehicles

	Atlas D/E/F	LV-3	LV-3A	LV-3B Mercury	LV-3C	SLV-3	SLV-3A	SLV-3 C/D	Atlas G	Atlas H	Atlas I	Atlas II/IIA
Main engine	MA-2/3	MA-3	MA-3	MA-3	MA-3	MA-3	MA-5	MA-5	MA-5	MA-5	MA-5A	MA-5A
Lift-off thrust (lbf) boosters and sustainer	368,000 or 392,000	392,000	392,000	392,000	392,000	392,000	431,300	431,300	438,800	438,800	474,000	490,000
Upper stage	None	None	Agena	None	Centaur	Agena	Agena	Centaur	Centaur	Solid	Centaur	Centaur
Main diameter (feet)	10.0	10.0	10.0	10.0	10.0	10.0	10.0	10.0	10.0	10.0	10.0	10.0
Booster height (feet)	69.5	69.5	69.5	69.5	69.5	69.5	79.3	79.3	83.5	69.5	83.5	92.2
Overall height (feet)	82.0	83.0	98.4	96.3	108.3	105.0	118.0	131.0	137.0	89.0	144.0	155.5
Maximum lift-off weight (pounds)	259,595	265,041	273,398	256,000	300,000	273,298	300,650	327,250	366,300	330,000	362,250	408,000 515,332
Payload into LEO (pounds)	1,750	1,750	—	3,000	—	—	—	—	—	4,400	—	—
Payload to GTO (pounds)	—	—	1,765	—	4,000	1,950	2,265	4,500	5,000	—	5,500	6,100

Source: See Mark Wade's *Encyclopedia Aeronautica* web site, http://www.austronautix.com/

the flight environment.[49] Despite all of the changes, Lockheed Martin went to great lengths to ensure the interfaces and flight environments provided to the payload did not change; the Atlas III utilizes the identical payload adapters, electrical and fluid interfaces, and payload fairings as the Atlas IIA/AS.[50]

Atlas IIIA

The Atlas IIIA uses an improved single-engine version of the Centaur upper stage with a stretched tank (5.5 feet) along with the new Atlas III booster. Design optimization has led to the elimination of over forty-five hundred individual parts from the Centaur, along with the incorporation of better seals and improved spark igniters. Guidance, tank pressurization, and propellant usage controls for both Atlas and Centaur are provided by the inertial navigation unit (INU) located on the forward equipment module of the Centaur. The Atlas IIIA has a payload capacity of up to 8,950 pounds.[51]

In a typical Atlas IIIA launch, the vehicle's two RD-180 thrust chambers are ignited shortly before liftoff, and preprogrammed engine thrust settings are used during booster ascent to minimize vehicle loads by throttling back during peak transonic loads that occur in the high dynamic pressure region while otherwise maximizing vehicle performance. Just over 2 minutes into flight, as the vehicle reaches an axial acceleration of 4 g, the engines begin to throttle back, eventually initiating a constant throttle rate to sustain acceleration at 5.5 g. Booster engine cutoff occurs approximately 3 minutes into flight and is followed by separation of the Centaur from the Atlas. For the first launch, these settings were 74 percent power for the first 5 seconds, 92 percent for the next 28 seconds, reducing to 64 percent for 30 seconds at max-q, then throttling back up to approximately 87 percent for the remainder of the flight. The performance of the RD-180 is so much greater than the MA-5 was that the vehicle reached first-stage velocity and latitude targets in only 3 minutes, compared with nearly 5 minutes required by the Atlas IIAS. As an interesting aside, at 92 percent power the RD-180 burns oxygen rich, consuming over 2,000 pounds per second—a level more than the combined rate used by all three Space Shuttle Main Engines during a shuttle launch.

The first burn of the single-engine Centaur lasts about 9 minutes, after which the Centaur and its payload coast in a parking orbit. Approximately 10 seconds after first burn ignition, the payload fairing is jettisoned. The second Centaur ignition occurs about 23 minutes into the flight, continues for about 3 minutes, and is followed several minutes later by the separation of the spacecraft from the Centaur.[52]

The Centaur and its RL-10 performed flawlessly on the first Atlas IIIA flight, but its failure on an earlier Boeing Delta III flight had resulted in the initial Atlas IIIA flight being delayed almost a year. This resulted in Lockheed Martin and International Launch Services losing three or four missions worth over $300 million. The first Atlas IIIA mission is estimated to have cost approximately $100 million, but Lockheed Martin projects the price will fall to roughly $85 million in the near future. It should be noted that Eutelsat most probably did not pay anywhere near full price for the first mission, given its development nature.[53]

Atlas IIIB

The single-stage Atlas IIIB booster is the same as used on the Atlas IIIA, and the primary change is the addition of an optional second engine to the Centaur upper stage. Using a two-engine Centaur, the Atlas IIIB has a payload capacity of up to 9,920 pounds.[54] The first Atlas IIIB was launched the morning of 20 February 2002.

Atlas V

Lockheed Martin is investing a significant amount of company funds, in addition to $500 million of Air Force funds, into the development of the Atlas V booster. This is, in essence, a completely new vehicle, sharing little but some payload interfaces and the RD-180 engine with earlier boosters. However, the successful first flight of the Atlas IIIA on 24 May 2000 has given Lockheed Martin a great deal of confidence on the ultimate success of the Atlas V. In fact, estimates are that that one success has removed about 80 percent of the risk associated with the first Atlas V flight, scheduled for mid-2002.

The Atlas V launch vehicle system is based on the Common Core Booster™ (CCB) powered by a single RD-180 engine. The CCB is 12.5 feet in diameter, 106.6 feet long, and contains 627,105 pounds of LO2 and RP-1 propellants. The major departure from previous designs is the replacement of the traditional pressure-stabilized propellant tanks with a new isorigid design that is structurally stable and does not need to be continually pressurized. The aluminum isorigid tank structure allows the elimination of propellant slosh baffles, simplifying production and lowering the cost. The tanks utilize one-piece spun domes and four barrel panels per tank section; this design approach eliminates approximately thirty-nine hundred components and thirty-nine thousand fasteners compared to the Titan IVB tankage. In addition, since it does not need to be pressurized at all times, ground handling of the new stage is easier than with earlier Atlas boosters. In April 2000 the first production tank underwent structural testing at Lockheed Martin Astronautics in Denver, Colorado. Flight hardware began to arrive at Cape Canaveral in mid-2001 for pathfinder operations at a completely rebuilt LC-41.[55] The first "booster on stand" operations were conducted during October 2001 using the first Atlas V flight vehicle (AV-001). This involved assembling the booster and mating the Centaur upper stage in the Vertical Integration Facility (VIF).

Due to the longer mission profile of the Atlas V (230 seconds for booster stage burn versus 186 seconds for Atlas III), the RD-180 engine is undergoing additional testing. As of April 2000 this testing was 35 percent complete, having completed 1,569 out of a planned 4,714 seconds of additional test firing. By the end of 2001, the RD-180 had accumulated over 27,500 seconds of total test time, including a 394-second test in July 2001 that certified the engine for Atlas V operations.

Newly developed Aerojet solid rocket motors (SRM) are available to augment the RD-180 for any given mission. These ground-lit motors contain approximately 95,000 pounds of propellant and utilize low-cost manufacturing and design techniques developed by Aerojet for their ICBM programs. From one to five SRMs can be added to each booster core; the SRMs will be certified by late 2002.

The Atlas V also uses an advanced fault tolerant avionics system incorporating dual Honeywell inertial navigation units. The new system has a design reliability of 0.99-plus as opposed to 0.987 for the Atlas IIA/AS system. It also allows for a more automated prelaunch check-out to improve overall reliability of the launch system.

Lockheed Martin is introducing a 14.9-foot usable diameter (17.5-foot outside diameter) Contraves payload fairing in addition to retaining the option to use the heritage 11-foot and 14-foot Atlas IIA payload fairings. The Contraves fairing is a composite-construction unit based on flight-proven hardware, while the two heritage fairings continue to be made of aluminum alloy. The heritage fairings are available in "extended" (31-foot) and "long" (34-foot) lengths; the Contraves fairing is available in 68- and 77-foot lengths, depending upon the needs of the payload. When combined with a standard Atlas payload fairing, the configuration is part of the Atlas V 400 series. The Atlas V 500 series uses the larger Contraves payload fairing derived from that used on the Ariane 5. Both Atlas V 400 and 500 configurations incorporate a stretched version of the Centaur upper stage (Centaur III), which can be configured as a single-engine Centaur (SEC) or a dual-engine Centaur (DEC).[56]

The Atlas V Heavy configuration was scheduled to use three CCB stages strapped together to provide capability necessary to lift the heaviest U.S. government payloads previously launched on the Titan IVB. This configuration was, in theory, capable of lifting 41,000 pounds into geostationary transfer orbit when equipped with an 86-foot-long payload fairing and a dual-engine Centaur.[57]

Fairly early on, Lockheed Martin determined that it was not commercially viable to build a heavy-lift infrastructure at both Vandenberg and Cape Canaveral, and the Air Force allowed Lockheed to forego to the Vandenberg facilities. When the Air Force competed the first twenty-eight EELV launches in 1998, Boeing won nineteen of them with their Delta IV vehicle, including the only two heavy-lift launches. This caused Lockheed Martin to reexamine continuing with the development of the Atlas V Heavy configuration. The Air Force, however, reminded Lockheed that under the EELV contract, a heavy-

Figure 3
Atlas V Configurations

Configuration	LEO 28 deg (in pounds)	LEO polar (in pounds)	Geosynch Transfer (in pounds)	Geosynch (in pounds)
Atlas V 401	12,500	10,750	5,000	N/A
Atlas V 501	10,300	9,050	4,100	1,500
Atlas V 511	12,050	10,200	4,900	1,750
Atlas V 521	13,950	11,800	6,000	2,200
Atlas V 531	17,250	14,600	6,900	3,000
Atlas V 541	18,750	15,850	7,600	3,400
Atlas V 551	20,050	17,000	8,200	3,750

lift vehicle had to be designed, even if it was uncertain that the vehicle would ever actually be built.[58] The Air Force eventually decided to concentrate all of its heavy-lift requirements on the Delta IV and allowed Lockheed Martin to discontinue development of this configuration.

The Atlas launch vehicle program is managed for the Air Force by the Atlas Program Office at the Space and Missile Systems Center, Los Angeles AFB, California. The Atlas launch vehicle is manufactured by Lockheed Martin Astronautics[59] in Denver, Colorado, with major components fabricated in San Diego, California (tanks), and Harlingen, Texas (booster and fairings). Commercial launch service contracts are managed by Lockheed Martin Commercial Launch Services, a division of International Launch Services located in San Diego.

Conclusion

In retrospect, the Atlas is probably best considered a marginal ICBM—its early basing modes, inaccurate guidance system, unproven warheads, pressure-stabilized tanks, and nonstorable propellants combined to make it a operational nightmare that probably would not have worked all that well in actual practice. Thankfully, we never had to use it and will never know.

But as a space launch vehicle there is no question that Atlas has made a mark for itself, and a great deal of money for its manufacturers. Looking back at early ICBM test footage, and a seemingly never-ending rash of Atlas vehicles blowing up on the launch pad, it is hard to imagine that the modern Atlas launch vehicle has an outstanding success record. It is perhaps even more difficult to imagine that somehow the engineers managed to get the four most important early Atlas launches—all carrying piloted Mercury capsules—off without a hitch. Atlas has proven to be one of the workhorse launch vehicles, competing head-to-head with Delta and Ariane in the world market. The string of thirty-nine straight successes with the Atlas II/IIA series prove the continued viability of the basic design.

Interestingly, the vehicle has long outlived its corporate parent. General Dynamics purchased the old Consolidated-Vultee company that began its development when Atlas was just a child. As Atlas entered middle age, the aerospace operations of General Dynamics were purchased by Martin Marietta and later merged into Lockheed Martin. The Atlas operations were a good investment.

But even Lockheed Martin realizes when a proven technology requires replacement, and the Atlas V is a very different vehicle, sharing essentially nothing with the ICBM that began the family over forty years ago. Gone are the pressure-stabilized tanks that made the ICBM possible. Gone also is the odd Rocketdyne stage-and-a-half engine, replaced by, of all things, an engine designed in Russia—the original target of the Atlas ICBM.

Atlas, in several versions, will continue to be a major U.S. launch vehicle for the foreseeable future. Eventually it might be replaced by some future reusable launch vehicle (RLV), but that was said thirty years ago when the Space Shuttle was being developed, and Atlas soldiers on. Thankfully, the weight of the world is no longer on its shoulders, only 5,000 pounds of communications satellite.

Notes

1. Several individuals graciously took time from their busy schedules to review this essay at various stages: Robert E. Bradley at the San Diego Aerospace Museum, Mark Cleary at the 45th Space Wing History Office, and Cynthia J. Thomas and Frank Watkins at Lockheed Martin.

2. Several good histories of Atlas have been written. See, for instance, John Clayton Lonnquest, "The Face of Atlas: General Bernard Schriever and the Development of the Atlas Intercontinental Ballistic Missile, 1953–1960," Ph.D. diss., Duke University, 1996. Also, Davis Dyer, "Necessity Is the Mother of Convention: Developing the ICBM, 1954–1958," *Business and Economic History* 22 (1993): 194–209 and *TRW: Pioneering Technology and Innovation since 1900* (Boston: Harvard Business School Press, 1998), pp. 167–94.

3. Jacob Neufeld, *The Development of Ballistic Missiles in the United States Air Force, 1945–1960* (Washington, D.C.: Office of Air Force History, 1990), pp. 1–2.

4. Wernher von Braun, "The Redstone, Jupiter, and Juno," in Eugene M. Emme, ed., *The History of Rocket Technology: Essays on Research, Development, and Utility* (Detroit: Wayne State University Press, 1964), pp. 108–9; also published in *Technology and Culture* 4 (Fall 1963): 452–55; John W. Bullard, "History of the Redstone Missile System," Hist. Div., Army Missile Command, Oct. 1965, pp. 135–51. The creation of the North Atlantic Treaty Organization in 1949 had provided a clear military need for a battlefield rocket.

5. *Jane's All the World's Aircraft, 1962–1963* (New York: McGraw-Hill, 1963), pp. 391–92; von Braun, "Redstone, Jupiter, and Juno," pp. 109–10; A. A. McCool and Keith B. Chan-

dler, "Development Trends in Liquid Propellant Engines," in Ernst Stuhlinger, Frederick I. Ordway, III, Jerry C. McCall, and George C. Bucher, eds., *From Peenemüünde to Outer Space: Commemorating the Fiftieth Birthday of Wernher von Braun* (Huntsville, Ala., 1962), 292.

6. On 17 March 1943, the Consolidated Aircraft Corporation merged with Vultee Aircraft, becoming the Consolidated-Vultee Aircraft Corporation. This name was often truncated to "Convair," although this did not become official until 29 April 1954, when Consolidated-Vultee Aircraft Corporation became the Convair Division of the General Dynamics Corporation. In between, Convair referred to itself alternately as CVAC or CONVAIR (all caps).

7. Both Hermann Oberth and Konstantin Eduardovich Tsiolkovsky had apparently proposed pressure-stabilized tanks earlier, but Karel Bossart was the first to actually construct such a design.

8. For a superb article on Bossert and the balloon tank, see Richard E. Martin, "The Atlas and Centaur 'Steel Balloon' Tanks: A Legacy of Karel Bossart," reprint by General Dynamics Corporation, 40th International Astronautical Congress paper, IAA-89-738, 1989.

9. Peter Alway, *Rockets of the World* (Ann Arbor, Mich.: Saturn Press, 1992), pp. 240–41.

10. This is still true of some IRBMs such as the SCUD and its derivatives.

11. Alway, *Rockets of the World*, pp. 241–42.

12. Project RAND (for Research and Development) was established under contract to the Douglas Aircraft Company on 1 October 1945. In February 1948, the Chief of Staff of the newly created U.S. Air Force wrote a letter to Donald Douglas that approved the evolution of RAND into an independent nonprofit corporation. On 14 May 1948, RAND was incorporated as a nonprofit corporation under the laws of the state of California.

13. It should be noted that there is a minority opinion that feels the Atlas is not a true stage-and-a-half vehicle since all engines are ignited on the ground.

14. Jay Miller, *The X-Planes: X-1 to X-31* (Arlington, Tex.: Aerofax, 1988), pp. 96–97.

15. The name "Atlas," also known as Project Atlas, was approved by the Air Force in August 1951. It is generally believed that the missile is named for the Greek god that "carried the world on his shoulders," but persistent stories circulate that the missile was named for the Atlas Corporation, which by this time had a controlling interest in Consolidated-Vultee.

16. Neufeld, *The Development of Ballistic Missiles*, pp. 55–56.

17. Ibid., pp. 50–100.

18. See John L. Chapman, *Atlas: The Story of a Missile* (New York: Harper and Brothers, 1960).

19. CEP is a means of averaging proximity to the target center—it is the radius of a circle within which half of the ordnance targeted for the center of the circle can be expected to fall.

20. A few sources indicate that the X-11 and X-12 designations were carried over to the Atlas A and Atlas B as built. People on the program at the time remember that the X-11 and X-12 designations had been applied to the test vehicles originally proposed for the five-engine version of Atlas but insist the designations were not used by the ultimate Atlas A and Atlas B test vehicles. Official documentation is inconclusive (some memos and flight logs use the designations; most do not). The last published version of Jay Miller's excellent *The X-Planes: X-1 to X-31* uses the designations for the vehicles that were flown, but the next revision will change this to indicate the designations were not used for the Atlas A and B and were limited to the never-built 12-foot-diameter, five-engine concepts. Deferring to the people that were actually there, I have elected not to use the designations here.

21. Initially the Atlas was designated B-65, but later the Air Force decided to reserve the traditional "bomber" series for piloted vehicles—the Atlas was redesignated SM-65 (for strategic missile). In reality the designation was seldom used, and Atlas variants were generally known by an alpha suffix (Atlas A, Atlas B, etc.). Various other designations were used by

Atlas over the years, such as PGM-16D (Atlas D), CGM-16D/E (Atlas D/E), and HGM-16F (Atlas F).

22. Ironically, in Russia Sergei Pavlovich Korolev was working on the competing R-7 ICBM—evidently both sides wanted to use the lucky number.

23. Initially this was simply WS-107A. The "-1" was added after development of the Titan was authorized as WS-107A-2.

24. Several missile projects were given designations in the traditional "bomber" series: the Martin B-61 Matador, Northrop B-62 Snark, Bell B-63 Rascal, North American B-64 Navaho, Consolidated-Vultee B-65 Atlas, Radioplane B-67 Crossbow, and the Martin B-68 Titan. All were subsequently redesignated either SM (strategic missile) or GAM (guided air-launched missile).

25. Miller, *X-Planes*, pp. 95–97.

26. Officially the Atlas consisted of the "Series A," "Series B," etc., but the "Series" was quickly dropped and the missiles became known as Atlas A, Atlas B, etc.

27. Miller, *X-Planes*, pp. 94–96.

28. Neufeld, *The Development of Ballistic Missiles*, pp. 222–44.

29. Ibid., pp. 235–50.

30. Many sources list this vehicle as an X-12. See note 21.

31. Miller, *X-Planes*, pp. 98–99.

32. Alway, *Rockets of the World*, p. 242.

33. S. P. Brown and G. F. Senn, "Project SCORE," *Proceedings of the IRE* 48, no. 4 (April 1960): 624–30.

34. Neufeld, *The Development of Ballistic Missiles*, pp. 205–6.

35. Ibid.

36. Chuck Hansen, *U.S. Nuclear Weapons: The Secret History* (Arlington, Tex.: Aerofax, 1988), pp. 107–8.

37. Ibid.

38. Neufeld, *The Development of Ballistic Missiles*, pp. 242–44.

39. Ibid., pp. 213–38.

40. Loyd S. Swenson Jr., James M. Grimwood, and Charles C. Alexander, *This New Ocean: A History of Project Mercury*, NASA SP-4201 (Washington D.C.: NASA, 1966), pp. 187–89.

41. Heinz Hermann Koelle, *Handbook of Astronautical Engineering* (New York: McGraw-Hill, 1961).

42. This created three different length boosters: the original ICBMs, LV-3x, SLV-3 variants, and Atlas H were 69.5 feet long; the SLV-3x variants were 79.3 feet long; and the Atlas G and Atlas I were 83.5 feet long. The Atlas IIA/AS would add a fourth variant at 92.2 feet long.

43. Dennis R. Jenkins, *Space Shuttle: The History of the National Space Transportation System—The First 100 Missions*, (Cape Canaveral, Fla.: Specialty Press, 2001), pp. 51–52.

44. This is a Roman numeral "one" and not a letter "I," although that would have logically followed Atlas H. However, "I" (and "O") are seldom used in official designations to avoid confusion with numbers.

45. Lockheed Martin fact sheet, undated.

46. Originally, the Atlas III was known as the Atlas IIAR (the "R" presumably representing the Russian engine).

47. RD-Amross is a joint-venture between NPO Energomash and Pratt & Whitney.

48. Craig Covault, "Atlas III Flight Tightens LockMart, Boeing Faceoff," *Aviation Week & Space Technology*, 29 May 2000, pp. 28–29.

49. Steve Sasso, "Evolution not Revolution: Lockheed's Atlas V," *Launchspace*, April 2000, pp. 30–31.

50. Ibid.

51. Covault, "Atlas III Flight," pp. 28–29.

52. Sasso, "Evolution not Revolution," pp. 30–31.

53. Covault, "Atlas III Flight," pp. 28–29.

54. Sasso, "Evolution not Revolution," pp. 30–31.

55. Ibid.; http://ilslaunch.com/stories/AtlasVUpdates/, accessed on 8 February 2002.

56. Ibid.

57. This payload fairing was actually larger than the original Atlas ICBM!

58. Robert Wall, "EELV Outlook Impacted by Weakness in Commercial Launch Market," *Aviation Week & Space Technology,* 3 July 2000, pp. S20–S22.

59. In May 1994, Martin Marietta acquired the Space Systems Division (primarily Atlas and Centaur) of General Dynamics Corporation. Lockheed and Martin Marietta merged in 1995 to form Lockheed Martin Corporation. Today, all missile programs (Atlas, Titan, Centaur, etc.) are concentrated within the Lockheed Martin Astronautics Company.

– 3 –

Delta

The Ultimate Thor

Kevin S. Forsyth

Introduction

Delta is one of the most enduring members of the original family of U.S. space launch vehicles and has long been known as "the workhorse of space." The vehicle originated with the U.S. Air Force's Thor, a medium-range weapon that became the first operational American ballistic missile. In 1960 NASA added flight-proven upper stages from the Vanguard project to create the Delta launch vehicle. Soon Delta was evolving through numerous updates and performance improvements that used dependable, "off-the-shelf" components. The payload capacity of the current vehicle is more than forty times that of the original, and after more than a third of a century of use, the Delta family of launchers has racked up one of the best flight records of any rocket in the world. Many of NASA's science programs have relied on Delta, and hundreds of experimental and operational communications, meteorological, and military satellites have been orbited using the vehicle as well.

Thor: From Ballistic Missile to Space Vehicle

The Thor missile program was initiated in December 1954, when United States Air Force (USAF) Headquarters issued a general operational requirement for a tactical ballistic missile (TBM) intended to travel a distance of between 1,150 and 2,300 miles to deliver a nuclear warhead with an accuracy of better than 2 miles CEP (circular error probable). As the United States had few enemies within 2,000 miles, the TBMs were expected to be placed in friendly nations well within range of strategic sites within the Soviet Union. The United Kingdom soon expressed interest in deploying TBMs on their soil. A TBM launched from the UK would be within range of Moscow, which was the intent for its range specification in the first place. (This proposal ran into difficulties since it appeared to conflict with U.S. policy on sharing atomic energy information, but Britain's interest spurred U.S. officials to hash out a legal and diplomatic work-around.)[1]

By July 1955 the program had changed the name of its objective from TBM to the slightly more descriptive intermediate-range ballistic missile (IRBM). During that summer, a small "technical task force" led by Cmdr. Robert Truax of the U.S. Navy, on loan to the USAF Ballistic Missile Division, and Dr. Adolph K. Thiel, a former engineer on the German A-4 (V-2) project who left the Redstone Arsenal to join Ramo-Wooldridge Corporation, prepared an IRBM design study. In late August, Truax and Thiel defined the basic parameters of the Thor:

- Eight feet in diameter and 65 feet long, dimensions that allowed it to be carried aboard a Douglas C-124 Globemaster cargo plane—a requirement for overseas deployment
- Range of 1,750 miles
- Maximum speed of 10,000 mph during reentry
- Gross vehicle weight of 110,000 pounds, fully fueled and loaded with warhead
- Use of "one-half" of a Navaho-derived Atlas booster engine for main propulsion (as Thiel put it, in those early days of engine development, "clearly, it is the only one available")
- Insistence on the AC Spark Plug all-inertial guidance system, with a radio-inertial backup (this setup was the inverse of that used in the early Atlas D and was less susceptible to enemy disruption)[2]

On 30 November 1955, representatives from three preselected companies—Douglas, Lockheed, and North American—met with Ballistic Missile Division officials. At this time they were issued a challenge, not to design a rocket per se, as Thiel and Truax had already seen to that, but rather to demonstrate the ability to create "a management team that could pull together existing technology, skills, abilities and techniques in 'an unprecedented time.'" The companies were given exactly one week to reply.[3]

Before the end of 1955, the Ballistic Missile Division had made its decision. The Douglas Aircraft Company was awarded a contract on 27 December as the prime contractor, responsible for building the airframe and integrating the other components into a workable missile. The Rocketdyne division of North American Aviation would provide the engines, AC Spark Plug the all-inertial guidance system, Bell Laboratories the backup radio-inertial guidance, and General Electric (GE) the nose cone/reentry shield. The Air Force assigned the XSM-75 designation to the new missile, and it would become the major element in Weapon System 315A (WS-315A), better known as Thor.[4]

Douglas engineers offered minor changes to the Truax-Thiel design study. One was tapering the fuel tank at the upper half of the vehicle, instead of using a symmetrical cylinder, to improve the aerodynamic profile. Also, while Ramo-Wooldridge preferred all-welded tanks, Douglas proposed bolted tank bulkheads, a simplified construction solution that proved adequate in testing.[5]

While Thor's airframe was a new design, most of its other major subsystems were derived from the Atlas intercontinental ballistic missile (ICBM) program. The main engine came almost directly from the Atlas MA-3 engine, which consisted of a pair of thrust chambers and nozzles fed by a single turbopump assembly. By deleting one thrust chamber and rerouting the plumbing for a more concise package, Rocketdyne created the MB-3. Thor also used the same reentry system (nose cone) as Atlas, and since Thor ultimately flew first, it was able to test nose cone designs created for Atlas.

Some common systems, particularly those in propulsion, turned into major headaches for both Thor and Atlas. The worst of these was the turbopump assembly that forced kerosene and liquid oxygen (LOX) into the combustion chamber. It had originally been designed for another application and was later politely described as "a marginal component which ultimately had to be significantly modified."[6]

Engine tests on the MB-3 began in March 1956 at the Rocketdyne test facility in the Santa Susana Mountains, northwest of Los Angeles. The test series was completed swiftly, with the first R&D engine being turned over to Douglas in June. The first flight engine was trucked to the Douglas assembly plant in Santa Monica, California, on schedule during September 1956.[7]

The following month the first Thor airframe, numbered 101, was flown aboard a C-124 cargo plane from the Douglas plant in Santa Monica to Patrick Air Force Base in Florida. The missile was prepared for flight at Complex 17, a site that itself had been built in much the same way as Thor—swiftly but with a basis in known technology. Located midway down the length of Cape Canaveral's southern beach, the launch pads were barely past their initial groundbreaking as late as February 1956. At that time Dr. Thiel had visited the Cape and realized there was no time to design a blockhouse for the complex, so the plans for the nearby Redstone complex were borrowed. By the time Thor 101 arrived, the blockhouse was ready, as was pad B—however, the crash course to build B had left pad A unfinished.[8]

Following numerous practice propellant loadings and test countdowns, and after a failed relay forced a month-long delay, Thor 101 left the pad on 25 January 1957. It did not travel far. A scant instant into the flight, contamination in the liquid oxygen supply caused a valve to fail, and the engine suffered "thrust decay." The Thor slid backward through the launcher ring and exploded on the deflector plate below, causing extensive damage to the pad and the support equipment it contained. The flight was later joked to have had a "six inch apogee."[9]

It would be three months until the next flight attempt. Thor 102 flew perfectly for 35 seconds until it was destroyed through no fault of its own. A miswired instrument led the range safety officer to believe the Thor was headed inland rather than out to sea, and he triggered the missile's self-destruct package. Though an obvious disappointment, the short flight gave confidence in the basic design. A month later, Thor 103 provided a lesson in the need for complex and comprehensive procedures. Its main fuel valve was defective, and over the course of several lengthy pauses in the countdown the fuel tank began to overpressurize. Unfortunately, the procedures then in use allowed the problem to go unnoticed. As the count resumed for a final attempt for the night, with 4 minutes still on the clock, the tank gave way and the missile was destroyed. Another three-month pause came before the launch of Thor 104, which flew for 92 seconds before losing control and breaking up.[10]

On 20 September 1957, just twenty-one months after Douglas began Thor construction, Thor 105 flew 1,100 miles downrange before splashing in the Atlantic Ocean. On-board telemetry provided information from dozens of sensors throughout the flight. After analysts factored out the heavy weight of the R&D instruments, they determined that the combat version of Thor could travel 1,500 miles, the distance specified in the contract; the flight was deemed a full-range success. By the end of the year, Thor had flown ten times and demonstrated a range of 2,700 miles. The AC Spark Plug all-inertial guidance system, omitted in the first flights and replaced by the simpler, backup radio guidance, had been integrated and also tested to full range.[11]

The IRBM test flight phases continued to progress, soon adding the GE nose cone and later a dummy payload that duplicated the size and mass of a nuclear warhead. Deployment of operational missiles to the UK aboard C-124 cargo planes began in August 1958.[12] Meanwhile, Thor took on the first of many upper stages that would ultimately lead to Delta.

Upper Stages for Thor

Thor 116 was the first to carry an upper stage—an Aerojet AJ-10-40 powered by a modified engine developed for use as the second stage of the Vanguard that used unsymmetrical dimethyl hydrazine (UDMH) for fuel and inhibited white fuming nitric acid as oxidizer. In this application it was named Able. Three Thor-Able vehicles were used to test the General Electric advanced re-entry vehicle destined for the Atlas ICBM, although the first of these, Thor 116, was lost on 24 April 1958 due to a turbopump failure in the main engine. Turbopump failures plagued early MB-3 engines—a phenomenon called "bearing walking," in which the bearings would move axially in their mountings, was exacerbated by low atmospheric pressure at altitude and was difficult to reproduce on the test stand. The next flight to suffer this fate would result in a significant disappointment.

Thor 127 was mounted with an upgraded AJ-10 stage, called Able I, beneath a small, fiberglass-cased solid motor built by Allegany Ballistic Laboratory. Known as the Altair X-248, this powerful but lightweight motor was developed as a backup upper stage for Vanguard. Atop this three-stage stack known as Thor-Able-Star was an 84-pound, spin-stabilized Air Force probe called Pioneer that had been designed to enter lunar orbit and take the first closeup images of the Moon with a simple television camera. The mission was launched on 17 August 1958, but 77 seconds into the flight the turbopump gave out and the vehicle exploded. America's first attempt to reach the Moon was lost.[13]

Another upper stage used with Thor was the Lockheed-built Agena; larger than Able and considerably more powerful, it had originally been developed as part of the Air Force's 1954 military space program. Thor-Agena was the booster for the Corona program, the United States' first spy satellite (also known by the cover name Discoverer). The Corona reconnaissance spacecraft was launched into polar orbit to take photographic swaths as it passed over the Soviet Union. Corona returned its exposed film in a heat-resistant "bucket" that was contained within the reentry vehicle on the nose. The package reentered over the Pacific Ocean, and after reaching a safe altitude, the "bucket" was ejected to be recovered in the air by a passing aircraft. Called "America's Secret Space Program," Corona suffered many of the same setbacks as the civil program in its early days but ultimately proved its worth. Discoverer 14 was snatched in midair by a C-119 cargo plane on 18 August 1960, providing the earliest photos of the USSR's Plesetsk rocket base. Later results returned evidence that allayed fears of both a "bomber gap" and a "missile gap" with the Soviet Union.[14]

NASA's First Launch Vehicle Contract

On 1 October 1958, the National Aeronautics and Space Administration (NASA) was born. It immediately became the prime agency for scientific space exploration, and in addition to its own Explorer program it also took control of several projects already begun by other agencies, such as Tiros (Television and InfraRed Observation Satellite). One project that fell into NASA's lap was the S-2, an Air Force satellite built to fly in an elongated orbit and study the newly discovered Van Allen radiation belts in detail. It also carried a crude television camera system. Renamed *Explorer 6*, it was placed aboard a Thor-Able III (yet another upgraded Able) and launched on 7 August 1959. It returned the first images of Earth ever taken from orbit.[15]

NASA had a charter to launch all nonmilitary payloads in the United States, so one of the agency's first chores was to assess its launch capability. In consultation with the Department of Defense Advanced Research Projects Agency, in January 1959 NASA released *The National Space Vehicle Program,* in which it expressed a need to consolidate the available fleet: "Our approach up to this time has been much too diverse in that we fire a few vehicles of a given configuration, most of which have failed to achieve their missions, and then call on another vehicle to take the stage. In this situation no one type of vehicle is tested with sufficient thoroughness and used in enough firings to achieve a high degree of reliability."[16] Four vehicles were proposed as the basis of the National Space Vehicle Program:

- Vega, an Atlas- and Vanguard-based rocket that ultimately never left the design stage because it turned out to be redundant in capacity to other launch systems
- Centaur, a powerful upper stage that would be the first to use the then-exotic propellant combination of liquid hydrogen and liquid oxygen
- Saturn I, a monster rocket with nine fuel tanks and a cluster of eight engines for the first stage, with an upper stage that would, like Centaur, use hydrogen fuel (this vehicle led to the development of the immense Saturn V, the Apollo moon rocket)
- Nova, another exercise in design that before cancellation prompted the development of the largest engine ever built, the F-1, which was used in the first stage of the Saturn V

One other vehicle was mentioned in NASA's launch needs assessment. Based on Thor-Able, it was called Delta and was termed "an interim general purpose vehicle." It was meant to be "used for communication, meteorological, and scientific satellites and lunar probes during 1960 and 1961," after which other vehicles, mainly Vega and Centaur, would replace it. Most important, "reliability rather than performance [was] to be emphasized by replacing or deleting those components of Vanguard and Thor-Able that have caused failures" while retaining those that had proven successful.[17]

In April 1959 the Goddard Space Flight Center, located in Greenbelt, Maryland, just outside Washington, signed NASA's first launch vehicle contract. Goddard commissioned the Douglas Aircraft Company to create, produce, and integrate twelve Delta vehicles. Milton W. Rosen, Director of Launch Vehicles and Propulsion in NASA's Office of Manned Space Flight, determined the name. It came from the radio code word for the fourth letter of the alphabet, since this would be the fourth modification of Thor for space flight after Able, Able-Star, and Agena. Thus the vehicle was variously known as Delta and Thor-Delta.[18]

The design of Delta was very similar to the Thor-Able-Star. The first stage

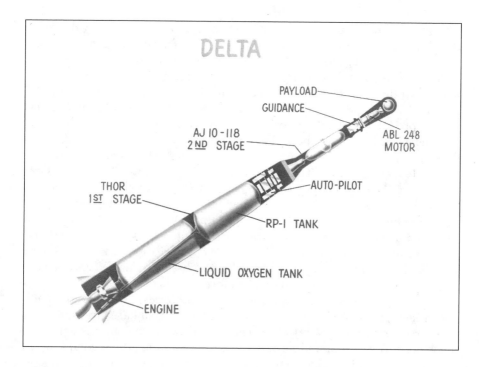

Image 3-1: The Thor-Delta launch vehicle concept in 1961. (NASA Photo)

was the Thor booster, modified as in Thor-Able to support upper stages rather than a nuclear warhead. It was powered by the Rocketdyne MB-3 Block I, which burned refined kerosene (Rocket Propellant One or RP-1) and liquid oxygen to produce 152,000 lbf. The engine was turbopump-fed, the pump being powered by a gas generator that also burned RP-1/LOX and which poured its exhaust out a long pipe mounted next to the regeneratively cooled engine nozzle. The engine was gimbal-mounted for steering in the pitch and yaw axes. The first stage also contained two vernier engines on its aft end, on either side of the main engine, to provide roll control.[19]

The Delta second stage consisted of the Aerojet AJ-10-118 engine, a 7,700-lbf engine that, as in Able, burned UDMH and nitric acid. This combination of propellants is hypergolic; that is, they spontaneously combust when mixed. This eliminated the need for an ignition system, as merely opening the valves would cause the engine to start. The engine itself was very simple in that it was pressure-fed, with no need for turbopumps to force propellants into the combustion chamber. Spherical tanks fed heated helium gas into the fuel and oxidizer tanks to keep them pressurized. A simple and reliable—if not overly powerful—configuration. The stage also carried a system that gave it a significant advantage in versatility over Thor-Able. When originally designed, the Vanguard second stage had contained a gas-jet attitude control system that allowed the stage to maintain or change its orientation in space following burnout of the AJ-10 engine. For simplicity's sake, this system had been deleted from the second stage as implemented in Vanguard and had remained missing from previous versions of Able. It would be restored to the newest version of Able that would become Delta's second stage.

The AJ-10 engine had been the source of numerous headaches during its development. Aerojet won the original contract for the Vanguard second stage by severely underbidding its main competitor, Bell Aircraft Corporation. Over its two-year development period Aerojet engineers struggled to accommodate the complex equipment the stage needed within the stringent weight restrictions. The engine package, exclusive of the thrust vector control actuator and its hydraulic tanks, wound up costing over $4 million to develop, around four times the amount Aerojet had bid.[20] Yet the cost overruns turned out to be a worthwhile investment, as the end result was a highly reliable stage that continues to fly in an enlarged and updated form to this day.

Aboard the second stage was the Bell Laboratories BTL-300 radio guidance system, a vacuum tube device that relied on ground-based transmissions to determine the rocket's trajectory. The guidance system controlled all steering actions for the first and second stages, then oriented the second stage in preparation for third-stage separation. A spin table was mounted on the forward end of the second stage, and, just before third-stage ignition, two small solid motors would fire to spin up the third stage and payload to over 100 rpm. The unguided third stage would then be released from the table and fired, its spinning motion providing gyroscopic stabilization as well as dampening out any thrust asymmetry in the propellant as it burned.

In a repeat performance, the Delta third stage was the ABL X-248, also known as Altair, that previously flew as the "Star" of Thor-Able-Star. Altair was a major innovation when it was developed as an upgrade to the Vanguard third stage. This cylindrical solid motor provided an average of 2,800 lbf for a duration of 38 seconds, yet it weighed just 500 pounds thanks to a lightweight, wound-fiberglass casing.[21]

The Original Twelve Flights

The first Deltas were able to lift 650 pounds into low-Earth orbit of 115 to 230 miles altitude and could have sent 100 pounds to geostationary transfer orbit (GTO). The vehicle was used to launch twelve payloads that brought a wide range of new discoveries in communications, weather, the Sun and its effects on Earth, and interplanetary space.

The use of satellites as communications relay stations was understood on the hypothetical level for many years before the rocket age began. Before he gained fame as a science fiction writer, Arthur C. Clarke wrote his now-famous essay describing how three satellites in geosynchronous orbit, about 22,300 miles above the earth, could provide worldwide communications coverage.[22] By the early 1960s, with the rudimentary guidance systems available, analysts were not so optimistic. With little experimental evidence on the ability of a satellite to perform accurate station keeping, it was assumed that some nineteen geosynchronous satellites, randomly spaced, would be needed to guarantee continuous communications between ground stations 3,000 miles apart. At lower altitudes hundreds of satellites could be necessary, so they would have to be extremely inexpensive to build and launch.[23]

One low-cost path was the passive reflector. As opposed to an active repeater, which carries microwave receivers, transmitters, and the power supply to run them, a passive reflector is essentially a mirror in orbit, reflecting transmissions back to the ground without any form of amplification or directional control. Many forms of passive reflectors were discussed including metallic spheres (with or without etched holes to save weight), mesh spheres, gravity-stabilized spherical segments, and even clouds of tiny magnetic dipoles (wires). This last concept runs counter to modern concerns about orbital debris, but some tests were conducted by the Air Force, amid controversy and with mixed results.[24]

Echo, the payload destined to fly on the first two Delta vehicles, was a simple sphere made of 0.5-mil thick aluminized mylar that measured 100 feet in diameter when fully inflated. Echo carried a set of small, solar-powered transmitters for telemetry but otherwise was a giant, silent silver balloon, held rigid only by internal pressure as its aluminum coating did not provide any structural support.

The launch of Echo 1 took place on 13 May 1960, when at 9:16 A.M. Greenwich Mean Time (GMT), the new Delta launch vehicle leapt from Pad 17A and arced out over the predawn Atlantic. The Thor first stage operated well and separated from the modified Vanguard second stage, but the attitude control system failed during second-stage burn. The vehicle was destroyed, and Echo 1 was lost.[25]

The return to flight of Delta took one day less than three months. Another early morning launch occurred on 12 August, and everything aboard Delta 2 went without a hitch. Three stages burned in succession to place the payload in a nearly circular 1,035-mile, 47-degree inclination orbit. The inflation system, having been tested a number of times on suborbital ballistic flights, operated well and the balloon filled without snagging. Echo 1A became the first passive communications satellite in orbit.[26]

Using two frequencies (960 and 2,390 MHz), signals were successfully bounced off Echo 1A. During orbits that passed over the continental United

States, two-way voice links were temporarily established between Bell Laboratories in Holmdel, New Jersey, and the NASA station at Goldstone, California. Some transmissions from the United States were even received in England. But soon micrometeorite impacts caused Echo 1A to gradually deflate, and as it shrank the mylar wrinkled with a resultant loss in effective reflectivity. The experiment in passive reflection gave good results, and even generated data on atmospheric density and solar pressure, but did not appear to be an adequate solution to the communications dilemma.

The follow-up Echo 2 was launched aboard a Thor-Agena B on 25 January 1964. By this time, experimental flights with active repeaters had been promising enough that Echo 2 would be the last passive reflector flown, and this flight's primary purpose wound up being an investigation into large spacecraft dynamics. The use of the more powerful Agena second stage meant a larger, 135-foot-diameter sphere could be lofted. Echo 2 was made of a three-ply laminate, a 0.35-mil mylar sheet sandwiched between two 0.2-mil aluminum sheets. When the sphere was slightly overinflated the skin "rigidized," or took on a permanent set, thus preventing wrinkling when internal pressure was lost.[27]

Although technology would pass by passive reflectors, both Echo 1A and Echo 2 demonstrated ground station tracking procedures that would come in handy with subsequent active satellites. Due to their great size and shiny finish, they appeared more brightly in the night sky than any stars and may still hold title as the most-seen manmade objects in space. Echo 1A reentered the atmosphere on 24 May 1968. Echo 2 followed on 7 June 1969.[28]

In mid-1958, the Advanced Research Projects Agency began work on an experimental meteorological satellite system that would ultimately become the Television and Infrared Observation Satellite. Cognizance of the program transferred to NASA on 12 April 1959, though Tiros 1 would remain manifested on one of Delta's forerunners, an Air Force Thor-Able II. It was launched on 1 April 1960, six weeks before the first Delta. All subsequent Tiros, Tiros Operational System (TOS), and Improved TOS satellites have flown on Delta vehicles. Five of these were in NASA's first contract of twelve launch vehicles.[29]

Tiros 2, launched on Delta 3 (23 November 1960), included a magnetic attitude control system, a simple coil around the outside lower portion of the satellite. By energizing the coil, the spacecraft could react with Earth's magnetic field to tilt its spin axis. Later flights would expand on this system for control in other axes.[30] Tiros 2 took thirty-seven thousand pictures and performed for ten months, well beyond its ninety-day design life. Its images were used to select proper weather conditions for the May 1961 Mercury flight of Alan B. Shepard Jr.

Tiros 3 (Delta 5 on 12 July 1961) carried two wide-angle cameras and three infrared scanners. It transmitted observations of all six of the major

Image 3-2: This three-stage Thor-Delta booster was launched from Cape Canaveral on 25 March 1961, carrying a 75-pound payload, the F-14 magnetometer-plasma satellite. (NASA Photo no. 61-Delta 4-12)

hurricanes in the 1961 season: Anna, Betsy, Carla, Debbie, Esther, and Frances. In September, Tiros 3 was credited with the discovery of Esther, two days before the hurricane was observed by conventional means. This was a first in space history.[31]

Tiros 4 (Delta 7 on 8 February 1962) began a program of ice reconnaissance that compared its results with those from airplanes, ships, and ground stations. Tiros 5 (Delta 10 on 19 June 1962) and Tiros 6 (Delta 12 on 18 September 1962) led off a period of overlapping observation for much of 1963, allowing almost continuous coverage of many regions. Forecasters could not have been more pleased.

Many of the satellites in one of NASA's longest-lasting programs, the Explorer series, was also launched on Deltas. *Explorer 10*, also known as P14, was the lightest primary payload ever launched aboard Delta, a mere 78 pounds. The extreme light weight was necessary to allow Delta 4 to propel *Explorer 10* into a highly elliptical orbit—some 138,000 miles at apogee, or almost half the distance to the Moon. In order to save weight, *Explorer 10* had no solar cells for power generation and was limited to the lifetime of its on-board batteries. Battery power lasted only fifty-two hours, at which point *Explorer 10* was still ascending in its first orbit and was about 42.3 Earth radii in altitude. Though brief, the mission returned good scientific data and was deemed successful by all.[32]

Explorer 12, flown 16 August 1961 on Delta 6, was the first of four Energetic Particle Explorers (EPE), spin-stabilized and solar powered spacecraft. EPE 1 measured cosmic ray particles, charged particles trapped in the Van Allen belts, solar wind protons, and magnetic fields in both Earth's magnetosphere and interplanetary space. To accomplish this, EPE 1 was (like *Explorer 10*) launched into a highly elliptical orbit, one which took it through both the lower and upper Van Allen belts and beyond. EPE 1 ceased transmitting on 6 December 1961, apparently as a result of failures in the power system.[33]

The Orbiting Solar Observatories were the mainstream workhorses of NASA's solar astronomy flight program throughout the 1960s. A fleet of eight spacecraft were built and flown from 1962 to 1971, and many were launched on Deltas. Their purpose was to extend observations beyond those possible with research balloons and sounding rockets. As Earth's atmosphere is opaque to many regions of the electromagnetic spectrum, sensors must be placed at high altitude—or better yet, in orbit—in order to receive much of the Sun's radiation.

The first of these was OSO 1, launched aboard Delta 8 on 7 March 1962. The package could carry 2, 3, or 4 telescopes of various types, with an emphasis on spectrographs and image forming spectroheliographs, "the most powerful tools in the astronomer's workshop."[34] The low altitude (345 miles) and inclination (33 degrees) of the OSO 1 orbit meant that during a portion of each orbit it would be occulted by the earth. Solar astronomers were able to correlate the total x-ray flux of the Sun with individual solar flares as well as whole regions of activity. The fact that solar flares cause short-term variations in x-rays was an unexpected discovery. OSO 1 also helped to interpret the changes in Earth's atmosphere observed by Ariel 1, such as the upper atmosphere density profile and the ionospheric electron density profile.[35] Following the loss of its second tape recorder on 15 May 1962, OSO 1 sent real-time data until May 1964, when its power cells failed.[36]

Ariel 1 was the first international satellite, a joint effort between Goddard Space Flight Center and the British National Committee of Space Research. Goddard was responsible for design and fabrication of the basic spacecraft structure, power supply, and other support systems such as thermal control and data storage. (Of course, Goddard also supplied the Delta vehicle and provided launch services.) The United Kingdom designed and built all flight sensors and their associated electronics.[37]

The spacecraft was originally intended to fly aboard a Scout, and its size reflected the restrictions of that rocket's 25.7-inch-diameter payload shroud. In 1960 and 1961, Scout suffered several launch failures, and Ariel was moved to Delta. The transfer was not without difficulty. Atop the Delta second stage and spin table was a separation system consisting of four curved metal panels, or petals, that held the third stage firmly in place. When explosive bolts severed the ring that retained the petals, they would fan open like a flower blooming to release the third stage and its payload. In its folded launch configuration, Ariel's sensor booms extended far enough aft that the petals would have struck them as they opened. The solution was to fabricate a magnesium cylinder, called a dutchman, to act as an adapter between the third stage and spacecraft. It moved Ariel forward by about 13 inches. Loath to waste space, Goddard engineers placed vibration and contamination sensors inside the dutchman to monitor the payload environment during flight.[38]

After its launch date slipped by about a month due to minor problems with the Delta vehicle, Ariel 1 was successfully launched on 26 April 1962 aboard Delta 9. Almost immediately, a failure in the common electrical system of the three Lyman-alpha ultraviolet radiation detectors caused that experiment to be permanently lost. Following the third-stage burn, Ariel 1 was spinning at about 160 rpm, a rate necessary to stabilize the third stage. By design, its speed would be slowed down using yo-yo de-spin weights prior to deployment of the solar panels and equipment booms. However, a foil shroud meant to protect the spacecraft from third stage outgassing caused that motor to overheat. The heat melted nylon tie-downs and/or RTV bonded tie-down fittings, and starting about 13 minutes into the flight various solar paddles and inertia and experiment booms deployed unevenly and prematurely. Fortunately, stringent design specifications kept Ariel 1 from losing any of its appendages to the unanticipated stresses, the spin rate ultimately was brought down to marginally above nominal, and the mission was able to continue to fruition.[39]

In the early morning of 9 July 1962, about 9:00 A.M. GMT, Thor missile number 195 launched from Johnston Island, some 800 miles southwest of Hawaii. At its tip was a Mk 4 reentry vehicle containing a W49 thermonuclear warhead. When it reached an altitude of 248 miles, the warhead detonated with a yield of 1.45 megatons. It was the Starfish Prime event, one of several high-altitude nuclear effects tests conducted by the United States as part of Operation Dominic-Fishbowl.[40]

Ariel was 4,000 nautical miles away when the Starfish Prime device detonated. Almost all its experiment sensors were instantly saturated, though for the most part they recovered and continued to function normally. Scientists received data on the intense but short-lived ionization effects of Starfish Prime and were able to study the artificial radiation belt produced by the device.[41] However, excessive radiation significantly degraded the performance of Ariel's solar panels and damaged certain other electrical subsystems, primarily semiconductors, as well. Beginning three days after the Starfish Prime event, and continuing for the remainder of its life, Ariel suffered intermittent system shutdowns, resulting in about one-third of the expected data return after 12 July. The tape recorder failed soon after that, and receipt of data was on again, off again until program termination on 9 November 1964.[42]

With the exception of Sputnik, no early satellite is more widely known than Telstar. It is often considered to have ushered in the era of satellite communications. Contrary to many sources, Telstar was not the first active communications satellite, a distinction that belongs to the U.S. Army's Project SCORE, nor the first operational communications satellite—Telstar was an experimental prototype. It was, however, the first commercial satellite in history, launched by NASA aboard a Delta but owned and operated by the American Telephone and Telegraph Company (AT&T).[43]

The same day of its launch (10 July 1962 on Delta 11), Telstar 1 became the first satellite ever to relay television signals across the Atlantic Ocean, as well as fax, voice, and data. Though the transmissions were brief thanks to the low orbit, and experimental in nature, they had considerable public impact. The satellite "promised to tie together the ears and eyes of the world."[44] Telstar 1 even had an instrumental song named for it that became the first number-one record on the American charts by a British rock group (it ended up selling five million copies worldwide).[45] Telstar 1 operated for about six months with only minor difficulties. As its elliptical orbit took it through the inner Van Allen belt and a portion of the outer belt, it absorbed radiation damage that ultimately destroyed transistors in the command system. Radiation from the Starfish Prime event, which occurred one day before launch of Telstar 1, was also a factor.[46]

Delta Begins to Evolve

Eleven successful flights out of twelve attempts represented a major success for the Delta program. Total development cost including twelve vehicles and launch support was $43 million, only $3 million over the initial estimate.[47] Small to medium payloads continued to fill NASA's manifest and soon Goddard Space Flight Center ordered an additional fourteen vehicles from Douglas Aircraft Company, the first of many follow-up procurements.[48] Meanwhile, the Air Force's Thor vehicle was evolving, and as improvements showed flight worthiness they were incorporated into Delta without the need for developmental test flights. By November 1963, upgrades to all three stages were in place.

First came Delta A, which used the Block II version of the MB-3 main engine. The Block II had a liftoff thrust of 170,000 lbf versus 152,000 lbf in the Block I. Two A model vehicles were flown with continuations of the Energetic Particle Explorers, EPE 2 and EPE 3.

Delta B took the A model and added an upgraded AJ10-118D second stage. Its propellant tanks were lengthened by 3 feet, and a higher energy oxidizer was substituted. In addition, the vacuum-tube radio guidance system was replaced by more robust hardware using transistors and semiconductors, though the Bell Labs system remained radio-inertial. With Delta B, the Delta program went from "interim" to "operational" status.

Delta C went a step further from B and replaced the third-stage Altair motor with Altair 2—the ABL X-258 motor developed as the Scout fourth stage. Altair 2 was a significant improvement over its predecessor. Just 3 inches longer and 10 percent heavier, it kicked out 65 percent more total thrust.[49]

Along with continuing scientific studies (Explorer, Tiros, OSO, to name but a few), significant progress in satellite communications was carried aboard the new Delta vehicles. The first Delta B, flight 15 (13 December 1962), launched Relay 1 into a medium-altitude orbit. The 172-pound craft was NASA's first active communications satellite and was in many respects a close sibling to AT&T's Telstar. Both experimental repeaters flew in low-to-medium orbits, were spin-stabilized, and had magnetic torquing coils for orientation. NASA used the same ground stations for Relay as did Telstar, leasing the Andover facility from AT&T.

Relay was a right prism just over 4 feet tall, tapered along its top half to fit neatly within the Delta payload envelope. Its omnidirectional receive and transmit antennas were located on a post at the tapered end, and command and telemetry antennas protruded from the opposite end. A redundant pair of transponders each produced 10 watts of output, though only one could be used at a time. When the primary transponder's power supply failed on orbit, the backup was used to success. In addition to transmission experiments, Relay

Image 3-3: A Delta rocket lifts off from Cape Canaveral carrying NASA's eighth Orbiting Solar Observatory (OSO) satellite in 1975. (NASA Photo no. GPN-2000-001328)

1 also measured radiation and returned results on the effects of radiation on electronics. Relay 1 lasted beyond its one-year design life when its shutoff timer failed to operate.[50]

As propulsion technology advanced, it became possible for a Delta B to place a 200-pound satellite into geosynchronous transfer orbit, a highly elliptical orbit with a perigee of 115 miles and apogee of 22,100 miles. Though the payload could reach the altitude of a geosynchronous orbit, it would not have sufficient velocity to remain there.

The next Delta B launch was of the first Syncom—NASA's third major communications satellite experiment after the passive Echo and the low-orbit active Relay. Syncom would be placed into geosynchronous orbit, and inaugurated the basic spacecraft layout and flight profile that would be used for most geosynchronous satellites launched on Delta. It was a basic spin-stabilized cylinder 29 inches in diameter. The curved face of the cylinder was covered in solar panels. An internal equipment shelf carried the requisite hardware to receive at 8 GHz and transmit at 2 GHz. A 2-watt traveling wave tube amplifier broadcast through a toroidal beam antenna; though one-fifth the power of Relay, the shaped beam focused the energy on a plane coincident to the earth vector. The system had the capacity to carry several two-way voice channels or a single television channel.[51]

Opposite the equipment shelf and antennas, Syncom carried the device that enabled geosynchronous orbit. Permanently attached to the airframe and substantially concealed within the solar panels was a Thiokol Star 13B solid motor. Its lightweight aluminum casing meant the 105-pound motor was 87 percent propellant by mass.[52] By lighting at apogee of the transfer orbit, the motor could impart enough energy to place Syncom in geosynchronous orbit. A motor used in this fashion, usually permanently attached to the spacecraft, is known as an apogee kick motor (AKM).

Syncom 1 was launched from Pad 17B late in the evening of 13 February 1963. Delta 16 placed it into its proper transfer orbit, and the AKM fired at the appropriate time, but during the burn Syncom 1 was crippled, likely from the explosion of a nitrogen tank used for attitude control once on orbit. Syncom 1 never reached operational orbit and the mission was lost.

Syncom 2, an identical mission, launched five months later aboard Delta 20. The vehicle performed flawlessly, the AKM placed the spacecraft into geosynchronous orbit, and Syncom 2 became the first satellite maneuvered to a specific longitude (55 degrees west). This mission profile was at the limit of performance for the Delta B. The vehicle did not have the power to eliminate the orbital inclination caused by launching from a site north of the equator, so Syncom 2's final orbit was inclined by 33 degrees. This meant that over the course of each day the spacecraft would trace an elongated figure-eight pattern over Earth that extended 33 degrees to the north and south of the equator. While data, telephone, fax, and video signals were successfully transmitted through Syncom 2 and the flight was a success, the orbital motion continued to require steerable antennas.[53]

Even with the upgraded third stage of Delta C, the vehicle still did not have sufficient power to follow a zero-inclination trajectory. More thrust was needed during first stage flight, but the MB-3 main engine was nearing the limits of its design. Another piece of flight-proven hardware was needed.

The Castor solid motor, generating a thrust of 53,000 lbf, was developed by Thiokol as the second stage of the all-solid Scout rocket. Douglas created the Thrust-Augmented Thor (TAT) by strapping three Castor motors to the aft section of the missile. Nose cones were added to the motors for aerodynamics and the nozzles were canted 7 degrees outward so their thrust vectors would pass through the vehicle centerline or roll axis. All three motors were lit at liftoff and were jettisoned, following burnout, to fall into the ocean. Castor booster motors were introduced in the TAT-Agena D, an Air Force vehicle used to loft the new, larger reconnaissance satellites of the Corona program.[54]

Using a TAT first stage, Delta C became Delta D, also known as the Thrust-Augmented Delta. The first Delta D (flight 25) launched Syncom 3, which was identical in size, shape, and capacity to the previous Syncoms. On 19 August 1964, Syncom 3 became the first satellite to enter geostationary orbit.[55] It was stationed almost directly above the intersection of the equator and the International Date Line in time for the 1964 Olympic Games in Tokyo, Japan. Syncom 3 relayed daily live coverage of the Olympics, publicly demonstrating that communications satellites are the sole practical means of transmitting live television over transoceanic distances. The new state of the art in broadcast communications had arrived.

One other Delta D flew, and ushered in the start of operational transatlantic satellite service. Less than eight months after Syncom 3 NASA launched Intelsat 1, commonly known as "Early Bird," the first satellite of the International Satellite Communications Consortium.

Early Bird was built on the same airframe design as Syncom 3 but had been upgraded to provide 240 telephone channels. At the time, the many submarine cables that had been expensively laid across the floor of the Atlantic could provide only 412 channels. With the placement of Early Bird at 28 degrees west longitude (later 38 degrees), the era of transatlantic cables came to an end.[56]

Thrust Augmented Improved Delta

The Delta E was introduced in 1965; capable of delivering an additional 100 pounds of payload to GTO over Delta D. It was frequently referred to as the Thrust Augmented Improved Delta, or TAID. All three stages and the booster motors were altered.

The strap-on augmentation became Castor II motors, a lengthened version of Castor with similar thrust but longer burn time. These would remain the standard augmentation for several years to come. The first stage received the latest MB-3 Block III main engine that featured improved reliability, but only increased liftoff thrust by 2,000 lbf.

Beginning above the first stage, the TAID silhouette differed considerably from earlier Deltas. The new AJ10-118E second stage used the same hypergolic propellant combination as its predecessor, but its tanks were widened from 33 to 55 inches, doubling its propellant capacity and burn time. With additional helium tanks to keep the main tanks pressurized, the 118E was given essentially unlimited restart capability. This improved Delta's versatility by allowing for a wide range of launch and orbital trajectories.

Two options were available for the third-stage motor: the Altair 2, or the FW-4D developed by United Technology Corporation for the Air Force. The two motors had similar specifications. Vehicles which used the FW-4D were designated Delta E1, the first sign that Delta variations might become more complex than a simple letter sequence could provide (see table 1).

The second stage's power allowed for heavier payloads, while its width required a larger payload fairing. The replacement fairing had previously flown aboard Agena upper stages and greatly increased payload space.

The first Delta E launched GEOS 1 (Geodetic Earth Orbiting Satellite) on 6 November 1965. The twenty-ninth in the Explorer series, it was instrumented with optical beacons, laser reflectors, a radio range transponder, Doppler beacons, and a range and range rate transponder. These were used to locate observation points (geodetic control stations) in three dimensions within 33 feet of accuracy, and to define the structure of the earth's irregular gravitational field. It was the first successful active spacecraft of the National Geodetic Satellite Program.[58]

In 1966, a complex of two Thor launch pads at Vandenberg Air Force Base that had supported the base's first missile launch (on 16 December 1958) were converted for use by Delta and renamed Space Launch Complex 2 (SLC-2, or "Slick 2"). The site was well suited for launching payloads into polar orbits. The first west coast Delta launch was on 2 October 1966, when the third Tiros Operational System meteorological satellite entered a 101-degree (polar) orbit.

A mixture of E, E1, C, and C1 models were flown from 1965 to 1968. Many of the missions were continuations of NASA projects: Orbiting Solar Observatories, Interstellar Magnetic Particle Explorers, and the Tiros satellites that had begun to evolve into the Tiros Operational System (to be administered by the fledgling Environmental Science Services Administration). Four spacecraft (*Pioneer 6, 7, 8,* and *9*) were launched into heliocentric orbits to study solar wind and cosmic rays beyond the influence of Earth's magnetic

Vehicle	Intro-duced	Stage	Major Modifications
Delta	1960	1 2 3	Modified Thor, MB-3 Block I engine AJ10-118 (Vanguard-derived propulsion system) Vanguard X-248 motor (Altair)
Delta A	1962	1	MB-3 Block II engine
Delta B	1962	2	AJ10-118D—tanks lengthened, higher energy oxidizer used
Delta C	1964	3	Scout X-258 motor (Altair 2) or FW-4D (Delta C1)
Delta D	1964	0	3 Thor-developed Castor I motors
Delta E	1965	0 1 2 3 F	Castor II motors MB-3 Blk III engine DSV-3E-3 (AJ10-118E)—propellant tanks widened Scout X-258 motor (Altair 2) or FW-4D (Delta E1) Agena-developed payload fairing
(Delta F)			Delta E1 without augmentation motors—never flown
Delta G	1966	3	None—used a special reentry vehicle for biological recovery (flights 43 & 51 only)
(Delta H)			Delta G without augmentation motors—never flown
Delta J	1968	3	TE-364-3 (Star 37D)
(Delta K)			Cryogenic upper stage design study—never flown
Delta L, M, N	1968	1 3	Long Tank Thor—tanks lengthened, cylindrical RP-1 tank L: FW-4 M: TE-364-3 N: none
M-6, N-6	1970	0	6 Castor II

Table 1
Delta Variants

and gravitational fields. Some thirty-five years after it was launched, *Pioneer 6* continued to send data to Earth, having traveled more than 18 billion miles.

Four Delta E1 launches were for Intelsat, and established the Intelsat II constellation of second-generation communications satellites. Intelsat II was similar in form to Early Bird (Intelsat 1) and, like that satellite, was built by Hughes Aerospace. The solar-cell-covered, spin-stabilized cylinder was twice the size and drew twice the solar power of Early Bird, and was equipped with an advanced antenna design developed by Hughes that permitted direct contact with a number of ground stations simultaneously. Like Early Bird, the Intelsat II satellites transmitted 240 two-way voice channels. In all, four were flown aboard Delta E1s. A pair of satellites was stationed over the Atlantic while the other pair covered the Pacific, though a failure in the apogee kick motor of "Atlantic 1" left it in a nonsynchronous orbit that limited its usefulness.

Two launches during this period were of special note: flights 43 and 51. These were Biosatellite 1 and Biosatellite 2. The vehicles were G models, essentially two-stage Delta Es.[59] The payload consisted of a capsule containing live specimens of frog eggs, fruit flies, wheat seedlings, and bacteria. The capsule was part of a special reentry vehicle, complete with built-in retrorocket for the deorbit burn, and would be caught in midair over the Pacific, using the method the Corona project had perfected. Upon recovery scientists could assess the effects of weightlessness and radiation. Biosatellite 1, launched 14 December 1966, was not recovered due to a failure in the retrorocket system, but Biosatellite 2, launched 7 September 1967, was successfully recovered in midair by the Air Force several days later and provided the first scientific data about basic biological processes in space.

Finally, on the Fourth of July 1968, the sole Delta J launched and placed *Explorer 38* in a highly inclined orbit to measure the intensity of celestial radio sources. The only change to this vehicle versus the Delta E was the use of the new Thiokol Star 37D motor as a third stage. The Star 37D was considerably larger, heavier, and more powerful than the FW-4D.[60]

The Long Tank and "Super Six"

In 1968, the Long Tank Thor first stage was introduced. This new stage changed the fuel tank (the upper half of the stage) from a tapered section to a cylinder; both fuel and oxidizer tanks were lengthened, for a total increase in stage length of over 14 feet. Total capacity increased by nearly 49,000 pounds of propellant while the dry weight of the stage only increased by 820 pounds.[61] Three Delta variations were available, all identical in external appearance and differing only in the choice of third stage. Delta L used the FW-4D motor, Delta M used the more powerful Star 37D, and Delta N was a two-stage version. A handful of vehicles launched with an extra trio of Castor II booster motors; these were known as "Super Six" models and increased the geosynchronous payload capacity by over 25 percent to 1,000 pounds.

By 1970 Delta was launching more than half of NASA's unmanned spacecraft each year. These included the earliest models of the Improved Tiros Operational System, though these were soon to be operated by the National Oceanic and Atmospheric Administration (NOAA). Other continuing NASA series were Orbiting Solar Observatories, sun-orbiting Pioneers, and Inter-

Table 2 Delta Designations	
First Digit: Type of First Stage and Augmentation	
0	Long Tank, MB-3 engine, Castor II motors (1968 baseline)
1	Extended Long Tank, MB-3 engine, Castor II motors (1972)
2	Extended Long Tank, RS-27 engine, Castor II motors (1974)
3	Extended Long Tank, RS-27 engine, Castor IV motors (1975)
4	Extended Long Tank, MB-3 engine, Castor IVA motors (1989-90)*
5	Extended Long Tank, RS-27 engine, Castor IVA motors (1989)†
6	Extra Extended Long Tank, RS-27 engine, Castor IVA motors (1989)
7	Extra Extended Long Tank, RS-27A engine, GEM 40 motors (1990)
8	Delta III Tank, RS-27A engine, GEM 46 motors (1998)
Second Digit: Number of Augmentation Motors	
3, 4, 6, or 9	Solid rocket motors
Third Digit: Type of Second Stage	
0	AJ10-118F (Aerojet) (Titan 3C Transtage derivative) (1972)
1	TR-201 (TRW)‡ (1972)
2	AJ10-118K (Aerojet) (1982)
3	RL10B-2 (Pratt & Whitney) (1998)

planetary Monitoring Platforms. They returned ever-expanding data on a wide range of space phenomena, and all of these programs were very successful. In addition, NASA orbited six satellites of the Intelsat III series for the International Satellite Communications Consortium.[62] By this time, Delta had established itself as a highly reliable vehicle. Despite a pair of unrelated launch failures in 1969, after eighty launches only six Deltas had failed to orbit—a cumulative success rate of 92 percent.[63]

A Busy Year of Changes

Over its lifetime, Delta has been the recipient of almost continual changes, whether hardware upgrades, new fabrication techniques, or improved ground support and processing. Even so, 1972 stands out as a momentous year for Delta. Over the course of the year, launches incorporated seven major up-

Table 2 (cont'd) Delta Designations	
Fourth Digit: Type of Third Stage	
0	No third stage
2	FW-4D (1968)§
3	Star 37D (TE-364-3) (1968)
4	Star 37E (TE-364-4) (1972)
5	Star 48B (TE-M-799) (PAM-D derivative) (1989)
6	Star 37FM (TE-M-783) (1998)
Dash numbers: Fairing Size	
–8	8-foot cylindrical (standard on "Straight 8"; discontinued)
–9.5	9.5-foot (standard on Delta II)
–10	10-foot metal skin-and-stringer (discontinued)
–10C	10-foot composite
–10L	lengthened 10-foot composite (in development)

Note: An example: Delta 7925-10. Extra Extended Long Tank, RS-27A engine, nine GEM 40 motors, AJ10-118K second stage, Star 48B third stage, 10-foot fairing.

* 2 missions.

† 1 mission.

‡ The first three flights are exceptions: Delta flights 92, 94, and 95 used the DSV-3P-4 engine, an AJ10-118F variant.

§ Use of theFW-4D was discontinued after creation, but before inception, of the four-digit system.

grades that, when complete, improved the payload capacity of Delta by about 40 percent, the largest single increase until the advent of Delta III.

By this time it was becoming obvious that the old letter designation system was inadequate to the task of describing a Delta's ever-changing components. Therefore, in 1972 the McDonnell Douglas Corporation (the product of a 1967 merger between McDonnell Aircraft Company and Douglas Aircraft Company) initiated a numerical designator system. This system, still in use today, uses four digits to describe, respectively, (1) the first stage tankage and engine, (2) the number of booster motors (usually a multiple of 3, with one early and several recent exceptions), (3) the second-stage engine, and (4) the third-stage motor (see table 2).

The hardware transition began with Delta 88, the first to fly with the Universal Boat Tail. This was a strengthened version of the first-stage aft structure, the primary purpose of which is as an attachment between the main engine and the liquid oxygen tank. The beefed-up framing prepared Delta for the 1974 introduction of the Rocketdyne RS-27 main engine. The new tail section also provided mounting points for up to nine solid booster motors. (However, being an N model, Delta 88 only used three Castor II motors for thrust augmentation.)[64]

The following flight was the first to use the four-digit system, and the model 0900 of Delta 89 was a substantially different vehicle than those that preceded it. Delta 89 carried a full complement of nine Castor II motors. Of these, six were lit at liftoff, with the remaining three ignited about the time the first set burned out and were jettisoned.[69] It also used the latest version of the Aerojet second stage, the AJ10-118F, which had been originally developed for the Titan III Transtage. This engine burned a higher-powered combination of hypergolic propellants: nitrogen tetroxide oxidizer and Aerozine-50 fuel (a 50-50 combination of hydrazine and unsymmetrical dimethyl hydrazine). The increase in thrust over the 118E version was more than 1,400 lbf.[70]

Delta 89 was also the first to utilize a brand-new flight controller, called the Delta Inertial Guidance System (DIGS). DIGS replaced a radio guidance system that, while simpler in terms of hardware, had a number of undesirable requirements and constraints. Among the benefits of DIGS:

- Eliminated an extensive and costly field support operation (tracking stations) required by the radio guidance system
- More than doubled spacecraft injection accuracy for two-stage missions
- Improved mission flexibility (the radio guidance system had boost profile and attitude constraints for antenna pointing, unnecessary in the self-contained DIGS)
- Consolidated functions of several previously separate hardware devices, thus simplifying the vehicle's ground check-out procedure

The DIGS hardware consisted of two separate boxes, an airborne guidance computer built by Delco Electronics and a Hamilton Standard inertial measurement unit made up of three strapdown integrating gyroscopes and three strapdown linear accelerometers. The DIGS built-in software performed both preflight and in-flight functions. During preflight checks, it performed system alignment, engine slew checks, and a dynamic closed-loop simulated flight test. In the flight mode DIGS performed all flight events and controls from engine start through spacecraft (or third-stage) separation. An added benefit of DIGS was that it used the same flight software regardless of vehicle configuration or mission profile, using preset program variables to account for a myriad of possibilities. For example, in its first two flights DIGS controlled a two-stage, nine-booster Delta to place ERTS 1 into sun-synchronous polar orbit, and a three-stage, six-booster Delta to place IMP 7 into a highly elliptical, near-escape orbit.[71]

ERTS 1, for Earth Resources Technology Satellite, is better known as Landsat 1. The satellite carried a television camera to obtain visible and near-infrared images and a multispectral scanner for radiometric images, Through 1974 Landsat 1 transmitted over 100,000 images covering 75 percent of the earth's surface; it continued to function into 1978. Later Landsat models Incorporated Thematic Mappers, which were designed to achieve higher image resolution, sharper spectral separation, improved geometric fidelity, and greater radiometric accuracy and resolution. In all, seven Landsats have been built and flown through 1999; all but one were successfully launched on Delta vehicles. Though it began as an experimental program, Landsat is now frequently referred to as "the central pillar of the national remote sensing capability."[72]

The flight of IMP 7, on Delta 90, introduced two more upgrades as a 1604 model. The first stage used the Extended Long Tank Thor, a 10-foot extension of the propellant tanks. The tanks themselves were built in an innovative new process. Previously, Thor tanks had used a typical airframe structure: an internal skeleton of stringers wrapped by and welded to an aluminum skin. This skin was 0.094 inch thick and was crisscrossed by integral ribs in a 3-inch square pattern, so that the inside of the tank resembled a waffle.

The new process, known as isogrid construction, was first developed by McDonnell Douglas for Delta in 1970. The tank walls were fabricated out of 0.5-inch-thick aluminum alloy plate, brake-formed into curves and welded into 8-foot-diameter cylinders. Then the interior walls were machined away so that the remaining material formed a thin skin with integral stiffening ribs in an equilateral triangular pattern. Though the isogrid ribs had twice the depth of those of the waffle, they were also twice as slender, and the larger spacing (4.619 inches on a side) of the isogrid meant fewer ribs. Not only was the skin only 0.050 inch thick (a reduction by nearly half), but most important, it was strong enough that most of the internal stringers could be eliminated from the design. The resulting isogrid structure was considerably lighter, stronger, and simpler to fabricate than a traditional airframe.

This flight also carried as its third stage the Thiokol Star 37E, a substantial improvement in total thrust over its 37D predecessor. The two motors were essentially identical, except that the 37E added a 14-inch cylindrical section between the two hemispheres of the 37D case.

The final capstone to Delta's 1972 improvements was the launch of Delta 92 carrying Telesat-A (known on orbit as Anik 1), Canada's first communications satellite. A Delta 1914, it was the first "Straight-Eight" vehicle, so named because it formed a continuous, 8-foot-diameter cylinder from its Universal Boat Tail to the nose cone of its brand-new, all-metal fairing. The second stage was fitted with a "mini-skirt" truss structure with an 8-foot thrust ring that attached between the fairing and a new, cylindrical interstage. The interstage was permanently affixed to the top of the first stage during stacking. It concealed the second-stage engine and much of its tankage and provided an improved first-to-second stage separation system.[73]

Thus in a single year, additional augmentation motors, improvement in second- and third-stage thrust, lengthening and lightening of the first-stage tanks, and the smoother aerodynamics of the Straight Eight, were all added to Delta. They combined to increase Delta payload capacity by about 1,200 pounds to low-Earth orbit and 400 pounds to geostationary transfer orbit.

Delta and the Dawn of Domestic Satellite Communications

After fourteen years and ninety-nine flights, ninety-one of which were successful, Delta was about to receive a new main engine. The Rocketdyne MB-3 engine had served Delta well, never suffering the catastrophic thrust problems that had recurred several times during Thor development. Lessons learned from MB-3, along with its sibling MA-3 of Atlas, had come in handy for Rocketdyne when NASA awarded the company an engine contract on 11 September 1958. The new engine, designated H-1, would be clustered in groups of eight to power the Saturn I and IB rockets. Using existing components and propulsion systems, Rocketdyne had successfully static tested the H-1 before the end of 1958. During the next ten years H-1 thrust would increase from 165,000 lbf to 205,000 lbf while many of its components were simplified for increased reliability. Rocketdyne ultimately built 322 engines for the Saturn program, after which it was a relatively simple matter to reconfigure the H-1 engine for use in Delta. In that application, the engine was designated RS-27.[74]

With the first flight of the RS-27 main engine in 1974, NASA had its standard Delta for the next five years in the 2914 vehicle. It didn't take long, however, for another variant to arrive on the scene.

On 16 June 1972, the Federal Communications Commission (FCC) issued a decision that would have far-reaching effects on the launch industry. The FCC began to allow competition in the private line field of domestic satellite communications. RCA Global Communications (Globcom) was the first to leap at the chance by renting transponder space on Canada's government-owned Telesat system, using the Anik satellite.[75] Anik 1 was Canada's first communications satellite. The Anik A series, three of which flew on Delta, were spin-stabilized HS-333 satellites developed by Hughes Aircraft Company, direct descendants of Syncom, Early Birdd, and Intelsat II. The HS-333 had typical specifications for Delta-launched communications satellites at the time: a 12-channel C-band transponder mounted with a reflector on a de-spun equipment shelf, atop a spinning cylinder of solar cells producing 300 watts of power, in a package weighing just under 1,300 pounds at launch including apogee kick motor. Another fleet of HS-333s was the Westar series, owned by Western Union. Westar 1, launched 13 April 1974 aboard Delta 101, was the first U.S. domestic satellite launched after the FCC decision. (As required by the FCC, RCA Globcom moved its service from Anik to Westar 2 in May 1975.)

RCA Globcom's use of rented transponders was an interim solution until their own line of satellites, known as RCA Satcom, could be launched. As designed, the RCA satellite advanced the state of the art but also shed light on a gap in NASA's launch vehicle capability. Satcom was a three-axis-stabilized craft carrying twenty-four C-band transponders, double the bandwidth of the HS-333. Unfortunately, its weight of 2,000 pounds made it too heavy for the Delta 2914, which could only lift 1,593 pounds to GTO. The next larger vehicle, the Atlas-Centaur, was overpowered for the task with a capacity of 4,200 pounds. RCA considered using an Atlas in a tandem launch of two spacecraft, but the risk of losing two satellites at once, combined with the inability to cost-effectively loft a replacement, precluded that choice.[76]

Instead, RCA, NASA, and McDonnell Douglas met in joint management negotiations and worked out a better solution. A new Delta, the 3914, would be built using Thiokol's powerful Castor IV motors for first-stage thrust augmentation. In addition to the standard "reimbursable launch" contract with NASA (which retained the sole charter for launching nonmilitary payloads within the United States), RCA signed "a unique 'user's contract'" with McDonnell Douglas, which paid a prorated portion of costs incurred by McDonnell Douglas in developing the Delta 3914. Thus RCA became the first private company to fund a launch vehicle development directly through the users fee" while NASA incurred no additional out-of-pocket expenses.[77]

The upgrade was not as simple as replacing the 2914's Castor II motors with Castor IVs. The new motor, first developed for the Athena H vehicle, was over 13 feet longer than the Castor II, weighed two times as much, and had almost twice the average thrust and burn time. Several changes were needed:

- The increased length of Castor IV required a forward attachment point on the LOX tank. An internal ring truss was added to the tank, and the stringers in the Universal Boat Tail were strengthened as well.
- The LOX tank antislosh baffle system was reconfigured to correspond to the revised natural frequencies of a Castor IV vehicle.
- With only a single mounting point, the Castor II had a simple separation system: an explosive bolt fired to release a thrust ball at the lower end of the motor, and aerodynamic drag combined with vehicle acceleration to pull the motor away. Wind tunnel tests on the Castor IV demonstrated the need for spring-loaded thrusters at the attach points in order to force the motor safely away from the core vehicle.

- The heavier motors necessitated more powerful motor hoists at the launch complex. Pad 17A was refurbished and some of its lower level platforms were redesigned. (Pad B's upgrade came some time later, as the first 3000-series flight from there did not occur until 1982.)
- Range safety invoked more stringent arming protocols, requiring the ability to remotely safe and arm the Castor IV motors. Because of the longer burn time, the down-range drop zone for motor casings was revised.

The more powerful motors also required a new staging sequence. A 2914 would light six motors at liftoff, three in flight, then drop all nine motors at once. To keep the acceleration of the 3914 within the same limits, only five motors were started at liftoff. When those were depleted, three motors were jettisoned followed by the other two motors one second later. (Computer simulations had predicted a danger of casing collision if the long motors were dropped all at once.) Then the remaining four motors were lit. This method of dropping spent motors soon after burnout allowed Delta 3914 to carry as little dead weight as necessary. The 5-4 sequence was used for about six years, after which time spacecraft were designed to withstand greater acceleration and a 6-3 sequence was resumed. Nine-motor flights continue to jettison motors in groups of three.[78]

The 3914 model would not become the standard Delta for another four years, but its first flight, that of RCA Global Communications' Satcom 1 on 13 December 1975, would make telecommunications history. Satcom 1, with some pinch-hitting by Westar 1, spurred the cable television industry to unprecedented heights with the help of a company known as Home Box Office (HBO).

Evolution Continues—but Production Ceases

In the late 1970s McDonnell Douglas Astronautics Corporation (MDAC) continued to evolve Delta, motivated in part by the realization that the Space Shuttle would take longer to become operational than originally planned. Flight 147 in December 1978 was the first to use DRIMS, the Delta Redundant Inertial Measurement System. Developed by MDAC, DRIMS replaced the Hamilton Standard inertial measurement unit that had been introduced with DIGS in 1972 (the guidance computer remained the same). It used three dry-tuned gyroscopes on orthogonally skewed axes and four accelerometers. The latest gyro technology and complete redundancy on all six axes of motion made DRIMS both reliable and cost-effective.[79]

In 1982 another second-stage became available, as Aerojet returned with the AJ-10-118K, the ultimate increase in tank size for that engine. The Aerojet AJ-10 and the TRW TR-201 had similar thrust and were used alternately on vehicles through 1988. (Ultimately the AJ-10-118K won out and is slated to be the standard Delta second stage well beyond the year 2000.)

An outgrowth of the shuttle development program was the Payload Assist Module (PAM), also known as the Thiokol Star 48B. PAM was a 15,000-lbf solid kick motor developed for use on the Space Shuttle to push satellites from low orbit to their final destinations. With attachment fitting alterations it became PAM-D (Delta) and carried a special distinction in the Delta program. Though it attached atop the Delta second stage and used a spin table like Delta third stages, PAM-D was considered to be part of the payload. Hence a 3910/PAM was, according to official documents, a two-stage Delta carrying a "perigee kick motor." As a result, some Delta flights were technically suborbital, as in the cases of SBS-A and -B.[80]

Even with the many improvements to the vehicle, the Delta launch manifest began to thin in the 1980s. Once the shuttle became operational, NASA preferred to manifest nearly all its payloads aboard that launch system. As NASA focused on the shuttle, ignoring the expendable launch market, satellite companies began to express greater interest in flying aboard Space Shuttles rather than expendable launch vehicles (ELV). Thus NASA's view that the ELV market was shrinking became self-fulfilling. McDonnell Douglas production lines slowed to a halt, and by 1985 it seemed that Delta would be permanently shelved as all U.S. payloads were to fly aboard the Space Shuttle.[81]

The period 1985–87 proved disastrous for U.S. space flight, due not only to the tragic loss of *Challenger* and its crew. During that time every major U.S. expendable launch vehicle—Titan, Atlas-Centaur, and Delta—suffered at least one failure. Delta 178 (3 May 1986) succumbed to an electrical short during its first-stage burn that caused its main engine to shut down. The powerless vehicle performed a full 360, shredding its fairing and payload, before range safety destroyed it. NASA scrambled to find sufficient vehicles to carry the sudden backlog of payloads whose owners preferred not to wait until the shuttle returned to flight. Three partially completed Deltas, in storage at the McDonnell Douglas factory in Huntington Beach, California, were rushed into operational service.[82]

Delta II: A New Beginning

It was the loss of *Challenger* that ultimately brought new life to the Delta program. For many years, the U.S. Air Force had been developing satellite-based navigation. (The earliest test, Transit 1B, had flown aboard a Thor-Able-Star in 1960 to become the world's first navigation satellite.) In the 1980s the Air Force was ready to deploy the operational Global Positioning System, or GPS. The space segment of GPS would consist of two dozen NAVSTAR II (*Navigation Signal Timing and Ranging*) satellites in six orbital planes, each orbiting the earth every twelve hours.[83] Users with GPS receivers could receive, decode, and process NAVSTAR signals to gain 3-D position, velocity, and time information.

The Air Force had been planning to put NAVSTARs in orbit using the Space Shuttle but was not willing to wait until the shuttle program could return to flight and clear its backlog of payloads. Therefore, it ran a competition to develop a Medium-Class ELV, and McDonnell Douglas won with a growth version of its 3920/PAM-D. It was called Delta II.

Delta II uses the longest first-stage tankage to date, called the Extra Extended Long Tank Thor. Total height of the vehicle is 125 feet, while its diameter remains 8 feet. The initial Delta II model, the 6925, used Castor IVA booster motors and a new, 9.5-foot diameter payload fairing. It also used the PAM-D motor, now officially considered to be the Delta third stage. The first NAVSTAR II was launched from Cape Canaveral on 14 February 1989. In all, nine NAVSTAR satellites flew on 6925 vehicles.

An added benefit of the Air Force contract was that it enabled McDonnell Douglas to successfully market Delta to commercial customers. The first commercial Delta lofted a communications satellite owned by Great Britain on 27 August 1989. Launches for Indonesia and India came in 1990, with many more to follow.

The current incarnation of Delta II, the 7000 series, first flew on 26 November 1990. Two major improvements occurred in the first stage. One was the newest version of the RS-27 main engine. The RS-27A has a wider nozzle, increasing the expansion ratio from eight to one to twelve to one. This makes the engine considerably more efficient at altitude, where it attains a thrust of 237,000 lbf. The Castor booster motors were also replaced. Built by Hercules Aerospace (now part of Alliant Techsystems), the new motors have graphite-epoxy casings, a substantial increase in propellant mass ratio over the steel casings of Castor motors. Each booster motor contributes about 109,500 lbf. Later vehicles also received a new avionics package. The RIFCA (redundant inertial flight control assembly) replaced the functions of both the DRIMS inertial unit and the separate guidance computer with a single flight box. RIFCA simplified installation and integration of the flight control system.

The Delta II earned the title of "workhorse of space" during the 1990s. In this decade the number of flights averaged over eight per year, reaching a crescendo of twelve launches in 1998. NASA launched over a dozen dedicated payloads aboard Delta IIs. Among these were some of the most dramatic robotic missions in NASA history.

Near Earth Asteroid Rendezvous (NEAR), launched 17 February 1996, was sent on a looping path to near-Earth asteroid 433 Eros. Arriving in December 1998, NEAR's main engine failed to complete its burn, and an intended rendezvous became a flyby. Flight engineers rewrote the engine's control software, recovered quickly, and brought NEAR in for a second attempt. On 14 February 2000, NEAR became the first satellite to enter orbit around an asteroid. Renamed in honor of the late asteroid researcher Eugene Shoemaker, NEAR Shoemaker has since altered its orbit several times while poring over

Image 3-4: At Launch Complex 17, Pad A, on Cape Canaveral Air Station, a Delta II is launched on 7 November 1996, propelling the *Mars Global Surveyor* to the red planet. Assembly of the three-stage rocket with its complement of nine strap-on solid-rocket motors began more than three months earlier with erection of the Delta II first stage. (NASA Photo no. KSC-96PC-1244)

Eros with its six specialized instruments. A complete surface map was pieced together by 27 June 2000.[84] On 12 February 2001, NEAR Shoemaker became the first spacecraft in history to land on the surface of an asteroid as it survived an intentional crash landing on Eros, a mission-ending task for which it was never designed.

NASA returned to Mars with its Mars Surveyor program in 1997. The adventurous flight schedule listed an orbiter and lander sent during each window of opportunity, about every two years, culminating with a sample return mission in 2005. The first pair, *Mars Global Surveyor (MGS)* and *Mars Pathfinder,* launched on separate Delta 7925 rockets on 7 November and 4 December 1996, respectively. Due to their different trajectories, *Mars Pathfinder* arrived first, plunging through the thin Martian atmosphere on the Fourth of July 1997. The lander reached the surface shrouded in an inflatable air-bag system that protected the payload while it bounced through a rock-strewn, dry river wash. The air bags deflated and an innovative tetrahedral petal design opened to expose *Pathfinder's* base station and its three solar panels. The station returned spectacular binoptic panoramas of the surrounding landscape, as well as atmospheric and meteorological data, during its eighty-three-day mission (nearly three times its design life). What captured the public's imagination, however, was a 24.3-pound (on Earth; about 9 on Mars) microrover called *Sojourner* that arrived strapped to one petal of the lander. Millions of people from around the world overloaded the Jet Propulsion Laboratory web site, downloading daily updates as the nimble little rover trucked around a field of comically named boulders, rocks, and pebbles. *Sojourner* moved from rock to rock semiautonomously, pressing its main instrument (an alpha proton x-ray spectrometer) against the rocks to determine their compositions. It operated for twelve times its design lifetime of seven days and is believed to have still been functioning when the main lander, through which it communicated, ran out of battery power.[85]

The publicity of *Pathfinder* utterly obscured the many successes of its sister ship, *Mars Global Surveyor. MGS* reached Mars in September 1997, using its main engine to enter a highly elliptical orbit. Over the next several months, a series of risky aerobraking maneuvers were used to circularize the spacecraft's orbit about 640 miles above the planet. Since it began full time mapping operations in April 1999, *MGS* has overcome solar array and high gain antenna problems to return tens of thousands of images with the greatest resolution ever seen of Mars.[86]

Other NASA missions flown aboard Delta and active in 2000 have proven just as spectacular as those described above, but do not necessarily receive the same public recognition without pretty pictures to return. Extreme Ultraviolet Explorer (EUVE) completed a map of the sky in the extreme ultraviolet range of the spectrum, and eight years after its launch is now a platform for guest observations. WIND has surpassed its five-year design lifetime and contin-

Image 3-5: The Mars Polar Lander awaits its launch opportunity aboard a Delta II 7425 at Launch Complex 17B at Cape Canaveral, Florida, in 1999. Note the use of only four solid-rocket motors around the core stage. (NASA Photo no. KSC-99PP-0002).

ues to make detailed observations of the solar wind, Polar is observing Earth's polar regions to study the effects of the solar wind on the ionosphere and magnetosphere, and Advanced Composition Explorer (ACE) is providing continuous solar observations from a Lagrange orbit a million miles from Earth.

Deep Space 1 tested with promising results a dozen new, experimental technologies, including a xenon ion engine and an autonomous navigational system that enables the spacecraft to determine the most appropriate trajectory to a specified location with minimal oversight by ground controllers. The ion engine holds the record for the longest operation of a propulsion system in space, logging 195 days on 8 August 2000. Flight controllers performed an unprecedented rescue mission after DS-1's star tracker failed, rewriting the spacecraft's on-board computer software to use its wide-angle camera for navigation. Deep Space 1 then triumphantly pursued an extended mission that went well beyond its original design. On 22 September 2001, DS-1 passed within 1,200 nautical miles of comet Borrelly at a relative speed of about 37,000 mph. Amazingly, not only did the unshielded spacecraft survive the potential impacts of cometary dust as it passed through the coma, it returned the highest-resolution images of a comet to date.[87]

The future of NASA-Delta collaborations appeared rosy as of the end of 2001. Launched 7 February 1999, Stardust will encounter comet Wild-2 in 2004. The spacecraft will collect particles from the comet's tail, then return the samples to Earth. If it succeeds, it will be the first opportunity for scientists to analyze directly cometary debris. *Genesis,* launched 8 August 2001, will do the same with particles of the solar wind. Future payloads, now in assembly or planning stages, include a number of specialized Earth observers, additional comet encounters, the first Mercury orbiter, and the final element in the NASA Great Observatories Program, the Space Infrared Telescope Facility. Despite the loss of both 1998 Mars Surveyor payloads to management errors and design flaws, the Mars Surveyor program will continue with an orbiter that launched in 2001 and a pair of rovers that will each launch on its own Delta in 2003. The rovers, 330-pound siblings of *Sojourner,* will land using *Pathfinder's* revolutionary design of air bags surrounding tetrahedral petals.

NASA has manifested these many payloads aboard Delta Medium (7925) vehicles and the newer Med-Lite capacity (with three or four booster motors). The choice of launchers is ideally suited to the payloads and statistically a safe bet. The Delta rocket has celebrated over forty years of reliable launch services. In 289 flights through the end of 2001, only 15 were total failures, a success rate of 94.8 percent. This rate improved over the years: only one Delta II has failed in 100 flights, for 99 percent success.[88]

The Next Evolution: Delta III and Delta IV

These 289 flights include the first three flights of the most powerful Delta to date. Thirty-eight years of continual development increased Delta's payload capacity slowly and gradually, but in 1998 this trend changed with the introduction of Delta III, an "upgrade" of the Delta II that is in a different payload class entirely. Delta III, known to Boeing planners as Delta 8930, can lift 8,400 pounds to GTO, over twice the capacity of Delta II.[89]

It accomplishes this by means of a number of major changes. Most significant is the new Delta III second stage, powered by a Pratt & Whitney RL10B-2 engine derived from the RL10 engine that has been the basis of the Centaur stage for over thirty years. (As ever, this Delta hardware upgrade comes from flight-proven technology.) Burning liquid hydrogen and liquid oxygen, the RL10B-2 has a world-record specific impulse rating of 462 seconds and is the first use of a high-energy cryogenic engine in a Delta.

In addition, new booster motors provide 25 percent more thrust, and three of the nine motors are equipped with thrust vector control for improved maneuverability. As in the first stage of the Delta II, a Rocketdyne RS-27A provides main engine thrust. It is fed by the same Extra Extended Long Tank Thor oxygen tank, but the fuel tank has been shortened and increased in diameter to improve control margins. The RIFCA avionics system and launch operations infrastructure are also identical to those of Delta II. A new 4-meter diameter composite payload fairing tops the assembly.[90]

Delta III's maiden launch occurred on 27 August 1998. Like the very first Delta in 1960, Delta flight 259 carried a real payload, the Hughes HS-601 communications satellite Galaxy-X, rather than a test article. Unfortunately, it also suffered the same fate. About 72 seconds after liftoff, attitude control was lost and the rocket broke apart.

Boeing immediately convened an investigative review board. It quickly exonerated the common systems of Delta II so that vehicle's schedule would not be affected by the failure. Within two months, the board announced its conclusions. The report centered around a 4-Hz roll mode, one of fifty-six roll modes in the control system, caused by the three air-lit solid booster motors rocking back and forth on their mounts in unison. As the vehicle flew and shed weight, this rocking motion became increasingly significant, and the control system attempted to counteract by oscillating the nozzles of the three ground-lit thrust vector control (TVC) boosters. The blow-down hydraulic system of the TVC nozzles, which ordinarily would have a margin of about half its capacity, depleted all of its hydraulic fluid in less than 20 seconds. At 10 miles altitude and 6.6 miles down range, the vehicle flew through a wind shear. The unassisted main engine lacked the control authority to fight the shear and the vehicle yawed severely and broke apart, triggering its self-destruct package.[91]

For Boeing, the software revision to account for the unexpected roll was relatively simple when judged against the 132 total "action items" returned by the review team. These resulted in numerous changes to in-house analysis, hardware testing, and systems integration. The review also uncovered a potential problem with deployment of the second-stage nozzle extension that could manifest in the cold of space under adverse power conditions. Engineering changes were implemented to improve reliability.[92]

The second Delta III flight came on 4 May 1999. The first stage ably demonstrated that Boeing's control software redesign was successful, and the second-stage nozzle extension deployed without a hitch. The RL10 engine burned for 8 minutes and then entered an 8-minute coast phase. During that first burn an anomalous shock exceeding 25 g was registered but did not affect the flight. The engine restarted for what was meant to be another 2 minutes, 42 seconds. Just 3.4 seconds into its second burn, however, the stage was rocked by another severe shock and automatically shut down, stranding the Orion F3 satellite in an eccentric, useless orbit.

Another review board was convened, focusing on the RL10B-2 engine. The Atlas-Centaur program, which uses similar RL10A-4 engines in the Centaur stage, was also put on hold.[93] Investigators determined a seam in the engine's combustion chamber had ruptured, leading to a catastrophic loss of pressure. A recently modified brazing process used by Pratt & Whitney to unite the four segments of the chamber was to blame, as it tended to leave an excessive number of air pockets along the seams. P&W again changed its brazing technique, and was delivering combustion chambers of excellent quality by October 1999. Meanwhile, Boeing worked to incorporate design modifications to reduce any potentially adverse vehicle-induced loads, which may have been a contributing factor.[94]

Delta III had passed return-to-flight evaluation by midyear 2000, but financial difficulties encountered by several satellite communication firms meant that no contracted commercial payloads were ready to fly. Boeing, anxious to prove the new vehicle to potential customers who might deem it unreliable and send their payloads to other launch providers, decided to launch the first test payload in Delta history. DM-F3 (Delta Mission–Flight Three) carried a payload that simulated the mass and center of gravity properties of a typical communications satellite, but aside from accelerometers and a telemetry package it had no on-board systems. As an afterthought, the spool-shaped mass simulator was painted with black-and-white patterns to provide a calibration target for use in optical imaging experiments by the Air Force and the Colorado Center for Astrodynamics Research.[95]

DM-F3 was launched on 23 August 2000 and was a complete success. Unlike a typical Delta II mission, where the second stage shuts down on command, the second stage of the Delta III burned to propellant depletion. This was as expected, as most Delta III payloads will have liquid propellant apogee engines that are more able than solid motors to compensate for the wider margin of insertion error a depletion flight incurs. The final orbit was somewhat lower in apogee than originally predicted, but was declared nominal for the atmospheric conditions on launch day.[96]

The future of Delta is literally bigger than ever with the advent of Delta IV. As with Delta II, it was spurred by a U.S. Air Force request for proposals, this time for an Evolved Expendable Launch Vehicle (EELV). The EELV, part of a $2 billion modernization by the Pentagon, was envisioned as a family of rockets capable of launching small, medium, and large military payloads. It is intended to replace the current Atlas, Delta, and Titan vehicles with a system that is up to 50 percent more efficient in terms of turnaround and on-pad time. In response, McDonnell Douglas submitted an original proposal in 1995 and began the Delta IV program in 1997.[97]

The Delta IV vehicle is based on an entirely new first stage called the Common Booster Core (CBC). A CBC is 16.25 feet in diameter, more than twice that of the Delta II, and stands taller than a complete Delta III. Its tanks are built in the Delta style of brake-formed isogrid skins joined using a friction stir welding technique recently developed for Delta II.[98]

Powering the CBC is the first large engine to be designed since the Space Shuttle Main Engine (SSME) was developed in the 1970s. The Rocketdyne RS-68 is the world's most powerful hydrogen-oxygen engine. Based on lessons learned from the SSME, it produces 650,000 lbf at sea level (compared to 393,800 pounds for an SSME at 104 percent throttle), but thanks to a simplified gas generator cycle it has 93 percent fewer parts. In typical fashion, the engine is gimbaled for pitch and yaw control, but unlike in previous Delta vehicles the RS-68 sends its turbopump exhaust through vectorable nozzles for roll control, eliminating the need for separate vernier engines.[99]

On 16 October 1998, the Air Force awarded Boeing with the lion's share of the EELV contract, ordering nineteen vehicles worth an estimated $1.38 billion. (In the interest of competition and corporate investment, a separate contract was awarded to Lockheed Martin for nine of that company's Atlas V EELVs.) Five different configurations were approved, all of which use the Pratt & Whitney RL10B-2 engine in the second stage.[100] The Delta IV-Medium carries a Delta III second stage with stretched tanks, and a standard Delta III 4-meter composite fairing. These are attached to the CBC with a tapered interstage. Three Medium-Plus models add pairs of Alliant GEM-60 strap-on booster motors, upsized versions of the Delta II GEM-40 and Delta III GEM-46 motors. A larger second stage, with a 5-meter diameter hydrogen tank, is also available, which along with a 5-meter composite fairing presents a cylindrical profile from stem to stern. The various Medium models can lift from 9,285 to 14,475 pounds to GTO.

The Delta IV-Heavy uses three CBC stages, strapped together side by side, to loft 28,950 pounds into geosynchronous transfer orbit. In flight, the central CBC will throttle back to 60 percent of its rated power during the period of maximum dynamic pressure, while the booster CBCs will maintain full thrust. Just under 5 minutes into the flight the booster CBCs will be jettisoned as the central CBC resumes 100 percent power and burns for another 80 seconds. The larger size 5-meter second stage will then carry the payload to orbit. With an extended fairing and special attachment fittings the IV-Heavy will have the option to provide dual payload capability for significant savings over two separate launches.[101]

To produce the Delta IV, Boeing has built a gigantic manufacturing facility at Decatur, Alabama, encompassing 1.5 million square feet of floor space. Innovative production tools include large, custom jigs for tank welding and a moving line for final assembly. The moving assembly line, common in the automotive and electronics industries, has never before been used for launch vehicles products of this size. Each CBC, two and a half stories high and over 135 feet long, creeps along the line at about half an inch per minute as workers add major components and perform check-out operations.[102]

Common Booster Cores are too large to travel aboard trucks along America's highways as readily as Delta II and III stages do, so Boeing has commissioned a new river- and seafaring vessel dubbed *Delta Mariner*. The *Delta Mariner*, with the capacity to carry a full Delta IV-Heavy package (three CBCs, an upper stage, and fairing), will travel the Tennessee-Tombigbee Waterway to the Gulf of Mexico on its way to the East Coast or West Coast launch sites. At Cape Canaveral, Boeing has converted Space Launch Complex 37, an historic site that launched eight unmanned Saturn I and IB rockets before it was mothballed in 1972.[103] At Vandenberg, work has begun on SLC-6, a site first built for the Air Force's Manned Orbiting Laboratory that was later intended to support polar-orbiting Space Shuttle launches.[104]

At either launch site, a new processing method represents a major departure for Delta IV. Since 1960, all Deltas have been vertically stacked on the pad, each stage or booster hoisted into place via cranes in the mobile service tower. Even with today's streamlined processing, the resulting minimum on-pad time for a Delta II remains about twenty-four days. Delta IV will be assembled prone in a Horizontal Integration Facility near the launch complex that contains enough space for up to four single-core vehicles, or one triple-core (Heavy) and two single-core vehicles. Meanwhile, payloads will be encapsulated in their fairings at clean room sites several miles away. The complete vehicle will then be transported to the pad and erected, after which the payload/fairing combination will be hoisted into place. Boeing expects horizontal integration will reduce on-pad time to seven to ten days, and hopes to launch some fifteen to eighteen vehicles per year (versus record years of twelve launches for Delta, in 1975 and 1998).[105] The first flight of Delta IV is expected in 2002 from Cape Canaveral, and 2003 from Vandenberg.

Delta IV has been designed for low cost and reliability, two touchstones of the Delta program since its inception. Its manufacture utilizes many of the techniques perfected by Douglas, McDonnell Douglas, and Boeing over four decades, and in flight the vehicle will be guided by the same high-precision RIFCA avionics system used by Delta II and III. Boeing images even depict Delta IV in the same light blue paint scheme that has become a virtual trademark of Delta. Nevertheless, with its new tankage, main engine, choice of fuel, erection process and launch sites, it could be argued that Delta IV is a Delta in name only, and a separate family. Thus Delta III may prove to be the last major upgrade to the original, 8-foot-diameter Thor vehicle. A NASA planning document of 18 August 2000 lists the Delta III as the launcher for three upgraded GOES weather satellites, but these will not launch until 2002 at the earliest. (No Delta IV launches for NASA have been manifested.) In the meantime, and for several years to come, the Delta II will continue to be NASA's small workhorse in the quest for further scientific knowledge of Earth, Mars, the Sun, the solar system, and humanity's place in the universe.[106]

Conclusion

For more than forty years the Delta family of launch vehicles has had a profound effect upon American society. It was the first vehicle used to launch communications satellites into orbit, and it remains the vehicle of choice for commercial payloads. Those commercial satellites placed into orbit by Deltas have revolutionized American culture, bringing the world closer through instantaneous global communications of voice, moving picture, and data. Those core commercial activities were funded in no small part by American businesses who have made satellite communications a profitable concern. More than any other, Delta has made possible the opening of space for Americans.

Notes

1. Jacob Neufeld, *The Development of Ballistic Missiles in the United States Air Force, 1945–1960* (Washington, D.C.: Office of Air Force History, 1990), p. 121.

2. Julian Hartt, *The Mighty Thor: Missile in Readiness* (New York: Duell, Sloan, and Pearce, 1961), pp. 41–43.

3. Ibid., pp. 48–51.

4. Neufeld, *The Development of Ballistic Missiles*, pp. 147–48. Hartt claims 28 December (p. 59).

5. Hartt, *Mighty Thor*, pp. 60–62.

6. Robert L. Perry, "The Atlas, Thor, Titan, and Minuteman," in Eugene M. Emme, ed., *The History of Rocket Technology: Essays on Research, Development, and Utility* (Detroit: Wayne State University Press, 1964), pp. 151–52.

7. Hartt, *Mighty Thor*, pp. 80–81.

8. Ibid., p. 73.

9. Ibid., p. 108; Marc C. Cleary, "The 6555th: Missile and Space Launches Through 1970/Chap. III, Sec. 4," available on line at http://www.patrick.af.mil/heritage/6555th/6555fram.htm, completed in 1991.

10. Hartt, *Mighty Thor,* pp. 113, 116, 128.

11. Ibid., pp. 132–35.

12. Neufeld, *The Development of Ballistic Missiles,* pp. 222–24.

13. Andrew J. LePage, "Operation Mona: America's First Moon Program," available on line at http://www.spaceviews.com/1998/04/article1a.html, first posted in 1998.

14. See Curtis Peebles, *The Corona Project: America's First Spy Satellites* (Annapolis, Md.: Naval Institute Press, 1998) and Dwayne A. Day, John M. Logsdon, and Brian Latell, eds., *Eye in the Sky: The Story of the Corona Spy Satellites* (Washington, D.C.: Smithsonian Institution Press, 1998).

15. Andrew J. LePage, "The Early Explorers," available on line at http://www.spaceviews.com/1999/08/article2a.html, posted in 1999.

16. NASA, in consultation with the Advanced Research Projects Agency, *The National Space Vehicle Program,* 27 January 1959, summary in John M. Logsdon, gen. ed. *Exploring the Unknown: Selected Documents in the History of the U.S. Civil Space Program,* vol. 4, *Accessing Space,* NASA SP-4407 (Washington, D.C.: NASA, 1999), p. 95.

17. Ibid.

18. Helen T. Wells, Susan H. Whiteley, and Carrie Karegeannes, *Origins of NASA Names,* NASA SP-4402 (Washington, D.C.: NASA, 1976), p. 12. "The National Launch Vehicle Program" refers to the vehicle simply as Delta, avoiding the military designation of Thor. However, many sources, including those at NASA Goddard, continued to call the new rocket Thor-Delta well into the 1970s.

19. Early Delta models included four small, triangular tail fins for directional stability. Wind-shear test flights on Thor determined them to be unnecessary, and they were subsequently deleted from Delta with the addition of booster motors.

20. Constance McLaughlin Green and Milton Lomask, *Vanguard: A History,* NASA SP-4202 (Washington, D.C.: NASA, 1970), chap. 4.

21. Mark Wade, "Encyclopedia Astronautica/X-248," available on line at http://www.friends-partners.org/~mwade/engines/x248.htm, posted 2000.

22. Arthur C. Clarke, "Extra-Terrestrial Relays: Can Rocket Stations Give World-Wide Radio Coverage?" in John M. Logsdon, gen. ed., *Exploring the Unknown: Selected Documents in the History of the U.S. Civil Space Program, Volume III, Using Space,* NASA SP-4407 (Washington, D.C.: NASA, 1998), pp. 16–22.

23. Leonard Jaffe, "The Current NASA Communications Satellite Program," in N. Boneff and I. Hersey, eds., *Proceedings of the XIIIth Congress of the International Astronautical Federation: Varna 1962* (New York: Springer-Verlag, 1964), p. 835; Andrew J. Butrica, ed., *Beyond the Ionosphere: Fifty Years of Satellite Communication,* NASA SP-4217 (Washington, D.C.: NASA, 1997). From nineteen satellites at 22,300 statute miles altitude, the curve passes through forty satellites at 5,000 miles, and four hundred satellites at 1,000 miles.

24. This was Project West Ford. NASA Scientific and Technical Information Division, *Significant Achievements in Space Communications and Navigation 1958–1964,* NASA SP-93 (Washington, D.C.: NASA, 1966), pp. 10–11; Donald C. Elder, *Out From Behind the Eight-Ball: A History of Project Echo* (San Diego: Univelt, 1995).

25. Steven J. Isakowitz, *International Reference Guide to Space Launch Systems* (Washington, D.C.: American Institute of Aeronautics and Astronautics, 1991), p. 204.

26. Jaffe, "Current NASA Communications Satellite Program," p. 840. Echo 1A is frequently referred to as Echo 1, ignoring the first flight attempt. It should be noted that Echo's designation is nonstandard. NASA's usual custom was to give spacecraft letter designations before launch and change the letter to a number after a successful launch. NASA's penchant for roman numerals tended to lead to more confusion. For example, one document notes that Tiros *I* (Tiros 9 when on orbit) is not to be mistaken for Tiros I (1 on orbit, or Tiros A before flight). This document uses arabic numerals for flight num-

bers and reserves use of roman numerals for program series (e.g. Intelsat III, NAVSTAR II, Delta II).

27. Jaffe, "Current NASA Communications Satellite Program," pp. 839–41.

28. Mike Evans, "JPL Mission and Spacecraft Library/Quicklook: Echo 1, 1A, 2," available on line at http://leonardo.jpl.nasa.gov/msl/QuickLooks/echoQL.html.

29. NASA Scientific and Technical Information Division, *Significant Achievements in Satellite Meteorology, 1958–1964*, NASA SP-96 (Washington, D.C.: NASA, 1966), p. 8. See also Arthur Upgren and Jurgen Stock, *Weather: How It Works and Why It Matters* (Cambridge, Mass.: Perseus Publishing, 2000).

30. Ibid., pp. 22–23.

31. Ibid., pp. 38–39, 60.

32. NASA Goddard Space Flight Center, "National Space Science Data Center Master Catalog/P14," available on line at http://nssdc.gsfc.nasa.gov/cgi-bin/database/www-nmc?61-010A, posted in 1996.

33. NASA Goddard Space Flight Center, "National Space Science Data Center Master Catalog/EPE-A," available on line at http://nssdc.gsfc.nasa.gov/cgi-bin/database/www-nmc?61-020A, posted in 1996.

34. Henry J. Smith, "Solar Astronomy," in *Astronomy in Space*, NASA SP-127 (Washington, D.C.: NASA, 1966), p. 14.

35. Ibid., p. 16.

36. NASA Goddard Space Flight Center, "National Space Science Data Center Master Catalog/OSO-A," available on line at http://nssdc.gsfc.nasa.gov/cgi-bin/database/www-nmc?62-006A, posted 1998.

37. NASA Goddard Space Flight Center, *Ariel I: The First International Satellite, Experimental Results*, NASA SP-119 (Washington, D.C.: NASA, 1966), pp. 3, 9–12.

38. Ibid., p. 72.

39. Ibid., pp. 72–74.

40. Carey Sublette, ed., "The High Energy Weapons Archive—A Guide to Nuclear Weapons/ U.S. Nuclear Testing/Operation Dominic," available on line at http://www.enviroweb.org/issues/nuketesting/hew/Usa/Tests/Dominic.html, posted 1997.

41. NASA Goddard Space Flight Center, *Ariel I*, pp. 23, 56.

42. Ibid., p. 100.

43. Bell Telephone Laboratories, "Preliminary Report, Telstar I, July–September 1962," in Logsdon, gen. ed., *Exploring the Unknown* 3:89–90; David J. Whalen, "Billion Dollar Technology: A Short Historical Overview of the Origins of Communications Satellite Technology, 1945–1965," in Butrica, ed., *Beyond the Ionosphere*, 95–127; Daniel Glover, "NASA Experimental Communications Satellites 1958–1995/Telstar," available on line at http://roland.lerc.nasa.gov/~dglover/sat/telstar.html.

44. Ibid.

45. Cub Koda, "AMG All-Music Guide/The Tornados," available on line at http://allmusic.com/cg/x.dll?p=amg&sql=B5672, posted 2000.

46. Glover, "Telstar."

47. J. Kork and W.R. Schindler, "The Thor-Delta Launch Vehicle: Past and Future," in Michal Lunc, ed., *Proceedings of the XIXth Congress of the International Astronautical Federation: New York 1968* (New York: Pergamon Press, 1970), p. 5.

48. NASA Office of Scientific and Technical Information, *Launch Vehicles of the National Launch Vehicle Program*, NASA SP-10 (Washington, D.C.: NASA, 1962), p. 10.

49. Kork and Schindler, "Thor-Delta Launch Vehicle," p. 10.

50. Daniel Glover, "NASA Experimental Communications Satellites 1958–1995/Relay," available on line at http://roland.lerc.nasa.gov/~dglover/sat/relay.html.

51. Ibid.

52. Cordant Technologies, "STAR Performance & Summary Chart," available on line at http://www.thiokol.com/StarPerf.htm posted 2000.

53. NASA Scientific and Technical Information Division, *Significant Achievements in Space Communications*, pp. 24–25.

54. W.M. Arms, *Thor: The Workhorse of Space—A Narrative History* (Huntington Beach, Calif.: McDonnell Douglas Astronautics, 1972), table A-1.

55. "Geosynchronous" refers to an orbit with a twenty-four-hour period, that is, synchronous with the earth's rotation. A "geostationary" orbit is a geosynchronous orbit with an inclination of 0 degrees, such that it appears to remain stationary over one point on the earth's equator.

56. Jonathan McDowell, "Geostationary Orbit Catalog," available on line at http://hea-www.harvard.edu/QEDT/jcm/space/logs/geo.date, posted 1999.

57. Based on Isakowitz, *International Reference Guide*.

58. NASA Goddard Space Flight Center, "National Space Science Data Center Master Catalog/GEOS-A," available on line at http://nssdc.gsfc.nasa.gov/cgi-bin/database/www-nmc?65-089A, posted 1996.

59. Design studies and unused versions caused some "letter leapfrogging" in the roster. *F* was an *E* without augmentation motors, but was never flown. *G* was a two-stage *E*. *H* was a *G* without augmentation, also never flown. *I* was skipped to avoid confusion with the roman numeral *1*. *J* thus became the next designation used, while *K* is rumored to have been a design study on the use of a cryogenic upper stage—a use that finally arrived, thirty years later, in the Delta III.

60. NASA Office of Space Science, *Mission Operations Report: Dynamics Explorer* (Washington, D.C.: NASA, 1981), 22.

61. Kork and Schindler, "Thor-Delta Launch Vehicle," p. 9.

62. Intelsat III was built by TRW and upgraded the spacecraft to fifteen hundred voice channels. It was the first global satellite communication system, and the first operational satellite system to use a de-spun antenna structure. A de-spun antenna allows a satellite to spin for stability while keeping its antenna pointed at Earth, thus avoiding the waste of transmitter power that an omnidirectional broadcast incurs. (Intelsat IV and later versions outgrew Delta and were launched on Atlas-Centaur rockets, among others.)

63. Charles R. Gunn, "The Delta and Thor-Agena Launch Vehicles for Scientific and Applications Satellites," in L. G. Napolitano, ed., *Proceedings of the XXIst Congress of the International Astronautical Federation, Constance 1970* (Amsterdam: North-Holland Publishing, 1971), 912–13.

64. Robert J. Goss, "Delta Vehicle Improvements," in *Significant Improvements in Technology*, NASA SP-326 (Washington, D.C.: NASA, 1973), p. 12.

65. Two missions.

66. One mission.

67. The first three flights are exceptions: Delta flights 92, 94, and 95 used the DSV-3P-4 engine, an AJ10-118F variant.

68. Use of the FW-4D was discontinued after creation, but before inception, of the four-digit system.

69. P. Plush, "The Delta 3914," *RCA Engineer*, June/July 1976, p. 67.

70. Goss, "Delta Vehicle Improvements," p. 12; "Propulsion Systems," *Aviation Week & Space Technology*, 8 March 1971.

71. Kenneth I. Duck, "Delta Launch Vehicle Inertial Guidance System (DIGS)," in *Significant Improvements in Technology*, pp. 14–15.

72. Ed Sheffner, "Landsat Program Chronology," available on line at http://geo.arc.nasa.gov/sge/landsat/lpchron.html, posted 1998. For a general history of Landsat, see Pamela E. Mack, *Viewing the Earth: The Social Construction of Landsat* (Cambridge, Mass.: MIT Press, 1990).

73. Goss, "Delta Vehicle Improvements," p. 13.

74. Roger E. Bilstein, *Stages to Saturn: A Technological History of the Apollo/Saturn Launch Vehicles*, NASA SP-4206 (Washington, D.C.: NASA, 1996), pp. 97–99.

75. E.D. Becken, "Satellite Communications," *RCA Engineer*, June/July 1976, p. 40.

76. John R. Pierce, *The Beginnings of Satellite Communications* (San Francisco: San Francisco Press, 1968), p. 5–12.

77. J.E. Keigler, "RCA Satcom—Maximum Communication Capacity per Unit Cost," *RCA Engineer*, June/July 1976, p. 50.

78. Plush, "Delta 3914," p. 67.

79. Paul D. Engelder, "DRIMS—A Redundant Strapdown IMU for Booster Guidance and Control," *Proceedings of the IEEE 1980 National Aerospace and Electronics Conference* (Dayton, Ohio: IEEE, 1980).

80. NASA Office of Space Transportation Operations, *Mission Operation Report: SBS-B/Delta Launch* (Washington, D.C.: National Aeronautics and Space Administration Report, 1981), p. 13.

81. House Committee on Science and Technology, *Delta Launch Vehicle Accident Investigation: Hearing before the Subcommittee on Space Science and Applications*, 99th Cong., 2d sess., 5 June 1986.

82. Ibid.

83. Nine NAVSTAR I satellites were launched as an experimental constellation in the 1970s aboard Atlas rockets. NAVSTAR II was the first set of operational satellites and was followed by IIA and IIR versions.

84. Johns Hopkins University Applied Physics Laboratory, "Near Earth Asteroid Rendezvous Mission," available on line at http://near.jhuapl.edu/, posted 2000.

85. Jet Propulsion Laboratory, "Mars Pathfinder," available on line at http://mpfwww.jpl.nasa.gov/default.html, posted 1999.

86. Jet Propulsion Laboratory, "Mars Global Surveyor," available on line at http://mars.jpl.nasa.gov/mgs/index.html, posted 2000.

87. Jet Propulsion Laboratory, "Deep Space 1," available on line at http://nmp.jpl.nasa.gov/ds1/, posted 2001.

88. Koreasat 1, launched on a Delta II on 5 August 1995, has been deemed a "partial success." One air-lit solid booster failed to detach, a first for Delta. Due to the increased weight, the satellite was injected into a lower-than-intended orbit; its on-board motor had to fire longer to reach geostationary orbit. This maneuver used about half of the propellant available for station keeping and shortened the useful life of the satellite.

89. The Boeing Company acquired McDonnell Douglas on 1 August 1997.

90. The Boeing Company, "Delta III Launch Vehicle," available on line at http://www.boeing.com/defense-space/space/delta/delta3/delta3.htm, posted 2000.

91. The Boeing Company, "Boeing Completes Delta 3 Failure Investigation, Changes Rocket's Control Software," news release available on line at http://www.flatoday.com/space/explore/stories/1998b/101998v.htm, posted 19 October 1998.

92. Florida Today Space Online, "Transcript of Boeing News Briefing on Delta 3 Return-to-Flight," available on line at http://www.flatoday.com/space/explore/stories/1999/040299d.htm, posted 2 April 1999.

93. The Titan IV program was not scheduled to use a Centaur upper stage for the remainder of 1999 and was not significantly affected.

94. The Boeing Company, "Delta 269 (Delta III) Failure Report," available on line at http://www.boeing.com/defense-space/space/delta/delta3/d3_report.pdf, posted 16 August 2000.

95. Justin Ray, "Demonstration Flight Ordered for Boeing's Delta 3 Rocket," available on line at http://spaceflightnow.com/delta/d280/index.html, posted 22 August 2000.

96. The Boeing Company, "Boeing Delta III Mission Data Shows Flight Success," available on line at http://www.boeing.com/news/releases/2000/news_release_000824h.html, posted 24 August 2000.

97. McDonnell Douglas, "Delta Rockets Advance Toward New Millennium," *Inside Delta,* May/June 1997, p. 2.

98. The Boeing Company, "Gaining Momentum in 1999: Delta IV Hardware and Facilities Take Shape," *Inside Delta,* January 1999, p. 4.

99. McDonnell Douglas, "Teammate Spotlight: Boeing North American Rocketdyne," *Inside Delta,* May/June 1997, p. 3.

100. The Delta IV-Small, a proposed version with an AJ-10 second stage and a Delta II 10-foot composite fairing, has been eliminated from planning documents.

101. The Boeing Company, "Delta IV Payload Planners Guide," sec. 1, p. 1–8, available on line at http://www.boeing.com/defense-space/space/delta/delta4/guide/index.htm, posted October 1999.

102. The Boeing Company, "Moving into 2000: Team Promotes Rocket Roll for Delta IV Assembly," *Inside Delta,* June 1999, p. 4.

103. The Boeing Company, "Redesigning Both Coasts," *Inside Delta,* July 2000, p. 6. Issues of *Inside Delta* from April 1998 onward are available on line at http://www.boeing.com/defense-space/space/delta/insidedelta.htm.

104. This is the fifth incarnation of a "star-crossed" complex that has supported only one successful mission in more than thirty-five years. See Roger Guillemette, "The Curse of Slick Six: Fact or Fiction?" available from Florida Today Space Online at http://www.flatoday.com/space/explore/special/slc6/slc6.htm, posted 1999.

105. The Boeing Company, "Delta IV Payload Planners Guide," sec. 6, p. 23, 30–35.

106. NASA Headquarters, "NASA Launch Services Manifest: Proposed Flight Planning Board 8/18/00," available on line at http://extranet.hq.nasa.gov/elv/IMAGES/ls.pdf, posted 2000.

– 4 –

Titan

Some Heavy Lifting Required

Roger D. Launius

Introduction

The first true space launch vehicles developed within the United States emerged from the Department of Defense's (DoD) ballistic missile programs of the 1950s, and as a result, most of the current U.S. space launch capability rests on the shoulders of the investment made in technology for the ballistic missiles of the cold war. The Titan launch vehicle family was developed by the United States Air Force to meet its medium ballistic missile lift requirements in the latter 1950s and early 1960s. It has been modified since that time into a heavy-lift space launcher for a broad range of military and scientific payloads.

Background

The development of the Titan family of boosters began in October 1955, when the United States Air Force (USAF) awarded the Glenn L. Martin Company a contract to build an intercontinental ballistic missile (ICBM). The missile was ordered as a backup to the Atlas ICBM being developed by Convair—when the Air Force had some nagging doubts about some of the technology being incorporated into Atlas. The new missile became known as the Titan I, the nation's first two-stage ICBM. It was also the first designed to be based in underground silos. The Air Force deployed fifty-four Titan Is followed by fifty-four improved Titan IIs. The first Titan II ICBMs were activated in 1962, and modified Titan IIs were selected to launch NASA's Gemini spacecraft into orbit during the mid-1960s. As a result of arms and nuclear-reduction treaties, the Titan II weapon system was deactivated during the mid-1980s; the mis-

Table 1
Titan Family Launch Vehicles*

Launch Vehicle/ Prime Contractor	Initial Operational Capability	Thrust of Launch	Specific Impulse	Payload to Orbit	Estimated Launch Price (Then-Year Dollars)
Titan I–II ICBM/ Martin Company	1961	430,000 lbf	280 s	4,000 pounds	Noncommercial
Titan IIIC-E/ Martin Company	1965	2,400,000 lbf	271.6 s	24,000–30,400 pounds, depending on configuration	Noncommercial
Titan 34D/ Martin Marietta	1982	500,000 lbf	280 s	33,800 pounds	Noncommercial
Titan II SLV/ Martin Marietta	1988	474,000 lbf	296 s	4,200 pounds	34 million
Titan III/ Martin Marietta	1989	2,600,000 lbf	271.6 s	11,000 pounds	130–150 million
Titan IV/ Martin Marietta	1989	3,300,000 lbf	285.6 s	39,000 pounds	196–248 million

*Source; Steven J. Isakowitz, Joseph P. Hopkins Hr., and Joshua B. Hopkins, *International Reference Guide to Space Launch Systems,* 3d ed. Reston, VA: American Institute for Aeronautics and Astronautics, 1999

siles were removed and the silos were destroyed. As a result of this decommissioning, several of the old Titan II ICBMs found use as space launchers in the post-*Challenger* era of space launch operations.

Both alone and with upper stages (notably Agena), the Titan II also launched quite a few nonmilitary payloads. The last Titan-Agena flew in 1987,[1] although newer versions of the Titan booster appeared thereafter. Various Titan III and IV systems employed a diversity of upgraded stages, strap-on boosters, and advanced subsystems. Collectively, this family of launchers has significantly affected U.S. access to space since the earliest years of the space age.

The Origins of Intercontinental Ballistic Missiles

Just after World War II the armed services undertook studies determining the feasibility of intercontinental ballistic missiles (ICBM) as strategic weapons that could deliver nuclear warheads to targets half a world away. Competition was keen among the services for a mission in the new "high ground" of space, the military importance of which was not lost on the leaders of the world. In April 1946 the U.S. Army Air Forces issued Consolidated-Vultee Aircraft (Convair) a study contract for an ICBM.[2]

There was, unfortunately, a lack of genuine progress on ballistic missile development between 1945 and the mid-1950s. An extreme conservatism about ballistic missiles reigned within the government through most of the Truman administration. Like the "experts" who once denied that aircraft could ever play a useful military role, critics of the embryonic missile proposals questioned both their feasibility and utility—sometimes questioning the good sense of their supporters. For instance, no less a figure than Vannevar Bush—godfather of the National Defense Research Council of World War II and advocate of research and development efforts in support of national security objectives—ridiculed all notions of ballistic missiles. Bush claimed that they would be both "astronomically" expensive and would require the development of impossibly sophisticated technologies to guide them to targets. In his view this, complicated by the obvious impossibility of creating an effective ballistic missile warhead, ensured that they would never achieve reality.[3]

Despite a lack of progress on ballistic missile development between 1945 and the mid-1950s, international events conspired to force the DoD out of its lethargy. The Soviet Union's detonation of a nuclear weapon in 1949, the fall of China to communist insurgents under Mao Tse-tung in 1949, and the invasion of South Korea by the communist north Korean regime in 1950 fundamentally altered national security policy. There were other difficulties and problems in early military space efforts as well. Interservice rivalry certainly was the most important, as every service fought for a mission in the arena even as they failed to support aggressive development efforts for launchers. Austere budgets, without "frills" like missiles, constituted another.

With relatively little progress made in the development of ballistic missiles, until the advent of the Eisenhower administration, interestingly enough it was a new "economy drive" in the DoD that provided impetus for the development of ICBMs. Determined that defense expenditures could and should be reduced, the Pentagon created a Guided Missiles Study Group (under its Armed Forces Policy Council) to recommend means for cutting defense costs. Secretary of Defense Charles E. Wilson in his 16 June 1953 directive creating the review committee specified that "a continuous effort should be made to standardize on one missile for production and use by all military departments, wherever, within the employment limitations of each type of missile, standardization appears to be practicable."[4]

This group discovered evidence of enormous delays in ICBM development and, in response, urged the creation of the Air Force Strategic Missiles Evaluation Committee (popularly known as the "Teapot Committee") to delve more deeply into the matter. Under the leadership of Prof. John von Neumann, this committee reviewed the status of the missile program and concluded that new warhead developments plus advances in rocket technology made an intercontinental missile immediately feasible. That conclusion, along with a series of implementation recommendations, reached Trevor Gardner, Air Force Assistant Secretary for Research and Development, in the first quarter of 1954. Enthused about the potential of the proposal, Gardner and von Neumann secured the active support of the Air Force Chief of Staff, Gen. Nathan F. Twining, and Secretary of the Air Force Harold E. Talbott. The resulting development effort became the Atlas program, the United States' first ICBM. The "stage-and-a-half" Atlas missile received high priority from the White House and hard-driving management from Brig. Gen. Bernard A. Schriever, a flamboyant and intense Air Force leader. Overseen from the newly established Western Development Division (WDD) in Los Angeles, California, the first Atlas was test fired on 11 June 1955, and a later version became operational in 1959.[5]

The Decision to Build Titan

As Atlas was being developed, on 12 January 1955 the Air Force accepted a recommendation coming from several sources—von Neumann's committee, Project RAND, and Bernard Schriever—that a second ballistic missile, a two-stage design, be developed as a hedge against any possible delays with the Atlas program. Using many of the same subsystems, this new ICBM would be an entirely different design and would have a significantly greater range and payload capacity. In addition, the Air Force determined that this new missile should be transportable overland by truck and thereby eliminate many of the logistics bottlenecks encountered in the Atlas program.[6]

Bernard Schriever emphasized the necessity of undertaking this second ICBM not only to ensure against a failure in the Atlas program but also to incite a sense of competition. He believed that Convair, the prime contractor for the Atlas launcher, had grown complacent in its development work because it was, in effect, "the only game in town." This would change if a second program began. Moreover, Schriever thought the technological challenges warranted a second project, believing it wise to sponsor an alternative configuration and staging approach with a second source. He wrote, "It is possible that such an approach might provide a design substantially superior with the availability of future component development and thus would provide a chance for great advancement even with a late start. In line with this thinking, it is presently believed that the second design should be oriented around greater technical risks which might offer dramatic payoffs."[7] Schriever believed, however, that only one of these two missiles would actually become operational. Either one of the development efforts could fail and be canceled, he concluded, or one of the ICBMs might turn out clearly superior to the other. Either way, only the better of the two missiles would actually enter active service.

Beginning in January 1955 the Air Force began the process to develop an alternate to the Atlas. The Air Research and Development Command (ARDC), the arm of the Air Force charged with weapons system development, formally submitted a proposal for this alternative ICBM to Air Force Secretary Harold E. Talbott, and on 28 April 1955 Talbott approved the concept. Accordingly, on 6 May 1955 ARDC solicited proposals for the new missile from Bell Aircraft, Douglas Aircraft, General Electric, Lockheed, and the Glenn L. Martin Company. In the solicitation for WS 107A-2, the ICBM that was later named Titan, ARDC asked that "the weapon system be capable of launching missiles from bases within the continental United States carrying thermonuclear warheads with a desired weight of 3,000 pounds to ranges of 5,000 nautical miles with a circular error probable of five nautical miles or less . . . and must have the capability to strike a retaliatory blow again any attacking enemy in a minimum amount of time."[8] Three companies responded to the solicitation and based on a recommendation from a source selection board the prime contract for WS 107A-2 went to the Martin Company because of its superiority in several engineering and management categories.[9] On 27 October 1955 the Air Force signed a letter contract with Martin to design, develop, and test a two-stage missile with a gross weight of approximately 225,000 pounds. This would lead to the final deployment of the Titan ICBM. While the final contract would not be signed until 22 January 1957, this October 1955 decision set the Air Force on the path of developing the Titan launch vehicle. The Martin Company then set out on development efforts for Titan at a new plant built in the Rocky Mountain foothills near Littleton, Colorado.[10]

Building Titan I

Designated SM-68 (Strategic Missile-68), the Titan I was 10 feet in diameter and 90 feet long. Martin engineers designed a Titan first stage that included a lightweight but self-supporting skin made of 2014 aluminum—an alloy of copper and aluminum—supported by structural members contained in the propellant tank walls. A semi-monocoque design, this first stage proved exceptionally important to the overall success of the Titan system and provided significant future growth capability.

The first stage of the Titan was powered by engines manufactured by the Aerojet-General Corporation that used liquid oxygen (LOX) and RP-1. The LR87-AJ-1 engine that went into the Titan I first stage had a dual thrust chamber and would deliver 300,000 lbf (sea level) for 140 seconds of flight. Its basic features included regeneratively cooled thrust chambers and a gas generator

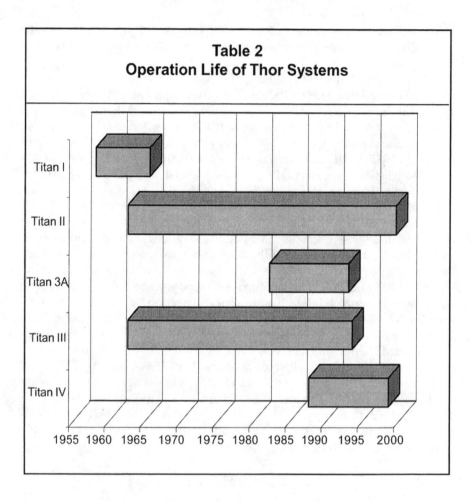

Table 2
Operation Life of Thor Systems

Titan I	
Titan II	
Titan 3A	
Titan III	
Titan IV	

1955 1960 1965 1970 1975 1980 1985 1990 1995 2000

assembly to power the propellant turbopumps. It achieved constant turbine speed, and the resultant stable propellant flow, by metering main engine propellants to the gas generator that powered the propellant turbopumps.

The second-stage engine for the Titan I also came from Aerojet-General. The LR91-AJ-1 engine, as expected of a second-stage engine, could be started at high altitude and offered 80,000 lbf for 155 seconds of operation. The only major difference between these two engines was that the LR91-AJ-1 included an ablative thrust-chamber skirt to ensure proper operation during the larger expansion ratio of the engine at ignition high in the atmosphere.

Engine tests began at Aerojet-General in mid-1956, and the first full-duration firings began in March 1957. Following successful tests by the manufacturer, Martin accepted delivery of the first engine for mating tests in

November 1957. More advanced versions of these two engines, designated the LR87-AJ-3 and the LR91-AJ-3, became the standard engines for the Titan I, and the first of these were delivered for configuration tests in March 1959 and flight tested for the first time on 13 May 1960.[11]

To ensure delivery of the nuclear device on target, the Titan I used a guidance system developed by the Bell Laboratories. With this system, ground-based radar tracked a missile transponder; acquiring trajectory data and making corrections as the missile flew toward the target. The ground controllers changed the flight path by sending pulse-coded radio signals to a three-axis reference system. A similar system was also used by Atlas. This did not work as effectively as intended, and it was vulnerable to enemy jamming and antimissile attack, so in 1959 the Titan program (and Atlas) went to a new inertial guidance system developed by American Bosch and MIT. This system was installed in the Titan ICBMs beginning in the autumn of 1962.[12]

The Glenn L. Martin Company manufactured a total of 163 Titan I vehicles, of which 62 were test articles and 101 were intended as operational ICBMs, although only 54 were actually deployed into silos. The Titan I flight test program was divided into Series I, II, and III. Twelve flights were programmed for each of the first two series, and 45 Series III flights were anticipated to complete the program. Series I flights were designed to test the Titan's first stage and explore the problem of starting the second stage's rocket engine at altitude. On Series II flights, the second stage's guidance system was operated in conjunction with the Titan's control system, and those flights served the additional purpose of testing the missile's nose cone separation mechanism. Series III flights validated the performance of the Titan I production prototype. To save time, Series I and II tests were run concurrently with considerable overlap in the flights.[13]

The first Titan I arrived at Cape Canaveral Air Force Station, Florida, on 19 November 1958, and the first successful Titan I launch took place on 6 February 1959. Successive launches occurred on 25 February, 3 April, and 4 May 1959. Although isolated performance discrepancies occurred during those flights, they still met virtually all of their test objectives. Encouraged by those results, Martin anticipated a swift test phase. Unfortunately, the next two launches failed dramatically. On 14 August 1959, as the Titan I gained thrust prior to its liftoff from Complex 19, its tie-down bolts exploded early, and one of the umbilicals generated a "no-go" signal to the ground-support equipment's flight controls as the missile lifted off the pad prematurely. The "no-go" signal prompted an automatic engine shut-down signal from the flight controls, and the missile lost all thrust. It fell back through the launcher ring and exploded, and the umbilical tower was damaged in the ensuing fire. The next launch, on 12 December 1959, was equally discouraging. The missile's first-stage destruct package ruptured the fuel tank 4 seconds after launch, and the second stage fell back on the pad and exploded. The next test flight did clear the launch complex on 5 February 1960, but the missile exploded 52 seconds after liftoff. Two other missiles experienced second-stage problems during their flights downrange on 8 March and 8 April 1960. On the other hand, one Titan I missile met all of its test objectives on 2 February 1960, and another landed in the Ascension Island range as planned on 28 April. Six consecutive missile flights to Ascension successfully followed between 24 February and the end of June 1960.[14]

After additional launches of the Titan I from Vandenberg Air Force Base, California, in 1960 and 1961, the Titan I ICBM force was deployed in six squadrons beginning in 1962. These were what was called the HGM-25A configuration and were based in hardened silos. These missiles then were raised to the surface after fueling and launched toward the enemy. At least 18 miles separated each of three launch complexes controlled by a hardened launch control center. Each Strategic Missile Squadron, in turn, operated three launch control centers. On alert, each squadron controlled nine missiles from three separate launch control centers (3x3–silo lift). The Titan I did not remain on alert long, for on 24 May 1963, Air Force leaders decided to phase out the missile by June 1965 in favor of follow-on systems.[15]

Second Generation: Titan II

Even as the Titan I program proceeded toward deployment, serving as the proving ground for many design and structural advances in multistage rocket technology, the Air Force began development of the Titan II in 1962. As the Air Force deactivated the Titan Is, 54 Titan II ICBMs would be placed in operational silos at three different locations around the country. Although close to Titan I in configuration, the Titan II made twenty-two configuration changes over the course of the test and evaluation period from that of the earlier ICBM. Of those, three represented significant alterations from the Titan I. Most obviously to the casual observer; the Titan II's second-stage diameter was stretched to 10 feet. Second, the overall missile length grew from 98 to 103.4 feet when mounting the reentry vehicle. Along with this the Martin Company increased skin thickness and beefed up the superstructure. Finally, the Titan II used a different staging sequence than its predecessor. In the Titan I, staging was triggered by the detection of low liquid level in the first-stage tanks. After a few seconds, explosive bolts separated the two stages and solid rockets ignited to push the second stage away from the first along a guide rail. A timer then ignited the second stage. In Titan II this complex system was simplified. Dispensing with explosive bolts, solids, and guide rails, the Titan II second stage ignited while still attached to the first; its exhaust vented outside the rocket until enough thrust had been generated to push the first stage away and to carry the vehicle on into orbit.[16]

In addition to these structural changes, the Titan II used a different propellant combination than its predecessor—A-50 hydrazine (Aerozine) and N204

Table 3
Titan II ICBM Units

Location	Unit	Unit Activated	Config-uration	On Alert	Off Alert
Lowry AFB, CO	724th Strategic Missile Squadron (SMS)	1 February 1960	3x3 silo-lift	18 April 1962	15 February 1965
Lowry AFB, CO	725th SMS	1 August 1960	3x3 silo-lift	10 May 1962	17 February 1965
Ellsworth AFB, SD	850th SMS	1 December 1960	3x3 silo-lift	28 September 1962	4 January 1965
Beale AFB, CA	851st SMS	1 February 1961	3x3 silo-lift	8 September 1962	4 January 1965
Larson AFB, WA	568th SMS	1 April 1961	3x3 silo-lift	26 September 1962	4 January 1965
Mountain Home AFB, ID	569th SMS	1 June 1961	3x3 silo-lift	16 August 1962	17 February 1965

Table 4
Titan II ICBM Specifications*

Titan Elements	Fully Assembled Missile Dimensions	
Stage 1 (including interstage structure, engines)	70.17 feet	
Stage 2	19.54 feet	
Reentry Vehicle Adapter	3.74 feet	
Mark 6 Reentry Vehicle	10.17 feet	
Total	103.4 feet	
Diameter	10 feet	
Fully Loaded Nominal Missile Weight	**Stage 1**	**Stage 2**
Airframe Empty	9,583 pounds	13,479 pounds
Oxidizer	160,637 pounds	37,206 pounds
Fuel	83,232 pounds	20,696 pounds
Ordnance, lubricants	1,43 pounds	743 pounds
Total	327,119 pounds	
Engine Thrust		
Stage 1 LR87-AJ-5 (sea level)	430,000 lbf	
Stage 2 LR91-AJ-5 (vacuum)	100,000 lbf	
Range		
Mark 6 Reentry Vehicle	6,350 miles	
Circular Error Probable	>1.1 miles	

* Source: David K. Stumpf, "Titan II: Sword and Plowshare," *Quest: The History of Spaceflight Quarterly* 7 (Spring 1999): 29–35.

nitrogen tetroxide. These propellants became the standard for all of the liquid-fueled configurations of the Titan family of boosters. The rocket's first-stage engine, the LR87-AJ-5, generated 430,000 lbf at sea level. Its second-stage engine, the LR91-AJ-5, was designed to provide 100,000 lbf at high altitudes. As an ICBM, the Titan II could boost a 4,500-pound nuclear device to a range

Table 5
Major Modifications to the Titan II during Flight Test Program*

Change	ICBM Frame Numbers Modified
1.Changes belly bands	Vehicles 1–9
2.Beefed-up waffle and skins	Vehicles 11–33
3.One-piece conduit, stage 1	Vehicles 1–9
4.Built-up conduit stage 1	Vehicles 11–33, 3A
5.Built-up cone fuel tank, stage 1	Vehicles 1–9, 11–16
6.One-piece cone fuel tank, stage 1	Vehicles 17–33, 3A
7.Oxidizer dome support, stage 1 forward	Vehicles 1–9, 33–33, 3A
8.Weld land area increase	Vehicles 2–9, 11–33, 3A
9.Interstage riveting	Vehicles 14–16
10.R/V adapter, Martin (Mk. 4)	Vehicle 11
11.R/V adapter, GE (Mk. 6)	Vehicles 1–9, 12–33, 3A
12.Translation rockets	Vehicles 11–12, 21–33, 3A
13.Spectoradiometer	Vehicles 24–29, 31, 33
14.Scientific passenger pod	Vehicles 24–25, 29, 31, 33
15.Malfunction detection system	Vehicles 24–25, 29, 31–33
16.40-foot staging cable	Vehicles 25, 29, 31–33
17.Air duct, stage 1 engine component	Vehicles 24–25
18.External camera pod	Vehicles 25, 32–33, 3A
19.Steel feed line (suction)	Vehicles 1–9
20.Aluminum feed line (suction)	Vehicles 11–33
21.Beefed-up transport section	Vehicles 32–33
22.Internal camera pod	Vehicles 29, 32–33, 3A

* Source: David K. Stumpf, "Titan II: Sword and Plowshare," *Quest: The History of Spaceflight Quarterly* 7 (Spring 1999): 29–35.

of 9,750 miles or an 8,000-pound reentry vehicle 6,350 miles. It was also able to lift approximately 4,200 pounds into a low-Earth circular orbit. With the addition of strap-on solid rocket motors (Graphite Epoxy Motors) to the first stage, the payload capability could be increased to 7,800 pounds to low-Earth polar orbit.[17]

The first full-duration firing of the Titan II's first-stage engine took place in March 1961 and throughout that summer other tests progressed through production engine tests and verification. The second-stage engines, however, ran into difficulties with ignition combustion instability. In essence, the problem was that thrust chamber pressure on the second-stage engine cycled through over 200 pounds per square inch at 25,000 cycles per second. This created a

Image 4-1: A Titan II ICBM clears its silo during a test launch from Vandenberg Air Force Base, California, on 16 February 1963. (USAF Photo)

vibration that acted like an ultrasonic saw on the thrust chamber. To solve the problem, Aerojet General engineers installed baffles with an oxidizer injector for thin film cooling to break up the instability long enough for the initiation of smooth combustion.

In addition, the second-stage engine's gas generator failed repeatedly during high-altitude start-ups. Gas generators used fuel and oxidizer to generate high-pressure gas for powering the turbopumps during flight. Solid-propellant cartridges provided the initial high pressure for spinning the turbines until the gas generators could take over. Unfortunately, this system failed to operate as intended and only after months of testing did Aerojet General, Martin, and Air Force engineers come up with a solution. They determined that this problem had to do with the lack of pressure contained in the gas generator at 250,000 feet altitude, and to solve it the engineers fabricated a rupture disc fitted over the roll nozzle at the end of the second-stage gas generator exhaust. This trapped the air pressure at launch in the generator until the solid-propellant start cartridge ignited in the upper atmosphere, and fortunately no more anomalies occurred with this part of the Titan II system.[18]

Formally, the Titan II R&D effort began on 16 March 1962 and ended after thirty-two flights on 9 April 1964.[19] Because these missiles were liquid-fueled, however, the Air Force decided not to deploy as many of them as it did the solid-fueled Minuteman developed during the same period. The storable fuels of the Titan II were still harder to handle than solids, and the ICBM therefore required more time to prepare for launch than the soon to be deployed Minuteman, which could be stored indefinitely already fueled and ready for flight. Accordingly, by the end of the 1960s the Air Force's ballistic missile fleet consisted of one thousand Minuteman and fifty-four Titan IIs on alert. Deactivation of the Titan II ICBM system began in July 1982, and the last missile was taken from its silo at Little Rock Air Force Base, Arkansas, on 23 June 1987. Deactivated missiles were then put in storage at Norton Air Force Base in San Bernardino, California, although most were subsequently moved into long-term storage at Davis-Monthan Air Force Base in Tucson, Arizona.[20]

The Titan II also had an unusual role in the human space-flight realm in the 1960s. To gain experience in spacecraft rendezvous, docking with another vehicle on orbit, and astronaut operations outside the capsule, NASA devised Project Gemini. Hatched in the fall of 1961 by engineers at Robert Gilruth's Space Task Group in cooperation with McDonnell Aircraft Company, Gemini started as a larger Mercury Mark II capsule but soon became a totally different proposition. It could accommodate two astronauts for extended flights of more than two weeks. It pioneered the use of fuel cells instead of batteries to power the spacecraft and incorporated a series of modifications to hardware. Its designers also toyed with the possibility of using a paraglider being developed at Langley Research Center for "dry" landings instead of a "splashdown"

Image 4-2: A Titan II ICBM lifts off from Cape Canaveral during a test flight in 1964. (USAF Photo)

Table 6
Deployment of Titan II

Location	Unit	Unit Activated	Confiruration	On Alert	Off Alert
McConnell AFB, KS	381st Strategic Missile Wing (SMW)	29 October 1961	18 launch complexes	5 July 1963	27 May 1986
Davis-Monthan AFB, AZ	390th SMW	1 January 1962	18 launch complexes	31 March 1963	21 May 1984
Little Rock AFB, AR	308th SMW	1 April 1962	18 launch complexes	16 May 1963	6 May 1987

in water and recovery by the U.S. Navy. But most important for this story, the system was boosted to orbit by the newly developed Titan II launch vehicle.[21]

Two-time Gemini astronaut Pete Conrad remembered the sensations of riding the Titan II into orbit during the Gemini program in a 1998 interview:

> At the time it was chosen for Gemini duty, the Titan II was the most powerful rocket in America's inventory. An improved version of the Air Force's Titan I, it first flew successfully in March 1962. Unlike Project Mercury's Redstones and Atlases, as well as the later Saturn Vs, the Titan used room-temperature propellants called hypergolics, which ignite on contact. At T-minus-0, an electrical signal set things in motion by igniting two small cartridges in the Titan's two first-stage Aerojet engines. The gas from the cartridges started the Titan's turbopumps spinning, which in turn forced both fuel and oxidizer into the engine's combustion chambers. When combined, the hypergolics emitted only a relatively small white flame and a rosy cloud rather than the burst of orange flame and billowing smoke of Mercury and Apollo launches. Four seconds later, if all systems remained go, the bolts that anchored the Titan to the launch pad exploded and the Titan was on its way.

During the Gemini program, the Titan undertook twelve flights—all but two of them with astronauts aboard—without a single failure.[22]

"Man-rating" the Titan II ICBM, however, was not straightforward. George E. Mueller, NASA Associate Administrator for Manned Space Flight, offered a synopsis of the major modifications necessary for the Titan II to make it acceptable for human space flight:

1. Structural modifications were made to the transition section above the second stage for attaching the spacecraft to the launch vehicle.
2. The AB inertial guidance system was removed and the General Electric MOD III-G installed. In addition, a Three Axis Reference System (TARS) was required for flight attitude.
3. By adding a tandem actuator system, a second hydraulic power supply, a second autopilot and redundant electrical power system, the failure probability in the flight control system was lowered by at least two orders of magnitude.
4. Rechargeable space system batteries replaced weapon system batteries in the electrical system.
5. A Malfunction Detection System (MDS) was installed to provide the astronauts with a detection system for noting malfunctions in order that an abort or escape action could be taken before a catastrophe occurs. Signals were provided in the spacecraft indicating pressure in fuel and oxidizer tanks, engine and thrust-chamber pressure, staging signals, excessive attitude rate changes, and range safety officers' actions.

Image 4-3: A Titan II carried *Gemini 12* into orbit from Launch Complex 19 at Cape Canaveral on 11 November 1966. The human-rated version of the Titan was externally indistinguishable from the ICBM. (NASA Photo no. KSC-66-21991)

6. Since the spacecraft has its own maneuverable engines, the vernier and retro-engines were removed.
7. The Titan II engine program was redirected to solve performance reliability and the longitudinal oscillation or "POGO" problem and combustion instability problems. The net result of this effort was an improved Titan II engine system that was "man-rated."[23]

Problems with the Gemini program abounded from the start. The Titan II had longitudinal oscillations, called the "pogo effect" because it resembled the behavior of a child on a pogo stick. Overcoming this problem required engineering imagination and long hours of overtime to stabilize fuel flow and maintain vehicle control. The fuel cells leaked and had to be redesigned. NASA engineers never did get the paraglider to work properly and eventually dropped it from the program in favor of a parachute system like the one used for Mercury. The Atlas-launched Agena target vehicle—intended as a rendezvous target for later Gemini flights—also suffered delays and cost overruns. All of these difficulties shot an estimated $350 million program to over $1 billion. The overruns were successfully justified by the space agency, however, as necessities to meet the Apollo landing commitment.[24]

NASA worked hard on these problems and, by the beginning of 1964, seemed to be on track to fly. The first operational mission took place on 23 March 1965 when astronauts Gus Grissom and John W. Young took the system on a "shakedown" flight. The second mission, flown in June 1965, stayed aloft for four days, and astronaut Edward H. White II performed the first American extravehicular activity (EVA), or spacewalk.[25] Eight more missions followed through November 1966. Despite problems encountered on all of them, the program succeeded in giving the astronauts and engineers of NASA enormous amounts of experience on long-duration missions, rendezvous and docking of two spacecraft, and EVAs. All of these skills were necessary to complete the Apollo Moon landings within the time constraints directed by the president.[26]

Beginning in the immediate post-*Challenger* era the decommissioned Titan II ICBMs that had been in storage for several years found a second use as space launch vehicles after refurbishment. The Martin Marietta Astronautics Group was awarded a contract in January 1986 to refurbish, integrate, and launch fourteen Titan II ICBMs for government space launch requirements. The Titan II, therefore, became a medium-lift space launch vehicle carrying payloads for the Air Force, NASA, and the National Oceanic and Atmospheric Administration (NOAA). Tasks involved in converting the Titan II ICBMs into space launch vehicles (SLVs) included

- modifying the forward structure of the second stage to accommodate a payload;

Table 7
Titan II Space Launch Vehicle Characteristics

Primary function		Launch vehicle used to lift medium-class satellites into space
Builder		Lockheed-Martin Astronautics
Launch site		Vandenberg AFB, CA
Stage 1	Length	70 feet
	Diameter	10 feet
	Engine thrust	474,000 lbf (vacuum)
Stage 2	Length	24 feet
	Diameter	10 feet
	Engine thrust	100,000 lbf (vacuum)
Guidance		Inertial with digital computers
	Subcontractor	Delco Electronics
Payload fairing	Diameter	10 feet
	Length	20 feet
		Skin and stringer construction—Tri-Sector Design
	Subcontractor	Boeing
Liquid rocket		Refurbished Titan II ICBM engines
	Subcontractor	Aerojet Tech Systems
Date deployed		September 1988

- manufacturing a new 10-foot-diameter payload fairing with variable lengths plus payload adapters;
- refurbishing the Titan's liquid rocket engines; upgrading the inertial guidance system; and developing command, destruct, and telemetry systems;
- modifying Space Launch Complex 4 West (SLC-4W) at Vandenberg AFB, California, to conduct the launches; and
- performing payload integration.

The Air Force successfully launched the first Titan II space launch vehicle from Vandenberg AFB on 5 September 1988. Several other Titan II SLVs have been used to launch satellites into orbit. For example, the Ballistic Missile Defense Organization's *Clementine* spacecraft was launched aboard a Titan II in January 1994. Initial reports were that this probe discovered water on the Moon in November 1996. Other payloads have included the USAF Defense Meteorological Satellite Program (DMSP) and NOAA weather satellites. Lockheed Martin Astronautics Corporation launched eight Titan II Space Launch Vehicles between 5 September 1988 and 19 June 1999 with a success record of 100 percent. All of the twelve Titan II SLVs that have been launched altogether have been part of the fourteen former ICBMs refurbished for space launches. The DoD had no plans to "rehab" additional Titan IIs after it completed the launch of the fourteen originally available.[27]

The Titan III Heavy Booster

The third major member of the Titan family of launchers was the Titan III. In 1961 representatives from the DoD and NASA studied the projected future heavy-lift requirements of space flight and found that none of the existing systems satisfied the needs they saw in the coming decade. As a result, on October 1961 the interagency Large Launch Vehicle Planning Group recommended that "the 120-inch diameter solid motor and the Titan III launch vehicle should be developed by the Department of Defense to meet DoD and NASA needs, as appropriate in the payload range of 5,000 to 30,000 pounds, low-Earth orbit equivalent." The principal reasons offered for proceeding with the Titan III included the following:

- The anticipated large number of DoD missions during this decade (the cost per launch was substantially lower than for Saturn-based vehicles)
- The importance to DoD of having a launching system not dependent on the use of cryogenic propellants
- The vehicle's greater flexibility (by virtue of the way the vehicle's building blocks can be combined)
- Easier logistics and training for DoD due to prior experience with the Titan II
- The development of large solid motor technology would be part of the development effort and cost[28]

The beginning efforts to build this system were underway by the end of 1962, with the Titan III intended as a modular vehicle that could satisfy a wide variety of heavy-lift requirements. In essence, it was an attempt to expand the payload lifting capability of a proven system. Following approval of this new vehicle, Martin Marietta began development and fabrication at its Colorado facility. Its engineers beefed up the structure and, most significant, added strap-on solid-propellant boosters to enhance the payload launch capability. At the same time, an additional stage that could maneuver on orbit, called a transtage, was developed to increase the versatility of the booster.

A key component of the later Titan III program was its use of solid rocket motors to increase payload life capacity. In May 1962 the United Technologies Corporation received a contract from the Air Force to produce the solid motor that would provide the initial propulsion for the Titan III launch vehicle.

United Technologies engineers moved out smartly and on 20 July 1963 successfully test fired a five-segment prototype of the motor that boosted a whole range of Titan III configurations. Beginning in 1965 these new launch vehicles successfully put into orbit a number of Department of Defense reconnaissance and communications satellites as well as NASA spacecraft.[29]

While the Air Force enthusiastically supported the development of the Titan III in the early 1960s, NASA initially believed it competed with the Saturn I and IB for heavy-lift payloads and pronounced it redundant. In no small measure, the Air Force's enthusiasm was based on the projected use of the Titan III as the launch vehicle for the Dyna-Soar military astronaut program.[30] Milton Rosen, Director of Launch Vehicles and Propulsion for NASA's Office of Manned Space Flight, described the issue in a memorandum to his boss, Brainerd Holmes, late in 1961:

> The TITAN III and the Saturn C-1 are competitive in orbital performance. The TITAN III, alone, has some escape capability which is enhanced by addition of a fourth stage. The Saturn C-1 has an appreciable escape capability through the addition of a third stage. One major difference is that the TITAN III core has a 10-foot diameter and only with difficulty could carry large diameter payloads. The Saturn C-1, on the other hand, has an 18-foot diameter and could be provided with a third stage of similar diameter, for example, the following combination [S-I—S-IV—S-IV]. Escape payloads presently planned by NASA for Centaur utilize the full 10-foot diameter of that vehicle. Future escape payloads, requiring greater launch vehicle capability, fall in the diameter class of 12 to 18 feet. Launch vehicle requirements for these payloads can be met by the Saturn C-1.[31]

At the same time, space observers calculated that a Titan IIIC would cost $24 million to procure and launch, while each Saturn IB cost an estimated $55 million. Carrying 23,000 pounds to low-Earth orbit, the Titan IIIC delivered its payload at a cost per pound of about $10,000. The Saturn IB cost about $15,000 per pound to deliver its 37,000 pound payload. Accordingly, the Titan III was easy for the DoD to justify on economic grounds. The Saturn I and IB went by the wayside after the Apollo Soyuz Test Project.[32]

Even without economic justification, however, the DoD would have pressed for the Titan III because of two important missions envisioned for it. The first of these, the X-20 Dyna-Soar program, had begun on 11 December 1961 and required the Titan IIIC to launch this military orbital spaceplane.[33] This winged, recoverable spacecraft did not possess as large a payload as NASA's capsule-type spacecraft and was always troubled by the absence of a clearly defined military mission. Accordingly, in September 1962 Defense Secretary Robert S. McNamara questioned whether Dyna-Soar represented the best expenditure of funds. This resulted in numerous studies of the program, but in 1963 McNamara canceled the program in favor of a Manned Orbiting Laboratory

(MOL). This military space station, along with a modified capsule known as Gemini-B, would be launched into orbit aboard a Titan IIIM vehicle that used seven-segment solids and was "human-rated." As an example of the seriousness with which the Air Force pursued the MOL program, which was later canceled, the third Titan IIIC test flight boosted a prototype Gemini-B (previously used as GT-2 in the Gemini test program) and an aerodynamic mockup of the MOL laboratory into orbit. It was as close as MOL would come to reality. The new military space station plan ran into numerous technical and fiscal problems, and in June 1969 the Secretary of Defense, Melvin R. Laird, informed Congress that MOL would be canceled.[34]

The Titan III is generally thought of as a Titan II with two big strap-on solid-propellant boosters, although the distinction between II and III was not the presence of the strap-ons but the fact that III was specifically designed for space missions. There have been a number of different versions of the Titan III. The Titan IIIA, first launched on 1 September 1964, was essentially the same as the Titan II that was used to launch Gemini except it featured a stretched stage 1 and stage 2 core and a Transtage upper stage (small liquid stage) and used an inertial guidance system. The follow-on Titan IIIB had an Agena upper stage instead of the Transtage and used a radio guidance package. The Titan IIIC became the general heavy-lift booster for the Air Force, and was the first to use strap-on solid rocket motors, in this case each made of five propellant segments. The IIIC made its first flight in June 1965. The Titan IIIM was intended to be a "man-rated" version of the Titan IIIC using seven-segment solid rocket motors to launch the MOL military space station; development was canceled before its first flight, although the basic solid rocket motor design would return on the Titan IV. The Titan IIID and IIIE were minor variations of the IIIC, in the first case deleting the upper stage entirely, and in the latter replacing the Transtage with a Centaur. The Titan IIID began flying in 1971, and the IIIE began flying in 1974 under the auspices of NASA and was used to launch a variety of planetary probes, including the Viking missions to Mars and the Voyager missions to the outer planets.[35]

Separately, but related, the Titan 34B was essentially a Titan IIIA (no solids) with a stretched stage 1, while the Titan 34D used 5.5-segment solid rocket motors. The Titan 34D was designed to provide a launch capability that filled the gap between existing expendable launch vehicles (ELVs) and the Space Shuttle, and it began flying in 1982.[36] Various upper stages, notably Transtage and Centaur, also found use. The Titan 34D in its East Coast version replaced the Titan IIIC and could be fitted with the Inertial Upper Stage (IUS) as well as the Transtage, although this was apparently only done once, in 1982 for a military satellite launch. A Titan 34D-7 was also designed, using the 7-segment solids from the IIIM, but instead evolved into the Titan IV.[37]

There were numerous other missions for a heavy-lift vehicle, however, and Titan IIIs sent eighty-two military and NASA satellites into orbit between

Image 4-4: This test launch of the Manned Orbiting Laboratory (MOL) program, a military space station, took place on 3 November 1966 from Cape Canaveral Air Force Station's Launch Complex 40. The program was eventually canceled before an operational vehicle was launched. (USAF Photo)

1965 and its phaseout in 1982. Titan IIIEs boosted such demanding missions as Viking, which landed on Mars, and Voyager, which flew past Jupiter, Saturn, Uranus, and Neptune.[38]

Indeed, the Titan III became a mainstay for heavy DoD payloads. In part this was because of the stubbornness of DoD officials, who resented NASA's bid to make the Space Shuttle the sole heavy-lift vehicle in the U.S. fleet of launchers. In the late 1970s and early 1980s representatives of these two governmental entities slugged it out over what they called "assured access to space" and how best to achieve it. What that meant, in reality, was that the DoD wanted to have capability to reach low-Earth orbit with critical national security spacecraft, especially reconnaissance satellites, at any time they chose. NASA leaders insisted that the Space Shuttle could achieve this reality, but DoD officials were skeptical and held out for a mixed fleet that included both the Space Shuttle and several types of expendable launch vehicles—including the Titan family of launchers. NASA would have preferred that the Titan assembly line be shut down and that the DoD rely on the Space Shuttle to launch all of its payloads. The DoD, in part because of its fear of losing turf to NASA and in part for very sound technical reasons, refused to place all of its launch requirements on the shuttle.[39]

Accordingly, well before the *Challenger* accident of January 1986, top Air Force and National Reconnaissance Office (NRO) officials, led by Secretary of the Air Force Edward C. Aldridge, had become concerned about a policy of total dependence on the Space Shuttle for launching intelligence satellites and other critical national security satellites. After a bitter fight with NASA, in 1985 Aldridge succeeded in convincing the White House and Congress to approve procurement of a limited number of what were called Complementary Expendable Launch Vehicles (CELV) as backups to the shuttle. Martin Marietta in 1985 won the competition to provide this capability with a more powerful variant of its Titan 34D launcher—the Titan IV. This contract also allowed Martin Marietta to keep the Titan production line open.

When the post-*Challenger* opportunity to enter the commercial launch market appeared, those most closely involved with Titan were able to convince the top executives of Martin Marietta to develop a commercial variant of the Titan III booster called Commercial Titan. This was a more powerful—and more expensive—launcher than either Delta or Atlas, and could be commercially viable only if it could schedule two payloads on the same launch, an approach pioneered by the Ariane launcher of Europe. However, Martin Marietta was not able to find many customers willing to fly at the same time, and in June 1989 announced that it would no longer attempt to market a single launch to two customers. This had the effect of removing the Commercial Titan from the marketplace, since the price of a launch was $130–150 million, too high for a single payload. Only four Commercial Titans were launched, all between 1990 and 1992. The commercialization of the Titan family had thus failed. The system might have been phased out of the U.S. fleet of launchers at that point except that government organizations, the only users with both the exorbitant funds required for launch and with heavy enough payloads, stepped forward to purchase several Titan IV launches.

Not only DoD contracts but also NASA procurements kept alive the Titan line of launch vehicles. Because of the *Challenger* accident, NASA could no longer rely solely on the Space Shuttle fleet for all of its launches. Therefore, on 14 May 1987 NASA announced its intent to procure from U.S. industry launch services using expendable launch vehicles. NASA Administrator James C. Fletcher said, "NASA's purpose in seeking expendable launch services is to lessen dependence on a single launch system, the Space Shuttle. Expendable launch vehicles will help assure access to space, add flexibility to the space program, and free the shuttle for manned scientific, shuttle-unique, and important national security missions." NASA indicated its future intent to purchase the equivalent of up to five Delta launches per year and two Atlas-Centaur or Titan III launches per year. Of these first ELV launches from NASA, four were Titan III orders and all of the payloads had been previously booked on the Space Shuttle. Since the shuttle fleet was grounded, however, ELVs represented NASA's only possible method of placing them in orbit.[40]

The Titan IV Heavy-Lift Launcher

When first conceived, the Titan IV was the largest space booster envisioned ever to be required by the Air Force. The vehicle was designed to carry payloads equivalent to the size and weight of those carried on the Space Shuttle, and it would rely principally on launches of reconnaissance and Milstar satellites to sustain its operations since there were no commercial payloads of those large sizes. The Titan IV consisted of two seven-segment solid-propellant motors, a liquid-propellant two-stage core, and a 16.7-foot-diameter payload faring. The system included a cryogenic wide-body Centaur upper stage, and also could be flown with an IUS or no upper stage at all. The overall length of the system was 204 feet when flown with an 86-foot payload faring.

The Titan IV's core consisted of an LR87 liquid-propellant rocket that featured structurally independent tanks for its fuel (Aerozine 50) and oxidizer (nitrogen tetroxide). This would minimize the hazard of the two mixing if a leak should develop in either tank. Additionally, the engines' propellant could be stored in a launch-ready state for extended periods. The use of propellants stored at normal temperature and pressure was intended to eliminate delays and give the Titan IV the capability to meet the strict launch windows required for military satellites. The second stage consisted of an LR91 liquid-propellant engine attached to an airframe, like stage 1.

The Titan IV–Centaur combination was capable of placing 10,000-pound payloads into geosynchronous orbit 22,300 miles above the Earth. The sys-

Table 8
Titan IVB Launch Vehicle Characteristics

Launch Weight		Approximately 1.9 million pounds
Stage 0 (SRMU)	Length	12.4 feet
	Diameter	10.5 feet
	Propellants	88% hydroxyl terminated polybutadiene
	Motor thrust	1,700,000 lbf per motor
	Weight	770,673 pounds
	Contractor	Alliant Techsystems
Stage 1	Length	86.5 feet
	Diameter	10 feet
	Propellants	hypergolic liquid—Aerozine-50 (hydrazine and unsymmetrical dimethy-hydrazine) fuel and nitrogen tetroxide oxidizer
	Motor thrust	551,200 lbf (full-duration average)
	Engine	Aerojet LR87-AJ-11
	Contractor	Lockheed Martin Astronautics
Stage 2	Length	32.7 feet (bottom of engine nozzle to top of forward skirt)
	Diameter	10 feet
	Propellants	hypergolic liquid—Aerozine 50 and nitrogen tetroxide
	Motor thrust	106,150 lbf (full duration average)
	Engine	Aerojet LR91-AJ-11
	Contractor	Lockheed Martin Astronautics
Centaur upper stage	Length	33 feet
	Diameter	10 feet
	Motor thrust	20,500 lbf (22,000 lbf for RL10A-4-1)
	Engine	Pratt & Whitney RL10A4 or RL10A-4-1
	Contractor	Lockheed Martin Astronautics
Payload fairing	Length	56–86 feet
	Diameter	200 inches aluminum isogrid construction, trisector design
	Contractor	The Boeing Company
Cost	$433.1 million	Titan IVB with Centaur upper stage
	$432 million	Titan IVB with Inertial Upper Stage

tem could also place 39,100 pounds into a low-Earth orbit at 28.6 degrees inclination or 31,000 pounds into a low-Earth polar orbit.

A Solid Rocket Motor Upgrade (SRMU) developed by Alliant Aerospace Company and Lockheed Martin Astronautics increased the launch capability of the new Titan IVB beginning in the early 1990s. Designed to take advantage of proven, off-the-shelf technologies, the SRMU system provided a 25 percent increase in performance and heavier lift capability than the boosters of its predecessor. The SRMU was a three-segment, 10.5-foot-diameter solid rocket motor that was approximately 112 feet tall and weighed over 770,000 pounds. With the SRMU, the Titan IVB Centaur was capable of placing 12,700-pound payloads into geosynchronous orbit and 47,800 pounds into a low-Earth orbit.[41]

In the sixteen-month period between June 1992 and September 1993, Alliant completed development and ground tested five full-scale SRMUs. The first SRMU-equipped Titan IVB was launched in February 1997. A second flight in 1997 launched the *Cassini* spacecraft and the *Huygens* probe on an international mission to study Saturn's moons, rings, and environment for four years.

The Air Force also contracted with Lockheed Martin to produce forty-one Titan IVs by the end of fiscal year 1999. That contract, valued at more than $12 billion, was originally awarded to the former Martin Marietta Launch Systems Company in February 1985. Titan IVs have only been launched only from Launch Complexes 40 and 41 at Cape Canaveral Air Station, Florida, and from Space Launch Complex 4E at Vandenberg AFB. The first Titan IV was launched on 14 June 1989 from Launch Complex 41 at Cape Canaveral Air Station. Ten years later that same complex saw its final liftoff (an Air Force Titan IVB) on 9 April 1999—the complex was being converted to launch Atlas V vehicles as part of the Evolved Expendable Launch Vehicle (EELV) program.

The Titan IV was not successful as a commercial launch vehicle, however, for the European Ariane booster containing similar technology provided lower-cost access to space than could be achieved by the Titan IV. Eventually, the evolutionary nature of modern technology rendered this Titan system obsolete, and the Air Force contracted with Lockheed Martin to develop a new system known as the Atlas V as part of the EELV program. In all, twenty-five Titan IV rockets have been launched since the first one flew in 1989. Of those, twenty-two were the Titan IVA series space launch vehicle and three were the newer, more powerful Titan IVB configuration. The Titan IVB is to fly fourteen additional missions over the next several years and then cease operations. The new EELVs, consisting of Lockheed Martin Atlas Vs and Boeing Delta IVs, are scheduled to become operational in 2003.[42]

Titan Family Mission Performance

As a family of launch vehicles, the Titan has proven a reliable method of placing spacecraft into orbit. There has been a relatively significant transformation in launch reliability from the beginning of the space age until the present, and the Titan family has mirrored that change. From the dawn of the space age, U.S. launchers have achieved a reliability rate of 85 percent. But an excellent success rate of all launchers of greater than 90 percent reliability has

Table 9	
Titan IVB Specifications	
Primary function	Launch vehicle used to lift heavy satellites into space
Builder	Lockheed Martin Astronautics
Power Plant	Stage 0 consists of two solid-rocket motors; Stage 1 uses an LR87 liquid-propellant rocket engine; and Stage 2 uses the LR91 liquid-propellant engine. Optional upper stages include the Centaur and Inertial Upper Stage.
Guidance	Delco Electronics ring laser gyro Guidance system manufactured by Honeywell.
Cost	Approximately $500 million, depending on launch configuration
Launch sites	Cape Canaveral AFS, FL, and Vandenberg AFB, CA
Thrust	Solid-rocket motors provide 1,700,000 lbf per motor at liftoff. First-stage provides an average of 548,000 lbf and second stage provides an average of 105,000 lbf. Optional Centaur upper stage provides 33,100 lbf and IUS provides up to 41,500 lbf.
Length	Up to 204 feet
Lift capability	Can currently carry up to 47,800 pounds into a low-Earth orbit, up to 12,700 pounds into a geosynchronous orbit when launched from Cape Canaveral; and up to 38,800 pounds into a low-Earth polar orbit when launched from Vandenberg. Using an inertial upper stage, the Titan IVB can transport up to 5,250 pounds into geosynchronous orbit.
Maximum liftoff weight	Approximately 2.2 million pounds
Date deployed	June 1989

been achieved in the last twenty years. It is probably most beneficial to focus on the causes of these failures with the intention of seeking to predict probabilities. In U.S. launcher failure histories fully two-thirds of the catastrophes in liquid-propulsion launchers resulted from subsystem failures other than engines. Taking those subsystems into consideration in design and testing are critical to future success, as is the ability to remove and adequately test subsystems absent in the rest of the launcher.[43]

The Titan family has achieved exactly the same reliability of U.S. launchers as a whole, 85 percent, for the whole of the operational existence of the booster. Without question, launch reliability has also improved over time. The

Titan I achieved a launch reliability of only 64 percent, and the Titan 34 had a reliability of 75 percent. These were unacceptable for almost everyone. On the other hand, the Titan IIs achieved an 87 percent, the Titan IIIs 91 percent, and the Titan IVs an 87 percent launch reliability rate. The failures of each Titan system are recounted in the table on pages 176-177.

The End of the Titan Launcher

Even though the Titan IIIs and IVs were not competitive for commercial launches from the beginning of their careers as commercial launch vehicles, the U.S. government used them extensively for large payload deliveries until the mid-1990s. For example, in the fiscal year 1995 Department of Defense appropriation, Congress included language that "terminates the Titan IV program after completion of the current contract," which delivered forty-one launchers to the DoD and provided for a total of forty launches (with one spare) through 2003. These forty-one Titan IVs were originally to be procured for $12 billion, but the cost increased by more than $10 billion over the life of the procurement. This legislation called the Titan IV "excessively expensive" and foreswore the continuation of this launcher on grounds of cost. It did note that the "U.S. government, primarily the Air Force, has 125 medium- and heavy-lift launch vehicles currently under contract as follows:

- 61 Delta II medium lift vehicles for various Air Force and NASA satellites;
- 9 Atlas II medium lift vehicles for the Defense Satellite Communications system;
- 14 Titan II medium lift vehicles for the Defense Meteorological Satellite Program; and
- 41 Titan IV heavy lift vehicles for the Defense Support Program satellite, MILSTAR, and classified payloads."

To date, the legislative record added, 50 of the 125 space vehicles under contract had been launched.[44] Congress carried this analysis even further in its appropriation language: "It is the Committee's belief that the expenditure of $1 billion to $2 billion could allow termination of the Titan IV heavy-lift vehicle, make current medium-lift vehicles more cost effective, and save billions of dollars over the next twenty years. For example, after the turn of the Century, if the average recurring launch cost of a new vehicle were $300 million, the development costs would be repaid after only two launches since the comparable Titan IV average launch costs are projected to approach $1 billion." Congress then went further, directing the DoD to phase out the Titan IV after 2003 and to begin immediately closing down one of the three Titan IV launch complexes.[45]

Table 10
Titan Family Launch Failures[*]

Titan I	
15 May 1959	Titan I B-4 exploded during static testing
3 July 1959	Titan I B-3 exploded during static testing
14 August 1959	Titan I B-5 exploded on the pad
8 March 1960	Titan I second-stage failure
8 April 1960	Titan I second-stage failure
1 July 1960	Titan I J destroyed 295 feet above pad
28 July 1960,	Titan I J destroyed 210-miles range
3 December 1960	Titan I J elevator collapsed, leading to explosion
Titan II	
27 November 1962	Titan II failure
19 September 1980	Titan II exploded in silo
5 October 1993	Landsat 6 Star-37XFP-ISS kick-motor malfunction
Titan 34	
24 April 1981	Titan 34B jumpseat 6 partial failure
28 August 1985	Titan 34D KH-11 no. 7 Stage 1 propellant feed system failure forced premature engine shutdown
18 April 1986	Titan 34D KH-9 no. 20 solid booster exploded 16 seconds after lift-off just over pad due to deteriorationof solid motor thermal insulation
2 September 1988	Titan 34D USA 31 broken transtage pressurization feed lines prevented the geosynchronous orbit apogee burn from taking place
14 March 1990	Intelsat 6 F-3 second stage failed to separate due to a wiring error in the stage separation electronics, stranding the payload in low Earth orbit

[*] Mark Wade, "The Wrong Stuff—A Catalogue of Launch Vehicle Failures," *Encyclo-*

John Egan, chief executive of Egan International, was a vocal critic of the Titan IV and made this point about the vehicle: "The Titan 4 is my favorite launch vehicle. It's the only one that violates every rule of economics. The last one coming off the assembly line is more expensive than the one before. . . . It's not Lockheed Martin's fault. The intelligence community likes to build their payloads on time of the rocket."[46] The high personnel and monetary

Table 10 Titan Family Launch Failures*	
Titan IIIA	
1 September 1964	Titan IIIA transtage 1 failure
Titan IIIB	
26 April 1967	OPS 4243 stage 2 engine lost thrust
16 February 1972	jumpseat 2 failure
20 May 1972	OPS 6574 failure
26 June 1973	OPS 4018 failure
Titan IIIC	
15 October 1965	OV2-01 partial failure
26 August 1966	IDCSP (8) ... IDCSP (14) failure
6 November 1970	IMEWS 1 partial failure
20 May 1975	DSCS II-05 partial failure
25 March 1978	DSCS II-09 destroyed by range safety
Titan IIIE	
11 February 1974	sphinx Failure
Titan IVA	
2 August 1993	NOSS 19 exploded 101 seconds after lift-off
12 August 1998	Mercury ELINT booster pitched over 40 seconds after launch due to guidance system loss of heading after power interrupt and was destroyed by range safety
Titan IVB	
9 April 1999	DSP F19 IUS first and second stages failed to separate
30 April 1999	Milstar 2 F1 Centaur software programming error

pedia Astronautica, 9 April 2000.

costs of procuring and launching a Titan IV, compared to other space launchers, are noted in table 11.

While the phaseout began in 1995, the impetus to end Titan IV operations accelerated suddenly following three successive failures of the Titan IV between August 1998 and April 1999. In its first twenty launches, Titan IV scored nineteen successes. The twenty-first flight, the only one that was not for the

Table 11 Launch Vehicle Comparisons[*]			
Comparison of Launch Costs by Vehicle[47]			
Launch Vehicle	Pounds to Low Earth Orbit	Cost per Launch (Fiscal Year 1993 dollars)	Cost per Pound Fiscal Year 1993 dollars
Titan II	2,000–4,000	40–45 million	10,000–22,500
Delta II	5,000–11,000	45–50 million	4,090–10,000
Atlas II	12,000–18,000	60–70 million	3,333–5,833
Titan IV	30,000–50,000	500-540 million	10,000–12,000

Comparison of Personnel Required for Launch Operations by Vehicle		
Space Launch Vehicle	Size of Launch Crew	Days on Launch Pad
Ariane IV	About 100	10
Delta II	300	23
Atlas-Centaur	300	55
Titan IV	More than 1,000	100

Comparison of Launch Base Range Operations	
Spaceport	Personnel Required for Each Launch
Kourou Space Center	About 900
Cape Canaveral AFS (excluding NASA)	11,000
NASA Kennedy Space Center	18,000

[*] Source: John T. Correll, "Fogbound in Space," *Air Force Magazine*, January 1994.

Air Force, launched the NASA/European Space Agency *Cassini* spacecraft toward Saturn in 1997. Three more Titan IVs were launched in 1997 and 1998 before a failed launch attempt on 12 August 1998. On that morning, a Titan IVA rocket launched from Cape Canaveral went out of control and exploded. After a thorough investigation, the Titan IV resumed flight in April 1999. However, its next two missions were also failures when the upper stages on the Titan failed to boost their payloads from low-Earth orbit to the target altitudes. These three failures caused the complete loss of three military satellites costing more than $3 billion to build. "When we have three failures in a row of

Image 4-5: The seven-year journey to Saturn for the *Cassini* space probe began with the lift-off of a Titan IVB–Centaur carrying the *Cassini* orbiter and its attached *Huygens* probe on 15 October 1997. In the foreground are the NASA vessels used to retrieve the Space Shuttle Solid Rocket Boosters. (NASA Photo no. KSC-97PC-1543).

any system...something is not right," said Brig. Gen. Randy Starbuck, commander of Cape Canaveral Air Station,[48] Florida, after the third failure on 30 April 1999.[49]

The first failure took place on 12 August 1998 and resulted in the loss of a highly classified NRO satellite. The accident investigation board found evidence that wire insulation damage existed in a wiring harness, resulting in at least one powered conductor with exposed wire that went undetected during prelaunch inspections and tests. After liftoff, the exposed wire intermittently shorted as vehicle vibration increased. These caused electrical interruptions that led to the guidance computer issuing a succession of pitch down and yaw right commands. At this point, according to the report, "the vehicle's northernmost Solid Rocket Motor (SRM #1) separated from the core vehicle, leading to vehicle breakup. Once this occurred, the vehicle's self-destruct system activated in order to prevent uncontrolled flight." The cost of this Titan IVA-20 mishap was more than $1 billion. Those costs included the launch vehicle, satellite, and range support.[50]

A second failure of a Titan IV on 9 April 1999 carrying a Defense Support Program (DSP) spacecraft from Space Launch Complex 41 at Cape Canaveral Air Station occurred when the launcher's IUS first and second stages failed to separate. These DSP satellites have been a key part of North America's early warning systems since the early 1970s. In their 22,300-mile geosynchronous orbits, they provide early warning to the United States and its allies by detecting missile launches, space launches, and nuclear detonations. The satellite and launcher were worth more than $1 billion, and it looked more and more as if the Titan IV was a star-crossed system.[51] Then, the Titan IV had another spectacular failure less than a month later, when a $1.23 billion Milstar satellite never made it to its intended 22,300-mile-high orbit on 30 April 1999. Instead, it circled Earth in a lonely, lopsided orbit that reached no more than 3,100 miles high. In every case, the Titan IVs appeared to do their job during the first several minutes of the launch, but the upper stages needed to propel the satellite higher did not fire properly.[52]

The next Titan IV launch on 22 May 1999 performed as intended and placed in orbit an unusual NRO payload from Vandenberg AFB. The secret payload was probably a set of three electronic intelligence spacecraft that can collectively provide data with short time variations.[53] Then the Titan IV turned in another good performance by launching into geosynchronous orbit a $682 million DoD satellite on 8 May 2000. It demonstrated that reforms in management and quality control across multiple areas for the Titan system had been successful. After reassessments of project management and a shakeup at Lockheed Martin and in the Air Force, it appeared that the program was back on track. But the Titan IV program had only ten more missions scheduled as of mid-2000, all of them involving large national security payloads before launch operations transitioned to the new Delta IV and Atlas V Evolved Expendable Launch Vehicles in 2003.[54]

To punctuate the demise of the Titan as a launch system, on 14 October 1999 at exactly 10:05 A.M. Eastern Daylight Time, Dykon, Inc. of Tulsa, Oklahoma, demolished the 7-million pound Umbilical and Mobile Service Towers at Space Launch Complex 41 (SLC-41) at the Cape Canaveral Air Station. This launch complex, first built in 1965, had been used to launch Titan IIICs then refurbished to handle Titan IVs when they came on line in 1989. As one of three Titan IV launch complexes—one more was located at Cape Canaveral and another at Vandenberg AFB—SLC-41 was phased out of flight status. In its place, the Air Force contracted with Lockheed Martin to design and build both the Atlas V EELV and a new launch platform at the site of the old launch complex. The first phase of this switchover called for removing the old towers so that new launch towers could be constructed.[55]

Conclusion

For some forty years the Titan family of launchers was a mainstay of the American heavy-lift fleet. The United States currently has available the Titan II SLV and Titan IV series for launch operations, and these continue to provide significant heavy-lift capability for a variety of military and scientific needs. Now, however, in the earliest years of the twenty-first century, it is obvious that the Titan launch vehicle is going to be phased out of operations. Lockheed Martin continues to own the assembly line for this system, to provide overall program management, system integration, and payload integration for the program, and to build the first and second stages and the Centaur upper stage of the vehicle. Lockheed Martin, however, also owns the Atlas series of launchers, and the Atlas V is a core part of the U.S. government's plans for the development of the EELV concept.

At a basic level, the story of the Titan family of launchers demonstrates the close relationship between the federal government and the aerospace industry, fueled by the difficulty of the rivalry between the United States and the Soviet Union in the forty years since the end of World War II. Without that cold war, the Titans would never have been built as ICBMs and would never have found a space launch mission. The DoD's commitment to the development of huge reconnaissance and other military applications satellites required the continuation, and even upgrading of the Titan family, but they never had a commercial market. They were simply too big and expensive. It is clear that there is little room in future space payload projections for the heavy-lift capability provided by the Titans. Those that remain may well soon, as some already have, become museum pieces, relics of the cold war that spawned them and lessons for future generations as they pursue future space-lift capabilities.

Notes

1. See David K. Stumpf, *Titan II: A History of a Cold War Missile Program* (Fayetteville: University of Arkansas Press, 2000).

2. Robert L. Perry, "The Atlas, Thor, Titan, and Minuteman," in Eugene M. Emme, ed., *The History of Rocket Technology: Essays on Research, Development, and Utility* (Detroit: Wayne State University Press, 1964), pp. 143–55; John L. Sloop, *Liquid Hydrogen as a Propulsion Fuel, 1945–1959,*

3. Vannevar Bush, *Modern Arms and Free Men* (New York: Simon and Schuster, 1949), pp. 84–86.

4. Quoted in Robert L. Perry, *Origins of the USAF Space Program, 1945–1956,* Space Systems Division Supplement, vol. 5 (Andrews Air Force Base, Washington, D.C.: Space Systems Division, 1961), chap. 4. The background of the ballistic missile decision of 1954 is perhaps the best-documented event in Air Force history. The "most official" version is summarized in *Congressional Record* (Appendix), 2 September 1960, Rep. L.C. Arends, pp. A6642–A6645.

5. This story is told in Edmund Beard, *Developing the ICBM: A Study in Bureaucratic Politics* (New York: Columbia University Press, 1976); Jacob Neufeld, *The Development of Ballistic Missiles in the United States Air Force, 1945–1960* (Washington, D.C.: Office of Air Force History, 1990).

6. W.E. Greene, *The Development of the SM-68 Titan,* AFSC Historical Publications Series 62-23-1 (Wright-Patterson Air Force Base, Ohio: AFSC, 1962), p. 11–12; Bruno W. Augenstein, *A Revised Development Program for Ballistic Missiles of Intercontinental Range,* RAND Corp. Special Memorandum 21 (Santa Monica, Calif.: RAND, 1954), pp. 1–38; Neufeld, *The Development of Ballistic Missiles,* pp. 114–16, 120, 129; Stumpf, *Titan II,* pp. 14–15.

7. Quoted in Greene, *Development of the SM-68 Titan,* pp. 11–12.

8. Quoted in "Semi-Staff Study on Titan ICBM, 30 July 1960," n.p., Air Force Materiel Command Historical Archives, Wright-Patterson Air Force Base, Ohio.

9. The pioneer aviator had founded the Glenn L. Martin Company. It merged with Marietta in 1961 to become the Martin Marietta Corporation, and then with Lockheed Aircraft Company in 1994 to be named Lockheed Martin.

10. John T. Greenwood, "The Air Force Ballistic Missile and Space Program (1954–1974)," *Aerospace Historian* 21 (Winter/December 1974): 194; Stumpf, *Titan II,* pp. 15–16.

11. H.T. Simmons, "Martin's Titan Project," *Missiles and Rockets,* October 1956, p. 20; and William B. Harwood, *Raise Heaven and Earth* (New York: Simon and Schuster, 1993), pp. 300–306; Greene, *Development of the SM-68 Titan,* p. 106; M.J. Chuclick, "History of the Titan Liquid Rocket Engines," in Stephen E. Doyle, ed., *History of Liquid Rocket Engine Development in the United States, 1955–1980,* AAS History Series (San Diego: Univelt, 1992); L.C. Meland and F.C. Thompson, "History of Titan Liquid Rocket Engines," p. 2, AIAA-89-2389, paper presented at AIAA/ASME/ASEE 25th Joint Propulsion Conference, Monterey, Calif., 10–12 July 1989; Stumpf, *Titan II,* pp. 16–18.

12. Greene, *Development of the SM-68 Titan,* p. 114. An outstanding discussion of the intricacies of missile guidance and control may be found in Donald MacKenzie, *Inventing Accuracy: A Historical Sociology of Nuclear Missile Guidance* (Cambridge, Mass.: MIT Press, 1990).

13. "History of the Air Force Missile Test Center," 1 January–30 June 1957, pp. 205–6, Air Force Historical Research Agency, Maxwell Air Force Base, Ala.

14. "History of the Air Force Missile Test Center," 1 July–31 December 1959, pp. 37, 174, 175, 265; Marvin R. Whipple, "Index of Missile Launchings by Missile Program, July 1950–June 1960," 15 December 1960, pp. 12-2, 12-3; "History of the Air Force Missile Test Center," "6555th Test Wing (Development) History," 21 December 1959–31 March 1960, "Organization" and "Mission," Air Force Historical Research Agency.

15. Stumpf, *Titan II*, p. 29.

16. David K. Stumpf, "Titan II: Sword and Plowshare," *Quest: The History of Spaceflight Quarterly* 7 (Spring 1999): 29–35.

17. Joseph J. Mihal Jr., "A Survey of Commercially Available Expendable Heavy Lift Launch Vehicles," M.S. thesis, Naval Postgraduate School, 1994, p. 5; Steven J. Isakowitz, Joseph P. Hopkins, and Joshua Hopkins, *International Reference Guide to Space Launch Systems*, 3d ed. (Reston, Va.: American Institute for Aeronautics and Astronautics, 1999), p. 264.

18. Stumpf, "Titan II: Sword and Plowshare," pp. 31–33; Barton C. Hacker and James M. Grimwood, *On the Shoulders of Titans: A History of Project Gemini*, NASA SP-4203 (Washington, D.C.: NASA, 1977), pp. 140, 168.

19. David F. Stumpf, "Titan II: Research and Development, Part II," *Quest: The History of Spaceflight Quarterly* 7 (Summer 1999): 40–45.

20. Neufeld, *The Development of Ballistic Missiles*, p. 239; "Titan II—Facts," 2000; Air Force Space Command, Public Affairs, "Titan II," March 1999.

21. James M. Grimwood and Ivan D. Ertal, "Project Gemini," *Southwestern Historical Quarterly* 81 (January 1968): 393–418; James M. Grimwood, Barton C. Hacker, and Peter J. Vorzimmer, *Project Gemini Technology and Operations*, NASA SP-4002 (Washington, D.C.: NASA, 1969); Robert N. Lindley, "Discussing Gemini: A `Flight' Interview with Robert Lindley of McDonnell," *Flight International*, 24 March 1966, pp. 488–89; Barton C. Hacker, "The Idea of Rendezvous: From Space Station to Orbital Operations, in Space-Travel Thought, 1895–1951," *Technology and Culture* 15 (July 1974): 373–88; Barton C. Hacker, "The Genesis of Project Apollo: The Idea of Rendezvous, 1929–1961," *Actes 10: Historic des techniques* (Paris: Congress of the History of Science, 1971), pp. 41–46; Hacker and Grimwood, *On the Shoulders of Titans*, pp. 1–26.

22. D.C. Agle, "Riding the Titan II," *Air & Space/Smithsonian*, August/September, 1998.

23. George E. Mueller, "Summary Learning from the Use of Atlas and Titan for Manned Flight," 21 December 1965, NASA Historical Reference Collection, NASA History Office, NASA Headquarters, Washington, D.C. (hereafter cited as NASA Historical Reference Collection).

24. Grimwood and Ertal, "Project Gemini," pp. 393–418; Grimwood, Hacker, and Vorzimmer, *Project Gemini Technology and Operations;* Lindley, "Discussing Gemini," pp. 488–89.

25. Reginald M. Machell, ed., *Summary of Gemini Extravehicular Activity*, NASA SP-149 (Washington, D.C.: NASA, 1968).

26. *Gemini Summary Conference*, NASA SP-138 (Washington, D.C.: NASA, 1967); Ezell, *NASA Historical Data Book* 2:149–70.

27. "Refurbished ICBM Launches QuikScat," *SpaceDaily*, 19 June 1999, NASA Historical Reference Collection; Lockheed Martin Astronautics, "Titan II—Facts," 2000; Air Force Space Command, Public Affairs, "Titan II," March 1999; David K. Stumpf, "Titan II: Space Launch Vehicle Program, Part III," *Quest: The History of Spaceflight Quarterly* 7 (Fall 1999): 44–47.

28. "Summary Report: NASA-DoD Large Launch Vehicle Planning Group," 24 September 1962, NASA Historical Reference Collection.

29. G.R. Richards and Joel W. Powell, "Titan 3 and 4 Launch Vehicles," *Journal of the British Interplanetary Society* 46 (1993): passim; J.D. Hunley, "The Evolution of Large Solid Propellant Rocketry in the United States," *Quest: The History of Spaceflight Quarterly* 6 (Spring 1998): 24.

30. McGeorge Bundy, Memorandum for the President, "Your 11 a.m. appointment with Jim Webb," 18 September 1963; Robert S. McNamara, Memorandum for the Vice President, "National Space Program," 3 May 1963, both in NASA Historical Reference Collection.

31. Milton Rosen, Director, Launch Vehicles and Propulsion, Office of Manned Space Flight, Memorandum to Brainerd Holmes, Director of Manned Space Flight, "Recommendations for NASA Manned Space Flight Vehicle Program," 20 November 1961, NASA Historical Reference Collection.

32. NASA, "Space Shuttle Economics Simplified," House Committee on Science and Technology, Subcommittee on Space Science and Applications, *Operational Cost Estimates: Space Shuttle*, 94th Cong., 2d sess., 1976.

33. As the weight and complexity of Dyna-Soar grew, it quickly surpassed the capabilities of the Titan II and was switched to the Titan III. Just before the program was canceled it looked like weight growth had outclassed even the Titan IIIC, and plans were being made to use Saturn IBs or other boosters.

34. Roy F. Houchin III, "Air Force–Office of the Secretary of Defense Rivalry: The Pressure of Political Affairs in the Dyna-Soar (X-20) Program, 1957–1963," *Journal of the British Interplanetary Society* 50 (May 1997): 162–68; Matt Bacon, "The Dynasoar Extinction," *Space* 9 (May 1993): 18–21; Roy F. Houchin III, "Why the Air Force Proposed the Dyna-Soar X-20 Program," *Quest: The Magazine of Spaceflight* 3 (Winter 1994): 5–11; Terry Smith, "The Dyna-Soar X-20: A Historical Overview," *Quest: The Magazine of Spaceflight* 3 (Winter 1994): 13–18; Roy F. Houchin III, "Interagency Rivalry: NASA, the Air Force, and MOL," *Quest: The Magazine of Spaceflight* 4 (Winter 1995): 40–45; Donald Pealer, "Manned Orbiting Laboratory (MOL), Part 1," *Quest: The Magazine of Spaceflight* 4 (Fall 1995): 4–17; Donald Pealer, "Manned Orbiting Laboratory (MOL), Part 2," *Quest: The Magazine of Spaceflight* 4 (Winter 1995): 28–37; Donald Pealer, "Manned Orbiting Laboratory (MOL), Part 3," *Quest: The Magazine of Spaceflight* 5, no. 2 (1996): 16–23.

35. Frank H. Winter, *Rockets into Space* (Cambridge: Harvard University Press, 1990), p. 92; Isakowitz, Hopkins, and Hopkins, *International Reference Guide to Space Launch Systems*, pp. 291–92.

36. Mihal, "Survey," p. 6.

37. Isakowitz, Hopkins, and Hopkins, *International Reference Guide to Space Launch Systems*, pp. 291–92.

38. Edgar Ulsamer, "Space Shuttle, High-Flying Yankee Ingenuity," *Space World*, June 1977, pp. 18–23; *Aerospace America*, July 1991, pp. 34–38.

39. Hans Mark, *The Space Station: A Personal Journey* (Durham, N.C.: Duke University Press, 1987), pp. 172–73, interview with Hans Mark, 16 April 2000.

40. NASA Press Release 87-76, "NASA Plans Use of Expendable Launch Vehicles," 15 May 1987.

41. Alliant Techsystems web site, accessed 10 July 2001, http://www.atk.com/aerospace/descriptions/DataSheets/Titan.htm.

42. Mihal, "Survey," pp. 7–8; Isakowitz, Hopkins, and Hopkins, *International Reference Guide to Space Launch Systems*, pp. 446–70; Lockheed Martin Astronautics, "Titan IV—Facts," 2000; Air Force Space Command, Public Affairs, "Titan IV," March 1999; "LockMart: A Billion Dollar Titan," SpaceDaily, 12 October 1999, NASA Historical Reference Collection; "Titan Accident Report Released," SpaceDaily, 18 January 1999, NASA Historical Reference Collection.

43. B. Peter Leonard and William A. Kisko, "Predicting Launch Vehicle Failure," *Aerospace America*, September 1989, pp. 36–38, 46; Robert G. Bramscher, "A Survey of Launch Vehicle Failures," *Spaceflight* 22 (November–December 1980): 51–58.

44. U.S. House of Representatives, "Department of Defense Appropriations Bill, 1995," Committee Report, 103d Cong., 2d sess. (PL 103-562), 27 June 1994.

45. Ibid.

46. Quoted in "Space Shots," *Space News*, 11 December 1995, p. 22.

47. Launch costs vary with circumstances—and circumstances definitely vary. Estimates similar to those on this chart, however, are used extensively within the space community to compare the four main U.S. expendable launchers. When a payload cannot be accommodated on Atlas II, it is an enormous jump in cost to put it on Titan IV. See, John T. Correll, "Fogbound in Space," *Air Force Magazine*, January 1994.

48. The Air Force seems very confused over whether it wants to call its facility at Cape Canaveral an "air station" or an "Air Force station" and has changed the designation several times over the past decade. In either case, it is the same spit of land adjacent to the NASA Kennedy Space Center.

49. Marcia Dunn, "Another Air Force Satellite Fails," AP news story, 1 May 1999, NASA Historical Reference Collection.

50. "Titan Accident Report Released," SpaceDaily, 18 January 1999, NASA Historical Reference Collection; Robert Wall, "Titan IVA Explosion Linked to Wiring Flaw," *Aviation Week & Space Technology*, 25 January 1999.

51. "Titan Team Ready for Last Launch," SpaceDaily, 7 April 1999, NASA Historical Reference Collection; Craig Covault, "Titan IVB/IUS Mission Failure Rocks Military Space Program," *Aviation Week & Space Technology*, 19 April 1999; Craig Covault, "Titan Mission Probe Eyes IUS Separation Problem," *Aviation Week & Space Technology*, 3 May 1999. For an excellent history of the DSP program, see Jeffrey T. Richelson, *America's Space Sentinels: DSP Satellites and National Security* (Lawrence: University Press of Kansas, 1999).

52. Craig Covault, "Titan, Delta Failures Force Sweeping Reviews String of Accidents will have Broad Program Impact," *Aviation Week & Space Technology*, 10 May 1999; Robert Wall, "Titan IV Flaws in Software," *Aviation Week & Space Technology*, 2 August 1999.

53. Craig Covault, "Titan Succeeds in NRO Flight," *Aviation Week & Space Technology*, 31 May 1999.

54. Craig Covault, "Titan IVB Flight Validates Quality Control Reforms Launch Success after Three Geosynchronous Failures Marks Pivotal Turnaround for USAF, Lockheed Martin, and the NRO," *Aviation Week & Space Technology*, 15 May 2000.

55. Dykon, Inc., "Demolition of Space Launch Complex 41, Kennedy Space Center, Cape Canaveral, Florida," March 2000, NASA Historical Reference Collection.

– 5 –

History and Development of U.S. Small Launch Vehicles

*Matt Bille, Pat Johnson,
Robyn Kane, and Erika R. Lishock[1]*

Introduction

The history of small American launch vehicles—those capable of lifting 1,100 pounds or less to low-Earth orbit (LEO)—is a surprisingly complex and interesting tale. Tracing the origins, development, and contributions of small launchers from the 1950s to the present illustrates the directions the U.S. military, civil, and commercial space programs have taken and sheds light on the roles of governmental and commercial developers in pushing back the space frontier. The cross-fertilization between military and civil small launch vehicle (SLV) programs was extensive in the beginning and has continued to the present day. In 2002 advancing technology is driving the U.S. space community to look more intensely at small satellites, which in turn creates a renewed interest in SLVs. In order to fully exploit this new interest, it is important to understand the history of SLV development and get a sense of where this field of space technology is headed.

Small launch vehicles have been part of the United States' space transportation fleet since the first orbital launches in 1958. Indeed, in 1958 and 1959, they were the only launch vehicles in existence for placing American satellites in orbit. After the initial burst of SLV development, the size and capability of U.S. launch vehicles steadily increased to accommodate the ever-increasing size and weight of the payloads they had to carry. With the national emphasis of space programs on manned flight and large-scale application satellites, there was almost no work done on new American SLVs from 1960 to 1988. Since then, there has been one solid success—the Pegasus—and several start-ups that have either failed or are still struggling to be born.

Background: Origin of the SLV

The Space Age may be said to have originated in the 1940s when warring nations, with Nazi Germany leading the way, began developing large, powerful rockets that demonstrated the possibility of satellite launch vehicles. While Wernher von Braun and his team were building their famous A-4 (more commonly known as V-2) missiles to attack England, they were well aware that future versions of these machines could put spacecraft into Earth orbit.

After the war, American and Russian experts who worked to take advantage of German technology came to the same realization. The United States won the biggest prize, obtaining the services of von Braun and most of his key personnel, but did not hasten to take full advantage of this coup. The U.S. Navy did sponsor a study of an Earth Satellite Vehicle Program as early as 1945. While this project went so far as to let design contracts for a two-stage launch vehicle, it was canceled in 1948 due to postwar budget cuts.

By 1955, when it became official the United States and the Soviet Union were seriously undertaking satellite programs, there were two existing areas of research from which to draw hardware and technology for rocket development. The first was in military missile design and development; the second was in the area of scientific sounding rockets, which in most cases also had military origins. These two sources drew on the same source of funding—the Department of Defense (DoD)—and the same body of engineering knowledge, although they were developing in different directions.

Origins: The Sounding Rockets

U.S. researchers had been firing small suborbital sounding rockets for decades before the true Space Age began. Researching the upper atmosphere was the impetus for Robert H. Goddard's pioneering work in liquid-fueled rockets. After World War II, scientific rocket activity intensified as new technology and ideas flowed from Germany. Given the situation prevailing in the war and during the Cold War which followed, it is no surprise that even the scientific rockets were all developed with military support. For example, by the time the war ended, the Army was already funding efforts at the Jet Propulsion Laboratory (JPL) to build large liquid-fueled rockets. The first result of this program, the WAC (Without Attitude Control) Corporal, began flying in 1945. This rocket had a thrust of 1,500-pound force (lbf)—normally boosted by a solid-propellant Tiny Tim military rocket motor—and was a developmental step on the way to the Corporal missile, but is best known for its scientific achievements.

The WAC Corporal's most famous role was in Project Bumper. The Bumper rocket was composed of a WAC Corporal mounted as a second stage atop an A-4 booster. On 24 February 1949, Bumper Round 5 reached a record altitude

Image 5-1: The WAC Corporal was the grandfather of all small launchers. Here a Bumper-WAC lifts off at the White Sands Proving Grounds in New Mexico on 24 February 1950. This two-stage rocket was intended to provide data for upper atmospheric research. On this date the vehicle, which employed a V-2 as the first stage with a WAC Corporal upper stage, obtained a peak altitude of more than 240 miles. (NASA Photo no. MSFC PA-3228)

of 250 miles. Project Bumper proved the possibility of staging large rockets and demonstrated the performance gain that could be achieved by using this technique.

From the WAC Corporal came the larger, more capable Aerobee. The Aerobee, which was intended to be a pure sounding rocket and not part of a missile program, was originally designed to replace the fast-dwindling supply of captured A-4s being used for high-altitude research. The Aerojet Engineering Corporation (now the Aerojet subsidiary of GenCorp) developed the rocket for the Applied Physics Laboratory (APL) at Johns Hopkins University with funding from the Navy Bureau of Ordnance and the Office of Naval Research (ONR). The first launch was on 14 November 1947. The Aerobee used a solid booster (18,000 lbf) and a liquid-fuel sustainer (4,000 lbf). An advanced version, the Aerobee-Hi, was proposed in 1952 and began flying in 1956.

The Naval Research Laboratory (NRL) also wanted a successor to the A-4. NRL engineer Milton Rosen had proposed a rocket program for upper-atmosphere exploration in 1945. He and his fellow engineers were delighted when the A-4s were captured, and the German rockets were put to good use in carrying instruments to unheard-of altitudes. As early as 1946, however, the NRL realized a successor vehicle would be needed in the near future. Rosen's team set their goal for the new rocket at "500 pounds of instruments to 500,000 feet."

Rosen and his colleagues designed a single-stage, fully steerable guided rocket weighing about 10,000 pounds. The Glenn L. Martin Company was selected to build a rocket called Viking based on Rosen's design. A contract for ten rockets was let on 21 August 1946 (an additional four rockets were purchased later). Reaction Motors, Inc. (RMI) was chosen to develop the engine, which used turbine-driven pumps and delivered a thrust of 20,000 lbf.[2] The Viking boasted several major advancements over the A-4—indeed, Rosen recalls the A-4 experience was mainly valuable for providing credibility to the idea of large rockets, and not for any technology. The most important advances were a superior control and orientation system, a gimbaled motor to provide steering, the use of integral tanks (in which the tanks are part of the rocket structure and use the fuselage skin as their outer walls) and the use of aluminum as the main structural material.[3]

Twelve Vikings were fired from 1949 through 1955. While the Viking program made significant contributions to our knowledge of the upper atmosphere, it made far greater ones to the development of rocket boosters. In 1954–55, the NRL was already studying the possibility of mating the Viking and Aerobee to produce a vehicle for high-speed reentry system tests. This configuration, the Viking M-15 (for Mach 15) was never built, but it became the basis of the satellite launcher used in Project Vanguard.[4]

Origins: The Missile Program

In 1950, von Braun and his group became the core of the Army's Ordnance Guided Missile Center at Redstone Arsenal, Alabama. (On 1 February 1956, the unit would acquire its more famous name, the Army Ballistic Missile Agency or ABMA). The von Braun team built on their A-4 technology and began designing larger, more capable rockets. The Korean War intensified the entire field of weapons development, and in July 1951 the all-out effort began to build a large guided missile—the Redstone.

The Redstone effort built on the captured A-4s, and on the experience gained by Project Hermes—an effort to assemble parts of captured A-4 into working missiles to allow engineers and technicians to gain practical experience. Beginning as early as 1944, the Army had sponsored work by General Electric to develop large rocket motors and guided missiles. This was expanded after the war to include Hermes, and was eventually joined by the von Braun team—soon the program began to manufacture both replacement parts and improved parts for the captured A-4s, eventually leading to essentially new missiles. The Hermes program ended in 1950, but it contributed to two future achievements: General Electric's X-405 engine for the Vanguard booster and the Hermes A-1 engine, built by the Rocketdyne Division of North American Aviation, which was the foundation for the engine used in the Redstone booster.[5]

The improved Rocketdyne engine used in the operational Redstone drew on the Hermes A-1 technology and a design developed for the Navaho intercontinental winged cruise missile. The Redstone's liftoff mass was 62,750 pounds, and the thrust at liftoff was 82,600 lbf. The vehicle stood 69 feet tall and was 6 feet in diameter. The propellants were liquid oxygen and ethyl alcohol, and guidance was provided by an inertial system controlling graphite exhaust vanes (the technique used in the A-4) and small aerodynamic rudders.

When the Redstone made its first test flight on 20 August 1953, it was easily the largest missile developed up to that point in the United States. The Redstone contributed to several subsequent programs, both military and civilian. It was the core of the Jupiter C (a.k.a. Juno I) launch vehicle for *Explorer I*. It was also the foundation for the larger Jupiter missile, which in turn became the satellite launcher called Juno II. The Redstone may be best known for its role in the Mercury program. Six Redstone launches from November 1960 to July 1961 carried Mercury capsules on suborbital flights. The last two of these carried Alan Shepherd and Gus Grissom, the first two Americans in space. Redstone space launch vehicles used an uprated engine, the Rocketdyne A-6, with a thrust of 91,000 lbf.

Almost before the Redstone began flying, thought was given to its suitability as a satellite launcher. In June 1954, the ONR hosted a meeting of the leading minds in rocketry and space exploration to review a proposal for an Earth satellite. One attendee was von Braun, who suggested the Redstone would make a serviceable first stage for a satellite launch vehicle. Encouraged by the support he received, von Braun produced the "Minimum Satellite Vehicle" proposal, dated 15 September 1954.[6] The launcher in this concept was the Redstone with clusters of solid-fuel Loki rockets added to form three upper stages. The satellite envisioned was truly minimal—it would weigh only about 5 pounds and be completely without instruments. The proposal, which eventually became known as Project Orbiter, specified launching from a Navy ship stationed at the equator to obtain the greatest possible boost from the Earth's rotation. The projected launch date was summer 1956, but the program was canceled in 1955 in a fateful and controversial decision.

Choosing the First SLV

On 29 July 1955, President Eisenhower publicly committed the United States to a satellite program as part of American participation in the International Geophysical Year (IGY). At the time, the DoD's Ad Hoc Group on Special Capabilities was already meeting to select a satellite and launch vehicle.[7] This eight-member panel was chaired by Homer Stewart of JPL and included representatives nominated by the Army, Navy, and Air Force. The Stewart Committee considered three options:

- An updated version of Orbiter, backed by the Army and ONR;
- A proposal from the NRL, using a new three-stage launch vehicle based on the Viking;
- An Air Force design launched by the yet-untested Atlas intercontinental ballistic missile (ICBM).

While the Army proposal had been updated from the 1954 original, the concept remained the same: a very small satellite, orbited by a Redstone-based booster augmented with clusters of small solid-propellant rockets. One change from 1954 was a recommendation to use scaled-down versions of a newer JPL-designed solid rocket, the Sergeant, in place of the original Loki.

The Committee issued its formal report on 4 August. The winner was the Navy proposal, which startled and infuriated the Army team.[8] After all, Redstone had been making test flights for two years, and its costs and capabilities were well known. The NRL's rocket would need considerably more development work, and their proposal contained very little cost data. Nevertheless, the Stewart Committee majority felt its three-stage rocket was likely to be cheaper—four times cheaper, in fact—and more reliable than the seemingly complex Orbiter booster.[9] In addition, the Committee thought that the Navy system (soon to be named Vanguard) offered more potential for growth. Finally, while development of the Vanguard booster presented a technical risk, it would take no resources from the military's missile programs.

Some modern historians have speculated that Vanguard was preferred because it had a less obvious military character. While the Vanguard selection did serve the government's agenda (codified in the classified National Security Council directive 5520) to establish "free transit of space" using a scientific satellite, there is no proof the Committee was explicitly charged with picking the "more civilian" satellite. It is true, however, that the satellite program's military aspects were downplayed as much as possible. Officially, Vanguard was "an IGY project in which the DoD is cooperating, rather than a DoD project."[10] Maj. Gen. John B. Medaris, commander of the Army's ballistic missile program, later referred to this attempt to separate military and civilian rocketry as "a costly and ridiculous division of the indivisible."[11]

In any event, the choice was made, and the Navy forged ahead. The NRL would design the Vanguard satellite, while the Glenn L. Martin Company was responsible for the launch vehicle. According to the contract, the Martin-built vehicle was to carry a payload weighing 21.5 pounds. John Hagen of the NRL was the project's director, and Milt Rosen became Technical Director.

The Contributions of Vanguard

Vanguard will always be remembered for the spectacular failure of its first orbital attempt in December 1957. This is unfortunate and very misleading. The Vanguard booster development effort was a highly successful program, especially in terms of the technology it developed and contributed to later launch vehicles.

There were a lot of growing pains on the way to these achievements. In selling its proposal, the NRL acknowledged the relative immaturity of its rocket but assured the Stewart Committee that the challenge was manageable and the risk fairly low. The first stage of Vanguard would be developed from the proven Viking, and the second stage would come from the equally mature Aerobee-Hi sounding rocket. Only the third stage would be new.

In practice, the challenges proved to be considerable. Late in 1956, the Vanguard satellite design was switched from a cone to a 20-inch sphere.[12] That required redesigning the second stage to give it a larger diameter. Many other changes were necessary to both the first and second stages to convert the existing designs for their new functions. The most obvious was a redesign of the first-stage engine. Viking's engine only had to lift a maximum of 15,000 pounds gross weight from the pad, while Vanguard was over 8,000 pounds heavier. Even more challenging was the required increase in velocity. The Viking never exceeded 5,000 feet per second (fps). Vanguard needed 25,000 fps. (Rosen insisted on a margin of safety, and almost 27,000 fps was achieved.)[13]

Partly as a result of these challenges, Vanguard produced a series of important design innovations. Increased use of aluminum (and, in the final version, fiberglass) produced an impressive mass ratio of 0.86, compared to a maximum of 0.80 for Viking. This improvement was quite a feat, considering Vanguard needed additional structural weight for staging, payload release, and other functions that did not apply to Viking. (The comparable figure for the A-4 was 0.69.) Another example of the inventiveness of Vanguard designers was the use of turbopump exhaust to power the first stage roll control jets.

The Vanguard third stage was to be the first large solid-propellant motor. To ameliorate the perceived risk, two contracts for completely different motor designs were let, one to the Grand Central Rocket Company and one to the Allegany Ballistics Laboratory (ABL). The ABL design used a fiberglass casing, another first for a large rocket, and an advanced double base propellant. While the Grand Central stage was used on the early flights, including the first two that reached orbit, ABL's X-248 allowed Vanguard to more than double its payload (to 52 pounds) and was used by several other launch vehicles.

As mentioned earlier, the Viking program consisted of twelve launches. There were actually fourteen Vikings, but the last two carried Vanguard designations—Test Vehicle (TV)-0 and TV-1. True Vanguard launches began with TV-2. By the time the Vanguard program ended with a final launch on 18 September 1959, twelve launches (not counting the two Vikings) had been attempted and three satellites had been placed in orbit.

The original plan was to fire Vanguard tests through TV-6 before making an official satellite attempt. However, the successful 4 October 1957 launch of Russia's *Sputnik I* placed the U.S. program in the political spotlight and created great pressure to compete in this new "space race." On 6 December 1957, TV-3 made the first Vanguard orbital attempt. The program's directors tried to make it clear this was indeed a test— it was the first launch of the full three-stage Vanguard booster configuration, and the second stage had never been flown at all. The decision to launch a satellite on TV-3 had been made in July, and was reaffirmed in the aftermath of *Sputnik I*. When Rosen was later asked whose idea that had been, he replied, "Everyone's." After *Sputnik*, "if we were going to launch a rocket, it was going to have a satellite on it."[14]

The desire to keep TV-3 low key was doomed by the political and media enthusiasm to match Sputnik. It is no exaggeration to say the entire world was watching on the cold December morning as the needlelike silver space vehicle, glistening in the Florida sunshine and the glare of publicity, was ignited. Two seconds after launch, the rocket burst into a spectacular fireball. To accentuate the disaster in a most ironic fashion, the satellite was blown clear and fell to the ground intact, its radio transmitters dutifully sending signals as if it were in orbit.

The second Vanguard orbital launch also failed, but the third, on 17 March 1958, placed the Vanguard 1 satellite in orbit. A second Vanguard was orbited in February 1959, although the upper stage imparted an erratic motion that spoiled some of the science results. The program's final success came when the 50-pound Vanguard 3 was orbited in September 1959.

Image 5-2: One of the earliest of the small launchers, this three-stage Vanguard (SLV-7) takes off from Cape Canaveral on 18 September 1959. Designated Vanguard III, the 50-pound satellite was used to study the magnetic field and radiation belt. In September 1955 the Department of Defense recommended and authorized the new program, known as Project Vanguard. The Vanguard vehicles were also used in conjunction with later boosters such as the Thor and Atlas. (NASA Photo no. MSFC 9139356)

The Jupiter-Juno Family

While the Vanguard was America's designated small launch vehicle, the Army continued to develop the Redstone into both a more capable missile and a reentry test vehicle. In a decision that would be vital to SLV development, the Jupiter missile program was approved in late 1955.

Jupiter C/Juno I

To pave the way for Jupiter missile development, Redstones were modified into a three-stage configuration called Jupiter C to test ablative reentry vehicles. The first Jupiter C flight, on 20 September 1956, became known as the "satellite that wasn't" because it could theoretically have placed a live fourth stage (which it was designed to carry) into orbit. The Jupiter C reentry test vehicle was very successful. It flew three times and demonstrated that the innovative concept of ablative shielding would protect a reentry vehicle on its journey down through the atmosphere. This contributed to all future missile and human space-flight programs.

Immediately after the *Sputnik I* launch, von Braun and Medaris urged the incoming Secretary of Defense, Neil McElroy, to allow the Army to proceed with a Jupiter-launched satellite. The Army's proposal, now called Project 416, envisioned launching four satellites at a total cost of $16.2 million.[15] On 8 November, McElroy told the Army to prepare for two launches in case a backup to Vanguard was needed, although approval for the actual launch was not yet given. Almost a month before this, however, General Medaris had (without proper authority) issued an order directing preparation of a Jupiter booster.

Despite its name, the Jupiter C's core vehicle was a modified Redstone, not a Jupiter missile. The tanks on the Redstone were longer than on the basic missile version, and the engine had been modified to a new configuration, called the A-7. Instead of alcohol, the A-7 burned liquid oxygen and hydyne, a mixture of unsymmetrical dimethyl hydrazine (UDMH) and diethyltrianine (DETA). This engine had a maximum (vacuum) thrust of 93,500 lbf.

In the satellite launching configuration, with a live fourth stage, the Jupiter C was redesignated Juno I, a name that never gained universal acceptance. Both the Jupiter C test vehicles and Juno I SLVs carried an unusual upper stage cluster. The system's ingenious design was driven by the limited size and power of the Redstone stage and the existing solid-fuel rockets. It used clusters of scaled-down Sergeant rockets, each six inches in diameter. Stage 2 was a ring of eleven such rockets, Stage 3 was a cluster of three, and Stage 4 (when carried) was a scaled-down Sergeant, or "baby Sergeant" as it was sometimes called. The guidance system, on the Redstone stage, was lost with stage separation, so the upper stages were spun to stabilize their trajectory.

The ingenuity displayed in making an orbital launch vehicle based on marginally capable rocket motors had the tradeoff of imposing severe conditions on the Juno's payload. The maximum acceleration imposed by the final solid-fuel stage was seventy times the force of gravity (70 g). The instruments, along with the entire satellite and upper-stage package, also had to survive spinning at up to 700 revolutions per minute (rpm) during the flight.

The spin rate was only possible because the upper section had been deliberately over-designed. Dr. Walter Downhower, *Explorer I*'s payload engineer, has explained that the reentry vehicle test program, for which this upper section was first built, only required a spin rate of 450 rpm for sufficient accuracy. JPL and Army engineers, hoping that someday their design would be used for a satellite and knowing a higher rate would be needed to provide the accuracy for orbital injection, built and tested the contraption for spin rates of up to 1,000 rpm. Without this overdesign, *Explorer I* could not have been launched, or at least would have taken months of additional work.[16]

Another difference between the Jupiter C/Juno I and the Redstone missile was the former's use of a different guidance system, the LEV-3. The LEV-3, based on the design used fifteen years before in the A-4, was installed to provide improved roll and yaw control. The LEV-3 was an off-the-shelf component (built by a company called Waste King Corporation, an amusing moniker for a government contractor) which, when combined with a Ford air-bearing gyro for pitch control and accelerometers, produced a guidance system equal in accuracy to the newly developed system built for Vanguard.[17]

There were six flights of the Juno I. The first, on 31 January 1958, was the most famous, placing *Explorer I* into orbit. There were five subsequent launches, the last being a unique five-stage configuration carrying NASA's Beacon 1 satellite. Three flights, including the Beacon 1 launch, failed, but the others placed two more Explorer satellites into orbit.

Juno II

The Jupiter missile, designated SM-78 by the Army, was a follow-on development to the Redstone. The diameter was increased to 8.8 feet and the liftoff thrust to 109,000 lbf. The fuel was kerosene instead of alcohol. Having succeeded in orbiting satellites with the Redstone-based Juno I, the Army and the new National Aeronautics and Space Administration (NASA) naturally examined what could be done with the Jupiter missile along the same lines. The result was the Juno II. The Juno II first stage was a modified Jupiter, with the most significant change being to the engine. The Rocketdyne S-3 used to power Jupiter missiles was modified to a more powerful configuration, the S-3D. The S-3D, with its 150,000 lbf, became the engine of choice for later Jupiter missiles as well as the Juno II.

The Juno II used the upper stage apparatus (sometimes called the "washtub") from the Juno I, but added a fairing that covered the assembly for streamlining. Juno II is best known for launching the *Pioneer 3* and *4* lunar probes, but it also made eight orbital launches. The first, with *Explorer 6*, was a failure. The second, on 14 August 1959, put the Beacon 2 satellite into orbit. Both of these were three-stage configurations, without the top baby Sergeant. The four-stage configuration launched six Explorer payloads from 1959 to 1961, but only three were successful. Premature shutdowns, control problems, and ignition failures caused the failure of the other three.

Project Pilot: The First Air-Launched SLV

In a then-classified and still largely unknown effort, the U.S. Navy in 1958 built the first air-launched SLV. This was Project Pilot or NOTSNIK ("NOTS" stood for Naval Ordnance Test Station, then the name of the Navy installation at China Lake, California; the "nik" was borrowed from Sputnik). The impetus for this program was *Sputnik I*. As recalled by Dr. Frank Cartwright, a group of NOTS physicists was watching *Sputnik I* track across the heavens when his colleague Dr. Howard Wilcox, head of the NOTS Weapons Development Department, said simply, "We ought to be able to do that."[18]

The idea caught on, and NOTS personnel set to work. After examining catapult-launched and ground-launched vehicles, the satellite team, which had only quasi-official status at best, settled on a rocket dropped from a fighter jet.[19] A formal proposal describing the air-launched approach was produced in February 1958.

To obtain more funding, NOTS officials in May 1958 promised Navy leaders they could put up satellites to monitor Project Argus, a series of nuclear tests set to begin in August. By using this approach, the NOTS team obtained approval to spend NOTS funds on an official basis, eventually totaling at least $4 million.[20] Unfortunately, this also created a schedule that appeared impossible. The NOTSNIK engineers had to finish the design work, then fabricate, assemble, and fly the entire system—satellite, instrumentation, and booster—in less than four months. NOTSNIK would essentially be tested by launching it into orbit.[21]

Constrained by the limited power of available solid-fuel rocket engines, NOTS engineers assembled a minimal five-stage SLV based on modified military rocket motors called HOTROCs. The first and second stages consisted of four motors grouped together. One pair was fired as the first stage, and the second pair, oriented at 90 degrees from the first pair, was fired as a second stage. The third stage was a single X-241 rocket motor. This was a variation of the X-248 developed by ABL for the Vanguard program. The fourth and fifth stages were small rockets built in the shops at NOTS. The tiny fifth-stage motor was actually mounted in the center of the 2-pound donut-shaped satellite. The entire launch vehicle weighed only 2,100 pounds, making it the smallest SLV ever built.

There were three launch attempts during July 1958 and three more in August. A post-mortem assessment by NOTS engineer Frank Knemeyer concluded that five of the six launches failed due to a variety of booster malfunctions.[22] The exception came on 22 August 1958, when a NOTSNIK rocket left its F4D Skyray carrier plane off the California coast and operated perfectly. The NOTSNIK team was ecstatic when a radio signal that may have been from the satellite's transmitter was picked up in Christchurch, New Zealand.[23] But there were no further signals. If the satellite was in orbit, its radio or battery must have failed.

No one knew—nor will we ever know—whether this shoestring project actually succeeded in putting a satellite into orbit. The Navy engineers themselves were divided in their assessments. The concept of an air-launched satellite booster, though, was to resurface three decades later. The Pegasus, a conceptual descendent of the primitive NOTSNIK, became the leading American SLV in the 1990s.

Thor and Its Progeny

In 1954, the Air Force issued a Request for Proposals for an intermediate-range ballistic missile (IRBM) capable of carrying a nuclear weapon 1,500 miles. The Douglas Aircraft Company proposed a design that used the warhead and guidance system already in development for the Atlas missile. This was accepted by the Air Force and named Thor.

Initially, the Air Force was in a race with the ABMA to secure funding for its IRBM program. The conflict ended with the 1956 "Wilson Memorandum," in which Secretary of Defense Charles Wilson stripped the U.S. Army of all missiles with a range of over 200 miles. ABMA continued Jupiter IRBM development, but the operational missiles would be turned over to the Air Force. With the Wilson Memorandum, the Air Force could develop both the Atlas and the Thor. Thus, Thor was given equal priority with the Atlas in December 1955, and the Western Development Division (WDD) became the missile's new sponsor at that time.

Because of the political climate, Thor was a high priority and "maximum risk" program. The first Thor missile was delivered to the USAF in October 1956, just ten months after the contract began. After four test failures, the first successful launch occurred on 20 September 1957.

The principal contractors for Thor were the Douglas Aircraft Company for the airframe and North American Aviation (Rocketdyne) for the engines. The missile weighed 111,400 pounds and was 62.5 feet long and 8 feet in diameter. It was propelled by a single rocket motor rated between 135,000 and 170,000 lbf. The inertial guidance used liquid-floating gyroscopes. Thor used gimbal-mounted motor exhausts like Vanguard, rather than the exhaust vanes employed on Redstone and Jupiter. However, the Thor, the Jupiter missile (not the Jupiter C/Juno I launcher), and the Juno II SLV shared essentially the same engine. The Rocketdyne S-3 and S-3D used on the Jupiter/Juno were very similar to the same company's LR79-7 for Thor.

Thor evolved through several three- and four-stage configurations that were used in many roles, including space exploration, military testing, and strategic reconnaissance. Thor's potential as a satellite launcher was recognized "right from the beginning."[24] Work on the missile and space launch versions overlapped—the first Thor SLV was fired while the Thor missile was still undergoing test launches. Table 1 lists the various Thor configurations.

The Thor Able 1 was a space launch vehicle consisting of a Thor booster and a Vanguard second stage, which at the time was essentially the only upper stage available. The first launch, on 17 August 1958, carried a lunar probe. Unfortunately, the vehicle exploded 77 seconds into flight. The mission was designated Pioneer 0. The next Thor Able 1 mission, *Pioneer 1*, was a partial success. The third stage produced insufficient thrust to achieve escape velocity, but the vehicle did travel over 70,000 miles into space. The final lunar probe, *Pioneer 2*, did not reach the moon because the booster's third stage did not ignite.

The next configuration, the Thor Agena, was a Thor booster with an Agena second stage—the entire Agena entered orbit. The Agena was powered by a Bell Aerospace Hustler 8048 motor, and so the vehicle was sometimes called the Thor Hustler. This configuration failed during its initial test on 21 January 1959, but successfully launched Discoverer 1, the first spacecraft in the Corona imagery intelligence program, the following month. Discoverer 1 was also the first satellite in polar orbit and the first to be launched from Vandenberg Air Force Base (VAFB), California. Between March and December, eight Discoverers were launched.

The Thor Able 2 series tested a new Bell Telephone Laboratory guidance system for the Titan ICBM and was also used in a failed attempt to launch the Navy Transit 1-A in September 1959. The Thor Able 3 space carrier launched *Explorer 6* in August. A Thor Able 2 launched the weather satellite Tiros I and a new configuration, Thor Able 4, launched *Pioneer 5* into a solar orbit between Venus and Earth.

The successful Thor Able Star (also known as Thor Epsilon) was introduced in April 1960. The Able Star stage used the Aerojet AJ-10–104 engine, which had a restart capability. This stage could carry 2.5 times as much propellant as the Thor Able second stage and was used to launch the Navy Transit navigation satellites (from Transit 1 in 1960 to Oscar 5 in 1965) and other Navy scientific spacecraft (such as SolRad). The vehicle achieved the first in-orbit re-ignition in April 1960 and the first double-, triple-, and sextuple-satellite launch.

The Thor Able Star also launched the Courier 1B (the Courier 1A launch had failed) communications satellite in October 1960. The Courier marked

Image 5-3: The launch of Thor–Able 5 launch vehicle with Tiros 1 (Television Infrared Observation Satellite) as its payload on 1 April 1960. (NASA Photo no. MSFC 9141923)

the one-hundredth launch of the Douglas Thor (military and scientific combined). At that time, Thor had boosted 60 percent of all U.S. satellites into orbit.

The first flight for the new Thor-Delta configuration occurred on 13 May 1960. The vehicle was named Thor-Delta because it was the fourth modification of the Thor (after Able, Able Star, and Agena). The second stage used an Aerojet General AJ-10-118[25] and the third stage the ABL-248 solid-propellant motor (the Vanguard third stage).

Thor-Delta was the first NASA-managed launch vehicle. It was called simply "Delta" at Cape Canaveral, while military rockets launched from Vandenberg AFB used the name "Thor." "Thor" was officially removed from the Delta name in 1972. Subsequent changes to the vehicle moved it out of the SLV class, and today's Delta is a powerful medium-lift launcher.

Thor-Delta's first mission, Echo 1, failed, but the second orbited Echo 1A, the first passive communications satellite. Some of the other spacecraft launched by the Thor Delta include the weather satellites Tiros 2 through 6, Ariel 1 (the U.S./UK astronomy satellite), Telstar 1 and 2, and *Explorers 10* and *12*.

The final Thor small launcher configuration was the Thor Altair (or Thor Burner 1). It was a space launch vehicle with a solid upper stage (the ABL X-248 developed for Vanguard). The Thor Altair was used between January 1965 and March 1966 to orbit six Defense Meteorological Satellite Program (DMSP) spacecraft. Later DMSP spacecraft were launched by the higher-performance Thor Burner 2/2A.

Note on the Origin of Solid-Fuel Boosters

While Vanguard and Juno used solid-fuel upper stages, very large motors, such as those used in the first stages, were initially all liquid-fueled. Solid motors produced less energy, and there were limits to how large they could be cast. Again, it was military missile work that spurred solutions to these barriers.

While the Hermes A-1 engine was an important step in liquid-fuel technology, the same program funded the Hermes A-2, which used solid fuel. Thiokol Chemical Corporation built the motor and fired it in December 1951. The A-2 was eventually canceled, but its motor powered four successful test flights in a vehicle called RV-A-10 in 1953. Thiokol also worked with JPL on the Sergeant program, developing the 15-inch-diameter Sergeant motor, which JPL scaled down to become the upper stages for the Jupiter/Juno series. Thiokol later scaled up this technology and, in 1958, won contracts to build all three stages for the Minuteman ICBM.[26]

Another impetus for solid-fuel engines arose from the Jupiter program. The Jupiter missile began its life as a joint Army-Navy development project. The Navy had never liked the idea of launching large liquid-fueled missiles from a ship, but up to that point, it seemed like the only option. However, new

Table 1
Thor Configurations

	Thor	Thor-Able	Thor-Able-Star (Thor Epsilon)
Purpose	IRBM	Able 0 was a reentry test vehicle Able 2 was mainly used to test the Titan GS Ables 1, 3, 4 were space carriers	DM21A used MB Block I DSV-6 (SLV-2) used MB Block II/III
Launch site	Varied	Cape Canaveral	DM-21A: Cape Canaveral DSV-6 (SLV-2): VAFB
Overall length (feet)	72	98.5	95
Stage 1 engine	Rocketdyne MB-1	Rocketdyne MB-1	Rocketdyne MB-3
Stage 1 propellant	LOX/RP-1	LOX/RP-1	LOX/RP-1
Stage 1 thrust (lbf)	170,000	170,000	172,000
Stage 2 engine	—	Aerojet General AJ10-42 (101A)	Aerojet General AJ10-104
Stage 2 propellant	—	IWFNA/UDMH	IRFNA/UDMH
Stage 2 thrust (lbf)	—	7,700	7,900
Stage 3 engine	—	Allegheny Ballistics Lab X-248 (X241)	—
Stage 3 propellant	—	Solid	—
Stage 3 thrust (lbf)	—	3,150	—
Stage 4 engine	—	Able 1: Thiokol Falcon solid rocket motor used as a retrorocket to correct lunar approach velocity in early Pioneer firings Able 3: Explorer 6 included an Atlantic kick rocket to lift the perigee (if needed)	—
Stage 4 thrust (lbf)	—	Able 1: 3,000 Able 3: 450	—
Burnout velocity	—	36,000	26,250

Table 1 Thor Configurations			
	Thor	**Thor-Able**	**Thor-Able-Star (Thor Epsilon)**
Capability	—	350 pounds to 300-mile orbit; 86 pounds to escape velocity	110 pounds to 385 miles
Number of flights	—	Able 0: 3 Able 1: 3 Able 2: 8 Able 3: 1 Able 4: 4	19 (11 DM-21A)
Successful flights	—	Able 0: 2 Able 1: 2 (Partial Failure (PF)) Able 2: 6 Able 3: 1 Able 4: 1	14 (7 DM-21A)
	Thor-Delta	**Thor-Agena A (Thor Hustler)**	**Thor-Altair (Thor Burner 1)**
	Subsequent versions simply referred to as "Delta"	—	—
Launch site	Cape Canaveral	VAFB	VAFB
Overall length (feet)	102	92	79
Stage 1 engine	Rocketdyne MB-3 (XLR79-NA-7)	Rocketdyne MB-3 (XLR79-NA-7)	Rocketdyne MB-3 (XLR79-NA-7)
Stage 1 propellant	LOX/RP-1	LOX/RP-1	LOX/RP-1
Stage 1 thrust (lbf)	170,000	170,000	170,000
Stage 2 engine	Aerojet General AJ10-118	Lockheed Agena A (Bell Aerospace Hustler 8048 engine)	Alleghany Ballistics Lab X-248 (Altair)
Stage 2 propellant	IWFNA/UDMH	IRFNA/UDMH	Solid (CTPB)
Stage 2 thrust (lbf)	7,900	15,500	3,100
Stage 3 engine	Allegheny Ballistics Lab X-248A5	—	—
Stage 3 propellant	Solid (CTPB)	—	—
Stage 3 thrust	2,800	—	—

	Thor-Delta	Thor-Agena A (Thor Hustler)	Thor-Altair (Thor Burner 1)
Table 1 **Thor Configurations**			
Stage 4	—	—	—
Burnout velocity (fps)	26,250 – 36,000	26,250	—
Capability	100 pounds to geosynch transfer	1,700 pounds to 100x1000-mile orbit	550 pounds to 300x510-mile orbit
Number of flights	12	16	6
Successful flights	11	11	5

advances in solid-fuel motors, especially from Thiokol, rekindled interest in development of an alternative. A solid-fuel Jupiter was designed, but was never actually built. Instead, the Navy withdrew from the Jupiter program in November 1956 and let development contracts for an all-solid missile called Polaris. Thiokol produced the first batch of six first- and second-stage development motors, with Aerojet General delivering the first production motors. The Polaris first stage provided the foundation for the first all-solid ground-launched SLV, called the Scout.

Scout: The Workhorse

No SLV had a longer or more productive career than NASA's Solid Controlled Orbital Utility Test (Scout).[27]

Scout was the world's first all-solid-fuel satellite launch vehicle to enter operational service. Prior to 1957, the PARD (Pilotless Aircraft Research Division) of the National Advisory Committee for Aeronautics (NACA) at Langley had been developing expertise in small rocket launch, guidance, automatic control, and telemetry. In addition, their missions had produced a significant amount of experimental data on hypervelocity performance and aerodynamic heating. In late 1957, the PARD engineers concluded that a four-stage combination of existing solid-fuel motors could provide orbital capability for a small payload, although there was some doubt in the space community as to whether any solid-fueled vehicle could reach orbit.

While PARD needed a four-stage system for reentry research, the vehicle cost was too high for a single project. In mid-1958, NACA asked Langley to prepare a Space Technology Program for a new space agency. In this plan, Scouts were listed as "small-scale recoverable orbiters" to be used in development of the Mercury capsule. The study was coordinated with the Air Force, which wanted a sounding rocket with higher performance than their Javelin. In October 1958, NASA came into existence, incorporating NACA (it would soon add the von Braun team from ABMA as well). Contracts were let for the Scout motors, guidance system, airframe, and launcher. Langley was the prime contractor, and both NASA and Air Force vehicles were procured through the Langley contract.

On 1 March 1959, NASA and the Air Force jointly announced the development of Scout, the "poor man's rocket." The aim of the Scout project was to produce a relatively inexpensive, reliable, solid-fuel rocket that could be used in standard configurations for NASA, DoD, and foreign payloads.[28] Vought Astronautics (later the LTV Aerospace and Defense Company, Missile Division, Dallas, Texas) manufactured it.

The standard Scout launch vehicle was a solid-propellant, four-stage booster system approximately 75 feet long. A fifth-stage velocity package could be added to increase the hypersonic reentry performance, enable highly elliptical orbits in deep space, or extend probe capabilities to the Sun. In contrast to most early launch vehicles, the Scout was a NASA development and not directly based on a missile system (although all the Scout's motors either came directly from military programs or were derived from motors developed for the military). Table 2 summarizes the general Scout configuration and motor sources. NASA renamed each engine for a star.

On 1 July 1960, the first four-stage Scout was launched from NASA Goddard Space Flight Center's Wallops Flight Facility (WFF) on Wallops Island, Virginia. The vehicle became operational in 1963. After more than three decades of service, the final Scout was launched on 8 May 1994 from Space Launch Complex 5 (SLC-5) at Vandenberg AFB, California.

In addition to NASA, Air Force, and Navy payloads, Scout was used by the European Space Research Organization and the governments of the Netherlands, France, Germany, Italy, and the United Kingdom. The Scout had three potential launch sites. The WFF was used for eastward launches; SLC-5 at Vandenberg AFB was used for high-inclination missions; and the San Marco platform, located at 2.9 degrees latitude off the Kenyan coast, was used for low-inclination missions.

Project officials attributed the Scout's high reliability to the standardization of launch and manufacturing procedures and the incorporation of off-the-shelf technology. Over the life of the program, the Scout was launched 116 times. This included 21 research and development (R&D) launches (with 11 successes) and 95 operational missions (with 91 successes).[30]

Configurations

The Scout was continuously upgraded during its lifetime, and each of the

		NASA	
Stage	**Engine**	**Designator**	**Comments**
1	XM-68 (Aerojet-General Corporation Jupiter Senior)	Algol	• Developed for the Joint Army-Navy-IRBM.[29] • Forerunner of the Polaris and Minuteman
2	TX-33 (Army Sergeant)	Castor	• Original configuration used Castor I • Later configurations used Castor II which was also used as the strao-ons for the Thor and Delta launch vehicles
3	ABL X-254	Antares	• Upgraded Navy Vanguard upper stage X-248 (Altair) • Not used outside the Scout program
4	X-248	Altair	• Used as an upper stage in the Thor-Delta, Thor Burner 1, and Atlas satellite launchers • Also used in the Javelin, Journeyman, and Shotput sounding rockets

Table 2
Scout Component Stages

stages was replaced at least once after the first version. Table 3 shows the various standard configurations.

Unlike NASA's other large expendable rockets, the Scout was assembled and the payload was integrated and checked out in the horizontal position. The vehicle was raised to the vertical orientation prior to launch, and it could be launched at any angle between straight vertical and 20 degrees from the vertical.

In all Scout versions, the Honeywell-built guidance and control (G&C) system provided attitude reference as well as the control signals and forces necessary to stabilize the vehicle in the pitch, yaw, and roll axes. Miniature rate gyros, in the inertial reference package, detected any deviation from the vehicle's programmed path and generated electronic signals to the steering system. The first stage was controlled by four fins containing moveable jet vane/fin-tip assemblies. The second and third stage used an "on-off" monopropellant (hydrogen peroxide) reaction control system. The fourth stage was unguided and spin-stabilized.

Image 5-4: The launch of a Scout space launch vehicle at the Wallops Flight Facility, Virginia. (NASA Photo no. MSFC 8904412)

Scout Evolution

Unlike the Vanguard and Thor, the Scout development was not a "time essential" or "maximum risk" undertaking. The program was well structured, and there were clear development phases followed by a "standard" vehicle design. The Scout's history has been divided by writer Andrew Wilson into eight phases, some of which overlap:[31]

I.	Development
II.	Prototype
III.	Recertification
IV.	Management
V.	Incentive Procurement
VI.	Award Fee Procurement
VII.	Continuing Program
VIII.	Final Production and Launches

There were ten missions in Phase I (the developmental phase). All the missions were identified with the "ST" (Scout Test) designation and, except for the last mission, used the X-1 configuration. During this phase, payload operation was a secondary objective. The first launch occurred on 18 April 1960 and is usually not included in the launch totals because it carried two dummy stages (stages 2 and 4). The other nine launches included four orbital, four probe, and one reentry mission. The only successful orbital mission during the development phase (*Explorer 9* on 16 February 1961) has the distinction of being the first all-solid rocket to orbit a payload (given that NOTSNIK was never officially credited with making orbit) *and* the first rocket to reach orbit from Wallops.

Phase II covered the period from April 1962 to November 1963, and included fourteen missions. The purpose of Phase II was to launch prototypes of the standard production vehicle and, hopefully, achieve a 90 percent success rate. Unfortunately, Phase II ended with only a 50 percent success rate. The main "lesson learned" from this phase was the need for close cooperation between the manufacturer and NASA. In response, design interfaces and checkout procedures were revised for the purpose of delivering to the launch site a certified vehicle essentially ready to launch.

Phase III should have been the fully operational phase, but instead, NASA and LTV concentrated on using it to improve the Scout's reliability. "Operational" or not, Phase III achieved a success rate of 93 percent (fourteen flights, one failure). Seven of the payloads were Explorer satellites. One suborbital flight carried a payload of two ion engines. This was the first successful ion engine test in space. Phase III also included the first launch by a foreign government; this was launched from Wallops by an Italian team on 15 December 1964.

For the final Phase III flight, Scout S-131R (10 August 1965), the name changed from the X series to the Scout B. It was an important flight because many changes had been made to the configuration simultaneously in an attempt to establish a new "standard launch vehicle." Other Phase III revisions included changing the second stage from the Castor I to the Castor II and the fourth stage from the X-258 to the FW-4S.

Phase IV was the first phase to use standardized vehicles (the X designation ceased). It covered November 1965 to June 1971. This phase had a remarkable 92 percent success rate for its twenty-two orbital and three reentry missions (twenty-five launches, two failures). The payloads included two Explorers, nine Navy Transit navigation satellites, and six Air Force OV3 environment research satellites. Also, the first launch from the San Marco platform occurred on 26 April 1967. The San Marco platform was located 3 miles off the Kenyan coast in international waters and was set up to meet the need for an equatorial launch site. The United States supplied training, launch vehicles, and tracking, while the Italian Space Agency provided the launch complex, launch crew, and satellite. There were a total of nine Scout launches from San Marco.

The Transit navigation satellites were the main series of military payloads orbited by Scout. Scout launched the final operational prototype, Transit 5C, on 4 June 1964. The first five operational Transits were launched on Thor Able Stars, but the sixth and all subsequent Transits were launched by Scouts.

Phase V was the "incentive procurement" phase because the contract for fifteen Scouts included funding for LTV to purchase the motors. The fourth and fifth phases overlapped, with Phase V covering April 1968 to December 1971. All fifteen flights launched during Phase V were successful. The launches used twelve Scout-Bs, one Scout-A, one Scout-D, and one X-5C. In the Scout-D configuration, the Algol 3 replaced the Algol 2C. This enabled the Scout to put a 400-pound payload into a 300–mile orbit, a 30 percent increase over Scout-B.

Phase VI, July 1971 to February 1975, achieved a perfect launch record in its fourteen orbital attempts. During Phase VI, the Scout-D was used extensively, the first Scout-E and Scout-A1 were launched, and "B1" vehicles were used instead of "B." The A1 and B1 used the Algol 2C instead of the Algol 3 first stage, while the "E" was the first to use the new Antares 3A third stage. Phase VI is notable for the large number of international launches included. During this period, Scout launches included satellites for ESRO (European Space Research Organization, the predecessor to the European Space Agency, or ESA), West Germany (Aeros), Italy (San Marco), France (Eole), and Britain (three satellites).

Phase VII began in 1975 and extended into the mid-1980s. For this phase, NASA contracted with LTV for fifteen vehicles, and the F1 and G1 configurations were introduced. Also, the number of consecutive Scout successes was

Table 3
Scout Configurations [32]

	A-1	B-1	D-1	E-1	G-1
	1961	*1962*	*1963*	*1964*	*1968*
Stage 1	Algol 1B/C (XM-68)	Algol 1C/D	Algol 2A	Algol 2A/B	Algol 2B
Stage 1 thrust (lbf)	105,850	127,000	127,000	127,000	127,000
Stage 2	Castor IA (TX-33)	Castor IA	Castor IA	Castor IA	Castor IIA
Stage 2 thrust (lbf)	64,300	64,300	64,300	64,300	58,000
Stage 3	Antares IA (X-254)	Antares IIA (X-259)	Antares IIA (X-259)	Antares IIA (X-259)	Antares IIA (X-259)
Stage 3 thrust (lbf)	13,500	21,000	21,000	21,000	21,000
Stage 4	Altair IA (X-248)	Altair IA (X-248)	Altair IA (X-248)	Altair IIA (X-258)	—
Stage 4 thrust (lbf)	3,000	3,000	3,000	5,000	—
Capacity (lbs to 115 miles)	130	167	192	227	—
Missions	8	6	10	15	1
Success	3 + 1 PF	2	7	14	1

capped at thirty-seven due to an Antares 2B nozzle failure on 5 December 1975.

A unique Scout mission for the military mission launched two inflatable balloon targets for the F-15/ASAT antisatellite weapon. The payload, USA-13/USA-14 (a.k.a. the Avco Integrated Target Vehicle) was launched on 13 December 1985. In the test, a kinetic kill vehicle would have been launched from a McDonnell Douglas F-15 fighter to fly a suborbital trajectory to impact the target satellite. Because Congress ended antisatellite testing after one flight against another target (an aging science satellite), the program was abandoned.[33]

Phase VIII began with Scout flight S-208C on 12 October 1984 and ended with the final Scout launch in 1994. This final period involved using the G1 configuration to launch several Navy and DoD satellites.

The Scout was an ideal launcher for the Pentagon's new "lightsat" programs. Two communication satellites, MACSAT 1 and 2, were orbited in May 1990 (S-212C). They were intended for experimental use as message relays but saw operational use in the 1991 Gulf War. Another satellite, REX (Radia-

Table 3 (cont'd) Scout Configurations					
	X-1	**X-2**	**X-3**	**X-4**	**X-5C**
	1965	*1966*	*1972*	*1974*	*1979*
Stage 1	Algol 2B	Algol 2B	Algol 3A	Algol 3A	Algol 3A
Stage 1 thrust (lbf)	126,750	126,750	106,000	106,000	106,000
Stage 2	Castor II[34]	Castor IIA	Castor IIA	Castor IIA	Castor IIA
	(TX-354)	(TX-354-3)	(TX-354-3)	(TX-354-3)	(TX-354-3)
Stage 2 thrust (lbf)	58,000	58,000	58,000	58,000	58,000
Stage 3	Antares IIA (X-259)	Antares IIA (X-259)	Antares IIA (X-259)	Antares IIB (X-259)	Antares IIIA (TEM-762)
Stage 3 thrust (lbf)	21,000	21,000	21,000	21,000	18,000
Stage 4	Altair II (X-258)	Altair IIIA (FW4S)	Altair IIIA (FW4S)	Altair IIIA (TEM-640)	Altair IIIA (TEM-640)
Stage 4 thrust (lbf)	5,000	6,200	6,200	6,200	6,200
Capacity (lbs to 115 miles)	270	315	400	425	465
Missions	12	30	15	1 E plus 2 Fs[35]	18
Successes	12	27 + 1 PF	15	1 E plus 1 F	18

tion Experiment, P89-1), was launched in 1991 (S-216C) to study the physics of the electron density irregularities that disrupt radio signals.[36]

The Scout also launched NASA's first Small Explorer (SMEX) mission, SAMPEX (S-215C, 3 July 1992). The Scout's second to last flight (S-217C, 25 June 1993) orbited the RadCal (P92–1) satellite for the USAF Space Test Program. Two of the Scout's final three missions orbited MSTI (Miniature Sensor Technology Integration) satellites sponsored by the Ballistic Missile Defense Organization (BMDO)/USAF to test miniature sensor technologies for missile detection and tracking and to gather background data on the Earth's atmosphere and terrestrial environment. Scout launched MSTI-1 (S-210C) in November 1992. MSTI-2 (S-218C) was the Scout's final launch on 5 May 1994.

The storied career of the Scout may not yet be over. While an initial attempt by LTV to commercialize the system failed, an Italian company called BPD Difesa e Spazio (Defense & Space) in 1998 proposed developing an enhanced version of the Scout. The "San Marco Scout" (later renamed as the VEGA for Advanced Generation European Vector) was to be an enhanced version of the Scout-G1 with more than triple the lift capacity (1,750 pounds to a 750-mile polar orbit).[37] A redesigned VEGA, with a greatly reduced heritage from Scout, was formally approved as an ESA joint program on 16 December 2000.[38]

Blue Scout

Blue Scout was an Air Force branch from the NASA launch program, using Scouts modified with off-the-shelf hardware. The program was designated the Hyper Environmental Test System (HETS) 609A. The three Air Force Scouts were contemporaries of the Scout X. Blue Scout 2 (XRM-89) was essentially an X-1. Blue Scout 1 (XRM-90) was essentially an X-1C (X-1 without the fourth stage).[39]

Blue Scout 1 had three flights. Flight D-3 was a ballistic firing, D-6 was destroyed after veering off course, and D-7 was a failed reentry test. Blue Scout 2 was almost identical to Blue Scout 1 but with a fourth stage. D-4 performed measurements in the Van Allen radiation belt. D-5 also carried thermal neutron detectors, ion and electron density probes, and high-energy radiation experiments. Although the instrument capsule was lost, it did telemeter some of its data.

The Mercury-Scout (MS-1) on 1 November 1961 was the final Blue Scout 2 flight. It was to provide a dynamic checkout of the Mercury tracking network. Immediately after launch, the vehicle developed erratic motions, and after 28 seconds the booster began tearing apart. It turned out that "cross-wiring" caused the failure; the pitch and yaw rate gyro connectors were transposed, so that yaw rate error signals were transmitted to pitch control, and vice versa.

The operational phase of Blue Scout began with two suborbital missions, flights O-1 and O-3, in 1961. Shortly after the program moved into the operational phase, however, production of Blue Scout 1 and 2 was canceled. The only Blue Scout satellite attempt was the failed MS-1.

There was, however, one subsequent program in which the Air Force launched Scouts procured by NASA. In the highly classified Program 417, which remained secret until 1998, the Air Force used Scouts to launch a series of small weather satellites. These satellites were to provide cloud-cover information about the Soviet Union for use by reconnaissance satellites and strategic bombers. For this program, NASA acquired the Scout components and provided them to the Air Force, which took responsibility for the booster assembly and use.[40]

These Scouts experienced significant reliability problems. The most serious failures concerned the Scout ABL-258 fourth stage, which the Air Force replaced with the Lockheed MG-18. Even with that change, the unreliability of the other stages (in particular the third stage ABL-259) impacted the vehicle performance. This program was a notable exception to the highly successful civil-military cooperation encountered in most SLV programs. Air Force officials were frustrated by what they perceived as NASA's lack of responsiveness to the military's requests for help in resolving the anomalies and in getting the Scouts delivered on schedule. One probable reason for the lack of cooperation was that only three senior NASA officials knew the classified program's true objectives and understood the Air Force's sense of urgency.

Program 417 included five launches between April 1962 and September 1963. One was an unqualified success; one was a partial success; three were failures. Even the two successful launches had booster anomalies.[41] Despite the many problems, the two successfully orbited satellites did provide important cloud-cover information, including coverage of the Caribbean during the Cuban Missile Crisis.

Another Air Force program that used Blue Scout was the UHF Emergency Rocket Communications System (ERCS). On 29 September 1961, Strategic Air Command (SAC) issued a SOR (Specific Operation Requirement) for the ERCS. The first test launch of an ERCS Blue Scout Junior from Vandenberg AFB on 31 May 1962 was successful. The Air Force declared the three Blue Scout Junior launch sites at Wisner, West Point, and Tekamah, Nebraska, operational on 11 July 1963. This gave SAC another means of ensuring reliable command, control, and communications. The Blue Scout Juniors carried UHF recorders with a prerecorded force execution message that could be transmitted to all units within line of sight of the apogee flight. A follow-on ERCS became operational on 10 October 1967 when technicians installed the system on a Minuteman II missile at Whiteman AFB, Missouri. This newer system vastly improved SAC's ability to transmit command and control messages to its forces and also made the Blue Scout Junior system obsolete. The original ERCS was deactivated on 1 December 1967.[42]

Interlude

While the Scout was modified and improved repeatedly, work on new U.S. SLVs essentially came to a standstill after 1960. It would not be until 1990 that a completely new U.S. SLV made it into orbit.

There was one "old new" vehicle used in this period. The venerable Redstone flew two types of space missions after giving birth to the Jupiter C. In addition to carrying manned Mercury capsules on suborbital flights, it was mated with two solid upper stages (an Antares-2 second stage and an Alcyone third stage) to produce a configuration called the Redstone Sparta. From late 1966 to late 1967, this SLV performed nine suborbital launches and one orbital launch. All these were from a new location, Woomera, Australia. The last Redstone Sparta, launched 29 November 1967, put up the Wresat satellite, the first payload ever orbited from the continent of Australia.

Pegasus Rises

By the mid-1980s, when military interest in "smallsats" began to rise again, the Scout was the only small booster in the nation's fleet. As Scout production wound down and flashes of renewed interest in small satellites began to appear, an opportunity was created for a new vehicle.

In 1987, Orbital Sciences Corporation (OSC) hatched an audacious idea to create a new and very different launch vehicle, the Pegasus. No one had attempted an air-launched SLV since the all-but-forgotten Project Pilot thirty years before.[43] The Pegasus developers did not base their concept on Project Pilot, drawing inspiration instead from the Vought ASM-135A, a two-stage suborbital antisatellite vehicle that had been fired successfully from an Air Force F-15 in 1985.[44] Hercules Aerospace Corporation (now Alliant Techsystems, part of ATK Thiokol Propulsion Company) joined Orbital in a joint venture to develop the launch vehicle. By mid-1988, they had a contract from the Defense Research Projects Agency (DARPA) for one firm launch and five options, enabling them to raise the last segment of the development cost, which was over $50 million.

Pegasus was designed as a three-stage, all-solid-propellant SLV. Hercules developed three new motors (with some heritage from the company's work on ICBM programs). These were innovate designs themselves, with graphite composite casings. The first stage nozzle was fixed, while the second and third stage nozzles could be swiveled. Cold gas thrusters near the front of the vehicle provided steering control and precise orbital injection.

The initial version of the completed SLV, called the Standard Pegasus, had a 22-foot-span delta wing on Stage 1, was 49 feet long, and weighed 41,000 pounds. It was nearly twenty times the weight of the only comparable design, the NOTSNIK launcher, but had over three hundred times the designed payload capacity. The wing, too, was graphite composite structure. The materials and computer design aids the Orbital engineers worked with couldn't have been imagined by their 1958 counterparts. (Nor could NOTSNIK designers have imagined an SLV with fifteen microprocessors on board.) Of the original model Pegasus, 94 percent of the structural weight was graphite composite, with aluminum a distant second at 5 percent and titanium making up the rest. The payload shroud was 72 inches long and 46 inches in internal diameter. The heaviest payload orbited using this configuration was 596 pounds.

Orbital President David Thompson predicted during the development effort that it would take sixteen to eighteen launches to recoup the Pegasus development costs. An estimate released in 1989 stated that the per-launch price would be between $6 and $7.5 million. DARPA paid $6 million each for the first two launches (although the agency also absorbed costs such as range support).[45]

The air-launched system offered several performance advantages over

Image 5-5: The Pegasus air-launched space booster is carried aloft under the right wing of NASA's B-52 carrier aircraft on its first captive flight from the Dryden Flight Research Center, Edwards AFB, California. The first of two scheduled captive flights was completed on 9 November 1989. Pegasus is used to launch satellites into low-Earth orbits. (NASA Photo no. EC89-0309-3)

ground launches. First, the aircraft launched the Pegasus at over 40,000 feet, where 75 percent of the atmosphere had already been left behind, significantly reducing the energy needed to reach orbit. Second, there was a slight gain in performance (1–2 percent) derived from the speed of the subsonic carrier aircraft. The trajectory was flatter than that used for a ground-launched vehicle, so less power was bled away in getting the SLV into the right attitude for orbital injection. Finally, these combined factors allowed for a vehicle that didn't have to withstand as much stress as a traditional ground-launched vehicle, so lighter construction could be used, reducing the overall vehicle weight. Orbital chief engineer Antonio Elias, who conceived the Pegasus and led the design project, estimated that, adding up all the advantages of air launch, "you end up with a 10–15 percent reduction in the total delta-V that the motors have to provide."[46]

Another advantage of an air-launch vehicle is its mobility. By September 2000, Pegasus had placed satellites in orbit from Edwards and Vandenberg Air Force Bases in California, the Shuttle Landing Strip at Kennedy Space Center, the Air Force's Skid Strip on nearby Cape Canaveral Air Force Station, and Wallops Flight Facility in Virginia. In one unique case, a Pegasus XL rocket assembled at Vandenberg Air Force Base was ferried to Madrid, Spain, where it was integrated with the Spanish MINISAT satellite, and then ferried to Spain's Canary Islands off the coast of Africa for launch. In October 2000, the booster added another launch site when a mission to launch a NASA satellite originated from the Kwajalein Missile Range in the central Pacific Ocean.

The first stage of the Standard Pegasus used an Orion 50S motor. It was 29 feet long, 50 inches in diameter, and burned for 72 seconds. The second stage used an Orion 50 motor and the third an Orion 38. The first two Pegasus stages were used thirteen times before being replaced by the more powerful motors of the Pegasus XL. The XL stages were first used in 1994. Details of all these motors are shown in table 4.

A fourth, liquid-fueled stage, the Hydrazine Auxiliary Propulsion System (HAPS) could be added to either version of Pegasus. With a thrust of 150 lbf and a burn time of 241 seconds, it could be used to nudge a payload into a higher orbit or provide more precise orbital injection.

The first Pegasus launch took place on 5 April 1990, orbiting the experimental Pegsat and a small military satellite. The second launch, fifteen months later, was the first flight using the HAPS. It was marred by a guidance error that placed the payload, DARPA's Microsat constellation, in an orbit much lower than planned, although the mission objectives were met. The third launch, in February 1993, resumed the pattern of success.

In 1991, the Pegasus captured an Air Force contract for a single launch plus numerous options, and in 1994 obtained a similar contract from NASA. These firmly cemented Orbital's position as the leading U.S. provider of SLV services. The costs of the Pegasus launcher soon rose well above the $7.5 million estimate from 1989. Costs of a Pegasus XL contract vary depending on the number of launches desired and the support and integration options included, but two single-launch NASA contracts in 1996 were for $12 million and $15 million.[47]

There were a total of nine flights of the Standard Pegasus from 1990 through 1998, eight of which were completely successful. Through August of 2000, there were twenty-one flights of the XL, with two failures and one partial failure. The most famous achievement of Pegasus was launching the ORBCOMM communications satellites. Five Pegasus missions from 1997 through 1999 lofted thirty-three of the 92-pound satellites, putting the world's first private LEO communications network in place.

The development work on Pegasus has led to its stages being used in two other launchers: Orbital's larger Taurus and the Orbital/Suborbital Program

Table 4
Pegasus Configurations[48]

	Standard Pegasus	Pegasus XL
Overall length (feet)	49.3	55.4
Stage 1	Alliant Techsystems Orion 50S	Alliant Techsystems 50S-XL
Stage 1 thrust (lbf)	109,000	132,500
Stage 2	Alliant Techsystems Orion 50	Alliant Techsystems 50-XL
Stage 2 thrust (lbf)	26,500	34,500
Stage 3	Alliant Techsystems Orion 38	Alliant Techsystems Orion 38
Stage 3 thrust (lbf)	7,800	7,800
Propellant (all stages)	HTPB solid	HTPB solid
Capacity	825 pounds to 125 miles	1,000 pounds to 125 miles
Missions	9	20
Successful missions	9	17

(OSP) launcher, a collaboration with the Air Force described below. Pegasus derivatives were also slated to carry NASA's suborbital X-43 hypersonic test vehicles. Unfortunately, on 2 June 2001 the first of the X-43A vehicles was lost when the modified Pegasus XL booster appeared to shed parts as it accelerated away from the NASA NB-52B launch aircraft over the Pacific Ocean off the California coast. The boost rapidly deviated from its planned trajectory and was destroyed by a combination of aerodynamic forces and the range safety system.

Capacity and Cost Trends

From the Vanguard and Juno I through Pegasus, SLVs have generally increased in capacity. Based on the cost of orbiting one pound of payload, they have also become more cost efficient, even though the total cost per mission of today's market leader, the Pegasus XL, is high by historical SLV standards. A comparison, with all numbers adjusted to Base Year (BY) 00 dollars, is shown in table 5.

Start-ups: Old and New

The Pegasus was the first new SLV to enter active service in decades, but it was not the only attempt to build such a booster.

Table 5
Launcher Cost Comparisons[49]

Launcher	Pounds	Millions of Dollars per Pound (in Base Year 00 Dollars
Redstone	100	.101
Juno I	37	.269
Juno II	92	.039
Thor-Able	265	.049
Thor-Able-Star (Thor Epsilon)	500	.038
Thor-Delta	460	.024
Thor-Agena (Thor Hustler)	1,875	.009
Thor-Altair (Thor Burner 1)	1,130	.015
Scout X-1	130	.102
Scout X-2	167	.079
Scout X-3	192	.069
Scout X-4	227	.057
Scout A-1	128	.100
Scout B-1	315	.041
Scout D-1	408	.032
Scout E-1 and F	425	.030
Scout G-1	465	.028
Standard Pegasus	575	.021
Pegasus XL	1,015	.031

On 5 August 1981, the first test model of a new liquid-fuel low-cost commercial rocket, the Percheron, was ready for launch. The experimental Percheron, built by a start-up company called Space Services, Inc. (SSI), exploded in a static test. The Percheron was meant to be a modular launcher, built of identical components 39 feet high and 4 feet in diameter. There was never a second launch, as the company failed to obtain the customers or retain the investors needed.[50]

SSI changed its design, looking instead to solid-fuel systems. On 9 September 1982, SSI successfully flew a single-stage suborbital test vehicle. Named

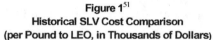

Figure 1[51]
Historical SLV Cost Comparison
(per Pound to LEO, in Thousands of Dollars)

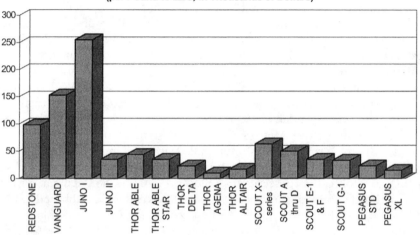

the Conestoga 1, the rocket was based on the Aerojet M56-A1 used as the second stage in Minuteman ICBMs.

However, SSI was unable to build on this accomplishment. In December 1990, SSI was purchased and became the Space Service Division of EER Systems. EER kept the Conestoga name and the modular concept, but switched to Thiokol solid motors. A mix of the Castor 4A and 4B, Star 37, and Star 48 motors were used. A family of boosters, from small to medium size, was envisioned, but the relatively large Conestoga 1620 (with a capacity of 1,950 pounds to LEO) was test-flown first. The first launch suffered a control failure and was destroyed after 45 seconds of flight. EER planned a second flight using its smaller Conestoga 1220 but was unable to raise adequate funding.

Pacific American Launch Systems, founded in 1982 by Gary Hudson, planned a family of launchers called the Liberty. The first to be built, the Liberty I, was a two-stage, liquid-fuel design, quite small by LV standards (75,000 lbf at liftoff). Each stage had a single engine. Stage 1 used LOX/kerosene and stage 2 N2O4/MMH. The whole vehicle was less than 56 feet high and was intended to cost $2.5 million to place a 500-pound payload in a polar LEO. The Strategic Defense Initiative Organization (SDIO) provided a small development contract, but Liberty had only reached the engine test stage when DARPA let the first light launcher contract to OSC for the Pegasus. The financial backers of Liberty dropped out, figuring the SLV market had been won by Pegasus before there was ever really a competition. About $2 million had been invested before the funding dried up and the company ceased operation in 1989.[52]

AeroAstro, one of the original "microspacecraft" development companies, explored entering the launch services market with its PA-X design. According to CEO Rick Fleeter, the company discovered it could lower the hardware costs for the launcher, but building a cheap microsatellite launcher proved difficult because overhead costs, such as range expenses, do not decrease with the size of the rocket involved. While engine tests were done, PA-X never attracted sufficient funding or demonstrated achieved economically feasibility. The project was suspended in 1995.[53]

A final entrant was MicroSat Launch Systems, which partnered with Canada's Bristol Aerospace in a venture called Orbital Express. Bristol is a leading builder of sounding rockets, and its motors were to be the building blocks of the SLV. Bristol had proposed its own SLV in 1984, but the Canadian government declined to provide the needed support. MicroSat was established in 1988 and joined with Bristol in 1989. The first rocket, the OV-10, was planned to carry up to 11 pounds into LEO for under $1 million. The OV-200 was to have a 200-pound capacity. By the end of 1990, these designs had evolved into a single vehicle, the Orbital Express, with a 300-pound capacity and a price tag of approximately $3.5 million. Several further evolutions of this design took place to meet the needs of prospective customers, but firm deals proved nearly impossible to close. In 1993, the only signed launch contract, with SDIO, was canceled by the customer. The project disintegrated and was never reconstituted.[54]

The ICBM Resurfaces

Many of the early SLVs had their origin in ICBM components. This concept had surfaced many times since. For example, the Minuteman ICBM's potential as a launch vehicle had been examined as early as 1963.[55] In the 1990s, with the retirement of the Minuteman II under arms control treaties, the idea of converting ICBMs to SLVs was reexamined.

The Air Force has funded development of small orbital launchers based on excess Minuteman II ICBMs, although the 1996 National Space Transportation Policy severely restricts when these can be used. Several orbital designs were produced by Lockheed Martin under the Multi-Service Launch System (MSLS) contract, but the contract expired without any such vehicles having been ordered. The replacement contract, known as the Orbital/Suborbital Program (OSP), was let to OSC. OSC launched the first OSP orbital vehicle in January 2000 and the second in July of the same year.

The OSP launcher, also known as Minotaur, incorporates the Orion 50XL and Orion 38 upper stages used on the Pegasus XL along with the Pegasus front section. This assembly is mated to the first two stages of a Minuteman II. The Minotaur is a highly capable launcher (up to 1,400 pounds to a 185-mile orbit at low inclinations), but, at a current cost estimate of $14 million per launch, does not significantly lower the per-mission costs of SLVs.[56]

Bantam: Building a Cheaper Booster

In 1997, NASA issued a request for studies to develop a new SLV called Bantam. The original request was for a rocket with a 330-pound capacity to LEO. It could be expendable or reusable, but it was to cost no more than $1.5 million per flight. This fitted well into NASA Administrator Dan Goldin's emphasis on "faster, better, cheaper" missions.

NASA provided $8 million in study money to Gencorp Aerojet, Pioneer Rocketplane, Summa Technology, and Universal Space Lines (USL). Except for the venerable Aerojet, all were new entrants in the launch vehicle business. Unfortunately, NASA's bold effort to look to new sources and new ideas failed to produce the desired results. USL's partner, Rocket Development Corporation, reported that a launcher the size of Bantam could not be operated commercially and therefore was not worth building, since the development costs could not be recouped by flying a small number of government-sponsored payloads.[57] No design submitted appeared likely to come close to the cost goal—around $3 million appeared to be the lowest cost achievable—and Bantam was refocused into a technology development effort in 1998.[58]

In April 1999, the concept resurfaced. At a meeting between Goldin and the Commander in Chief, U.S. Space Command, it was decided to try again to identify a low-cost approach to putting 220 pounds into orbit.

On the Horizon

As of this writing, the only company besides Pegasus actively promoting an American SLV with flight heritage is Coleman Aerospace Company. Coleman is in partnership with Israeli Aircraft Industries (IAI), and is proffering an American-built version of IAI's Shavit launch vehicle. The LK-1 booster can place a 975- pound payload into an equatorial orbit with a 430–mile apogee.[59] Since 1988, the Shavit launcher has succeeded in three of five attempts. Marketing, however, was severely hampered by U.S. State Department concerns about allowing a U.S. company to use a launcher with the Shavit's missile-related origins. Coleman hopes to revive marketing when a stretched version of the Shavit stages, with no missile heritage, is flight-proven in the near future.[60] Coleman is also exploring a small orbital launcher for NASA's Alternate Access to Station program based on its strategic target vehicles, which are extracted by parachute from a cargo plane and ignited at altitude.[61]

In addition to the fully operational Pegasus and the flight-ready LK-1, there are several development efforts and proposals for additional SLVs. The thrust of all these concepts is the same: despite NASA's skepticism about the Bantam study results, some designers remain convinced they can drastically reduce the costs of flying an SLV.

The most mature SLV project is the Microcosm Sprite. The Sprite's development is being funded by the Air Force Research Laboratory (AFRL) as part of a broader program to develop a launch vehicle family called Scorpius. The first orbital Sprite launch is planned in 2004. The goal of the program is a very simple, clean-sheet design that would place a 450-pound payload in low orbit for under $2 million per launch. The notional response time is as low as eight hours beginning with delivery of the payload.[62] Congress appropriated $6 million for this program for FY 01 and the same in FY02.

Assuming the Sprite is funded to completion, which depends on obtaining commercial customers as well as continuing DoD support, it may offer a significant new SLV capability. Even in a worst-case scenario, in which the Sprite missed both its cost and responsiveness targets by a considerable margin, the resulting vehicle would still be a major improvement in both areas over any other U.S. vehicle expected to fly in the near term.

The Air Force Research Laboratory's Space Vehicles directorate (AFRL/VS) has suggested what is, in essence, a return to the minimal air-launched vehicle concept used in Project Pilot. The AFRL proposal is to build an SLV that can be launched from an F-15 or F-22 fighter. The launcher would be a three-stage solid-fuel design, weighing 2,800–3,000 pounds, and able to place a 66-pound Microsat in a 310-mile orbit at an inclination of 55 degrees. Development cost is estimated at $200 million, with a $1 million recurring cost per booster.[63] At this writing, an advanced version of this concept, involving a high-performance, specialized aircraft as the first stage, is being studied at DARPA, where it is known as RASCAL (Rapid Access, Small Cargo, Affordable Launch.)[64] Other private efforts underway in 2002 included three balloon-launched concepts, being developed by JP Aerospace, the High Altitude Research Corporation (HARC), and Starhunter Corporation. Ground-launched systems were being pursued by Rocket Propulsion Engineering Company, whose Prospect LV-1 would carry 300–400 pounds to LEO, Wickman Spacecraft & Propulsion, which has a design for a 200-pound-payload booster, and Thurber Space Systems, which sought to build a 500-pound-class liquid-fuel booster.[65]

SLV Contributions, 1958–1999

The history of SLVs, especially the early vehicles, is often forgotten. Having retraced the record of such programs allows us to take a "big picture" look at what SLVs have contributed to the space programs of the United States.

In 1958 and 1959, SLVs were the only satellite launchers America had to pit against the heavy-lifting ICBM boosters of our Cold War adversary. U.S. engineers and technicians took up the challenge in a magnificent fashion. The Explorer and Vanguard programs struggled through setbacks and orbited the first U.S. satellite, the first solar-powered satellite, and the first important scientific instrumentation to be placed in orbit. They also made major contributions to the larger rockets that came after, with Vanguard contributing a host of technical innovations, plus entire stages, and Jupiter engine work leading eventually to the mighty Saturn system.

As the large boosters became available, the SLVs did not disappear. Instead, they proliferated, with the Thor and Scout joining the stable. These rockets filled the small but vital niches in the space program, orbiting military, scientific, and other satellites that continued to make major contributions to our knowledge of the universe and to satellite applications.

SLVs Today

The resurgence of interest in small satellites and small launchers in the past decade has borne considerable fruit. A new generation of "smaller, faster, cheaper" spacecraft is revolutionizing our approach to both Earth science and planetary exploration. The first private communications network in LEO, ORBCOMM, is using its small satellites to speed data around the globe. DoD is exploring microsatellites and their launchers for several applications in sensing, on-orbit servicing, and space object inspection.

Despite this, the 1990s market for microsatellite launch and the complementary small launch vehicles has not exploded as proponents hoped. Microsatellite launch is a classic chicken-and-egg problem. If there were a cheap, dependable small launcher on the market (Pegasus is dependable, but, at over $15 million per launch, too expensive for many missions), DoD and other users would be more likely to use microsatellites. On the other hand, if there was a steady demand for microsatellite launch services, SLV companies would be able to obtain financing and market such a vehicle.

The SLV market—indeed, the entire launch market—has changed drastically from the early days of SLVs. With the formation of ESA and the Ariane launch vehicle, European countries have moved towards larger satellites and away from U.S. launch vehicles. Other countries have developed their own

Table 6
U.S. Small Launch Vehicles

Name	First Launch Launch (or) End of Development)	Payload to LEO (pounds)	Origin
Vanguard	1958	50	Sounding rockets
Jupiter C/Juno I	1958	35	Redstone missile
NOTSNIK	1958	2	Small military rockets
Juno II	1959	100	Jupiter IRBM
Scout stages	1960	150	Polaris and Vanguard
Thor-Able Thor-Delta Thor-Altair	1958 1960 1965	330 500 550	Thor IRBM and Vanguard stages
Thor-Agena Thor-Able-Star	1959 1960	1,100 1,000	Thor IRBM
Standard Pegasus	1990	825	New design
Pegasus XL	1994	1,000	New design
Abandoned Startups			
Orbital Express	1993	300	Sounding rockets
Liberty	1988	500	New design
Percheron	1981	Not available	New design
PA-X	1995	22/750 (2 versions)	New design
Current Projects and Proposals			
Sprite		450	New design
LK-1		970	Israeli Shavit SLV
AFRL air launched		65	New design
Corvus		120–140	New design
Prospect LV-1		300–400	New design
Wickman Bantam		200	New design

Scout-class launchers (Japan in 1970, India in 1970, and Israel in 1988). Several reusable launchers are in prospect for the future, and their proponents hope these will make getting into space so cheap and routine they could, in theory, make the small expendable launcher obsolete.

Accordingly, while the microsatellite wave has created considerable interest in SLVs, they do not yet serve a large or stable market. Additionally, American SLVs face severe competition from the nations of the former Soviet Union. A July 1998 launch of two university microsats from a converted submarine-launched ballistic missile (SLBM) cost a reported $200,000.[66] A true renaissance, in which several new American SLVs reach flight status, will require entrepreneurs like Microcosm to prove their beliefs that SLVs can be built and launched far more cheaply than today's vehicles.

Conclusion

The history of American SLVs is a story of innovation, perseverance, and success. Despite the uncertain market situation, SLV proponents, with varying degrees of government support, are still forging ahead today. If small satellites truly become the ubiquitous vehicles predicted by their proponents, those who persevere will be in a strong position to capture an emerging market and continue the SLV's historical record of contributions to the exploration and exploitation of space.

Notes

1. Opinions expressed in this chapter are solely those of the authors. This study does not represent the views, policies, or plans of Analytic Services Inc. (ANSER), Booz Allen Hamilton, Titan-SenCom, any other corporation mentioned herein, the U.S. Air Force, the Department of Defense, or the United States government.

2. Milton Rosen, *The Viking Rocket Story* (New York: Harper & Brothers, 1955), pp. 25–27.

3. Milton Rosen, telephone interview with Matt Bille, 25 October 1999. Early in the Viking development project (then called Neptune), Rosen had met von Braun and discussed the planned design. Von Braun replied, "I don't think you'll get away with aluminum" (he felt the skin temperatures encountered would be too high for aluminum to take), and he was also skeptical about solving the complicated mathematics involved in steering with a gimbaled engine.

4. John P. Hagen, "The Viking and the Vanguard," in Eugene M. Emme, ed., *The History of Rocket Technology: Essays on Research, Development, and Utility* (Detroit: Wayne State University Press, 1964), p. 125.

5. On the postwar V-2 program at White Sands and Cape Canaveral, see U.S. Army Ordnance Corps/General Electric Co., "Hermes Guided Missile Research and Development Project, 1944–1954," 25 September 1959.

6. Wernher von Braun, *A Minimum Satellite Vehicle Based on Components Available from Missile Developments of the Army Ordnance Corps* (Huntsville, Ala.: Guided Missile Development Division, Redstone Arsenal), 15 September 1954.

7. Donald A. Quarles, "Ad Hoc Advisory Group on Special Capabilities" (Memorandum RD 263/7), 13 July 1955.

8. Office of the Assistant Secretary of Defense (Research and Development), *Report of the Ad Hoc Advisory Group on Special Capabilities*, (RD 263/9, Log No. 55–1067A), August 1955, p. 4.

9. Ibid., pp. 6, 15; Constance McLaughlin Green and Milton Lomask, *Vanguard: A History*, NASA SP-4202 (Washington, D.C.: 1970), pp. 49–50.

10. Walter McDougall, . . . *the Heavens and the Earth: A Political History of the Space Age* (New York: Basic Books, 1985), p. 130.

11. John B. Medaris, *Countdown for Decision* (New York: Putnam, 1960), p. 135.

12. Hagen, "Viking and the Vanguard," in Emme, ed., *History of Rocket Technology*, p. 129.

13. Rosen, interview with Bille, 25 October 1999.

14. Milton Rosen, telephone interview with Matt Bille, 6 November 1998.

15. Paul H. Satterfield and David S. Akeins, *Historical Monograph—Army Ordnance Satellite Program* (Army Ballistic Missile Agency, 1 November 1958), p. 62.

16. Walter Downhower, telephone interview with Matt Bille, 1 April 1999.

17. Raymond M. Nolan, "Details of Jupiter-C Guidance System Revealed," *Missiles and Rockets*, March 1958, pp. 183–84.

18. Frank Cartwright, telephone interview with Matt Bille, 16 November 1997.

19. China Lake Museum Foundation, *Secret City: A History of the Navy at China Lake* (video documentary), 1993; Mikey Strang, "Interview with Dr. Howard Wilcox, re: The NOTS Project," 25 May 1967, p. 1. Dr. Wilcox recalled that he suggested the air-launched concept; it has also been credited to Dr. Bill McLean, then Technical Director of NOTS.

20. Sewell, R.G.S. "Informal Thoughts on the NOTS Project" (memorandum), 2 September 1958.

21. Ibid.; Eugene Walton, "A Review of the Management of the NOTS I Project" (memorandum), Management Division, Naval Ordnance Test Station, August 1960; H.D. Parode and Special Writing Projects Branch, Naval Weapons Center, "Project Pilot: Informal Research Package," no date (cover letter by Parode is dated January 7, 1977).

22. Frank Knemeyer, "Astronautics Test Schedule" (NOTS internal memorandum), 18 March 1960.

23. It is commonly reported that further signals were received from this satellite, or that at least one other satellite may have made orbit. Frank St. George, who manned the NOTS tracking station, is certain he heard only one signal from one launch. Frank St. George, telephone interview with Matt Bille, 7 May 1999.

24. Julian Hartt, *The Mighty Thor* (New York: Duell, Sloan and Pearce, 1961), p. 137.

25. The current model is an AJ-10-118K, used by the Delta II, USAF Titan, and Japan's N-II.

26. E.S. Sutton, "From Polymers to Propellants to Rockets—A History of Thiokol," paper (AIAA 99-2929) presented at the 35th AIAA/ASME/SAE/ASEE Joint Propulsion Conference, 20–24 June 1999, Los Angeles, CA. Thiokol is now part of ATK Thiokol Propulsion Company.

27. Sometime during 1958, PARD's William Stoney, soon to be assigned overall responsibility for development of the new rocket, named it "Scout." Given engineers' propensity for acronyms, Scout has come to stand for "Solid Controlled Orbital Utility Test system." However, Stoney insists today that the various acronyms attached to the name "Scout"(even in official publications) have all been "after-the-fact additions." According to him, Scout was named in the spirit of the contemporary Explorer series of satellites with which the rocket was often paired. He and his colleagues gave no thought at the time to deriving its name from a functional acronym. See Helen T. Wells, Susan H. Whiteley, and Carrie E. Karegeannes, *Origins of NASA Names* (Washington, D.C.: NASA SP-4402, 1976), p. 20.

28. Andrew Wilson, "Scout—NASA's Small Satellite Launcher," *Spaceflight* 21, no. 11 (November 1979): 446–59.

29. PARD researchers were advisors to the Navy in the Polaris program.

30. Joseph Adams Shortal, ed., *History of Rocketry and Astronautics: Proceedings of the Twelfth, Thirteenth and Fourteenth History Symposia of the International Academy of Astronautics* (San Diego: Univelt, 1980).

31. Andrew Wilson, "The Scout Launcher . . . An Update," *Journal of the British Interplanetary Society* 34 (1981): 193–95; Wilson, "Scout—NASA's Small Satellite Launcher."

32. The five missions using a fifth stage include the letter "A," as in X-3A. The missions with an M2 fourth stage rather than the Altair use the letter "M" (X-2M and X-3M). Two X variants did not include a fourth stage and used a "C" designator (X-3C and X-5C).

33. Jonathan C. McDowell, "The Scout Launch Vehicle," *Journal of the British Interplanetary Society* 47 (1994): 99–108. The F-15 ASAT was initially tested against the orbiting Solwind P78-1 satellite, on 13 September 1985.

34. The Castor II had a lower thrust than the Castor I, but a longer burn time.

35. The F-1 is an E-1 without the small Alcyone 1A fifth-stage motor. Accordingly, the statistics for the E and F were combined, since stages 1–4 were identical.

36. A follow-on mission, REX II (P94-2), was launched by a Pegasus XL on 9 March 1996. SMC/TELM managed the program.

37. Marco Caporicci, "Vega: A European Small Launcher," *Reaching for the Skies*, no. 18 (Sept. 1998): 7–9: European Space Agency, "Successful Engine Test Firing for the Vega Small Launcher," Press Release, No. 23–99, Paris, 18 June 1999; Teal Group, *World Space Systems Briefing*, July 1998, Teal Group Corporation, Fairfax, VA.

38. *Space Daily*, "ESA Gives Green Light to New Rocket Program," 8 January 2001, http://www.spacedaily.com/news/booster-01a.html.

39. Joel Powell, "Blue Scout—Military Research Rocket," *Journal of the British Interplanetary Society* 35 (1982): 22–30; Eric W. Weisstein, "Blue Scout," 1996.

40. National Reconnaissance Office, *A History of Satellite Reconnaissance Volume II*, no date (declassified 1998). Scouts were delivered to Vandenberg in seventeen boxes from nine sources.

41. Note that this USAF program occurred in the same time frame as the Scout Phase II, which only achieved a 50 percent success rate.

42. State of Nebraska history on the web at http://www.esu3.k12.ne.us/districts/ralston/ms/slv1.html, accessed 10 July 2001.

43. There was also a proposal to use the North American X-15 research aircraft to carry a Blue Scout rocket. This would have used the X-15 as a first stage and the Blue Scout as an upper stage. This program was terminated before any hardware was produced.

44. Bob Richards, Orbital Sciences Corporation, interview with Matt Bille, 13 August 2001.

45. Bruce A. Smith, "Pegasus Air-Launched Test Vehicle Is Rolled Out," *Aviation Week & Space Technology* (14 August 1989): 36.

46. Craig Covault, "Commercial Winged Booster to Launch Satellites from B-52," *Aviation Week & Space Technology* (6 June 1988): 16.

47. Teal Group Corporation, *World Space Systems Briefing*, July 1998.

48. *Encyclopedia Astronautica*, http://www.rocketry.com/mwade/spaceflt.htm, updated 6 January 2002, (hereafter cited as *Encyclopedia Astronautica*); OSC Web site, www.orbital.com, accessed August 2001.

49. The cost per pound is based on the Flyaway Unit Cost of rockets that had at least one proven orbital success. All capacity and cost numbers are from *Encyclopedia Astronautica*; the figures were inflated/deflated using the Gross Domestic Product calculator found on www.jsc.nasa.gov/bu2/inflateGDP.

50. *Encyclopedia Astronautica*.

51. Data from Mark Wade, *Encyclopedia Astronautica, www.astronautix.com*, updated 6 January 2002.

52. Ibid.; Gary Hudson, e-mail to Matt Bille, 21 October 1999.

53. David Goldstein, AeroAstro, telephone interview with Matt Bille, 22 September 1999; Rick Fleeter, interview with Matt Bille, 22 August 2000.

54. Geoffrey V. Hughes, *The Orbital Express Project of Bristol Aerospace and MicroSat Launch Systems, Inc.* (Washington, D.C.: American Institute of Aeronautics and Astronautics, 1997).

55. National Reconnaissance Office, *A History of Satellite Reconnaissance Volume II*.

56. Scott Haskell, Leslie Doggrell Eugene Fossness, Dino Sciulli, Troy Meink, and Joseph Maly, "EELV Secondary Payload Adapter (ESPA)," *Proceedings of the 13th Annual AIAA/USU Conference on Small Satellites*, 23–26 August 1999, paper SSC99-X-2.

57. T.K. Mattingly, Rocket Development Corporation, telephone interview with Matt Bille, 2 November 1999.

58. Warren Ferster, "NASA Shifts Gears on Low-Cost Launcher," *Space News* (8–14 June 1998): 3.

59. Coleman Aerospace Company, "LK-1 Small Expendable Low Earth Orbit Launch Vehicle," 1999.

60. Frank Krens, Coleman Aerospace Company, telephone interviews with Matt Bille, 22 October 1999 and 8 August 2001.

61. Coleman Aerospace Company, "SELVS Mission Profile," http://www.crc.com/aerospace/spacelaunch.htm, accessed 25 July 2001.

62. James Berry, Robert Conger, and James Wertz, "The Sprite Mini-Lift Vehicle: Performance, Cost, and Schedule Projections for the First of the Scorpius Low-Cost Launch Vehicles," *Proceedings of the 11th Annual AIAA/USU Conference on Small Satellites*, 23–26 August 1999, paper SSC99-X-7; Jim Berry, Scorpius Chief Engineer, telephone interview with Matt Bille, 25 July 2001.

63. AFRL, "Launched on Demand Micro-satellites," concept submitted to SMC TPIPT, 1999; Kevin Bell and Ruth Moser, "Conceptual Designs for On-Demand Microsatellites and Air Launched Launch Vehicles," *Proceedings of the AIAA Space Technology Conference*, 28–30 September 1999.

64. DARPA RASCAL briefing, April 2001.

65. Greg Allison, High Altitude Research Corporation, interview with Matt Bille, May 2000; Joel Powell, JP Aerospace, interview with Matt Bille, 15 August 2001: Rocket Propulsion Engineering Company home page, www.rocketprop.com, accessed 20 August 2001; Chrishma Hunter Singh-Derewa, "New Launch Methodologies for the Micro-Millennia," *Proceedings of the 15th Annual AIAA/USU Conference on Small Satellites*, 13–16 August 2001, paper SSC01-X-5; "Thurber Space Systems," company brochure, 2001; Rocket Propulsion Engineering Company home page, www.rocketprop.com, accessed 20 August 2001.

66. University of Surrey Small Satellite home page, http://www.ee.surrey.ac.uk/SSC/SSHP/nano.html, accessed 29 August 2000.

6

Minuteman and the Development of Solid-Rocket Launch Technology

J.D. Hunley[1]

Introduction

The Minuteman missile was only possible because of a large number of technologies developed during and after World War II. Further technologies developed specifically for Minuteman have, in turn, helped to enable production of the large solid rocket motors for the Titan III and Titan IV space launch vehicles and the still larger Solid Rocket Boosters for the Space Shuttle. Thus, Minuteman was not only a major step forward in missile technology and our longest lasting strategic missile, it was a major contributor to launch-vehicle technology. This paper will trace the evolution in rocket technology that made Minuteman possible and will then discuss the ways in which Minuteman and other programs contributed to launch-vehicle technology. It will conclude with a brief discussion of the further advances necessary for the boosters used on the Titan launch vehicles and the Space Shuttles plus the eventual use of portions of Minuteman missiles themselves as stages for launch vehicles.

Castable Composite Propellants

One important strand in the complicated skein of technologies that led to Minuteman was the development of castable composite propellants.[2] During World War II, the vast majority of rockets produced for the war effort had employed extruded double-base propellants (consisting of nitrocellulose and nitroglycerine, hence the term double-base). These were limited in size by the nature of the extrusion process used at that time to produce them. In extrusion using a solvent, the process of production limited the size, and the elasticity was too low for bonding large charges to the motor case. With a solventless process, the immediate limitation was the size of the press used for extrusion, but this had its practical limits as well, and it yielded inferior physical properties and greater hazards of explosion than extrusion using a solvent. These factors created the need for castable propellants.

The first known castable solid propellant was a composite[3] developed by John W. Parsons of the rocket team at the Guggenheim Aeronautical Laboratory of the California Institute of Technology headed by Frank J. Malina. Parsons had taken chemistry courses at the University of Southern California and had worked as a chemist without having a degree. Under a series of contracts from the National Academy of Sciences, the U.S. Army Air Corps, and the U.S. Navy to develop jet-assisted takeoff (JATO) units to help aircraft in taking off under difficult conditions, such as short runways or carrier decks while carrying heavy loads, the rocket team had developed a restricted-burning propellant made from compressed black powder. Unfortunately, the propellant exploded upon ignition if stored for periods of time longer than a few hours. To solve this problem, in June 1942 the imaginative if unorthodox Parsons hit upon a castable propellant consisting of asphalt as a binder and potassium perchlorate as an oxidizer. The resultant mix, known as GALCIT 53 (after the acronym for the laboratory), did not have a particularly impressive performance compared, for example, with ballistite (a double-base composition) but retained its performance much better than compressed black powder at temperatures down to 40 degrees Fahrenheit (F).[4]

At even lower temperatures, however, GALCIT 53 cracked. It also melted in the tropic sun, and it was very smoky when burning. This limited the takeoff of second and follow-on aircraft using JATO units on a single runway because the smoke restricted visibility. Consequently, researchers at GALCIT, which became the Jet Propulsion Laboratory (JPL) in the course of 1944, began searching for an elastic binder with storage limits beyond GALCIT 53's extremes of -9 to 120 degrees F. In particular, a young engineer named Charles Bartley, who was employed at JPL from June 1944 until August 1951, began examining synthetic rubbers and polymers, eventually hitting upon a liquid polysulfide compound designated LP-2, made by the Thiokol Chemical Corporation for such purposes as sealing aircraft fuel tanks.[5]

Like many important innovations, LP-2 resulted from an initial discovery that was made inadvertently. In 1926, a physician named Joseph C. Patrick, who had found chemistry more interesting than medicine, had been seeking an inexpensive way to produce antifreeze from ethylene using sodium polysulfide as a hydrolyzing agent. Instead of antifreeze, his procedure yielded a synthetic rubber that a firm he helped found, Thiokol, marketed in the form of gaskets, sealants, adhesives, and coatings, since the polysulfide polymer was resistant to weather, solvents, and electrical arcing. Then in 1942, Patrick and an employee named H.L. Ferguson found a way to make the first liquid poly-

Image 6-1: The origins of solid-rocket booster technology may be traced to the first rocket-assisted airplane takeoffs, using Guggenheim Aeronautical Laboratory, Caltech (GALCIT)–developed solid propellant, 28-lbf, jet-assisted takeoff units (JATO) during World War II. This photo is of the first test of the JATO concept, conducted at March Field, California, on 12 August 1941. (NASA Photo)

mer that included no volatile solvent and yet could be cured to form a rubber-like solid. It found use during World War II in sealing not only fuel tanks but also gun turrets, fuselages, air ducts, and the like.[6]

Before learning of LP-2, Bartley and his associates at JPL had tried a variety of moldable synthetic rubbers to use as binders and fuels, including Buna-S, Buna-N, and neoprene. Neoprene had the best properties for use as a binder and burned the best of the lot, but molding it required high pressures. Like the extrusion process used with double-base propellants, this made the production of large propellant grains impractical.[7] Meanwhile, Thiokol chemists seeking new applications for their polymer had begun to send out technical data about it. At a meeting of the American Chemical Society, Bartley was asking about a liquid that would polymerize to a solid elastomer (a rubberlike substance). Dr. Frank M. McMillan, who represented Shell in the San Francisco area, told him about Thiokol's product, and Bartley acquired small quantities of LP-2 from Walt Boswell, the Thiokol Chemical Corporation's representative for the western United States.[8]

With encouragement from U.S. Army Ordnance and the U.S. Navy, Bartley—joined by John I. Shafer, a design engineer, and H. Lawrence Thackwell Jr., whose expertise lay in aircraft structures—began to develop a small test vehicle designated Thunderbird in 1947. This vehicle had a six-inch diameter and was intended to test whether polysulfide propellants could withstand the forces of high acceleration that a potential large launch vehicle would undergo. Bartley had already found that an end-burning grain of polysulfide propellant did not produce steady thrust but burned faster at first and then leveled off. He attributed this to accelerated burning along the case to form a convex cone, a hypothesis that he confirmed by quenching the flame partway through the burn.

To solve this problem of unsteady thrust, the three JPL engineers adopted a grain design used in Great Britain in the late 1930s that featured an internal-burning, star-shaped cavity. This design protected the case from excess heat because all of the burning was in the middle of the propellant grain. It also provided a constant level of thrust because as the internal star points burned away, the internal cavity became a cylinder with roughly the same surface area as the initial star. Bartley had read about the star design from a British report and had instructed Shafer to investigate it. He found that the government-owned, contractor-operated Allegany Ballistics Laboratory (ABL) in Pinto, West Virginia, had used the British design in the uncompleted Vicar rocket and a scaled-down version of it named the Curate. Using equations from the ABL report on the projects, Shafer developed a number of star designs beginning in 1947. Combining a polysulfide propellant with the star design and casting it in the case so that it bonded thereto, the team under Bartley produced a successful Thunderbird rocket that passed its flight tests in 1947–48.[9]

Meanwhile, in a parallel development, Caltech had conducted another rocket program during World War II in Eaton Canyon northeast of Pasadena. This had produced tactical rockets for the Navy using a dry extrusion process for double-base propellants (not composites). They all used external-burning grains, but in 1944 the Eaton Canyon project expanded into the Mojave Desert at what became the Naval Ordnance Test Station (NOTS), Inyokern. By the end of 1945, Caltech operations at Eaton Canyon had ceased, and all operations were at NOTS, carried on by the Navy. In early 1946, NOTS began successful flight tests with a Caltech rocket that had a 5.0-inch diameter and a length of 38 inches. Known as the "White Whizzer," this rocket, which was designed and built at NOTS when the programs there were still managed by Caltech, had an aluminum case and an internal burning charge. It became the prototype for Navy tactical rockets over the next twenty years, including the famous Sidewinder guided missile and folding-fin aircraft rockets developed at NOTS (Naval Weapons Center, China Lake after 1967), where the Navy estimated in 1968 that 70 percent of the airborne weapons in the armed services were developed. Both the White Whizzer and the Curate were significant because all the service rockets used by the United States in World War II had used external-burning charges and because the internal-burning charges were more susceptible to combustion instability, hence more difficult to employ successfully.[10]

While these rockets were being tested, another significant development for composite propellants was occurring—the replacement of potassium perchlorate as an oxidizer by ammonium perchlorate, which offered higher performance (specific impulse) and less smoke. Apparently, the Thunderbird used a propellant with the designation JPL 100, which contained a mixture of ammonium perchlorate and potassium perchlorate.[11] Already in 1947, however, JPL had developed a propellant designated JPL 118 that used only ammonium perchlorate as an oxidizer together with polysulfide as the binder. Although this propellant had yet to be fully investigated in 1947,[12] by mid-1948 it had been tested and showed that it had a specific impulse of 198 pounds of thrust per pound of propellant burned per second (lbf-sec/lbm).[13] This was still relatively low compared with a typical performance of double-base propellants of about 230 lbf-sec/lbm, but it was higher than the 185 lbf-sec/lbm for the asphalt-potassium perchlorate propellant.[14]

During 1942 Malina, Parsons, and other members of the GALCIT rocket group, including the famous aerodynamicist Theodore von Kármán, had formed the Aerojet Engineering Corporation to produce their JATO motors. Later known as Aerojet General Corporation, this firm began using ammonium perchlorate in its Aeroplex (polyester polymer) propellants in 1948 in order to increase specific impulse and reduce smoke. Funded by the Navy's Bureau of Aeronautics to develop a basic understanding of the production and employment of solid propellants, Aerojet increased the specific impulse of its ammonium perchlorate propellants to 235 lbf-sec/lbm, but its Aeroplex binder was not case bondable, leading the firm to switch in 1954 to a polyurethane propellant that was.[15]

In the interim, Thiokol sought to sell its polymer to Aerojet and another manufacturer of rockets (Hercules, on which see below), but both rejected it because its 32 percent sulfur content made it a poor fuel. Thiokol sales representative Joseph Crosby then asked the Army if it wanted to allow the promising developments with the Thunderbird rocket to end. The result in the spring of 1947 was Army Ordnance encouragement for Thiokol to go into the rocket business itself. In early 1948, Thiokol set up its budding rocket operations in a former ordnance plant in Elkton, Maryland. In April 1949, it moved to the Army's new Rocket Research and Development Center at Redstone Arsenal in Huntsville, Alabama.[16]

The Falcon Missile

Under contract to the Army, Thiokol produced a T-40 motor intended for use as a JATO unit. As a propellant, it used JPL 100 (rechristened T-10) with both potassium perchlorate and ammonium perchlorate as oxidizers in a polysulfide binder; it had a specific impulse of 190 lbf-sec/lbm. Unlike the Aerojet propellants of that day, it featured a case-bonded motor design. Also in 1949, Thiokol designed the T-41 motor for the Hughes Aircraft Company's Falcon missile under development for the U.S. Air Force. This was a shorter version of the Thunderbird motor whose development was started at Elkton and moved to Huntsville, where a larger version called the T-42 evolved from it.[17]

According to Air Force officer (later Colonel) Edward N. Hall, who was highly instrumental in promoting the development of solid propellants, the Falcon tactical (air-to-air) missile contributed "quality control techniques for rubber-base propellants, design data for case-bonded grains, [and] aging characteristics of rubber-based propellants."[18] As indicated by the narrative above, its contributions were not solely those of Thiokol but were shared by JPL in an early example of technology transfer. In October 1947, Charles Bartley from JPL was present at a meeting of Thiokol personnel, representatives of Army Ordnance, and the Navy's Bureau of Aeronautics to discuss the kind of work Thiokol was expected to do in the further development of polysulfide propellants. The next day, Bartley met again with Thiokol personnel to relate JPL's experience with polysulfide-perchlorate propellants.[19]

In about January or February 1949, a trip report by a Thiokol employee discussed a visit to JPL's Solid Rocket Section, of which Bartley was the chief. It covered such matters as grinding ammonium and potassium perchlorate, mixing it and the liquid polymer in a vertical mixer, pouring the propellant, the grease used to extract the mandrel to create the internal cavity, preparation of the liner for the combustion chamber, igniters using black or a special igniter powder, and testing. Also reported was a visit to Western Electrochemical Company, which supplied the perchlorate. The document concluded with some recommended changes in Thiokol's operating procedures at Elkton as a result of the trip.[20]

As helpful as JPL's assistance was to Thiokol, however, the kinds of contributions Hall mentions seem to have come significantly from work done at Thiokol's Elkton and Huntsville plants. Thiokol discovered that the size of perchlorate particles was important in motor operation and propellant castability, so it introduced a micromerograph to measure particle size. To reduce absorption of moisture by the perchlorates, it installed air conditioning in the grinding rooms. It determined the optimal mixing time for the propellant and replaced a barium grease JPL had used to extract the mandrel from the middle of the cast propellant after it cured with a Teflon coating. This latter step was necessary not only because the grease-affected part of the propellant had to be sanded after extraction of the mandrel but also because the grease had enfolded into the propellant, causing weak areas. Thiokol also introduced a "temperature-programmed cure cycle," pressurized curing, and a method of casting that eliminated propellant voids resulting from shrinkage and air bubbles.[21]

While these developments were occurring at Thiokol, JPL was wrestling with a test vehicle named the Sergeant, not to be confused with the later missile of that name. Army Ordnance had authorized its development as a sounding rocket with a diameter of 15 inches—about twice as large as any existing solid-propellant motor. Designed with an extremely thin (0.065-inch) steel case and a star-shaped perforation, it was expected to attain an altitude of up to 700,000 feet while carrying a 50-pound payload. Static tests with a thicker case in February 1949 showed that a polysulfide grain of that diameter could function without deformation. But when the JPL researchers (including Bartley, Shafer, and Thackwell) shifted to the thinner case, the result was twelve successive explosions, the last on 27 April 1950. At this point, JPL Director Louis G. Dunn canceled the project for the sounding rocket and cut back all solid-propellant work at the laboratory to basic research. The researchers soon determined that the cause of the explosions was a chamber pressure that was too high for the thin case and points on the star configuration that were too sharp, promoting cracks. An easy solution would have been a more conservative (thicker) case and rounded points on the star. With Dunn's cancellation of the project, Thackwell took his knowledge of solid-propellant rocketry to Thiokol's Redstone Division in Huntsville, Alabama.[22]

The RV-A-10 Missile

In the meantime, Thiokol had teamed up with General Electric in the Hermes project to produce a solid-propellant rocket much larger than the Sergeant sounding rocket. Project Hermes, sponsored by the Army, began in November 1944 with an Army contract awarded to General Electric to study the German V-2 liquid-propellant missile. Starting in 1946, sixty-seven V-2s were fired at White Sands, New Mexico. Meanwhile, in 1950 work began at Redstone Arsenal on the Hermes A-1, a smaller, modified version of the V-2. At the same time, Army Ordnance initiated a solid-propellant rocket initially known as the A-2.[23]

The initial requirements for the A-2 were to carry a 500-pound warhead to a range of as far as 86 miles, but the requirements changed to a payload weighing 1,500 pounds, necessitating a motor with a diameter of 31 inches. Thiokol became the contractor for the motor, and development started in May 1950. By December 1951, the program had successfully completed a static test of the 31-inch motor. Then, between January 1952 and March 1953, there were twenty more static tests at Redstone Arsenal and four flight tests of the missile at Patrick Air Force Base, Florida. In the course of the program, the missile came to be designated the RV-A-10. During the testing, project engineers encountered problems with nozzle erosion and combustion instability that they had not anticipated.[24]

The four flight tests achieved a maximum range of 52 miles (on flight one) and a maximum altitude of 195,000 feet (flight two) using a 0.20-inch-thick motor case and a propellant grain employing a star-shaped perforation with broad tips on the star. The propellant was designated TRX-110A, which included ammonium perchlorate as the oxidizer. The propellant took advantage of an Air Force–sponsored project (MX-105) entitled Improvement of Polysulfide-Perchlorate Propellants that had begun in 1950 and issued a final report in May 1951, written by Thiokol employees. Meanwhile, however, on test motor No. 2 a propellant designated T-13, which contained a polysulfide designated LP-33 and ammonium perchlorate, achieved a specific impulse at sea level of 196.2 at 80 degrees F but also demonstrated combustion instability. This led to the shift to TRX-110, which had a slightly lower specific impulse but no combustion instability.[25]

Thiokol had arrived at the blunter-tipped star perforation as a result of photoelastic studies of grains performed at the company's request by the Armour Institute (later renamed the Illinois Institute of Technology). This, together with a thicker case wall than JPL had used on the Sergeant sounding rocket, eliminated the problems with cracks and explosions that the Sergeant test vehicle had experienced.[26] However, TRX-110 proved not to have sufficient initial thrust. The solution to this problem was to change the size of the ammonium perchlorate particles from a mixture of coarse and fine pieces to one of consistently fine particles, which yielded not only higher initial thrust but also a more consistent thrust over time—a desirable trait. Meanwhile, the Thiokol-GE team gradually learned about the thermal environment to which the RV-A-10 nozzles were exposed. Design of the nozzles evolved through subscale and full-scale motor tests employing a variety of materials and techniques for fabrication. The best materials proved to be steel with carbon inserts, while a roll weld with finish machining proved superior to casting or forging as the method of production for the nozzle itself.[27]

Another problem encountered in fabricating the large grain for the RV-A-10 was the appearance of cracks and voids in the grain when it was cured at atmospheric pressure, probably the cause of a burn-out of the liner on motor No. 2. The solution proved to be twofold: (1) Thiokol poured the first two mixes of propellant into the motor chamber at a temperature 10 degrees F hotter than normal, with the last mix 10 degrees F cooler than normal, and (2) Thiokol personnel cured the propellant under pressure with a layer of liner material laid over the propellant to prevent air from contacting the grain. Together, these two procedures eliminated the voids and cracks.[28]

With these advances in the state of the art of producing solid-propellant motors, the RV-A-10 became the first known solid-propellant rocket motor of such a large size—31 inches in diameter and 14 feet 4 inches long—to be flight tested by that date (February–March 1953). Among its other firsts were scaling-up the mixing and casting of polysulfide propellants to the extent that more than 5,000 pounds of it could be processed in a single day; the routine use of many mixes in a single motor; the use of a tubular igniter rolled into coiled plastic tubing (called a jelly roll) to avoid the requirement for a heavy closure at the nozzle end to aid in ignition; and the first use of jet vanes inserted in the exhaust stream to provide thrust vector control of a solid rocket. As recently as December 1945, the head of the Office of Scientific Research and Development during World War II, Vannevar Bush, had stated, "I don't think anybody in the world knows how [to build an accurate intercontinental ballistic missile] and I feel confident it will not be done for a long time to come." Many people even in the rocket field did not believe that solid-propellant rockets could be efficient enough or burn long enough to serve as long-range missiles. The RV-A-10 was the first rocket to remove such doubts from at least some people's minds. Arguably, it provided the technological basis for the entire next generation of missiles from Polaris and Minuteman to the large solid boosters for the Titan IIIs and IVs and the Space Shuttle,[29] although many further technological developments would be necessary before they became possible (including significant improvements in propellant performance).

The Sergeant Missile

The first major application of the technologies developed for the RV-A-10 was the Sergeant missile, for which JPL began planning in 1953 under its ORDCIT contract with the Army's Ordnance Department. JPL submitted a proposal for a Sergeant missile in April 1954, and on 11 June 1954 the Army's Chief of Ordnance programmed $100,000 in fiscal year 1955 research and development funds for developing Sergeant. At the same time, he transferred control of the effort to the commanding general of Redstone Arsenal. Using lessons learned from the liquid-propellant Corporal missile, JPL proposed a co-contractor for the development and ultimate manufacture of the missile. In February 1956, the Sergeant Contractor Selection Committee unanimously chose the Sperry Gyroscope Company (a division of Sperry Rand Corporation) for this role, based on JPL's recommendation and Sperry's capabilities and experience with other missiles, including being prime contractor for the Sparrow I air-to-air missile system for the Navy. Meanwhile, already on 1 April 1954, the Redstone Arsenal had entered into a supplemental agreement with the Redstone Division of Thiokol to work on the solid-propellant motor for the Sergeant, and the program to develop Sergeant began officially in January 1955.[30]

This is not the place for a detailed history of the Sergeant missile. It took longer to develop than was originally planned and was not operational until 1962, by which time the Navy had completed the much more capable Polaris

A1 and the Air Force was close to fielding the much more significant and successful Minuteman I. JPL Director Louis Dunn had warned in 1954 that if the Army did not provide for an orderly research and development program for the Sergeant, "ill-chosen designs . . . [would] plague the system for many years." In fact, the Army did not provide consistent funding and then insisted on a compressed schedule. This problem was complicated by differences between JPL and Sperry and by JPL's becoming a NASA instead of an Army contractor in December 1958. The result was a missile that failed to meet its in-flight reliability of 95 percent. While it met a slipped ordnance support readiness date of June 1962, it was classified as a limited-production weapons system until June 1968. On the other hand, it was equal to its predecessor, Corporal, in range and firepower while being only half as large and requiring less than a third as much ground-support equipment. Its solid-propellant motor could be gotten ready for firing in a matter of minutes instead of the hours required for the liquid-propellant Corporal. An all-inertial guidance system on board Sergeant made it virtually immune to known enemy countermeasures, whereas Corporal depended on a vulnerable electronic link to guidance equipment on the ground.[31]

The Sergeant motor was a modification or direct descendant of the RV-A-10 motor. The latter (in the TRX-110A formulation) had ammonium perchlorate as an oxidizer, and the TP-E8057 propellant for the Sergeant motor (designated JPL 500) also used that oxidizer and an LP-33 liquid polymer in addition to p-quinone dioxime curing agent and smaller amounts of magnesium oxide as a reinforcing agent, sulfur as a curing accelerator, and nylon tow as another reinforcing agent. Its specific impulse was about 185 lbf-sec/lbm. It employed a five-point star grain configuration, used a case of steel at slightly more than half the thickness of the RV-A-10 case, and a steel nozzle with a graphite nozzle-throat insert.[32]

Polaris

While Sergeant was in development, the Navy had gotten the Polaris missile developed and operational in a shorter period of time, although the technology it employed goes back to some of the same sources that contributed to the Sergeant. It also, however, brought some new players to the table. Polaris came along at a time when interest in ballistic missiles among the military services had increased significantly and with that interest came rivalry over roles and missions. The full story of these developments is too complicated and nuanced to be covered in full here. But in all three of the contending services—the Army, Navy, and Air Force—there were both proponents and opponents of large ballistic missiles. One development that gave hope and ammunition to the proponents was a prediction in the spring of 1953 by the Air Force Scientific Advisory Board that by the end of the decade the United

States would have developed a 1,500-pound thermonuclear warhead with a 1-megaton yield (i.e., equal to the explosive power of a million tons of TNT). This was half the weight and fifty times the power of the initial warhead proposed for the liquid-propellant Atlas missile, thereby reducing not only the size of the missile required to launch the warhead but also the accuracy required for its guidance system. This was welcome news at a time when the United States was becoming increasingly concerned that the Soviet Union would develop a long-range ballistic missile before the United States did.

Because of this concern, President Eisenhower asked the Office of Defense Mobilization Science Advisory Committee within the Executive Office of the President to examine how science and technology could protect the country against a surprise attack. The committee created the Technological Capabilities Panel, better known as the Killian Committee because its chairman was James Killian, the president of Massachusetts Institute of Technology. In its report of February 1955, the Killian Committee recommended that the National Security Council recognize the intercontinental ballistic missile (ICBM) as a very high priority. It also argued that the United States develop both land- and sea-based intermediate-range ballistic missiles (IRBM) capable of carrying warheads 1,500 miles (as opposed to the 5,000-mile-plus range of ICBMs). This recommendation sparked rivalry among the three services as to which one(s) should develop the IRBM. In November 1955, the Joint Chiefs of Staff recommended that the Air Force develop a land-based IRBM (the liquid-propellant Thor) and the Army and Navy jointly develop the liquid-propellant Jupiter.[33]

Placing a missile at sea, especially beneath the surface on board a submarine, made it mobile, hard to find, and easy to disperse, hence difficult for the enemy to destroy. It also removed the missile a long distance from U.S. cities, thereby protecting them from first strikes or countermeasures.[34] However, adapting a missile for launch from a submarine required engineering changes that would delay development at a time when the perceived threat was great. The Army was willing to undertake these changes because its position in the roles-and-missions struggle was weak and the Navy offered funds to pay for the adaptations. For the Navy, however, the idea of a liquid-propellant missile aboard ship was anathema.[35] The propellants for Jupiter were the cryogenic (extremely cold) liquid oxygen and the hydrocarbon RP-1 (kerosene).[36] In 1949, the Navy had conducted tests to see what would happen if an accident occurred with a liquid-propellant missile. The thermal shock from the liquid oxygen split the supporting I-beams of a mocked-up steel warship and cracked steel plates.[37]

Adding to the Navy's aversion to liquids was the fact that cryogenic propellants like liquid oxygen had to be loaded just before launch so that little of the material boiled away. This was a real problem in the confined spaces of a submarine and was a problem even for ground-launched missiles since speed

Image 6-2: A Peacekeeper missile being test launched from its underground silo. The Peacekeeper is the most recent of all U.S. intercontinental ballistic missiles, and it was built using solid-rocket technology. With the end of the cold war, the United States revised its strategic policy and has agreed to eliminate the multiple reentry vehicle Peacekeeper ICBMs by the year 2003 as part of the Strategic Arms Reduction Treaty II. (USAF Photo)

of launch is critical for any potential use of missiles as a counterthreat to a perceived enemy. Even later, when earth-storable liquid propellants became available, they were not practical on a submarine, where the pitching and rolling of the vessel created intolerable stresses on piping, valves, and fittings in the missile's engine and could cause leaks in propellant tanks. In addition, the low density of liquid propellants necessitated larger missiles than would conveniently fit on a submarine.[38]

For these reasons, and because most of the technical specialists assigned to the Navy's Special Projects Office (SPO) for a fleet ballistic missile came from solid-propellant-research backgrounds, the Navy made no secret of its plan to convert to a solid-propellant missile as soon as advances in the technology permitted. This did not worry the Army because most of its experts foresaw no likelihood of that happening in the near future.[39] Among the unresolved technical problems were low performance (specific impulse), unstable combustion (also a problem with liquid-propellant rockets), the lack of the ability to ensure that combustion terminated as soon as the precise desired velocity had been achieved (to assure accuracy), control of the thrust vector (for steering), reliability of nozzle materials (given the heat and corrosive chemicals to which they were exposed), materials and technology for the construction of large combustion chambers, and warhead size.[40]

Despite the daunting nature of this list, solutions to all of these problems yielded rapidly to research and development on a variety of fronts. Early in January 1956, the Navy approached the Aerojet General Corporation and the Lockheed Missile and Space Division for help in producing a solid-propellant missile. Capt. Levering Smith, who as a commander at NOTS had participated in developing a 50-foot solid-propellant missile called "Big Stoop," was brought into the Special Projects Office to direct their efforts, supported by studies at NOTS. Partnering with the Navy's Special Projects Office, the contractors came up with a design designated the Jupiter S (for solid) that had enough thrust to carry a standard nuclear warhead the requisite distance. It would do so by clustering six solid rockets in a first stage and having one in the second stage. The Jupiter S, however, would be 44 feet long and 10 feet in diameter. An 8,500-ton vessel could carry only four of them compared to sixteen of the later Polaris missiles. Consequently, the Navy and its contractors continued their search.[41]

One key contributor to the solution was the Atlantic Research Corporation (ARC), a firm founded on 24 January 1949 in Virginia to "work in chemistry and chemical engineering," as Dr. Arthur Sloan—one of its two founders—put it. The Navy's Bureau of Ordnance awarded the fledgling firm a contract in 1949 to develop a polyvinyl chloride fuel and binder that led to the Arcite family of propellants. Later, even before its contacts with Aerojet and Lockheed, the Navy contracted with Atlantic Research to seek a way to improve the specific impulse of solid propellants. Two young engineers working for Atlantic Research—Keith Rumbel and Charles B. Henderson, both chemical engineers trained at the Massachusetts Institute of Technology—had undertaken theoretical studies of how to increase the performance of solid propellants in 1954. In the course of their research, they became aware that other engineers, including some from Aerojet, had calculated that the performance of solid propellants could be increased by including aluminum powder among the ingredients. But the calculations of the other researchers had shown that once the amount of aluminum in the propellant exceeded 5 percent of the total weight, the performance of the propellant would again decrease. The other researchers—basing their calculations on contemporary theory about the nature of the chemical reactions that occurred when the propellant burned and doing those cumbersome calculations without the aid of computers—achieved incorrect answers and abandoned the investigation of aluminum as an additive that would significantly increase specific impulse (although it did serve to dampen some kinds of combustion instability).

Rumbel and Henderson refused to be deterred by the alleged 5-percent limit. As Henderson commented in an interview in 1997, they did not have the bad advice that beyond the 5-percent point, the aluminum in the propellant would not burn. Consequently, they tested polyvinyl chloride with significantly more aluminum in it, finding that the performance increased as they added more aluminum. They fired small grains of the propellant into an enclosed chamber filled with helium and found that as the chamber pressure increased, they needed to increase the amount of oxygen in the products of combustion as both aluminum and carbon combined with the oxygen. They also found that the energy content of the propellant had to be high enough to ignite the aluminum, which apparently needed a thin coating of aluminum oxide on its exterior to melt before it could ignite, necessitating a high flame temperature. A further requirement was that the aluminum particles had to be small to burn properly. Thus, a lack of false assumptions (which passed for theory at that time) and good empirical technique enabled Rumbel and Henderson to achieve an important technical breakthrough. They began their laboratory work on aluminum in 1955, with the first two static firings of rocket motors occurring in January 1956. The grains they used were 5 inches in diameter and 14 inches long with a star perforation. Their propellant achieved specific impulses significantly higher than the RV-A-10 or the Sergeant missile. Soon after this, in early 1956, Aerojet verified the results of Atlantic Research's firings in the latter's test stands by actually launching a 100-pound rocket (by weight).[42]

The addition of aluminum essentially provided sufficient performance for Polaris, but that missile's development could succeed only if the other technical problems could be solved. Fortunately, the Navy was not the only service working to solve them. At Wright-Patterson Air Force Base (AFB) in Ohio, Maj. Edward N. Hall had first become acquainted with missile technology when he led an Air Force Propulsion Group through German factories at the infamous underground Work Camp Dora near Nordhausen in Thuringia where prisoners of war produced German V-1 and V-2 missiles as well as turbojet and piston engines for fighters. He later assisted in the division of captured missile equipment between the United States and Great Britain. Still later, he became assistant chief of the Non-Rotating Engine Branch in the

Power Plant Laboratory at Wright-Patterson AFB. There, he became intimately aware of the limitations of existing solid propellants when he and others at the Power Plant Laboratory sought a solution to a problem with takeoff lengths of the B-47 swept-wing jet bomber and sought a solution through use of jet-assisted[43] takeoff (JATO) units. As a consequence, in late 1951 the Air Force—through the Non-Rotating Engine Branch—began to study and fund the development of solid-propellant rocketry. Its goals were not limited to assisted takeoff but looked forward to "the use of solids for long-range ballistic missiles." Using funds acquired in "various semi-legitimate ways over several years," Hall and the others "established a very low-profile project with several contractors in the rocket field" to develop reliable thrust-termination devices, lightweight cases, nozzles able to withstand the hot gases expelled from the combustion chamber, thrust-vector-control mechanisms with low drag, and high-performance propellants.[44]

Details of this project are sketchy—understandably so in view of the "semi-legitimate" funding. Clearly, some work was done out of Wright-Patterson AFB before the Air Force's Western Development Division (WDD, in the Air Research and Development Command) began its operations in a former schoolhouse in Inglewood, a suburb of Los Angeles, on 5 August 1954. In Inglewood, Hall (now a lieutenant colonel) was the Chief for Propulsion Development in the effort to develop the liquid-propellant Atlas (and later the Titan and Thor) missiles, but he continued his work on solids as well. In December 1954, he invited major manufacturers in the solid-propellant industry—Aerojet, Thiokol, Atlantic Research, Phillips Petroleum, Grand Central Rocket (founded by Charles Bartley), and Hercules Powder—to discuss prospects for solids.[45]

The outcome, apparently, was the Air Force's Large [Solid] Rocket Feasibility Program (AFLRP). Evidently, Hall instituted this program because in 1956, when he became program director for the liquid-propellant Thor missile, his boss, Gen. Bernard A. Schriever, agreed that he could keep solid-propellant studies among his responsibilities even though he gave up his other propulsion duties. Meanwhile, according to an Aerojet account, the AFLRP got started. An Aerojet history written by a number of former employees of the firm characterizes this program as "a three way, one year, basic design and development competition starting in September 1955 between" Aerojet and the two solid-rocket company groups that "were viewed as being Air Force oriented, Grand Central/Lockheed, Phillips Petroleum/Rocketdyne." (Lockheed purchased Grand Central in 1960–61, and Rocketdyne eventually took over the rocket operations of Phillips Petroleum, with which the Air Force had contracted in 1952 to develop low-cost JATO units and get the Air Force involved in the solid-rocket business. Aerojet's affiliation was primarily with the Navy at this point.)

Specific companies looked at different technologies. According to the Aerojet account, that company's polyurethane was the best available propellant, so Aerojet drew the assignment to look at what was believed to be the least favorable case material with respect to its strength-to-weight ratio, aluminum. The competition took place among six major technologies—the production process for the propellant, its performance, concepts for thrust vector control and thrust termination, ignition, and case material. Aerojet achieved favorable results in all six areas, and the Air Force dropped its two competitors, bringing Thiokol and Hercules into a competition with Aerojet for the second year of the program. Again, Aerojet won the competition, in part because of two innovations.

The first, credited to John Fuller of Aerojet, involved changing the parameters for launch and trajectory of a missile to match the structural characteristics of solid rockets, which were very different from those of liquid rockets. Advanced thrust-vector-control and thrust-termination technologies available by then permitted rapid and controllable changes in a missile's pitch plus the ability to cut off the thrust rather precisely so as to control the missile's velocity. The existing liquid-propellant intermediate-range and intercontinental ballistic missiles (IRBMs and ICBMs, respectively) had fragile structures that limited axial accelerations and pitch maneuvers to small and gradual increments. The liquid rockets were also forced to go "through the regime of maximum aerodynamic pressure [at] essentially zero pitch angle." These limitations on the Air Force's Atlas and Thor systems resulted from "extensive development effort." But complying with them wasted a great deal of propellant. With the much sturdier cases of solid rockets, Aerojet recommended higher initial accelerations and much more rapid "kick over" in pitch to a much higher angle with the vertical. These changes, attributed to Fuller's "inspiration" and "elegant analysis," yielded an increase in range of more than 20 percent over the practices required at the time for liquid-rocket operations. Fuller and George Henry of Aerojet "developed a set of partial differential equations" and performed the analysis on the clumsy electro-mechanical desk calculators of the day, but it "took two weeks to convince Space Technology Laboratories (STL— the Air Force's [systems engineering] agency) of the validity of this approach." Once STL agreed, however, the procedure became, according to Aerojet, "the basic type of maximum range trajectory for all solid rocket ICBMs."[46]

The second major innovation by Aerojet consisted of its contributions to the development and continued improvement of polyurethane propellants themselves. This was a complicated process. It began under a small Navy, rather than Air Force, Nitropolymer Program funded by the Office of Naval Research about 1947 to develop high-energy binder systems for solid propellants. The principal contractor was Aerojet, with Purdue and Ohio State universities getting separate research contracts. Aerojet never had more than seven or eight professional chemists working on the program, together with a few technicians. Although they succeeded in synthesizing a number of new high-energy compounds, in every case the process required heating the components to high temperatures—not a safe practice with potentially explosive compounds. Then Rodney Fischer found "an obscure reference in a German patent" indicating "that iron chelate compounds would catalyze the reaction of alcohols with isocyanates to make urethanes at essentially room temperature." Other metallic chelate compounds worked, but the Aerojet team used mainly ferric acetylacetonate as a catalyst. This discovery basically began the development of polyurethane propellants in many places besides Aerojet.

Meanwhile, in 1949 at the suggestion of Dr. Karl Klager (then working for the Office of Naval Research in Pasadena), General Tire (which had acquired Aerojet in 1944–45) began work on foamed polyurethane, leading to two patents held by Klager, Dick Geckler and R. Parette. In 1950, Klager went to work for Aerojet, and by 1954 he headed its solid-propellant development group. After 1956 when Polaris was started—with official approval coming in December of that year—Klager's group decided to minimize the percentage of oxidizer in the propellant by including some oxidizing capacity in the binder.

In April 1955, Klager and his group found out about the work at Atlantic Research with aluminum and began adding it to the propellant. As they began scaling the propellant size up to the dimensions for Polaris, however, cracks in the grain led to explosions. Reasons for the cracking and solutions to the problem were not fully forthcoming until the 1970s, but Klager and his group were able to duplicate the explosions on a laboratory scale and make corrections to the mixing procedures that convinced the Navy it was safe to continue with the development, as proved to be the case. There were further developments and multiple contributors, but the upshot was a successful propellant for Polaris A1. For his work in developing both the grain and the propellant, Klager received the Navy's Distinguished Public Service Award in 1958, and Marvin Gold, also of Aerojet, received the Navy's Civilian Meritorious Public Service Citation for his leadership of the research, development, and scale-up of nitrochemistry employed in the Polaris missile.[47]

The propellant Aerojet developed for both stages of Polaris A1 was a cast, case-bonded polyurethane composition with a large percentage of ammonium perchlorate and a significant amount of aluminum for stage 1, less aluminum and more ammonium perchlorate for stage 2, featuring a six-point internal burning star configuration. With an almost identical expansion cone angle of about 35 degrees for each of four nozzles in each of its two stages, Polaris featured a higher specific impulse for stage 1 than had previously been achieved in a composite solid rocket and a still higher one for stage 2, the latter being elevated in part because of the reduced atmospheric pressure at the altitudes over which it fired in the missile's trajectory. Overall, Polaris A1 was only 28.5 feet long and 4.5 feet in diameter as compared with Jupiter S's 44 and 10 feet respectively.[48]

The combustion chambers (cases) for both stages of Polaris A-1 consisted of rolled and welded steel, which had been modified in accordance with stringent Aerojet specifications that had resulted from extensive metallurgical investigations. Each of the four nozzles for stage 1, and apparently also stage 2, included a steel shell, a single-piece throat of molybdenum, and an exit-cone liner. Steering for the missile was provided by jetavators designed by Willy Fiedler of Lockheed, the prime contractor for Polaris. A German, Fiedler had worked on the V-1 program during World War II and had developed the concept of the jetavator while working for the Navy at Pt. Mugu Naval Air Missile Test Center. He had patented the idea and then adapted it for Polaris. Jetavators rotated into the exhaust stream of stage 1's four nozzles and deflected the flow (so-called thrust vector control) to provide pitch, yaw, and roll control. Jetavators in stage 2 were similar to those in stage 1 but featured different materials.[49]

Although the jetavators proved to be "one of the more difficult engineering achievements" in Polaris A1 and "succeeded very well" in the long run, they posed serious problems during static tests of the stage-1 motor. As initially designed, the jetavators consisted of solid rings with spherical inside surfaces. They were attached to the nozzles and rotated into the exhaust stream to effect steering. To prevent exhaust gases from passing between each jetavator and the nozzle to which it was attached and thereby hitting the aft end of the motor, there was a blow-back seal between the inside surface of the nozzle and the inside surface of the jetavator. With this arrangement, the jetavators tended to crack early in the firing sequence because of the ignition or mechanical and thermal shock to which they were exposed. The jetavator bearings also had a tendency to overheat. There was a further problem with aluminum oxide in the exhaust stream that collected on the blow-back seal and impaired jetavator movement.

The solution the Polaris team devised included a redesign to better absorb the structural loads and thermal stresses to the jetavators. Then the team had to eliminate the blow-back seal to avoid the "sticking" caused by the buildup of aluminum oxide. Finally, to avoid overheating of the aft end of the motor during flight tests as a result of exhaust blow-back (a condition known in the parlance of the missile developers as "hot foot"), and to preclude aerodynamic base heating from the speed at which the missile flew through the atmosphere, the team had to add external insulation.

To obtain the data needed to make these design changes, the Navy and Lockheed called upon the Naval Ordnance Test Station at Inyokern, California; the Air Force's Arnold Engineering Development Center at Tullahoma, Tennessee, with its high-altitude test chambers; and the sled tracks at Edwards Air Force Base, California. Test flights at Cape Canaveral also contributed data and led to the first fully successful flight on 20 April 1959.[50] These tests show that despite interservice rivalries, which were intense, Polaris, like other programs, could not have succeeded without a great deal of interservice cooperation to solve common national problems. It also helps to illustrate the wide variety of organizations and agencies that were necessary for successful rocket and missile development. It was by no means a parochial undertaking. Principal credit for the success of Polaris, of course, belonged to the Navy and its Special Projects Office headed by Admiral Raborn, with then-Captain Levering Smith as technical director after June 1957, and L.T.E. Thompson, first technical director of NOTS, as chief scientist.

As noted already, equally critical with steering the missile was termination of the thrust. For the missile to deliver the warhead to its target area, not only did it have to be pointed in the right direction during powered flight, it also had to travel at the right speed. Once it reached the correct terminal velocity, its second stage motor had to cease firing immediately. This could be achieved in liquid-propellant missiles simply by terminating the flow of propellants. For solids, the task was more difficult. After considering various alternatives, the Polaris team elected a technology that opened six vents (called thrust-reverser termination ports) in front of the second stage. These had plugs that were pyrotechnically removed at the proper velocity, permitting exhaust gases to escape and halt the acceleration so that the missile with its warhead would travel on a ballistic path to the target area.[51]

To control both the jetavators and the thrust-termination system of course required a guidance system. Even before the Polaris program got under way as a separate Navy effort, the Special Projects Office had contracted with the Massachusetts Institute of Technology's Instrumentation Laboratory to look at ship stabilization for the Jupiter missile. The Instrumentation Lab was also designing accelerometers and gyroscopes for the Air Force's liquid-propellant Thor missile, which ultimately used an all-inertial guidance system featuring an analog on-board computer. So the Navy contracted with the Instrumentation Lab to design an inertial system for the Polaris even before that missile had a name. Since Polaris was smaller than Thor, the Polaris guidance system had to be smaller as well, even though launching it from a moving submarine made its guidance and navigation tasks more complicated than for a missile launched from a fixed site on land.

Because the Polaris team decided to use a digital computer for the guidance system at a time when digital computers were slower and larger than their analog counterparts, what really sold the Navy on the Instrumentation Lab was the development by two of its mathematicians—Richard H. Battin and J. Halcombe Laning Jr.—of a method called Q-guidance for the Atlas ICBM. This enabled much of the computation for guidance (the Q matrix) to be done well before the missile was fired, leaving only a limited amount of calculation to be done by the computer on the missile. Another advantage of working with the Instrumentation Lab was that—at least in the view of the Special Projects Office's technical director, Adm. Levering Smith—the lab's fluid-floated gyroscope would be easier to adapt to the high accelerations of solid motors than other kinds of gyros. This development illustrates the technology that sometimes transferred even from liquid- to solid-propellant programs.

Since the Instrumentation Lab did not have a production capability, the Navy selected the General Electric Company to build the guidance system. The resultant guidance system featured a 2.5-inch-diameter inertial rate-integrating gyroscope (IRIG) with a pendulous integrating gyro accelerometer (PIGA) using the same gyroscope. Three of the IRIGs at right angles to one another kept a platform stable. On the platform were three PIGAs in a predetermined orientation. Supporting the platform were three gimbals made from beryllium that permitted movement of the platform within limits. The onboard computer constituted arguably the first fully transistorized, digital differential analyzer used in a missile. Unlike a general-purpose digital computer, it was limited in its capabilities to performing the few calculations needed for solving the differential equations employed in Q-guidance, allowing it to be quite small. The entire "black box" with its inertial components and electronics weighed in at a mere 225 pounds—"the smallest all-inertial guidance system developed at that time." At that relatively early date, the system caused innumerable problems with its transistors, testing, wiring, gyroscopes, and core memory. But it was ready on schedule. Its mean time between failures was problematical, but it was satisfactory for the roughly 2 minutes of guided flight. It guided a preprogrammed turn during the period when the first stage was providing the propulsion and then adjusted the trajectory, directed the thrust termination in the second stage and then the separation of the reentry vehicle that would carry the warhead in a real launch. In fact, on 6 May 1962 the USS *Ethan Allen* (SSBN-608) actually launched a Polaris A1 missile from a submerged position in the Pacific. The reentry vehicle landed "right in the pickle barrel" and detonated its nuclear payload during Operation Dominic (Frigate Bird Event). Earlier than that, however, the USS *George Washington* (SSBN-598) had launched the first functional Polaris missile on 20 July 1960, with a second missile launched three hours later. Both launches were successful over the 1,200-nautical-mile range set for Polaris A1. The fleet then deployed the missile on 15 November 1960.[52]

Minuteman

As noted already in part, Air Force Col. Ed Hall "deserves most of the credit for maintaining interest in large solid-rocket technology [during the mid-1950s] because of the greater simplicity of solid systems over liquid systems," as Aerojet's Karl Klager has stated. Hall's efforts leading to the Air Force's Large Rocket Feasibility Program "contributed substantially to the Polaris program." But the Air Force's own large solid missile, the Minuteman, did not formally begin its development until the Air Force secured final Department of Defense (DoD) approval for it in February 1958, over a year later than Polaris.[53] The reasons for this are complex.

On the one hand, although the Navy was not united in support of a fleet

ballistic missile, opposition to it was perhaps muted by the fact that not supporting it would clearly benefit the Air Force and thus hurt the Navy in the
interservice competition for roles and missions. Furthermore, the Navy was
clearly averse to a liquid-propellant missile for shipboard use, so support of a
solid was easier to obtain. Partly for these reasons, Adm. William F. Raborn,
who headed the Special Projects Office, had unusual support and authority to
proceed with Polaris development. He combined this with heterogeneous engineering to get the job of developing and deploying Polaris completed expe-
ditiously and effectively.[54]

On the other hand, although the Air Force had its own heterogeneous
engineers, those in the upper management of ballistic missiles at the Western
Development Division were already committed to liquid-propellant missiles.
Even if not ideally suited to their role, such missiles were less problematical
on land than they would have been at sea. General Schriever had comparable
authority and support to that enjoyed by Admiral Raborn,[55] but the heterogeneous engineers who were committed to a conversion to solid propellants for
ballistic missiles had a difficult job convincing him to support the change. No
doubt, the development of Polaris aided their cause by providing what Harvey
Sapolski has called "competitive pressure" in the continuing battle over roles
and missions of the armed services with respect to missile technology.[56]

But they nevertheless had their own battles within the Western Development Division and the Air Force to win support for what became Minuteman.
Besides Hall himself, the principal proponent for a solid-propellant ballistic
missile within the WDD was Barnet R. Adelman. He had earned bachelor of
science and master of science degrees in chemical engineering at Columbia
University and had done experimental work on both liquid and solid propellants at JPL before moving to a position as technical director of the Rocket
Fuel Division at Phillips Petroleum. He then became Director of Vehicle Engineering for the Ramo-Wooldridge Corporation, which had the title of systems
engineering and technical direction contractor for the Air Force at the Western Development Division.[57]

The details and precise dates of Hall's and Adelman's struggle within the
WDD and the Air Force to win acceptance of the solid-propellant ICBM that
became Minuteman are sketchy. According to one source, by March 1957 the
two men had agreed that technology for such a system had progressed sufficiently for it to be possible. In August of that year, Hall had turned over responsibility for the Thor missile to others and accepted an assignment from
General Schriever to develop the solid ICBM.[58] By this time, incidentally, WDD
had been redesignated (in July 1957) as the Air Force's Ballistic Missile Division (BMD). Apparently, General Schriever did not initially have much faith
in the idea of a solid ICBM because he wanted Hall simply to study the idea
whereas Hall insisted on a development effort. He was given no staff and only
a windowless office previously used for storage, but Adelman continued to

work with him, as did a few others from Ramo-Wooldridge, despite opposition from many in the firm.[59]

The specifics of Minuteman technology would continue to evolve, but the basic idea for the missile involved taking advantage of the reduced complexity of solids over liquids to cut the number of people required to launch it. Each Minuteman missile could be remotely launched, without a crew of people in attendance, from a control center using redundant communications cables, and maintenance was much lower than for liquid-propellant rockets. Although the Air Force considered using mobile Minuteman missiles, ultimately it decided to launch them exclusively from silos underground. Because solids were more compact than liquids, Minuteman I (according to one set of statistics) was only 54 feet long to Atlas's 82, Titan I's 98, and Titan II's 105 feet. Its diameter was a little over half that of the other three missiles (5.5 feet to 10 feet), and its weight was only 65,000 pounds to Atlas's 269,000, Titan I's 220,000, and Titan II's 330,000. This made the costs of silos much smaller than they might have been and substantially reduced the amount of thrust necessary to get the missiles launched. Thus, the initial costs of Minuteman were only one-fifth and the annual maintenance costs one-tenth those of Titan I. Moreover, a crew of two could launch ten Minuteman missiles whereas *each* Titan I required a launch crew of six.[60]

Although these facts were only projections in 1957–58, Hall and Adelman used this sort of information (among others) to convince higher Air Force and Department of Defense leaders to support Minuteman development. There was a series of briefings to this end. The principal one occurred on 8 February 1958 when Hall was in Washington in the company of Schriever and his deputy, Col. Charles H. Terhune Jr. Details of this trip differ among accounts, but what prompted the trip apparently was the issue of whether Polaris could serve as a land-based missile. At this point in time, Schriever evidently still was not sold on solids as alternatives to the liquid-propellant missiles he was developing, but he did accept the need for a viable alternative to Polaris as a land-based missile. According to Neal, who is not entirely reliable but was writing close to the time these events occurred and did interview the participants when their memories were fresh, Terhune brought Hall in to brief Schriever on the solid-propellant missile he was developing, which was not yet named Minuteman. Hall stressed the cost effectiveness and simplicity of the new missile, and Schriever decided to take Hall to Washington to present the case for Minuteman, as it then became.

Accounts agree that the two key briefings were of the Air Force Vice Chief of Staff, Gen. Curtis LeMay, who generally was opposed to missiles and favored bombers as the way to deliver nuclear weapons (or deter their use), and of Secretary of Defense Neil McElroy. According to Hall, LeMay's opposition to Atlas and Titan had stemmed from their vulnerability, the length of time it took to launch them, the number of people required for a launch, and the quantity of components that could fail and preclude a successful launch. Hall convinced him that Minuteman solved those problems. Subsequently Hall—and, according to him, LeMay—also impressed McElroy with the advantages of Minuteman, a name Schriever selected from several Hall had proposed.[61]

Even at this stage, some influential people in the Ramo-Wooldridge organization continued to oppose Minuteman. This is not the place to go into the details of the ensuing controversy beyond pointing out that Hall and his supporters in Ramo-Wooldridge were skillful enough heterogeneous engineers and sufficiently in command of the technologies involved in Minuteman to overcome the opposition. The upshot of these efforts was approval on 20 February 1958 of $50 million for research and development of the missile, an amount that was later increased at least partly through the efforts of Schriever, who now supported the missile.[62]

Soon after these developments, both Hall and Adelman left the Minuteman program. In 1958 Hall was transferred to an assignment as director of a North Atlantic Treaty Organization IRBM development program that eventually led to the French Diamant missile. The same year, Adelman left Ramo-Wooldridge to help found a new firm initially named United Research Corporation and then United Technology Corporation (UTC) that developed the solid rocket motors for the Titan III and Titan IV launch vehicles, which were important examples of technology transfer as well as further innovation.[63]

During the ensuing period from official approval of Minuteman in February 1958 to turnover of the first operational Minuteman I missile to the Air Force's Strategic Air Command in October 1962, there continued to be changes in program directors and the names of the Air Force organization responsible for Minuteman development. The immediate successor to Hall as program director was Col. Otto Glasser in 1959, but from August 1959 to 1963 the director was Col. and then Brig. Gen. Samuel C. Phillips. The Ballistic Missile Division split in two in 1961, with the Ballistic Systems Division retaining responsibility for Minuteman and moving to Norton Air Force Base near San Bernardino, California. Meanwhile, Ramo-Wooldridge's Guided Missile Research Division had been redesignated Ramo-Wooldridge Space Technology Laboratories in 1957, and Ramo-Wooldridge itself merged with a firm named Thompson Products to become Thompson Ramo Wooldridge (later shortened to TRW).[64]

Development of Minuteman involved a team relationship that initially used multiple approaches to arrive at individual technologies. The Ballistic Missile Division's contracts with Aerojet and Thiokol specified that they would work on all three stages of the missile. A contract with a third firm, Hercules Powder Company, assigned it to work on the third stage in addition to the other two firms. As they developed the relevant technologies, parallel development gave way to specific responsibilities. Thiokol won the contract for the first stage, Aerojet for the second, and Hercules for the third. The Space Technology Labs retained its role in systems engineering and technical direction, while the BMD selected The Boeing Company for the role of missile assembly and test (a role later expanded to include integration and check-out of operational facilities), North American Autonetics for that of guidance and control, and Avco for the reentry vehicle—to list only the major contractors. The BMD awarded all of the initial contracts in 1958, but award of the third-stage contract did not occur until the fall of 1960.

These companies and organizations all sent representatives to frequent program review meetings and quarterly gatherings of top corporate officials. At these meetings, participants examined progress and determined problem areas that required solution. Then, the relevant organizations arrived at solutions to keep the program on track.[65]

Although the basic conception and much of the technology for Minuteman I was in place by February 1958, many problems did remain. One particularly severe technical challenge involved materials for the nozzle throats and exit cones. The addition of aluminum to the propellant provided a high enough specific impulse to make Minuteman feasible, as it had done with Polaris, and it produced aluminum oxide particles that damped combustion instabilities. But the hot product flow degraded the nozzle throats and other exposed structures. It seemed that it might not be possible to design a vectorable nozzle that would last for the 60 seconds necessary for the operation of the missile. Resolving this problem required "many months and many dollars . . . spent in a frustrating cycle of design, test, failure, redesign, retest, and failure." The Minuteman team tried many different grades and exotic compounds of graphite, which seemed the most likely material, but all of them resulted in either blowouts or such severe erosion as to degrade engine performance. One solution proved to be a tungsten throat insert, a compromise because of its higher weight and increased cost. For the exit cones, the solution was the use of Fiberite molded at high pressure and loaded with silica and graphite cloth. It was only a partial solution because although it provided better erosion resistance than the graphite, it still was subject to random failures.[66]

Another problem area involved the vectorable nozzles. Although thrust vector control had been studied extensively and solved for Polaris by means of jetavators, flight-control studies for Minuteman had shown a need for nozzles in stage 1 that would vector the thrust 8 degrees from the plane of the missile's fuselage. Thiokol's first static test of the massive motor in 1959—the largest built to date—resulted in the ejection of all four nozzles only 30 milliseconds into the firing, which was well short of full-stage ignition. In October of that year, there were five successive explosions of the motors and their test stands, each with a different mode of failure. In January 1960, BMD halted first-stage testing. Its propulsion division chief at the time, Col. Langdon Ayers, accompanied by others from BMD and STL, went to the Thiokol plant to confer on the problem. Participants pinpointed two problem areas, internal insulation and nozzles, as masking other potential problems.

Two concurrent programs of testing ensued. Firings with battleship-steel cases tested movable nozzles, while flight-weight cases tested a single, fixed nozzle massive enough to sustain a full-duration firing. Thiokol solved the problem with insulation by summer, but the nozzle problem was not solved until the fall—dangerously close to the date set for all-up testing of the entire missile. Fortunately, Colonel Phillips had ordered Thiokol to begin manufacturing of the first stage except for the nozzle installation. This enabled installation of the nozzle as soon as its problem was resolved.[67]

A third major problem among many that could be mentioned involved launching from a silo. The first successful launch of the missile had occurred in February 1961, which then-General Phillips referred to as "December 63rd" since the launch had originally been scheduled for December 1960. It and two successive launches took place from a surface pad and were so successful that the team decided to advance the first launch from a silo to August 1961. The missile blew up in the silo. Fortunately, team members recovered enough of the guidance system from the wreckage to determine that the problem was not the silo launch itself, as critics contended, but quality control. Solder tabs containing connections had vibrated together, causing all of the stages to light at once. As Phillips said, the missile "really came out of there like a Roman candle." The team had been able to prepare the fifth missile for flight with the problem fixed by mid-November, when it had a successful flight.[68]

Previous testing of silo launches enabled this quick recovery. There had been critics in Ramo-Wooldridge from the beginning who had contended that it was not possible to fire the missile successfully from a silo. In early 1958 engineers from BMD and STL arranged for a program to develop the underground silos. It was divided into three phases, with the first two involving only subscale models of Minuteman, while the third tested full-scale models. Most of the testing took place at Edwards Air Force Base in the Mojave Desert about 100 miles northeast of BMD, but some of it was done on subscale wind-tunnel models by Boeing engineers in Seattle, Washington.

The testing at Edwards occurred at the rocket site, east of the main base on Leuhman Ridge, where a contingent from the Power Plant Laboratory at Wright-Patterson Air Force Base had moved in 1953 to begin testing rockets (at first, only small ones). Soon, however, it was testing Atlas, Thor, Titan, Bomarc, and Agena. By 1959, the entire Rocket Propulsion Branch of the Power Plant Laboratory had moved to Edwards. Heading the Minuteman testing at the rocket site was Maj. Rex Grey, who had previously tested Matador and Bomarc at Cape Canaveral, Florida. Initially, subscale silos tested issues such as heat transfer and turbulence. Already by November 1958, there had been fifty-six tests.

Meanwhile, in Seattle, Boeing engineers modeled the pressures the rocket's exhaust gases imparted to the missile and silo. They also examined the acoustic effects of the noise levels generated by the rocket motor on such delicate systems as guidance—a significant concern. Armed with such data, engineers at Edwards began one-third-scale tests in February 1959 and inaugurated construction of full-size silos. The full-scale tests that followed initially used a mock-up made of steel plate with ballast to match the weight and shape of the actual missile and only enough propellant (in the first-stage motor) to provide about 3 seconds of full thrust—enough to move the missile, on a tether, out of the silo and to check the effects of the thrust on the silo and missile. The tether was configured so that the missile would not drop back on the silo and damage it.

The silos were large, designed so that various-sized steel tubes could be inserted into them to simulate the size of a final silo. Sixteen-foot silos proved more than adequate, and the testing led to a twelve-foot diameter for the silo. Eventually, flight-weight missiles were launched from the silos. These tests ensured that the second silo launch at Cape Canaveral on 17 November 1961 was fully successful.[69]

There were other problems, but the Minuteman team overcame them and delivered the first missile to the Strategic Air Command for operational use in October 1962. This was almost exactly four years after the first contracts had been signed with industrial contractors to begin development of the missile. It was also a year earlier than initially planned because the schedule had been sped up.[70]

The missile that emerged from this process was the A model of Minuteman I (LGM-30A), later to be succeeded by a B model (LGM-30B). It was 53.7 feet long and consisted of three stages plus the reentry vehicle.[71] The first stage, produced and developed by Thiokol, featured a new propellant. Already in 1952, Thiokol chemists had begun searching for a polyhydrocarbon polymer that would have a higher fuel value than its polysulfide propellants. The laboratory director at Huntsville was Dr. W.F. Arendale, with the research being done by Dr. Dean Lowry and a young chemist named W.E. Hunter. Apparently, they used something of a trial-and-error methodology, informed by their knowledge of chemistry. First they tried polyisobutylene, then copolymers of isoprene and isobutylene, and finally a copolymer of butadiene and isoprene. All of these polymers proved resistant to the addition of functional groups that could be easily cured. Finally, in 1954 they developed a copolymer of butadiene and acrylic acid, polybutadiene-acrylic acid or PBAA that permitted higher concentrations of solid ingredients and greater fuel values than previous propellants. The carboxyl groups that the acrylic acid produced also reacted with a liquid epoxide resin to yield a cured polymer binder. Interestingly, the chemists had synthesized samples of the new material in 32-ounce Coke bottles because they proved to be the right size to fit in the homemade polymerization cabinet they used in characterizing the new propellant, and they were also inexpensive.

A further advantage of PBAA over Thiokol's polysulfide propellants was its higher hydrogen content. When Thiokol tried a composition of polysulfide polymers with 15 percent of aluminum, the hydrogen was insufficient to take full advantage of the aluminum to increase the specific impulse. With PBAA, a favorable reaction of oxygen with the aluminum generated significant amounts of hydrogen in the exhaust gases, reducing the average molecular weight of the combustion products (since hydrogen is the lightest of all elements). Most of the sulfur in the polysulfides, with its high atomic weight, was gone, replaced by hydrogen "heated to a higher flame temperature than the old polysulfides by the energetic combustion of aluminum to aluminum oxide." Despite the advantages, including a higher specific impulse, Thiokol delayed the transition to polybutadienes in some of its existing rocket motors because of the reliability and maturity of the old technology, but it did make the shift for Minuteman I.[72]

Thiokol did not produce the PBAA itself, except in research quantities. At the time, it did not yet have a pilot plant or production-size emulsion polymerization reactors. The Government Polymer Laboratories in Akron, Ohio, produced the early "scale-up" polymer samples. Further scaling up occurred at the BF Goodrich pilot plant, and production occurred in Louisville, Kentucky, at what was initially known as the National Synthetic Rubber Corporation but soon became the American Synthetic Rubber Corporation. Reynolds Aluminum supplied the aluminum particles, and the Shell Chemical Company produced the epoxy resin that served as the curing agent.[73]

The resultant propellant was a Class 2 material, meaning that it would burn rather than detonate, making it less dangerous to handle than detonable propellants. The oxidizer for an early version of this propellant, known as TP-H-8009, was ammonium perchlorate. A change in propellants occurred because the PBAA had a lower tear strength than the polysulfides. Thiokol solved that problem by adding 10 percent of acrylonitrile to the PBAA, creating polybutadiene-acrylic acid-acrylonitrile (PBAN), also produced by American Synthetic Rubber Corporation.[74]

For the mixing of the propellant, a 300-gallon mixer blended the constituents of the propellant in an automated process developed jointly by Thiokol and Toledo Scale Corporation. This process controlled the mixing of the premix, oxidizer, and epoxide-curing agent to very fine tolerances (e.g., a variation of no more than 2 pounds in 3,000 of oxidizer). The mixed propellant then had to have all trapped air removed to make the batch denser and preclude uneven burning resulting from voids created by air bubbles. Following this process, the propellant in a casting can enveloped in a hot-water jacket traveled to the casting pit. There, a motor case and mandrel for the internal cavity awaited it. The case was fitted with a vacuum-bell lid with a valve through which the propellant flowed for casting in the case, evacuated to very low pressure. This low pressure "drew" the propellant, which had the consistency of peanut butter at this point, into the motor case. It took thirteen batches of propellant, continuously fed, to fill the case. Then, with the near-vacuum removed by the circulation of air between the vacuum bell and the motor case, the propellant cured for about two days, gradually cooling to prevent uneven shrinkage, which could produce stress cracks that would lead to unprogrammed burning and possibly an explosion from excessive pressure.

The case consisted of double vacuum-melted steel exhibiting great strength. Three firms supplied the cases: Curtiss-Wright, Allison Division of General Motors, and General Electric. Each case had a forward dome that was welded to the central, cylindrical section and an aft dome that screwed to the cylinder. The cases consisted of six welded sections. Before having the propellant loaded, the cases underwent x-ray inspection of the welds and Magnaflux inspection for cracks.

Next, Thiokol bonded a premolded rubber liner about three-eighths of an inch thick to the inside of the motor case's head. It added aft-case insulation to preclude exhaust gases at temperatures of 5,500 degrees F from burning through the nozzle end of the motor case. Technicians also installed six inert, foamed-plastic slivers of very light weight in the case to take the place of the heavier propellant that would burn only at the very end of the thrust cycle and not contribute to the missile's range. This made the missile lighter, hence easier to launch. For the cylindrical case itself, a sprayed rubber liner covered the steel and insulated it from the heat of the burn; it also took care of the different coefficients of expansion of the propellant and the metal and helped bond the propellant to the case.

In the center of the case a Teflon-coated mandrel created the tapered, six-pointed, star-shaped cavity in the propellant once the latter had cured and the mandrel was removed. The Teflon prevented adhesion of the propellant to the mandrel and allowed ease of removal after the casting and curing was completed. The four movable nozzles for stage 1 vectored 8 degrees and were the largest vectorable nozzles yet used for a solid-propulsion rocket. The left and right nozzles moved up and down for pitch control, while the top and bottom nozzles moved sideways for yaw control. For control in roll, the lateral nozzles moved in opposite directions. A control unit produced by the Autonetics Division of North American Aviation as part of the guidance and control system operated the nozzles through a battery-powered hydraulic system. This system transmitted fluid pressure to hydraulic actuators.

On the outside of the case, a spray-head deposited a plastic insulation called Avcoat (made by Avco) to protect the motor from aerodynamic heating through ablation. To ignite the propellant, Minuteman I, Wing I featured a pyrogen pyrotechnic igniter. According to Air Force figures, the first-stage motor was 24.34 feet long, 65.7 inches in diameter. The total impulse delivered by the propellant was 10,324,400 pound-seconds (lbf-sec). This compared with 1,087,000 lbf-sec for the Sergeant missile, which was 16.275 feet long and had a diameter of 32 inches.[75]

For stage 2 of Minuteman I, Aerojet continued with its polyurethane propellant, including ammonium perchlorate for the oxidizer and aluminum powder to add to the specific impulse and control the burning rate. It used two slightly different propellant grains, a faster-burning inner grain and a slower-burning outer grain. The combination resulted in a conversion of the four-point, star-shaped internal burning cavity to a cylindrical cavity as the propellant burned, leaving no slivers of propellant that did not burn and contribute to the thrust of the missile. The combination also produced a near-neutral pressure-time curve over the period when the second stage was delivering thrust, producing the optimum range desired for the missile within the constraints of the existing design.[76]

The case for stage 2 was the same steel used for stage 1. It was made by machining ring-rolled forgings of the three center sections plus the forward and aft sections of the case, then welding them in a heat-treated, annealed condition, tempering them at 1,070 degrees F. It would have been possible to make the case from the glass, filament-wound material used for stage 3 (see below), but stage 2 was subjected to the most severe bending moments after launch, since it was the middle stage, and to provide the degree of stiffness it required, the glass-filament case would have had to be too heavy. Aerojet, like Thiokol, used three different suppliers of the case, but two of them were different from those in Thiokol's list. General Motor's Allison Division was the common supplier for both sets of cases, but Aerojet also used Avco's Lycoming Division and its own Downey Division. The second stage was 13.2 feet long and 44.3 inches in diameter.[77]

The aft closure bolted to the case, which included integral skirts so it could be attached to the interstage structures by Boeing at Air Force Plant 77 at Hill Air Force Base, Ogden, Utah, the stage itself being produced at Aerojet's Sacramento, California, plant. Inside the motor, Aerojet applied a lightweight, premolded, erosion-resistant rubber ranging in thickness from a tenth of an inch at the forward end to an inch at the rear where the exhaust gases posed the greatest danger of overheating the case. General Tire and Rubber (Aerojet's parent company) and a firm named Garlock supplied the insulation, with Aerojet personnel checking it to ensure it met the weight specifications and then vulcanized it to the inside of the case under pressure of nitrogen gas. As with stage 1, the external insulation was Avco's polyamide-epoxy resin, Avcoat.

To load the propellant, Aerojet positioned the case vertically in a vacuum bell located in a pit. The vacuum prevented any contaminants, moisture, air bubbles, and other external ingredients from entering the grain and causing problems when the motor ignited and burned. The 10,000-plus pounds of propellant were mixed in five batches and fed into the case from propellant "pots" suspended from cranes. It took four to six days for the propellant to cure. Aerojet personnel cast the outer grain first, with its own mandrel to keep the propellant in place. Then, they removed this mandrel and inserted the one for the inner grain. Once the grain had cured, Aerojet subjected it to a variety

of x-ray inspections, including one of the bond interface between the propellant and the case insulation, which was critical to the proper burning of the
propellant. These inspections revealed not only possible bond deficiencies but
also voids, cracks, porosity, and uneven mixture of fuel and oxidizer—any of
which would lead to rejection of the motor.

The four nozzles that were a feature of stage 2 as well as stage 1 were
designed by Bendix and supplied by that firm, Straza Industries, and Aerojet's
Downey Division. Aerojet attached them to the ends of the aft closure. The
nozzles featured forged tungsten throat inserts supported on a graphite backing about an inch thick. Pressure-molded Fiberite exit cones provided an expansion ratio of eighteen to one. The nozzles pivoted 6 degrees in pairs for
pitch, yaw, and roll control—about 2 degrees less than stage 1 and 2 degrees
more than stage 3. Aerojet cast propellant on the interior of the aft closure
both to provide additional thrust and to add insulation until the propellant
had burned.

Aerojet used a pyrogen, pyrotechnic igniter similar to the one it had used
on the Polaris motor. The device was essentially a small motor containing
about two pounds of grain similar to that used in the rocket motor. Bulova
supplied the safe and arm device, common to all three Minuteman I stages,
under an Air Force contract.[78]

For this stage, Aerojet used a planning system known as the Pilot Run
Item Master Schedule Committee (PRIMISCO), which was another inherit-
ance from Polaris to Minuteman. The PRIMISCO chair was responsible for
establishing a total master plan and a detailed plan. Representatives from the
project engineering office for each functional area of the stage plus the production control, manufacturing, quality control, facilities, and industrial engineering functions formed the committee. Milestone dates for transition from
research and development to production status included each tool and piece of
hardware needed for the stage. Stage-2 development and production actually
stayed ahead of this schedule, which was controlled by a computer that checked
the status of about twenty-six hundred items every two weeks and addressed a
memo to each program manager detailing the personnel who failed to achieve
their objectives. Unlike Thiokol and Hercules, who used Air Force production
plants, Aerojet built its production facility with company funds and consequently
went into production earlier than the other two companies.[79]

For stage 3, the Hercules Powder Company used the previously mentioned
glass-filament-wound case and a different propellant with a different core
pattern than those used for stages 1 and 2. These differences stemmed in significant part from the fact that Hercules came from a different tradition than
did Aerojet and Thiokol. It owed its origins to an antitrust suit against its
parent organization, E.I. duPont de Nemours & Company in 1912, that resulted in the creation of the one-thousand-employee explosives manufacturer
as well as a rival named the Atlas Powder Company. Both joined duPont in a

list of the United States' six largest explosives companies in 1913. Hercules alone produced more than 50,000 tons of smokeless powder during World War I before branching out into other products made from nitrocellulose, including plastics, lacquers, and photographic film. In the course of World War II, Hercules manufactured a great deal of double-base propellant for small rockets employed in air-to-air, air-to-ground, and antitank roles.

On 31 December 1945, Commander Levering Smith of later Polaris fame brokered the transfer of Allegany Ballistics Laboratory (ABL) from George Washington University's management to that of Hercules, with most of the employees staying on at the lab and several hundred Hercules employees transferring there from the firm's other operating locations. Hercules' earlier rocket propellants had been extruded, but after the war, the firm took over a technique for casting double-base propellants by treating nitrocellulose particles (known as casting powder) with a plasticizer and solvent (mostly nitroglycerin) plus a stabilizer, followed by heating and curing to create a homogeneous product that was formed and cured in a mold to form a much larger propellant grain than was possible under the extrusion process. Previously, ABL's largest extruded grains had been six inches in diameter. Its first cast, double-base grain was 16 inches in diameter, and scaling that up to much larger sizes seemed (and was) eminently possible. This process—which had resulted from the wartime research of Drs. John F. Kincaid and Henry M. Shuey at the National Defense Research Committee's Explosives Research Laboratory at Bruceton, Pennsylvania (operated by the Bureau of Mines and Carnegie Institute of Technology), and had been developed at ABL after Kincaid and Shuey were transferred there— enabled Hercules to produce a propellant grain that was not only as large as the castable, composite propellants Thiokol and Aerojet were making but also had a higher specific impulse (as well as a greater danger of exploding instead of burning as intended in a rocket motor).

Following development of a JATO unit designed to boost experimental vehicles, ABL under Hercules management developed rocket motors for a great many of the tactical rockets and missiles employed by the armed services in the postwar period. These included the Deacon rocket, Honest John artillery rocket, the Talos antiaircraft rocket, and the Nike and Terrier antiaircraft missiles.

The next step in the process was to add aluminum and ammonium perchlorate to the cast-double-base process to increase the performance of the propellant. At the same time, the process shifted from simply adding nitroglycerine and a stabilizer to the nitrocellulose to form a colloidal solid (solution casting) to slurry mixing the multiple ingredients and then casting them. ABL did not develop this process, known as composite-modified double base (CMDB), until 1958, evidently with the involvement of Kincaid and Shuey.[80]

Even earlier, Atlantic Research Corporation had developed a laboratory process for preparing CMDB. As manufactured, nitrocellulose is fibrous and not suitable to use as an unmodified additive to other ingredients being mixed to form a propellant. Dr. Arthur Sloan and D. Mann of ARC developed a process that dissolved the nitrocellulose in nitrobenzine and then separated out the nitrocellulose by mixing it with water under high shear (a process known as elutriation). The result was a compact and spherical series of particles of nitrocellulose with diameters from about 1 to 20 microns. These particles combined readily with liquid nitroplasticizers and crystalline additives in propellant mixers. The result could be cast into cartridge-loaded grains or case-bonded rocket cases and then converted to a solid with application of moderate heat. Sloan and Mann patented the process and assigned it to ARC.

Keith Rumbel and Charles Henderson then began scaling the process up to larger grain sizes and developing propellants in 1955. They developed two CMDB formulations beginning in 1956. Their propellants yielded specific impulses that were quite high compared to other solid propellants. As ARC's pilot plant became too small to support the firm's needs, production shifted to the naval facility at Indian Head, Maryland. Because the plastisol process developed by ARC was, according to Henderson, simpler, safer, and cheaper than other processes in existence at the time, Hercules and other producers of double-base propellants adopted ARC's method of production.[81] Meanwhile, in April 1958 the Department of Defense had begun work to expand the range of the Polaris missile from the 1,200-nautical-mile range achieved by its A1 missile to the 1,500 nautical miles originally planned for A1. The new missile, called Polaris A2, was originally slated to achieve the longer range through use of higher-performance propellants and lighter cases and nozzles in both stages, but the Navy's Special Projects Office decided to confine these improvements to the second stage, where they would have greater effect (since the second stage was already at a high speed and altitude when it began firing and did not have to overcome the weight of the entire missile and the full effects of the earth's gravity) and where risk of detonation of the high-energy propellant would not compromise the submarine. Hence, the SPO invited Hercules to offer ways to produce a second stage that would have higher performance.

Hercules was in a favorable position to do this because it had acquired the Young Development Laboratories of Rocky Hill, New Jersey, in 1958. Richard E. Young, the founder of the company, was a test pilot who had worked for the M.W. Kellogg Company on the Manhattan Project. In 1947, Kellogg had designed a winding machine under a Navy contract, leading to a winding laboratory in New Jersey that built a fiberglass nozzle. In 1948 it moved to Rocky Hill, New Jersey, where Young formed the development laboratories under his own name and filed the basic patents for winding. He and his firm moved from nozzles to cases, with Young becoming obsessed to solve the problem of strength-to-weight ratios in rocket motors, the key to which was lighter materials. Lighter weight cases could improve the mass fraction (relating the weight of the propellant to the weight of the entire rocket or stage), which become as important a measure of merit as specific impulse. Properly wound, fiberglass provided great strength combined with low weight, and in the mid-1950s ABL successfully tested small rockets and missiles featuring cases made of Young's Spiralloy material. On 18 September 1959, ABL's third-stage motor for Vanguard, featuring a fiberglass case, successfully placed a satellite in orbit, marking what was reportedly the first successful use of such a case in this application.[82]

For Polaris A2, Aerojet provided the first stage, and Hercules provided the second with a filament-wound case and a cast, double-base grain that contained ammonium perchlorate, nitrocellulose, nitroglycerine, and aluminum, among other ingredients. The grain configuration consisted of a twelve-pointed internal-burning star. It yielded a high specific impulse for a solid propellant. The missile had a successful test flight on 10 November 1960. The motor was 84 inches long and 54 inches in diameter, featuring four swiveling nozzles with 17.5-degree half-angle exit cones made of steel, asbestos phenolic, and Teflon with a graphite insert.[83]

The third stage for Minuteman I was somewhat smaller, being 84 inches long but only 37.5 inches in diameter. The case was a filament-wound material consisting of about 80 percent glass filament and some 20 percent epoxy resin. Hercules' (formerly Young's) plant in Rocky Hill, New Jersey, was not the only supplier of the cases. Another firm named Black, Sivalls and Bryson in Ardmore, Oklahoma, supplied some of them, and Hercules moved its case production to Clearfield, Utah, in the course of the Minuteman I production. The winding of the case was complicated because not only did stage 3 have four nozzles like the other two stages; it also featured four coated aluminum tubes for thrust termination, a feature that was exclusive to stage 3 in Minuteman I. The glass filament—fed by means of a spool and passing through an epoxy solution—was wound around a mandrel fitted with Buna rubber that was heat-resistant and became the internal insulation for the case. Once the case was completed, inspected, and tested hydrostatically, it went to Arrowsmith Tooling in Los Angeles, California, for installation of the aluminum mating rings for the nozzles. Arrowsmith technicians bonded the rings to the case skirt with epoxy resin. Arrowsmith also installed an operational "raceway" for the circuitry from the guidance section that would control nozzle actuation.

The propellant grain included two separate compositions (as did the other two stages). The composition used for the largest percentage of the grain resulted from Hercules' extensive research on high-energy propellants. It included a high explosive named HMX [tetrakis-(nitraza)cyclooctane] that had greater energy than ammonium perchlorate but was detonable, hence less safe. HMX was combined with ammonium perchlorate. Other ingredients included nitroglycerine, nitrocellulose, and aluminum. The second composition had the same basic ingredients as the first except for the HMX. It formed a horizontal segment at the front of the motor, whereas the composition with the HMX covered the aft portion of the motor plus a small portion in front. A hollow core ran through the center of the motor but stopped before it reached the segment containing the non-HMX composition. Described as a "core and slotted tube-modified end burner," it was roughly cone-shaped before tapering off to a cylinder. It produced a high specific impulse for a solid propellant.[84]

The nozzles had a housing with layers of various materials possessing different rates of conduction. The insert material was made of two substances with an insulation material located behind one of them to protect the moving parts of the nozzles. There was a composite exit cone with a contoured shape. A hydraulic control unit supplied by North American Aviation's Autonetics Division rotated the four nozzles in pairs up to 4 degrees in one plane to provide control in pitch and yaw.

The igniter was a unit designed by Hercules called a Pyrogen K-29D-6. It fired by igniting pellets that in turn ignited a small cast grain that began ignition of the main grain. Explosives, allowing the thrust to go out the side rather than the nozzles, opened the side ports for thrust termination.[85]

The inertial guidance systems for not only Minuteman I but also Minuteman II and III were provided by North American Aviation's Autonetics Division, which became the Autonetics Strategic Systems Division of Rockwell International by the time Minuteman III came along. The guidance system employed a stable platform with velocity measured by G6B4 gyroscopes and acceleration by a VM4A accelerometer. Like Polaris, Minuteman used a digital computer. The system was fully operational on the first flight of Minuteman I in February 1961.[86]

Stage 3 completed its preliminary flight-rating test in January 1961 and its qualification test in August 1962. Stage 2 was a month behind both of those dates, whereas stage 1 was on the same basic schedule as stage 3. The Minuteman I, Wing I became operational at Malmstrom AFB, Montana, in February 1963.[87]

It would be tedious and unnecessary to follow the evolution of Minuteman through all of the improvements that went into its later versions, but some discussion of the major changes is in order. Wings II through V of Minuteman I featured several improvements to increase the missile's range. This had been shorter than initially planned because of the acceleration of the Minuteman I schedule. This was not a problem at Malmstrom because it was so far north, but it would become a problem starting with Wing II. Consequently, for the Wing II and subsequent missiles, more propellant was added to the aft dome of stage 1, and the exit cone included contouring that made the nozzles more efficient. In stage 2, the material for the motor case was changed from steel to titanium. Titanium is considerably lighter than steel but also more expensive. Since, however, each pound of reduced weight yielded an extra mile of range, the advantages of titanium seemed worth the extra cost. The nozzles also were lighter and recontoured. Overall, the reduction in weight totaled only a fraction less than 300 pounds despite an increase in propellant weight. The increase in propellant weight yielded a range increase of 315 miles. There were no significant changes to stage 3.[88]

For Minuteman II, again the major improvements occurred in Aerojet's stage 2 (and in the guidance system, which will be discussed below). There had been problems with the nozzles and aft closure of stage 1 involving cracking and ejection of graphite. These were solved through an Air Force reliability improvement program for which details are lacking. There had also been problems with insulation burning through in the aft dome area of stage 3. Again, unspecified design changes inhibited the flow of hot gases in that region, solving the problem. Stage 2, however, featured an entirely new rocket motor with a new propellant, a slightly greater length and a substantially larger diameter, and a single fixed nozzle that did not swivel but used a liquid-injection thrust vector control system for directional control.[89]

The new propellant was carboxy terminated polybutadiene (CTPB), which, interestingly, was not developed by Aerojet but by other propellant companies. Some accounts attribute the development of CTPB to Thiokol, which first made the propellant in the late 1950s and made it into a useful propellant in the early 1960s.[90]

A history of Atlantic Research Corporation agrees that Thiokol produced the CTPB but attributes the solution of a curing problem to ARC.[91] It frequently happens in the history of technology that innovations occur to different people at approximately the same time. This appears to have been the case with CTPB, which Aerojet historians attribute to Phillips Petroleum and Rocketdyne without providing details.[92] These two companies may have been the source for the CTPB Aerojet used in Minuteman II, stage 2. Like Thiokol, in any event, Aerojet proposed to use MAPO as a cross-linking agent. TRW historians state that that firm worked with Aerojet to develop the stage-2 design.[93] The CTPB that resulted had better fuel values than previous propellants, good mechanical properties such as the long shelf life required for silo-based missiles, and a higher solids content than previous binders had offered. The propellant that resulted yielded a propellant specific impulse significantly higher than that used in stage 2 of Minuteman I, Wing II.[94]

Despite the fact that CTPB represented a significant step forward in binder technology, it remained less widely used than it otherwise would have because of its higher cost compared with PBAN and the emergence in the late 1960s of an even better polymer with both lower viscosity and lower cost known as hydroxyl-terminated polybutadiene (HTPB) that became the industry standard for newer tactical rockets thereafter. Karl Klager of Aerojet—who proposed it to NASA for use on the Astrobee D and F sounding rockets, on which it flew successfully—claims to have developed it. Even it did not replace PBAN for all uses, including that of the Titan and Space Shuttle solid rocket motors (on which see below), because PBAN could be produced for $2.50 per pound at a rate in the 1980s of four million pounds per year, a much higher rate than any other propellant could claim.[95]

Meanwhile, the second major change in the stage-2 motor for Minuteman II was the shift to a single nozzle with liquid thrust vector control replacing movable nozzles for control in pitch, yaw, and roll. Static firings had showed that the same propellants produced seven to eight points less specific impulse when fired from four nozzles than from a single nozzle. Examination of the reason revealed that with the four nozzles, liquid particles agglomerated in the approach section of the nozzles and produced exit cone erosion, changing the configuration of the exit cone in an unfavorable way. The solution was not only a single nozzle on Minuteman II's second stage but also the liquid-injection thrust vector control. The Navy had begun testing a freon system for thrust vector control for the second stage of Polaris A3 (third version of Polaris) in September 1961, well before the Minuteman II, stage-2 program began in February 1962. Metering valves injected freon into the exit cone of the nozzle. This created a shock pattern that caused the exhaust stream to deflect. The system was low in weight, insensitive to propellant flame temperature, and posed negligible constraints on the design of the nozzle. Early experimental work on this technique took place at the Naval Ordnance Test Station at Inyokern, California, which had been a leading developer of rockets since World War II. Aerojet, the Allegany Ballistics Laboratory, and Lockheed had all done analytical work to select the best locations for injectors, the most favorable injector, and the expulsion system. Despite this pioneering work by the Navy and its contractors, according to TRW historians, their firm still had to determine how much "vector capability" stage 2 of Minuteman II would require. TRW analyzed the amount of injectant that could be used before sloshing in the tank permitted the ingestion of air, and it also determined what the system performance requirements were. Since Aerojet was involved in the development of the system for Polaris, however, it seems likely that its involvement was also important. In any event, the Minuteman team also used freon as the injectant, confining it in a Vitron rubber bladder inside a metal pressure vessel. Both TRW and Aerojet studied the propensity of the freon to "migrate" through the bladder wall and become unavailable for its intended purpose. They found that only 25 of the 262 pounds of freon would escape, leaving enough to provide the necessary control in pitch and yaw. A separate roll-control system operated using a solid-propellant gas generator.

In addition to these changes, stage 2 of Minuteman II increased in length from 159.2 inches for Minuteman I to 162.32 inches in the new configuration. The diameter increased from 44.3 inches to 52.17 inches, resulting in an overall weight increase but a slight decrease in the propellant mass fraction from 0.897 to 0.887.[96]

A final significant improvement to Minuteman II was in its guidance system. Minuteman I had featured what J.M. Wuerth of the Autonetics Division of Rockwell International, designer of the Minuteman guidance and control system, has claimed was "the first mass production application of semi-conductors to high reliability military electronics." Minuteman II, he says, was "the first program to make major commitment" to semiconductor integrated circuits and "miniaturized discrete electronic parts." The integrated circuits made the computer more powerful, resulting in improved accuracy as a consequence of more refined computations of trajectory. The miniaturization reduced size. Hence, the Minuteman II computer had more than two and a half times as much memory as Minuteman I in roughly one-quarter the size. The new computer was also lighter and required less power. At the same time, the electronics exhibited a reduced vulnerability to nuclear effects.

A further improvement in the Minuteman II guidance and control system involved the substitution of beryllium for aluminum in the stable platform supporting the gyroscopes. Beryllium was very expensive and had an extremely toxic dust when machined, but its greater lightness and rigidity plus its less extreme thermal expansion made it far superior to aluminum. Additionally, the gas-bearing gyros were filled with hydrogen instead of helium, reducing drag and therefore drift for a further improvement in accuracy. Among other changes, Minuteman II's system also replaced the VM4A accelerometer on Minuteman I by a pendulous integrating gyro accelerometer from Draper Lab. It, too, employed gas-bearing gyros. The overall result was that Minuteman II was three times as accurate as Minuteman I—a requirement resulting from a shift in U.S. strategy from massive retaliation to flexible response necessitating that the missile be accurate enough to destroy an enemy missile in its silo. One disadvantage of the new guidance and control system was the infancy of the technology involved in integrated circuits. This resulted in their failure soon after deployment. Many had to be pulled out of service and replaced, but continued development solved the reliability problem and ultimately reduced the costs of system maintenance.[97]

Minuteman III featured multiple independently targetable reentry vehicles (MIRV) with a liquid fourth stage for deployment of this payload. This feature was not particularly relevant to launch vehicle development except that the added weight required for it necessitated higher booster performance. Stages 1 and 2 did not change from Minuteman II, but stage 3 became larger. Hercules did not win the contract for the larger motor. Instead, Aerojet got the award for design and development of the new third stage. Subsequently, Thiokol and a firm to be discussed below, the Chemical Systems Division of United Technologies Corporation, won contracts to build replacement motors. Stage 3 featured a fiberglass motor case, the same basic propellant Aerojet had used in stage 2 only in slightly different proportions, an igniter, a single nozzle that was fixed in place and partially submerged into the case, a liquid-injection thrust vector control system for control in pitch and yaw, a separate roll-control system, and a thrust-termination system.

Aerojet had gotten into the development of filament-wound cases quite early, first in Azusa and later in Sacramento. It produced most of the Minuteman fiberglass combustion chambers there but ceased filament winding in 1965. Meanwhile, Young had licensed Hercules' fiberglass technology to Black, Sivalls, and Bryson in Oklahoma City, and it became a second source for the Minuteman third-stage motor case. This instance and the three firms involved in producing the third stage illustrate the extent to which technology transferred among the contractors and sub-contractors for government missiles and rockets (see below).

The issue of technology transfer among competing contractors and the armed services, which were also competing over funds and mission jurisdiction, is a complex one about which a whole chapter, even a book, could be written. To address the subject briefly, there had been a degree of effort to exchange knowledge about rocket propulsion technology beginning in 1946, when the Navy and Army provided funding for a Rocket Propellant Information Agency (RPIA) within the Johns Hopkins University's Applied Physics Laboratory. The Air Force added its support in 1948, whereupon the RPIA became the Solid Propellant Information Agency (SPIA). After the launch of *Sputnik* by the Soviet Union in 1957, the Department of Defense's Advanced Research Project Agency and the newly created NASA began participating in SPIA activities. With the further development of rockets and missiles that ensued, it had become obvious by 1962 that a better channel for exchange of information was necessary. Hence, the DoD created an Interagency Chemical Rocket Propulsion Group in November 1962, with the name changing to the Joint Army/Navy/NASA/Air Force (JANNAF) Interagency Propulsion Committee in November 1969. This effectively promoted sharing of technology. In addition, "joint-venture" contracts, pioneered by Levering Smith of the Navy, often mandated the sharing of manufacturing technology between companies. These contracts served to eliminate the services' dependence on sole sources for a given technology that could be destroyed by fire or possible enemy targeting. It also provided for competitive bidding on future contracts.[98]

To return to the subject of Minuteman, according to the Air Force's Grey Books, at least, the propellant for Aerojet's third stage had less CTPB and more aluminum than stage 2. The grain configuration consisted of an internal-burning cylindrical bore with six "fins" radiating out in the forward end. This yielded a high vacuum specific impulse for a solid propellant. The igniter used black-powder squibs to start some of the CTPB propellant, which in turn spread the burning to the grain itself.

The 50-percent-submerged nozzle had a graphite phenolic entrance section, a forged tungsten throat insert, and a carbon phenolic exit cone. As compared with the 85.25-inch-long, 37.88-inch-diameter third stage of Minuteman

II, that for Minuteman III was 91.4 inches long and 52 inches in diameter. The mass fraction improved from 0.864 to 0.910, and with a somewhat greater specific impulse, the new third stage had more than twice the total impulse of its predecessor with a propellant mass that had grown in roughly the same ratio.[99]

The thrust vector control system for the new stage 3 was similar to that for stage 2 except that a different substance was used as the injectant into the thrust stream to control motion in the pitch and yaw axes instead of freon. Helium gas provided the pressure to insert the injectant instead of the solid-propellant gas generator used in the second stage. Roll control again came from a gas generator supplying gas to diametrically opposed nozzles. When both were operating, there was neutral torque in the roll axis. When roll torque was required, the flight-control system closed a flapper on one of the nozzles, providing unbalanced thrust to rotate the motor around its axial centerline.

To ensure accuracy for the delivery of the warheads, Minuteman had always required precise thrust termination for stage 3 as determined by the flight-control computer. On Minuteman I, the thrust-termination system consisted of four thick tubes integrally wound in the sidewall of the case and sealed with snap-ring closures to form side ports. Detonation of explosive ordnance released a frangible section of the snap ring, thereby venting the combustion chamber and causing a momentary negative thrust that resulted in the third stage dropping away from the postboost vehicle. The system for Minuteman III involved six circular shaped charges on the forward dome. Using high-speed films and strain gauges, the Minuteman team found that this arrangement worked within 20 microseconds, cutting holes that resulted in a rupture of the pressure vessel within 2 additional milliseconds. Subsequent examination of the case showed cracks that radiated from the edge of the holes. TRW used a computer code called NASTRAN to define the propagation of the cracks. It then determined the dome thickness needed to eliminate the failure of the fiberglass. Aerojet wound "doilies" integrally into the dome of the motor case under each of the circular charges. Testing of the resultant configuration showed that it had eliminated the cracking. Thereafter, the system worked to vent the pressure in the chamber and produce momentary negative thrust. This resulted in the third stage's dropping away from the postboost vehicle.[100]

Large Segmented Rocket Motors and the Titan Solid Rocket Motors

The developments of technology for Minuteman II and Minuteman III did not have immediate effects upon launch vehicle technology. Some of them would not do so until much later when stages of Minuteman II themselves became part of a launch vehicle. Meanwhile, however, a new firm had entered the list of those engaged in developing and producing large rockets, and the Air Force had initiated a program, later funded by NASA, to develop and test large, segmented solid rocket motors. Both developments contributed to the large solid-propellant boosters used on Titan III and Titan IV as well as the Space Shuttle, although both also owed significant technological debts to the Minuteman program.

The link to Minuteman was evident in the founding of the new firm, initially named United Research Corporation in 1958. The president of the new firm, which began in Los Angeles and then moved north to Menlo Park, was Lt. Gen. Donald L. Putt. He had retired from the Air Force in the position of Deputy Chief of Staff for Development and had had a significant role in assembling the team for Minuteman. He also had ties to JPL through Caltech, where he had earned a master's degree in aeronautical engineering with Theodore von Kármán as his mentor. Another major figure at United Research with even stronger ties to both JPL and Minuteman was Barnet R. Adelman, who was highly instrumental in founding the new firm and served in the beginning as its vice president, general manager, and director of operations. He subsequently succeeded Putt as the firm's president in 1962. We have already followed his role in Minuteman, and of course, he brought his knowledge of that missile's early development with him to United Research, which was subsequently renamed United Technology Corporation and then United Technology Center.

The new firm succeeded in recruiting a lot of knowledgeable people in the field of solid-propellant rocketry from several different firms, including many who had worked with the Air Force in the ballistic missile field. Among them was David Altman, who had done a lot of work on propellants and rocket combustion at JPL before moving on to Ford Motor Company's Aeroneutronic Systems. He remembers that in the mid-1950s, he had already had discussions with Adelman about segmented motors. Other key people in UTC came from Thiokol, Aerojet, Atlantic Research, and the Naval Ordnance Test Station at Inyokern, California, as well as Ramo-Wooldridge's Space Technology Laboratories. Early in 1960, Adelman stated that of the roughly forty "really top-grade men" in solid-rocket technology, UTC had recruited about ten of them.

Soon after its founding, United Aircraft Corporation bought one-third interest in the new firm, eventually becoming the sole owner. When United Aircraft assumed the name United Technologies Corporation in 1975, its solid-rocket division changed its name to Chemical Systems Division. In the remainder of this chapter, the solid-rocket division will be called UTC and CSD and the parent firm's name will be spelled out.[101]

Despite Adelman's early interest in segmented motors, Aerojet appears to have been the first firm to get involved with testing them. In 1957—before UTC was founded—Aerojet responded to indications from the Air Force that it might need very large solid rocket motors by cutting a 20-inch-diameter Regulus II booster into three segments, which the firm joined together using bolted flange joints. Using the same propellant and nozzle as in the Regulus, Aerojet successfully test fired the bolted booster, achieving identical performance to that of single-piece motors.

It had become obvious that motor sizes for solids would become too large to be transported (except by barge), so segmenting them clearly made a lot of sense. No doubt encouraged by Aerojet's preliminary successes, in about April 1959 the Powerplant Laboratory of the Wright Air Development Center in Dayton, Ohio, initiated a procurement effort for $495,000 that was carried out after the organization moved to Edwards AFB, California. It awarded a contract to Aerojet for a series of 100-inch-diameter motor firings. The overall goal was 20 million pound-seconds of total impulse, but the initial goal was 230,000 lbf for 80 seconds, which equates to 18.4 million pound-seconds of total impulse. Aerojet's first test with a segmented motor larger than Regulus successfully fired a 65-inch-diameter Minuteman first stage that was cut in half and rejoined with a lock-ring (also called lock-strip) joint configured similarly to subsequent joints. It had outer and inner lock strips held together by a lock strip key in the center and an O-ring inboard of that—both key and O-ring holding portions of the joint together and preventing leakage. The Minuteman test motor also featured a single nozzle instead of the four used in the first stage of the actual missile, and it used different propellant and grain designs to achieve the durations of firing that the program targeted. It had a graphite-block entrance cap and graphite throat on the nozzle, with an exit-cone half angle of 20 degrees to provide resistance to erosion from the hot, corrosive exit gases. This motor yielded 160,000 lbf for 60 seconds on 5 May 1961.[102]

Aerojet next designed, built, and tested a test-weight, 100-inch-diameter motor designated TW-1. Its technology was based on the tests with the Minuteman-sized motor. The chamber for the 100-inch motor, which consisted of three segments, was made of a rolled and welded barrel (cylindrical) section plus two welded and spun hemispherical sections for the head and the nozzle ends. The material was a steel alloy. Following tensile tests by the University of California at Berkeley, Aerojet stuck with the joint used in the tests of the Minuteman-sized motor. The company used the same basic nozzle design as in the previous tests. It successfully fired the TW-1 motor on 3 June 1961. There was erosion of the nozzle throat, but it was smooth. Dynamic measurements indicated that there should be no problem with adding segments to the motor. It yielded 450,000 lbf of maximum thrust and burned for over 45 seconds.

Following these successful tests, the goals of the program shifted to using lighter flight-weight motors, the first of which was designated FW-1. This motor had two center segments, with the follow-on motor (FW-2) having three. The program also included studies of thrust-vector-control systems. The principal objective of FW-1, however, was for the 100-inch-diameter motor to produce a total impulse of more than 30 million lbf-sec in 60 seconds, to demonstrate the integrity of the segmented joints, and to gather data on the progression of the flame within the grain during burning. Besides the two center segments, the motor had shorter forward and aft segments, including a refurbished center segment from TW-1 to show that it could be reused. Most of the rest of the motor consisted of new parts, but it used the same material, insulation, lock-strip joints, heat treatment, and methods of fabrication as TW-1. The nozzle assembly was the same as in TW-1, but it had new graphite, rubber, and silica cloth. This time, however, the technicians bonded together the graphite insert and entrance cap before bonding them to the nozzle. Also, they spread a layer of zirconium dioxide onto the surface to reduce the thermal shock. They reused the steel exit cone from TW-1, processing it as they had in the earlier test.

Using a new propellant that had a slightly higher solids content, the FW-1 motor fired on 26 August 1961. It was completely successful. Inspection after the event showed that the parts performed their functions and could be used again after refurbishing. There was one hot spot on the forward segment, but the case held up without burn-through. Analysis suggested that the cause for the hot spot was molten metal from the igniter reducing the thickness of the insulation, which could be prevented by insulating the igniter housing. The nozzle performed well. There was a circumferential crack in the graphite throat insert, but it may have resulted from quenching with water after the firing rather than from the firing itself.

Aerojet made the cases for the three center segments plus the fore and aft closures on the FW-2 motor out of steels stronger than that used in the previous motors. The segmented lock-strip joints were essentially the same as in previous motors except for the addition of tapered-wedge steel strips to prevent dynamic movement during ignition and firing. Technicians extended the entrance cap with a reinforced ring and bonded the throat together using graphite cement. Then they overwrapped it with asbestos felt that was impregnated with a resin. They spread zirconium oxide over the graphite throat insert. The outer steel shell of the exit cone had an inner liner. The thrust-vector-control system used an aqueous solution of sodium perchlorate that flowed at 63 percent strength through three banks of fixed orifices to provide three injection pressures.

The propellant used a polyurethane binder with still higher solids content than the FW-1 motor, including more ammonium perchlorate than FW-1 but slightly less aluminum. The test firing on 17 February 1962 was successful in terms of ballistic performance, yielding a total impulse of 47,679,000 lbf/sec for 88.1 seconds (essentially reaching the goal of 48 million lbf/sec for a duration of 90 seconds). All the components lasted through the higher temperature and greater length of the firing, and the motor parts proved to be reusable. The nozzle throat remained intact during the firing, but there was some surface spalling (breaking off of fragments) at 65.2 seconds into the firing. This caused the throat area to erode by 8.5 percent and pressure to drop by 10 pounds per square inch, making the performance of the insert only marginal. The thrust-vector-control system also was less than fully successful, failing to achieve the anticipated deflection rate for the motor thrust. On the other hand, the insulation of the igniter precluded damage to the insulation on the case as occurred in FW-1.

These tests, which completed Phase I of the Air Force's Large Segmented Solid Rocket Motor Program, demonstrated that large, segmented rockets would work in actual firing. This fundamental success led the Air Force to contract with the Aerojet General Corporation to continue the testing with FW-3 and -4 motors and perform further tests of thrust-vector-control systems. The FW-3 motor had five center segments with a new design for the aft closure assembly including the liner for the nozzle throat to accommodate 120 seconds of firing and 71 million lbf-sec of thrust. The thrust-vector-control system had two reactive fluids—sodium perchlorate and nitrogen tetroxide—with three injectant pressures. At 77 feet in length and a weight of 876,000 pounds, this was the largest solid rocket motor yet built.

FW-3 fired on 9 June 1962, achieving the predicted thrust at the 42-second mark of the firing. Shortly after that, the motor malfunctioned. There was a gas leak at the nozzle throat. It propagated around the aft closure near the joint and ejected the graphite throat at 45.3 seconds into the firing. This caused a significant decrease in pressure and thrust. As best analysts could determine, the graphite experienced compressive failure.

The FW-4 motor had the same configuration as FW-3 but had only two center segments instead of five. It used an improved graphite known as CFW for the nozzle insert with an improved design featuring a phenolic-resin-impregnated graphite tape that was overwrapped with a high-silica-phenolic-resin tape. Zirconium oxide again covered the throat. Two jet tabs provided thrust vector control in conjunction with a dual-fluid injectant system using sodium perchlorate and nitrogen tetroxide. Otherwise, the motor used parts refurbished from FW-3. Firing took place on 13 October 1962. The motor performed as expected until the 55.7-second point, when it apparently ejected the rear portion of the throat insert. Nine-tenths of a second later, there was evidence of a gas leak at the nozzle throat. It expanded circumferentially until the 83.3-second point, when a thrust surge ejected the remaining throat inserts, whereupon the thrust and pressure decreased rapidly. At 101.5 seconds, the motor ejected the exit tab assembly, and at 140 seconds the loss of thrust and pressure caused the propellant to stop burning. The liquid thrust-vector-control system had worked to some degree until the 46.3-second mark, when ejection of the exit cone disrupted it. However, it had not performed as well as expected. This showed that freon was still the best fluid for injection.

Even though all aspects of the 100-inch-diameter motor tests by Aerojet had not been successful, the firings had demonstrated that a segmented design was the solution for the transportation of very large rocket motors. The lock strip joint had also proved itself in the tests. But the failure of the nozzle-throat inserts and the thrust-vector-control systems showed that there was still work to be done in those areas. By the end of the tests, Lockheed and UTC had also experienced nozzle problems, and all of Aerojet's competitors had tested large, segmented motors. UTC, for example, had tested three motors measuring 90 and 87 inches in diameter and featuring either four or three segments (in those orders) during 1961. Lockheed Propulsion Company followed with three motors measuring at least 100 inches in diameter from 1962 to 1964.[103]

During 1961 while Aerojet and UTC were testing their early large, segmented motors, Harold W. Ritchey, who had become president of Thiokol, argued that a 3,000,000-lbf solid rocket could be developed within three years. He said that solid propulsion was the cheapest way to put payloads into orbit and the easiest solution to the high thrusts required for large launch vehicles. Aerojet's president, Dr. Ernest R. Roberts, who had previously managed research and development in the firm's solid-rocket plant, stated that no further proof was needed since Aerojet had already demonstrated 400,000 lbf. In the same year, John Crowley of Grand Central Rocket, soon to be acquired in full by Lockheed Propulsion, stated that solids offered the greatest hope of matching the Russian superiority in lift capability. And Barnet Adelman of UTC said segmentation offered the easiest way to achieve the sizes needed for heavy lift. Despite Aerojet's early work with segmentation, Adelman obtained a patent for segmented-joint design for such motors.

UTC, as noted, had already begun to develop large solid motors. With the assistance of Tom Polter from the Stanford Research Institute in Palo Alto, California, it began to develop a specific propellant for large segmented boosters, starting with the proven PBAN, used in stage 1 of Minuteman I, as Adelman would have known from his significant involvement in Minuteman development. The company's research with its own funds resulted in a successful test firing of the P-1 solid-propellant motor on 15 December 1960. The 87-inch-diameter motor yielded over 200,000 lbf for about 75 seconds, followed on 9 February 1961 by the P-1–2 with two middle segments, a forward and aft closure. It had a case made of a low-carbon steel plate known as H-11, a fixed nozzle with liquid thrust vector control (using nitrogen tetroxide), a length of 43 feet, 7 inches, and a diameter of 90 inches. It yielded 399,000 lbf for 79 seconds and was followed by the P-2 motor, which had only one center segment and was only 87 inches in diameter but produced about 500,000 lbf over a 75-second period. It had no thrust vectoring, but the nitrogen-tetroxide system used on P-1–2 was eventually used on the solid rocket motors of the Titan IIIC.

UTC's cases featured a taper of 1.22 degrees that the firm argued was effective in preventing erosive burning of the grain in long, internal-burning motors. During 1961, however, the firm abandoned this idea and agreed to construct its motor cases with zero taper. This permitted the segments to be interchanged, which provided an important advantage.[104]

Subsequently, UTC won a separate contract with the Air Force in May 1962 to produce the booster that would provide the initial propulsion for the Titan III launch vehicle. The Titan III featured storable liquid propulsion for the ensuing stages. This was a significant coup for so recent a firm, as its competitors were Lockheed, Aerojet, and Thiokol. Already on 20 July 1963 UTC successfully test fired a five-segment prototype of the motor that boosted a whole series of Titan IIIC, D, and E configurations. These launch vehicles successfully put into orbit a number of Department of Defense reconnaissance and communications satellites as well as NASA spacecraft beginning in 1965.[105]

The PBAN propellant used by UTC for the so-called zero stage of the Titan IIIs differed from other PBAN propellants in having greater toughness because of the addition of methyl nadic anhydride to the basic PBAN-ammonium perchlorate-aluminum propellant. It had a mass fraction of 85.3 percent and a configuration consisting of an eight-point star in the forward closure of the five-segment solid rocket motor, with the internal perforation becoming cylindrical through the rest of the motor including the aft closure. The motor was 84.65 feet long and 10 feet (120 inches) in diameter. The case was made of carbon bearing steel. The very large nozzle (covering 1,116 square inches at the throat) used carbon phenolic for the throat area. It provided an expansion ratio of 8.0 to 1.0. Given Aerojet's experience with nozzles, this so-called tape-wrapped carbon-phenolic throat was a major gamble for UTC, but it worked. A carbon tape was able to withstand the high temperatures of the exhaust gases, while the phenolic resin glued it together, yielding a throat area that remained structurally sound throughout the firing of the motor. This development was a major advance in large solid-rocket technology, and the resultant throat performed flawlessly on all Titans, paving the way for the shuttle boosters. The segments were joined by tongue-and-groove, clevis joints developed by engineers at UTC and another division of the parent firm, Pratt & Whitney. These five-segment solid rocket motors for the earlier Titans laid the groundwork for the five and a half-segment Titan 34D motors (1982) and the seven-segment Titan IV motors (1989) built by UTC (later CSD).[106]

To return from the specific Titan solid rocket motor developments (under a separate Air Force program designated 624A) to the Air Force's Large Segmented Solid Rocket Motor Program (designated 623A), in late 1962 the Air Force Rocket Propulsion Laboratory (AFRPL) at Edwards AFB began managing what in essence was a follow-on program to develop large solid-propellant motors that NASA and the DoD might use for space launch vehicles. The Air Force provided funding for 120- and 156-inch-diameter segmented motors and for further work on thrust-vector-control systems, with NASA providing funding for a part of the 156-inch and the follow-on 260-inch programs. Liquid thrust vector control, jet tabs, and jet vanes were all tested in the early stages of the follow-on program. In the process, Lockheed had developed a Lockseal® mounting structure for the nozzle to provide thrust vector control.

Thiokol later succeeded in scaling it up to the size necessary for the large motors, and it became known as Flexseal.®

Another critical technology to be developed in the program was case material for the 156-inch (and later, 260-inch) motors. A new material that seemed promising was maraging steel—a tough, strong steel that was low in carbon and high in nickel and that formed hardening precipitates as it aged. A great many organizations were involved it its testing, and the Aeronautical Systems Division of Air Force Systems Command had arranged a series of meetings in 1962–63 to exchange data resulting from the testing organizations, which included the Douglas Aircraft Company, North American Aviation, the Case Institute of Technology, Aerojet General, the Naval Weapons Laboratory, General Motors, the Mellon Institute, the Battelle Memorial Institute, The Budd Company, Lear Siegler, the Naval Research Laboratory, Thiokol, Lockheed, United States Steel, the Crucible Steel Company of America, the Army Materials Research Agency, the Frankford Arsenal, General Dynamics, The Boeing Company, the International Nickel Company, and others—a list that emphasizes the enormously diversified group of participants in the development of rocket technology and the interservice cooperation that often occurred despite countervailing interservice rivalry.

Applications for the maraging steel varied from submarines and hydrofoils to rocket cases, but the needs of the big booster programs were the driving forces in the urgency of the testing. Maraging steels with 18-percent were the primary focus of the testing because they had demonstrated greater fracture toughness than conventional steels with the same level of strength, permitting thinner (thus lighter) cases.[107]

Meanwhile, AFRPL had awarded Lockheed Propulsion Company a contract to design, produce, and test the first 120-inch motor designated 120-1. This motor had a single center segment and two rather large closures. Lockheed made the case from steel that was rolled and welded by Exelco of Silver Creek, New York. It contained 164,000 pounds of an unspecified propellant, a fixed nozzle with two separate liquid thrust-vector-control systems, one using nitrogen tetroxide and the other, freon. The nozzle throat used a bulk graphite material. A successful test firing took place on 12 May 1962. It lasted 123 seconds and yielded a maximum thrust of 350,000 lbf. In a demonstration of the possibility of reusing case segments, this one was reused by Aerojet in a subscale motor for a 260-inch test and then by Thiokol to test a hot-gas-injection thrust-vector-control system.

Lockheed Propulsion Company also conducted the first 156-inch-diameter motor test, although it was designated 156-3, on 28 May 1964. It, too, featured a single center segment plus fore and aft closures. Like most other 156-inch motors, it used a case made of rolled and welded 18-percent maraging steel, made by Exelco. The motor was 70.4 feet long. The nozzle throat consisted of tape-wrapped graphite phenolic and was the first nozzle of its size (37.25 inches) tested on a solid rocket motor. It had an expansion ratio of 6.0 to 1.0. The thrust vector control system involved four jet tabs mounted on the rear of the nozzle. The system produced 7 degrees of side force but caused thrust degradation in the motor. Components of the system also failed, necessitating a redesign. The motor contained 424,000 pounds of PBAN propellant with a solids loading of 84 percent. It produced 949,000 lbf in a test lasting 108 seconds.

After refurbishing, Lockheed reloaded the same case with a propellant loading of PBAN weighing 625,000 pounds with the same percentage of solids. This time, the jet tabs provided only 6.17 degrees of deflection, but the test on 30 September 1964 of motor 156-4 yielded a maximum thrust of 1,094,110 lbf, a record at that time. The firing lasted 142.8 seconds.

Meanwhile, AFRPL had awarded Thiokol a contract to design, produce, and test a heavier 156-inch motor employing a gimbal nozzle. The effort included three tests of 65-inch-diameter motors to evaluate possible gimbal nozzles. The motor and nozzle operated successfully except that the exit cone of silica phenolic cloth failed 30 seconds into the test. However, the 156-1 motor with 691,000 pounds of 86-percent-solids PBAN in an 18-percent maraging steel case 78.1 feet long and a nozzle gimbaled to 5 degrees did yield 1,471,000 lbf and burned for 126 seconds. A follow-on test of motor 156-2 by Thiokol took place on 27 February 1965. It had a 100.5-foot case of 18-percent maraging steel with two center segments, 758,000 pounds of PBAN loaded with 85 percent solids, and an external, fixed nozzle with no thrust vector control. It yielded 3,250,000 lbf, far exceeding previous motors, in a test lasting 58.7 seconds that was considered a success.[108]

The next step in the 156-inch motor testing was motor 156-5. This was a Lockheed flight-weight motor using 18-percent maraging steel like the previous 156-inch motors. It used a fixed nozzle submerged in the motor itself and featuring a 60.4-inch throat and a 6.45-to-1 expansion ratio. Nitrogen tetroxide provided 3.5 degrees of thrust vector control. The motor was 75.1 feet long and was built as if it were a first stage or booster for a launch vehicle or ballistic missile. On 14 December 1965, it fired for 55.25 seconds and produced a maximum thrust of 3,107,000 lbf—judged a complete success.

Lockheed followed this with motor 156-6, a flight-weight, monolithic (unsegmented) device built as if it were a second stage of a missile or launch vehicle. It was 33.3 feet long with a deeply submerged, ablative fixed nozzle having a 34.5-inch throat and a expansion ratio of 8.16 to 1. Thrust vector control of 4.3 degrees came from another nitrogen-tetroxide system. Like motor 156-5, it had a PBAN propellant, only it weighed 273,000 pounds to the 687,000 for 156-5. Both featured solids loading of 86 percent. On 15 January 1966, motor 156-6 fired for 65.0 seconds, yielding a maximum thrust of 1,025,000 lbf in a test judged to be completely successful.

The three final motors in the 156-inch series, numbered seven through nine, were designed, built, and tested by Thiokol. Motors 156-7 and 156-8 both had fiberglass cases, with the former being a monolithic, 21.1-foot motor, and the second a 58.3-foot motor with one center segment and fore and aft closures. Motor 156-7 was built as if it were a third stage and was fired on 13 May 1966 in a diffuser to simulate altitude. The motor performed as expected, but the nitrogen-tetroxide thrust-vector-control system produced only 0.77 degrees of thrust deflection instead of the design figure of 2.0 degrees. There was also a burn-through of the diffuser, but this was nevertheless the largest composite case fired up to that time.

Motor 156-8 had a case and nozzle developed by the Materials Laboratory at Wright-Patterson AFB, with the case built by B.F. Goodrich and the ablative nozzle manufactured by TRW. The nozzle was external (rather than submerged) and fixed with a 32.9-inch throat and a 7-to-1 expansion ratio. There was no thrust vector control. The propellant was 86-percent-solids PBAN, whereas Thiokol had used PBAA on the previous motor. On 25 June 1968, the motor fired for 118 seconds and yielded 1,089,000 lbf as expected, except that the tail-off lasted longer than expected. Following the firing, a hydroburst test of the case caused it to fail just short of 1,100 pounds per square inch (psi), which met the minimum safety factor of 1.25, the firing having produced 806 psi. The segmented case performed as designed. The nozzle's ablative material had been cured under pressure provided by high-tension shrink tape, and it performed in a way comparable to ablative nozzles cured under pressure produced by hydroclaves or autoclaves (devices for producing pressure).

Motor 156-9 returned to 18-percent maraging steel for its case, but under Air Force contract, Thiokol designed and built the nozzle with a Flexseal® bearing. Lockheed had competed to win this contract, but Thiokol got the bid despite Lockheed's development of the Lockseal® concept on which it had a patent pending. The case for this motor was monolithic, and the nozzle was submerged with a 34.5-inch throat and an expansion ratio of 8 to 1. The 33.4-foot motor contained 277,000 pounds of PBAN loaded with solids to 86 percent of the grain. On 26 May 1967, it fired for 77 seconds and yielded a maximum thrust of 983,000 lbf. The Flexseal® bearing performed as intended and was later used on many solid motors.[109]

In the program for the follow-on 260-inch motor, the AFRPL granted parallel contracts to Aerojet and Thiokol in 1963 to develop monolithic devices. It insisted that both contractors manufacture the truly enormous cases out of rolled plates of 18-percent maraging steel. On 1 March 1965, NASA and the DoD modified their agreement about responsibilities for the program, and NASA's Lewis Research Center in suburban Cleveland, Ohio (today the Glenn Research Center), took over full management and technical responsibility for the 260-inch program. Meanwhile, Thiokol had contracted with the Newport News Shipbuilding Corporation to build the case, which was much larger than any previously built for rockets but was only two-thirds the size of the highly stressed hull of the Polaris submarine. According to Aerojet sources, Thiokol had specified a relatively low-cost procedure for the manufacture of the case with heat treating to a slightly greater strength than Aerojet used but a consequent lower ductility. On 11 April 1965, the case failed during hydrotesting at 540 pounds per square inch (psi) of pressure, slightly more than half of the design strength of 1,040 psi. The cause was a preexisting flaw in the 0.75-inch-thick case in a membrane next to a longitudinal weld. This removed Thiokol from the program, ending its long series of successes in scaling up rocket motors.

Aerojet set more conservative standards in its contract with Sun Shipbuilding and Dry-dock of Chester, Pennsylvania. It required, by the account of its former employees, "much more complex, accurate, and sturdy weld position tooling, better plate preparation in the weld areas, a more forgiving multipass welding technique with extensive [ultrasonic] inspection, and heat treating to only 200 Ksi [compared with Thiokol's 250 Ksi]—which gave higher ductility." This approach was more expensive than Thiokol's, but none of Aerojet's chambers burst. Aerojet used a nozzle produced by TRW of Cleveland, Ohio, featuring a tape-wrapped carbon phenolic 71-inch throat cured under hydroclave pressure. The nozzle had an expansion ratio of 6 to 1. Aerojet had the first 260-inch case floated by barge from Pennsylvania to its Homestead Facility in Dade County, Florida, where it arrived during a hurricane but was undamaged when winds beached the barge. Aerojet mixed the 1,676,000 pounds of PBAN propellant using both batch and continuous mixers, resulting in no discernible difference in physical properties or performance between the results of the two different procedures. Goodyear Tire and Rubber Company developed a method of installing the insulation in the case on-site. The motor used an aft-end system of ignition with the igniter tethered so it would land in a safe place once it was ejected by the main gas flow from the motor. It was placed in a 150-foot-deep caisson sunk into the ground to provide a pit for casting and curing the propellant in the case and then to serve as a test stand, with the motor firing vertically. The firing took place on the night of 25 September 1965, with the burn lasting 113.7 seconds and producing 3,567,000 lbf. The flame was visible 30 miles away in Miami. The length of the motor was only 80.7 feet, about half the size needed to replace the Saturn booster, which was the potential application (never, of course, to be put into effect). As a consequence, this motor was designated 260–SL-1, the SL standing for short length. However, this and 260–SL-2 fired at design conditions using a burning rate and nozzle size appropriate for the full-length design. The similar SL-2 motor fired on 23 February 1966 with comparable results.

The 77-foot SL-3 motor had a partially submerged, 89-inch throat with an expansion ratio of 4.1 to 1. On 17 June 1967, despite having slightly less PBAN

84-percent-solids propellant (1,654,000 pounds), it yielded 5,884,000 lbf in a firing of 77 seconds—achieving a record for a single rocket propulsion unit that stood until at least 1994.

There was no direct application of the technologies from the 260-inch program, but Thiokol's participation in the overall Large Segmented Solid Rocket Motor Program, including the 120- and 156-inch motors, gave it both experience and access to designs, materials, methods of fabrication, and test results that contributed to the Shuttle Solid Rocket Boosters, although it also drew upon its experience with the first stage of Minuteman (and with a contract it won for the Poseidon C3, a successor of Polaris).[110]

The Space Shuttle's Solid Rocket Boosters

As these comments would suggest and as all the world knows, Thiokol won the contract for the solid rocket motors that formed the principal components of the Space Shuttle's twin Solid Rocket Boosters (SRBs), although United Space Boosters Inc. assembled, checked out, launched, and refurbished them for reuse. NASA's Marshall Space Flight Center had the responsibility for propulsion elements of the shuttles and had granted study contracts on 27 January 1972 for the SRBs to UTC, Aerojet, Lockheed Propulsion, and Thiokol for $150,000 apiece. On 16 August 1973, United Space Boosters Inc., a subsidiary of UTC, won the fabrication contract. Meanwhile, on 16 July 1973, Marshall issued a request for proposals to develop the solid rocket motors to Aerojet, Lockheed, Thiokol, and UTC. NASA selected Thiokol to design, develop, and test the motors on 20 November 1973, which Lockheed protested to the General Accounting Office (GAO). Until the protest was decided, NASA issued study contracts to Thiokol so it could continue its work. The GAO left it up to NASA to decide whether to reconsider the contract award, and the agency awarded a letter contract to Thiokol on 26 June 1974.[111]

The design of the solid rocket motors and the SRB as a whole was conservative, reflecting the prevailing approach of the Marshall Space Flight Center and the need to rate solid boosters, for the first time, for human flight (in the parlance of the day, man-rating). The case was made of the same steel used on the Minuteman missile and Titan solid rocket motors rather than the 18-percent nickel maraging steel used on the 156- and 260-inch motors. The SRB was 146 inches in diameter and 125.3 feet long, with each of four segments being 164 inches long and the fore and aft sections making up the rest of the length. The Ladish Company in Cudahy, Wisconsin, made the cases for each segment without welding by a process called rolled ring forging whereby a hole was punched while the metal was hot and then rolled to the 146-inch diameter. Initially the minimum thickness was 0.49 inches, but the thickness has varied with some motors having lightweight and others medium-weight rather than the standard weight. The differences in thickness were measured in thousandths

Image 6-3: Filling the ground with billows of smoke and steam created by the flaming solid rocket boosters (SRBs), the space shuttle *Atlantis* speeds toward space on mission STS-106, launched on 19 September 2000. The SRBs are the inheritors of the early solid-rocket program. (NASA Photo no. KSC-00PD-1263).

of inches, but they yielded savings of several thousands of pounds in weight, resulting in significant improvements in mass fraction, hence payload.

A separate firm, that of Cal Doran near Los Angeles, California, provided heat treatment to the case segments, after which they went farther south to Chula Vista near San Diego, where tang-and-clevis joints were added mechanically. The joint was significantly different from the one used on the Titan motors. The latter had used a single O-ring, and the outside portion of the joint pointed downward, discouraging the formation of moisture in the joint. For the shuttle SRB, the tang was added to the bottom of the casing, and it fit into the clevis on top of the adjoining section. There were 177 pins for each joint, and they fit through holes drilled in both the tang and clevis to hold the adjacent segments together but still permit disassembly and reuse. To protect the case and joints in the Titan SRMs, UTC had ensured careful manufacture and assembly of the internal insulation of the case. It had then used heating strips and a careful application of putty to prevent high-pressure gas "blow by" due to improper sealing of the O-rings. To "man-rate" the SRBs, the designers added a second O-ring, which proved inadequate in the *Challenger* disaster on 28 January 1986, when hot gases leaked from a joint and caused the external (liquid-propellant) tank to explode. The solution for the shuttle was principally a new "tang capture feature" that fit around the mating segments of the joints to keep positive pressure on the O-ring. Among other changes, a pressure-actuated flap called a J-seal replaced putty used in the earlier arrangement.[112]

For thrust vector control, Thiokol used the Flexseal® design it had scaled up from Lockheed's design for the 156-9 motor. It was capable of 8 degrees of deflection, which was necessary among other reasons for the shuttle to perform its now familiar roll downrange after lifting off the launch pad, a maneuver for which the thrust and gimbaling of the Space Shuttle Main Engines were not sufficient. A liquid-injection thrust-vector-control system such as that used on the Titan motors was also not usable in the shuttle because of the level of thrust and the sheer size of the SRBs, which would have required too much fluid.[113]

The propellant for the SRBs was the now-conservative and familiar PBAN with an 86-percent-solids loading, 16 percent aluminum, 69.7 percent ammonium perchlorate as oxidizer, 0.3 percent iron oxide as a burning rate catalyst, about 12 percent PBAN, and 2 percent epoxy curing agent. This yielded a propellant specific impulse of about 243 lbf-sec/lbm and a theoretical specific impulse of above 260 lbf-sec/lbm. The grain configuration was an eleven-point star in the forward-most segment tapering back to a large, smooth internal burning segment at the rear. With 1,108,600 pounds of propellant in the four segments, this configuration yielded 268,610,000 pounds of total impulse. This was the largest solid booster ever flown and provided 71 percent of the shuttle's thrust during the first 27.5 miles (75 seconds) of ascent into space.

The nozzle was partially submerged with a throat area of 2,327 square inches and an initial expansion ratio of 7.16 to 1. The nozzle throat was insulated with carbon cloth impregnated with phenolic resin, like the Titan solid rocket motors. Other portions of the nozzle, where the heat was less intent, used glass and silica phenolic.

Following *Challenger,* there were other changes to the SRBs, including an increase in length by 10 inches, but basically there was no change in the propellant, specific impulse, or grain configuration. Like the Titan IIIs and IVs, the shuttle has carried NASA spacecraft, military reconnaissance and communications satellites into space, performing a variety or roles from launch and repair of the Hubble Space Telescope and doing scientific experiments in the microgravity of its payload bay, to supporting the International Space Station. Both systems have also launched a number of space probes—for example *Viking 1* and *2* and *Voyager 1* and *2* for the Titan IIIE; *Magellan, Galileo,* and *Ulysses* for the shuttle.[114]

The SRBs for the shuttle have remained essentially unchanged since the improvements following *Challenger*. A team of Lockheed and Aerojet did design an improved advanced solid rocket motor under a 1989 contract, but before they could proceed to testing, Congress canceled the project in October 1993.[115]

The Titan IV Solid Rocket Motor Upgrade

There was, by contrast, a significant upgrade of the Titan IV solid rocket motors that, despite a lengthy and troubled period of development, did come to fruition. The new motors, under the Titan IV Solid Rocket Motor Upgrade (SRMU), were sponsored by the Air Force's Space and Missile Systems Division, located at Los Angeles Air Force Base, California, which also furnished the technical direction of the effort. The Phillips Laboratory at Edwards Air Force Base, previously known as AFRPL and since October 1997 redesignated the Propulsion Directorate of the Air Force Research Laboratory, along with The Aerospace Corporation, provided technical support. And Martin Marietta, the prime contractor of the Titan IV program and now part of Lockheed Martin, subcontracted with Hercules Aerospace Company (itself acquired by Alliant Techsystems in March 1995) in October 1987 to develop, qualify, and produce the improved solid rocket motor.[116]

Because the overall length of the SRMU was supposed to remain the same as the old solid rocket motor (although it actually increased from 111.9 to 112.5 feet), a 25 percent performance gain was achieved by increasing the principal diameter from 120 to 127.7 inches, decreasing the weight of the case by changing most of it from steel to a graphite-fiber epoxy composite, and adopting a higher-energy propellant, HTPB (discussed above). The graphite composite case evolved from technology in space, strategic, and tactical rocket and missile programs but especially from a filament-wound composite developed for the shuttle. The composite was never used on the shuttle because of concerns whether it could withstand the bending forces generated when the Space Shuttle Main Engine ignited and over various other technical and economic issues. Even though the forward dome and three cylindrical sections were the only composite portions of the SRMU case, with the joint rings and aft dome made of steel, the SRMU achieved the highest propellant mass fraction of any large space booster still in production—0.896 as compared with 0.86 for the SRM.

The HTPB propellant contained 88 percent solids. The grain configuration was a ten-slotted design in the forward segment with a tapered centerport in the center and aft sections. This delivered a standard sea-level theoretical specific impulse only a few lbf-sec/lbm higher than the older SRM, but the vacuum-delivered specific impulse was significantly higher than that for the SRM and for the redesigned shuttle SRB as well.

Like the Space Shuttle and unlike the SRM, the SRMU had a Flexseal®

omniaxial nozzle with an aft exit cone that could move 6 degrees in all directions (2 degrees less than the shuttle, to be sure). It had a tape-wrapped throat
with insulators.[117]

Because of *Challenger* and other failures of joints (notably a Titan 34D
explosion 8 seconds after liftoff in April 1986), the number of segments in the
SRMU was reduced to three and the joints were improved. The joints put on
at the launch site (called field joints) included both primary and secondary O-
ring seals that caused the gaps in the joint to close upon pressurization of the
motor. The pyrogen igniter was consumable with a composite case (instead of
steel in the SRM) and a CTPB propellant with a twenty-one-spoke grain.

The manufacturer subjected the motor to a high level of nondestructive
inspection including ultrasonic inspection of almost all of the bonds between
propellant and insulation and insulation and the case after the case was loaded
but before the segments left the plant and before they were mated at the launch
site. Despite this, on 1 April 1991 developmental motor PQM-1 exploded in its
first static test inside an environmental tower designed to control temperatures. It did so because of overpressure on the case resulting from a propellant grain deformation. A modification to the aft segment of the grain fixed
the problem, and the second developmental motor, PQM-1', had a successful
test on 29 June 1992. This was followed by successful tests of four qualification motors between 15 October 1992 and 13 September 1993.[118]

Finally, on 23 February 1997, the first Titan IVB (as the Titan IV with the
SRMUs was called) launched successfully from Cape Canaveral with a Defense Support Program missile early warning satellite as its payload. An Inertial Upper Stage built by Boeing carried DSP-18 with six thousand infrared
sensors to its geosynchronous orbit. *Aviation Week & Space Technology* pronounced the successful use of the new motor "the most significant advance in
U.S. solid-rocket propulsion since redesign of the Space Shuttle boosters after the *Challenger* accident 11 years" before. It also said that development of
the new Titan IVB took nine years (actually, almost nine and a half) and cost
about $1 billion.[119]

A second launch of the Titan IVB on 15 October 1997 sent NASA's Cassini
mission to Saturn on its way successfully from Cape Canaveral. A further
launch of a military reconnaissance satellite by a Titan IVB occurred successfully on 9 May 1998, but on 12 August 1998, an electrical short in the power
supply of a Titan IVA (with the older SRM motors as boosters) caused a catastrophic failure (unrelated to the SRM) that led to the grounding of all Titan
launch vehicles. The Titan IVB returned to flight on 9 April 1999, when it
launched another DMSP satellite from Launch Complex 41 at Cape Canaveral,
the last such launch from that complex, which was being turned over to
Lockheed Martin for that company's Evolved Expendable Launch Vehicle
(EELV). The EELV was an Air Force program to develop a family of launch
vehicles that would substantially lower the cost of putting U.S. payloads into

space. It was intended to replace the Titan IV in the next century. Still another Titan IVB launched a classified payload for the Air Force and the National Reconnaissance Office on 22 May 1999.[120]

The Minotaur

Perhaps less significant in some respects but a fitting topic for the conclusion of this chapter was the first use of Minuteman stages in conjunction with commercial upper stages to launch satellites. As the cold war was ending in the early 1990s, the Strategic Arms Reduction Treaty, ratified in July 1991, required the dismantling of 450 Minuteman II missiles and their destruction if they were not used for orbital launches, which had to be approved on a case-by-case basis by the Secretary of Defense. Under arrangements with the Air Force's Rocket Systems Launch Program, founded in 1972 to collect, store, convert, and use excess ballistic missiles as launch vehicles, a program called the Orbital/Suborbital Program let a contract with Orbital Sciences Corporation—a relative newcomer to the launch vehicle arena—in September 1997 to create a launch vehicle called the Minotaur that combined the first and second stages of a Minuteman II with the second and third stages of Orbital's Pegasus XL small launch vehicle. The contract, worth up to $206 million, provided for as many as six missions per year beginning in 1999 with options covering a total of twenty-four launches through 2004. At an estimated cost of $12.5 million the Minotaur could launch 1,400 pounds of payload to low-Earth orbit for a cost per pound of $8,929. This compared with a cost for the Pegasus XL of $16.25 million for 1,100 pounds of payload to low-Earth orbit or $15,000 per pound.[121]

Orbital Sciences began development of the initial Pegasus launch vehicle in 1987 in a joint venture with Hercules Aerospace in which Hercules had responsibility for three new rocket motors and a payload fairing, while Orbital provided all other mechanical and avionics systems, all software (ground and flight), a mechanism for dropping the three joined stages of the launch vehicle from a launch airplane, and overall systems engineering. The two partners evenly split the more than $50 million development cost. In July 1988, the Defense Advanced Research Projects Agency (soon to become ARPA) awarded Orbital an $8.4-million contract for one Pegasus launch vehicle and fixed-price options for five additional launches of Pegasus. This was the first completely new U.S. space launch vehicle designed since the 1970s. It was launched from a specially modified Lockheed L-1011 "mothership," although NASA's Dryden Flight Research Center launched the first six Pegasus vehicles from the same NB-52 launch aircraft that had launched the X-15 research airplanes. Since the launch aircraft carried the three stages to about 38,000 feet and a speed of about Mach 0.5 (half the speed of sound), Pegasus had about twice the performance of a similarly sized vehicle launched from the ground.

The first Pegasus vehicle, rolled out in August 1989 and launched by the

Image 6-4: The launch of a Minuteman II in 1992. Boeing continues to support the Minuteman ICBM program with various system upgrades and modernization projects. (Photo courtesy of The Boeing Company)

NB-52 on 5 April 1990, successfully placed two spacecraft into orbit—an ARPA/ Navy experimental communications satellite and a NASA Goddard Space Flight Center bus containing two experimental canisters and a payload-environment instrument package, both of which remained attached to the third stage of Pegasus. The remaining launches from the NB-52 followed through 1994, after which Orbital's L-1011 assumed launch duties. Pegasus XL, conceived in 1991, involved a lengthening of stages 1 and 2, permitting an increase in propellant of 24 and 30 percent, respectively. Its first launch on 26 June 1994 was also the first Pegasus launch from the L-1011. Initially the launch went well, but loss of vehicle control at 35 seconds into the flight led to flight termination. A second unsuccessful launch occurred in 1995, but on 9 March 1996 a Pegasus XL was successfully launched from its L-1011 and went on to deploy the Space Test Program Radiation Experiment II into orbit. On 2 July 1996, a Pegasus XL launched a Total Ozone Mapping Spectrometer, and on 8 September 1996, another Pegasus XL placed the Fast Auroral Snapshot Explorer, an astrophysics spacecraft, into orbit.

Both Pegasus and Pegasus XL featured an HTPB propellant with a vacuum specific impulse that varied from about 289 to 294 lbf-sec/lbm, depending on the stage. The HTPB had an 88-percent-solids loading. The cases were all graphite epoxy, and all three stages had carbon-carbon composite throat inserts in the nozzles and carbon-phenolic nozzles with graphite-epoxy overwraps. The stage-1 motor had a fixed nozzle, but stages 2 and 3 had Flexseal® nozzles, reflecting the transfer of that technology from Lockheed to Thiokol and then to other companies.[122]

Although initially scheduled for launch in 1999, the Minuteman-Pegasus hybrid did not finally launch until 26 January 2000 from Vandenberg Air Force Base, California. Payloads placed into orbit included the Air Force Academy's FalconSat, Stanford University's OPAL, Arizona State University's ASUSAT, the Air Force Research Laboratory's Optical Calibration Sphere Experiment, NASA Marshall Space Flight Center's Plasma Experiment Satellite, and Weber State University's Attitude Controlled Platform.[123]

Conclusion

This series of launches in some sense concluded an era in the history of solid-rocket technology, with the ensuing era beginning but awaiting definition. Both the Air Force and NASA were pursuing cheaper access to space as well as a reduction in rocket exhaust products that harmed the environment. What that would mean for solid-rocket technology is not yet clear. Meanwhile, the modest beginnings of solid-propellant rockets and missiles during and after World War II had evolved into enormously powerful space launch motors and equally sophisticated technologies that enabled them to provide the initial— and in some cases upper-stage—thrust for space launch vehicles. The number of firms and government entities that contributed to this evolution was truly staggering. While interservice rivalries among the military services in some ways stimulated and in other ways impeded these developments, there existed at the same time many examples of interservice cooperation that furthered the common cause.[124] There has also been a great deal of technology transfer among companies and other organizations, with the Flexseal® thrust-vector-control device and PBAN and HTPB propellants offering prime examples. The use of aluminum as a fuel has been a further example. Innovation has been extensive if usually poorly documented, and the result has been a number of conflicting claims for priority in developing new technologies that are extremely difficult to sort out.

Many readers of this chapter may lament that World War II and the cold war have been the principal stimuli for rocket development, with the threat (in some cases, realization) of the loss of human life the initial result. I regret it as much as anyone does, but it is an inescapable fact, as is some destruction of the environment every time a large rocket motor fires. On the other hand, a significant result has been the capability to launch communications and other satellites that have enhanced our knowledge of our world and the universe. The latter almost certainly would not have occurred in our lifetimes had it not been for the threats that war and the cold war provided because the funding for research and development would have been lacking. In any event, the development did occur for both good and evil, and the process was extremely complex, as this chapter has suggested.[125]

Notes

1. I would like to thank Charles B. Henderson, Wilbur C. Andrepont, Dr. Ross Felix, Dennis R. Jenkins, Col. Edward N. Hall, and Prof. Edward W. Price for their helpful comments on a draft of this chapter.

2. The term castable refers to a type of propellant that is prepared in a fluid form, poured into either a mold or a rocket's combustion chamber, and then cured until it becomes solid. Composite propellants are those that consist of separate particles of oxidizer, possibly a fuel, and other substances dispersed in an elastic matrix that serves as a binder and also as a fuel. These and other definitions can be found in [P.] Thomas Carroll, "A Brief Technological Primer on Solid Propellant Rockets" (Pasadena, Ca.: JPL/HN-12, 1971), p. 6.

3. See Clayton Huggett, C.E. Bartley, and Mark M. Mills, *Solid Propellant Rockets* (Princeton, N.J.: Princeton University Press, 1960), p. 125; W[illiam] A[lbert] Noyes Jr., ed., *Chemistry: A History of the Chemistry Components of the National Defense Research Committee, 1940–1946* (Boston: Little, Brown, 1948), pp. 103, 111–13.

4. See J.D. Hunley, "A Question of Antecedents: Peenemünde, JPL, and the Launching of U.S. Rocketry," in Roger D. Launius, ed., *Organizing for the Use of Space: Historical Perspectives on a Persistent Issue* (San Diego: Univelt, 1995), pp. 17–18, 22, and the sources cited there, including P. Thomas Carroll, "Historical Origins of the Sergeant Missile Powerplant," in Kristen R. Lattu, ed., *History of Rocketry and Astronautics* (San Diego: Univelt, 1989), pp. 123–25; Clayton R. Koppes, *JPL and the American Space Program: A History of the Jet Propulsion Laboratory* (New Haven, Conn.: Yale University Press, 1982), pp. 10–13; and J.W. Par-

Image 6-5: A payload launch vehicle carrying a prototype interceptor is launched from Meck Island in the Kwajalein Missile Range on 18 January 2000 for a planned intercept of a ballistic missile target over the central Pacific Ocean. The target vehicle, a modified Minuteman intercontinental ballistic missile, was launched from Vandenberg Air Force Base, California, at 6:19 P.M., Pacific Standard Time, and the vehicle carrying the prototype interceptor was launched about 20 minutes later. The intercept, however, was not achieved. The test was performed by the Ballistic Missile Defense Organization's National Missile Defense Joint Program Office. Defense and industry program officials will conduct an extensive review of the test results to determine the reason for not achieving an intercept. (DoD photo)

sons and M.M. Mills, "The Development of an Asphalt Base Solid Propellant," Air Corps Jet Propulsion Research GALCIT Project No. 1, Report No. 15 (Pasadena: California Institute of Technology, 1942), esp. pp. 1–4, 7–8.

5. Carroll, "Historical Origins," pp. 130–31; Koppes, *JPL,* pp. 13, 36; draft intvw. of Charles Bartley by John Bluth (then JPL oral historian), Carlsbad, Calif., 3–4 October 1995, pp. 14–24, 171. Thanks to John for furnishing me a copy as well as for his assistance in gathering copies of other JPL documents on a research trip to the JPL archives. A polymer is a compound consisting of many repeated, linked, simple molecules. Once it assumed the name of JPL, the former GALCIT organization retained its affiliation with Caltech but operated as a permanent installation working for Army Ordnance. In 1958, it transferred to the new National Aeronautics and Space Administration (NASA) in which, still operated by Caltech but now under contract to NASA, it functioned essentially as a NASA center.

6. E[rnie] S. Sutton, "From Polymers to Propellants to Rockets—A History of Thiokol," American Institute for Aeronautics and Astronautics (AIAA) paper 99–2929, presented at the Joint Propulsion Conference in Los Angeles, 20–24 June 1999, pp. 1–3; see also Sutton's "How a Tiny Laboratory in Kansas City Grew into a Giant Corporation: A History of Thiokol and Rockets, 1926–1996" (Chadds Ford, Pa.: privately printed, 1997), pp. 3–8 (cited hereafter as "History of Thiokol." I am indebted to Bob Geisler for providing a copy of the latter source and for organizing the session at the Joint Propulsion Conference where Ernie Sutton presented his paper.

7. A grain in this sense is simply the mass of solid propellant containing fuel and oxidizer, both of which are necessary in rockets to enable them to operate outside the atmosphere. Unlike jet engines, which burn fuel in the presence of atmospheric oxygen, rockets must carry their own supply of oxidizer, as an ingredient of the grain in the case of solid propellants or as either a separate propellant in liquid bipropellant systems or a component of a liquid monopropellant.

8. Sutton, "Polymers to Propellants," pp. 4–5; Bartley intvw., pp. 16–17; Carroll, "Historical Origins," p. 133. Sutton stated that the polymer Bartley acquired was LP-3, but Bartley and Carroll said it was LP-2. Sutton has since told me by e-mail that Bartley and Carroll were right.

9. Above two paragraphs are based on Carroll, "Historical Origins," pp. 132–38. Carroll has researched this matter thoroughly and covered it in much more detail than I have presented here. ABL is located across the Potomac River from Pinto, Maryland, and is 9 miles southwest of Cumberland, Maryland. At the time Vicar and Curate were developed, the laboratory was operated by George Washington University for the Office of Scientific Research and Development. See T.L. Moore, "Solid Rocket Development at Allegany Ballistics Laboratory," AIAA paper 99–2931, presented at the Joint Propulsion Conference in Los Angeles, 20–23 June 1999, pp. 1–5.

10. Coverage of the White Whizzer is based exclusively on comments provided to me on 4 August 2000 by Edward W. Price, who was heavily involved in the design, testing, and flying of the rocket. For background on the Eaton Canyon project and NOTS, see Charles H. Herty III and Francis A. Warren, "U.S. Double-Base Solid Propellant Tactical Rockets of the 1940–1955 Era," paper presented to the AIAA 9th Annual Meeting, Washington, D.C., 8–10 January 1973, unpaginated; J.D. Gerrard-Gough and Albert B. Christman, *The Grand Experiment at Inyokern: Narrative of the Naval Ordnance Test Station During the Second World War and the Immediate Postwar Years* (Washington, D.C.: Naval History Division, 1978); and "From the Desert to the Sea: A Brief Overview of the History of China Lake," on the Internet at http://www.nawcwpns.Navy.mil/clmf/hist.html.

11. See Joseph W. Wiggins, "The Earliest Large Solid Rocket Motor—The Hermes," paper presented at the AIAA 9th Annual Meeting and Technical Display, Washington, D.C., 8–10 January 1973, p. 345, where he states that the motor for the Falcon missile, built by Thiokol under subcontract to the Hughes Aircraft Company (which had the contract for the missile as

a whole), "was essentially a reduced length version of the JPL 'Thunderbird' Motor used for the first flight tests of case bonded, internal burning polysulfide propellant in 1947." The Falcon motor, he adds, contained both potassium perchlorate and ammonium perchlorate in "JPL 100L propellant." M. Summerfield, J.I. Shafer, H.L. Thackwell Jr., and C.E. Bartley, "The Applicability of Solid Propellants to High-Performance Rocket Vehicles," JPL ORDCIT Project Memorandum No. 4-17, 1 October 1947, p. 23, confirms this.

12. Summerfield et al., "Applicability of Solid Propellants," pp. 11, 23.

13. H.L. Thackwell Jr. and J.I. Shafer, "The Applicability of Solid Propellants to Rocket Vehicles of V-2 Size and Performance," JPL ORDCIT Project Memorandum No. 4-25, July 21, 1948, p.2. In this report, the two researchers concluded that solid-propellant rockets with JPL 118 as their propellant and a ten-point star configuration could equal the performance of the V-2 liquid-propellant German missile with only half its size and that solid rockets of greater length could double the range of the V-2.

14. L.G. Dunn and M.M. Mills, "The Status and Future Program for Research and Development of Solid Propellants," JPL Memorandum No. 4-5, 19 March 1945, p. 8; Raymond E. Wiech Jr. and Robert F. Strauss, *Fundamentals of Rocket Propulsion* (New York: Reinhold Publishing, 1960), p. 78. Neither of these sources gives the precise conditions of pressure and nozzle expansion ratio for these specific impulses, but presumably the figures are at least roughly comparable with the specific impulse of the JPL 118.

15. For the founding of Aerojet, see J.D. Hunley, "The Evolution of Large Solid Propellant Rocketry in the United States," *Quest: The History of Spaceflight Quarterly* 6 (Spring 1998): 24 and the sources cited there. On ammonium perchlorate, see Karl Klager and Albert O. Dekker, "Early Solid Composite Rockets," unpublished paper dated October 1972, pp. 13–24. Thanks to Klager for providing me with a copy. For the fact that aeroplex (not so named in the source) "cured into a very hard and brittle propellant grain which was not susceptible to case binding and needed trap supports," technical memoir by H.W. Ritchey [of Thiokol], c. 1980, p. 6, kindly supplied by Ernie Sutton.

16. Sutton, "Polymers to Propellants," pp. 6–7; Dave Dooling, "Thiokol: Firm Celebrates 30 Years of Space Technology," *Huntsville Times*, 22 April 1979, pp. 4–5, seen in the NASA Historical Reference Collection, NASA Headquarters, Washington, D.C. (hereafter cited as Historical Reference Collection), folder 010912, "II Thiokol."

17. Sutton, "Polymers to Propellants," pp. 9, 23; Sutton, "History of Thiokol," p. 58; Wiggins, "Hermes," p. 345.

18. Edward N. Hall, "Air Force Missile Experience," *Air University Quarterly Review* 9 (Summer 1957): 22–23.

19. Wiggins, "Hermes," p. 344.

20. Memorandum to L.F. Welenetz, n.d. Ernie Sutton generously sent me this and two other early Thiokol documents from his collection. On the front of this one is a note to Ernie indicating that the memo "must be a copy of a trip report of Glen Nelson's written in late January or early February 1949." On JPL's contributions, see also A.T. Guzzo, "Progress Report: Development of Standard Operating Procedures for Manufacture of Polysulfide-Perchlorate Propellants," Thiokol Corporation, Redstone Division Report No. 26-51, November 1951, p. 3, which states, "Early in 1948, when Thiokol initially attempted to produce polysulfide-perchlorate propellants on a pilot-plant scale, the work of the Jet Propulsion Laboratory furnished a valuable fund of basic information."

21. Sutton, "History of Thiokol," p. 28, which includes the quotation from physical chemist Harold W. Ritchey, whom he describes on p. 20 as "the most influential person in Thiokol's rocket history"; Guzzo, "Progress Report," pp. 3–8; Wiggins, "Hermes," p. 422.

22. Carroll, "Historical Origins," pp. 140–41; Hunley, "Evolution," pp. 25, 34 n. 22.

23. Sutton, "History of Thiokol," p. 29.

24. Wiggins, "Hermes," pp. 347–48, 355.

25. Ibid., pp. 357–62. LP-33 was a polymer that had better physical performance at low

temperatures than LP-3, the propellant that succeeded LP-2. However, LP-3 was less inclined to "creep" (deform) at high temperatures.

26. Sutton, "Polymers to Propellants," p. 11.

27. Wiggins, "Hermes," pp. 363, 388–90.

28. Ibid., p. 421.

29. Ibid., pp. 389, 396, 400–401. Wiggins does not mention the Titans and Space Shuttle,
but had he been writing later, no doubt he would have. Quotation taken from Sutton, "Polymers to Propellants," p. 10.

30. [Mary T. Cagle], "History of the Sergeant Weapon System" (Redstone Arsenal, Ala.:
U.S. Army Missile Command, [1963]), pp. 23–25, 31, 43–47, 97.

31. Ibid., pp. 64–65, 193–94, 199–200, 202–3; see also Koppes, *JPL*, pp. 63–66, 71–77,
which uses internal JPL sources different from those cited by Cagle.

32. Cagle, "History of the Sergeant," pp. 278–79; J.D. Hunley, "The History of Solid-
Propellant Rocketry: What We Do and Do Not Know," AIAA Paper 99-2925, presented at the
35th AIAA/ASME/SAE/ASEE Joint Propulsion Conference and Exhibit, 20–24 June 1999, p.
2; and see above in the text on the RV-A-10. Among the problems with Sergeant development
was combustion instability in the earlier motors. This was solved in the JPL 500 motor by
about 1958–59. See Cagle, "History of the Sergeant," pp. 104–7.

33. There is a multitude of sources for the developments summarized in the two paragraphs above. Two of the most useful ones are Edmund Beard, *Developing the ICBM: A Study
in Bureaucratic Politics* (New York: Columbia University Press, 1976) and Michael H.
Armacost, *The Politics of Weapons Innovation: The Thor-Jupiter Controversy* (New York: Columbia University Press, 1969). A valuable summary with much specific information is provided by John C. Lonnquest and David F. Winkler, *To Defend and Deter: The Legacy of the
United States Cold War Missile Program*, USACERL Special Report 97/01 (Rock Island, Ill.:
Defense Publishing Service, 1996), pp. 33–36, 41–42, 47–48. My thanks to Rebecca Cameron
for providing me a copy of this study, part of the Cold War Project sponsored by the Depart-
ment of Defense. For the ranges of ICBMs and IRBMs, see, for example, Harvey M. Sapolsky,
The Polaris System Development: Bureaucratic and Programmatic Success in Government
(Cambridge: Harvard University Press, 1972), p. 2. An interesting example of interservice
rivalry is provided by ABMA/BMD Memo for File, "Comment by Col. E. N. Hall (BMD-
ARDC) Regarding Successful JUPITER Launch," 12 June 1957, which quotes the Air Force's
Hall as saying about the launch, "I think it is a tragic thing for the country!" See also the
memo, Maj. Gen. J. B. Medaris for Gen. Gavin and Gen. Daley, 24 December 1957, about
Hall's efforts to "push" a "500–700 mile solid propellant missile," in which Medaris suggests
vigorous efforts for a medium-range ballistic missile on the part of the Army. Copies of both
documents from the Medaris Papers at the Florida Institute of Technology were generously
provided to me by Michael Neufeld. To the first statement, Colonel Hall commented about a
draft of this chapter, "I don't recall making this specific comment, but indeed it was a tragic
thing. In reality, there never was a Jupiter missile. It flew with a Thor engine which Medaris
called, 'ground support equipment.' The Jupiter program was a . . . nuisance to me. To provide the Army with engines for this misfit, I had continually to distort my development, test,
and production schedules." Letter, Hall to author, 15 July 2000.

34. Armacost, *Thor-Jupiter*, p. 106.

35. See esp. Sapolsky, *Polaris System Development*, pp. 22–24.

36. Frederick I. Ordway and Ronald C. Wakeford, *International Missile and Spacecraft
Guide* (New York: McGraw-Hill, 1960), p. 5.

37. James Baar and William E. Howard, *Polaris!* (New York: Harcourt Brace and Company, 1960), p. 14.

38. See Hunley, "Evolution," p. 25, and the sources cited there.

39. But cf. the memo by Medaris cited in note 32.

40. See esp. Sapolsky, *Polaris System Development*, pp. 26–29, 47; Baar and Howard, *Polaris!* p. 15.

41. Sapolsky, *Polaris System Development*, pp. 26–28; Philip D. Umholtz, "The History of Solid Rocket Propulsion and Aerojet," AIAA paper 99-2927, presented at the AIAA/ASME/SAE/ASEE Joint Propulsion Conference, 20–24 June 1999, in Los Angeles, unpaginated (fifth page).

42. The above three paragraphs are based on Hunley, "Evolution," p. 26, and the sources it cites (including a letter from and telephonic intvws. with Karl Klager); J.F. Sparks and M.P. Friedlander III, "Fifty Years of Solid Propellant Technical Achievements at Atlantic Research Corporation," AIAA paper 99-2932, presented at the 35th AIAA/ASME/SAE/ASEE Joint Propulsion Conference, 20–24 June 1999 in Los Angeles, pp. 1, 8; letter, Charles B. Henderson to me, 16 October 1997; my intvw. of Henderson, 15 October 1997. It was the patient explanations of both Klager and Henderson, taken together, that enabled a nonchemist to understand the technical details of this discovery. In a subsequent letter of 18 September 1998, Henderson commented on a correct but misleading statement in Hunley, "Evolution," p. 26, where I said that Atlantic Research "did not establish a six-person Missile Systems Division until 1960." As Henderson put it, "The Missile System Division was a separate adjunct to the rocket company. It purchased rockets from several rocket companies and assembled them into the four-stage Athena System. . . . The East Coast divisions of Atlantic Research had grown to maybe 800–1000 people by 1960 and were actively involved in developing tactical rocket systems." According to R.A. Fuhrman, "Fleet Ballistic Missile System: Polaris to Trident," von Kármán Lecture for 1978, paper presented at the AIAA 14th Annual Meting, February 1978, in Washington, D.C., p. 12, "Tests at the Naval Ordnance Test Station in 1955, which were confirmed by the Atlantic Research Corporation in 1956, demonstrated a significant increase in specific impulse obtained by the addition of finely divided aluminum to the propellant." This offers a variant to the explanation above for the discovery of aluminum as significant contributor to specific impulse, but it will require further research to discover the details and how they fit with the interpretation offered in the narrative of this chapter.

43. Although called "jet" these were in reality non-air-breathing rockets.

44. This account pieced together from the biographical data in Hall, "Air Force Missile Experience," p. 23; Roy Neal, *Ace in the Hole* (Garden City, N.Y.: Doubleday, 1962), pp. 49, 52–62; Edward N. Hall, "USAF Engineer in Wonderland, Including the Missile Down the Rabbit Hole," undated typescript generously provided by Colonel Hall, pp. 44–45; my telephonic intvw. with Hall, 13 November 1998 (transcript in the NASA Historical Reference Collection), pp. 3–4; and Hall, *The Art of Destructive Management: What Hath Man Wrought?* (New York: Vantage Press, 1984), pp. 31–38, quotations from p. 38. Hall had earned a bachelor of science degree in engineering from the City College of New York in 1935, a professional degree in chemical engineering (Ch.E.) from the same institution in 1936, and a master of science degree in aeronautical engineering from Caltech in 1948. For information on Work Camp Dora, see esp. Yves Béon, *Planet Dora: A Memoir of the Holocaust and the Birth of the Space Age* (Boulder, Colo.: Westview Press, 1997), trans. Yves Béon and Richard L. Fague, ed. with introd. by Michael J. Neufeld, pp. xix, 125, and passim. As Edward W. Price pointed out in a review of a draft of this chapter, 4 August 2000, this section should not be interpreted to suggest that Hall and the Air Force were the only innovators with vision in this period. Others were also making progress, and by 1951, the Air Force could have used JPL JATOs with ammonium-perchlorate oxidizers to solve the B-47 problem.

45. Neal, *Ace in the Hole*, pp. 64, 67; Hall, "Air Force Missile Experience," p. 23, for his title; Hunley, "Evolution," pp. 25, 27, for background on Grand Central and Hercules.

46. This account based principally on Bob Gordon et al., *Aerojet: The Creative Company*, self-published by the Aerojet History Group (Los Angeles: Stuart F. Cooper, 1995), pp. IV-15

to IV-16, all quotations from these two pages; but see also Neal, *Ace in the Hole,* p. 73; Klager and Dekker, "Early Solid Composite Rockets," p. 22; letter, Hall to author, 15 July 2000; and Hunley, "Evolution," p. 25, for some of the details. The account in Gordon et al., *Aerojet* (written in this case by Philip Umholtz but based on multiple contributions from those involved), discusses another innovation in the technology of rocket cases called aerowrap that "contributed significantly to Aerojet's favorable position" in the AFLRP, but since it was used only experimentally, I have not included it in the narrative. STL was initially an autonomous division of Ramo-Wooldridge, renamed in November 1957 from its Guided Missile Research Division. For its complex evolution, see Davis Dyer, *TRW: Pioneering Technology and Innovation since 1900* (Boston: Harvard Business School Press, 1998), pp. 198, 207, 227 ff., and 230 ff.

47. Gordon et al., *Aerojet,* pp. IV-98 to IV-103, IV-119, quotations on p. IV-99; short biography of Dr. Karl Klager that he generously provided; Hunley, "Evolution," p. 27 and the sources cited there; Sapolsky, *Polaris System Development,* p. 33, noting Secretary of Defense Charles E. Wilson's authorization of Polaris development in December 1956. However, as Philip Umholz wrote in *Aerojet,* already in July the Navy had given Aerojet a letter of intent to develop propellants specifically for Polaris.

48. Chemical Propulsion Information Agency's *CPIA/M1 Rocket Motor Manual* (hereafter cited as CPIA, *Rocket Motor Manual*); Hunley, "Evolution," p. 27; Fuhrman, "Polaris to Trident," pp. 11–16. Above in the narrative for the dimensions of Jupiter S.

49. K[arl] Klager, "Early Polaris and Minuteman Solid Rocket History," unpublished paper kindly provided by Klager, pp. 6–8, figure 9; Fuhrman, "Polaris to Trident," p. 12; Graham Spinardi, *From Polaris to Trident: The Development of US Fleet Ballistic Missile Technology* (Cambridge: Cambridge University Press, 1994), p. 52.

50. Fuhrman, "Polaris to Trident," pp. 12, 14; Spinardi, *Polaris to Trident,* p. 52.

51. Fuhrman, "Polaris to Trident," p. 13; Spinardi, *Polaris to Trident,* pp. 39, 53; Klager, "Early Polaris and Minuteman," pp. 7–8; comments of Edward Price on a draft of this chapter, 4 August 2000.

52. Spinardi, *Polaris to Trident,* pp. 42–45; Donald MacKenzie, *Inventing Accuracy: A Historical Sociology of Nuclear Missile Guidance* (Cambridge, Mass.: MIT Press, 1990), pp. 16–18, 22–25, 120–23, 147–49; Fuhrman, "Polaris to Trident," p. 15, both quotations from Fuhrman. Gyroscopes work in guidance by virtue of their inertia, which is the force that keeps a child's top upright as long as it is spinning. Since they change their orientation in a predictable way, they can signal to the computer when a missile or rocket is deviating from a predetermined course. But to do this accurately over a period of time, they must encounter very little friction. Floating a gyro in a liquid was one method of reducing friction, other methods including ball bearings of incredibly accurate roundness, gas, electric, and magnetic "bearings." Although radio guidance was arguably more accurate than inertial guidance, the military preferred the latter because it was self-contained on the missile and therefore less vulnerable than a radio guidance, which was dependant on an outside ground installation. This could either be destroyed by enemy missiles or its signals to the "friendly" missile could be jammed. On p. 315, MacKenzie defines a PIGA as "a gyroscope . . . supported in an unsymmetrical way so as to make it sensitive to acceleration, not just rotation." The PIGAs were critical elements in determining when to initiate thrust termination. The guidance systems for Polaris and Minuteman were not relevant to solid-rocket booster development, but I discuss them because they were significant in the history of guidance and are relevant to launch-vehicle guidance.

53. Klager, "Early Polaris and Minuteman," p. 14, quotations from this source; Hunley, "Evolution," p. 28 and the sources cited there; USAF Intercontinental Ballistic Missile System Program Office, "Minuteman Weapon System History and Description" (Hill AFB, Utah, May 1996), p. 17. The last source lists the February 1958 date but says the road map leading to Minuteman began in 1956. In his letter to me, 15 July 2000, Hall says that Secretary of

Defense McElroy approved Polaris and Minuteman the same day. This must have been some sort of supplementary approval for Polaris, however, because the literature on that missile is clear on its approval in December 1956 by McElroy's predecessor as Secretary of Defense, Charles E. Wilson. See, for example, Sapolsky, *Polaris System Development*, pp. 33–34, and Fuhrman, "Polaris to Trident," p. 4, where he gives the precise date of 8 December 1956.

54. See especially Sapolski, *Polaris System Development*, pp. 21–23, 31, 35–37, 45–47, 49–51, 55–56, 64–65, 74, 97–99, 149–50, 157–59; Fuhrman, "Polaris to Trident," pp. 2–3. On heterogeneous engineering, see John Law, "Technology and Heterogeneous Engineering: the Case of the Portuguese Expansion," in Wiebe E. Bijker, Thomas P. Hughes, and Trevor Pinch, eds., *The Social Construction of Technological Systems: New Directions in the Sociology and History of Technology* (Cambridge, Mass.: MIT Press, 1987), pp. 111–34, and MacKenzie, *Inventing Accuracy*, p. 28, where he defines the term as "the engineering of the social as well as the physical world"—in this instance, garnering support and funding for Polaris and encouraging the Polaris team to overcome obstacles and succeed in the physical engineering of the system.

55. Again, a convenient summary treatment of this issue is provided by Lonnquest and Winkler, *To Defend and Deter*, pp. 43–44. See also Beard, *Developing the ICBM*, pp. 188–93.

56. Sapolski, *Polaris System Development*, p. 39.

57. Biographical synopsis of Adelman provided by Karen Schaffer of Chemical Systems Division, the current name of the organization Adelman later helped found; my telephonic intvw. with Adelman, 23 February 2000. On Ramo-Wooldridge and its role, see W.S. Kennedy, S.M. Kovacic, and E.C. Rea, "Solid Rocket History at TRW Ballistic Missiles Division," paper presented at the AIAA/SAE/ASME/ASEE 28th Joint Propulsion Conference, Nashville, Tenn., 6–8 July 1992, pp. 1, 13, which recognizes on the latter page that Hall and Adelman were "the primary proponents of solid-propellant ICBMs and led a small group of engineers to monitor and direct the progress of the technology." As will be discussed below, Ramo-Wooldridge was one of the organizations that became TRW. On TRW, see also Dyer, *TRW*, which should be compared with Hall's less flattering descriptions of Ramo-Wooldridge in his interviews and writings, including *The Art of Destructive Management* and "USAF Engineer in Wonderland."

58. Kennedy, Kovacic, and Rea, "Solid Rocket History," p. 13; Neal, *Ace in the Hole*, p. 82. Both Adelman and Hall in their interviews have emphasized the unreliability of Neal's study, but it provides many details not available elsewhere and, if compared with other sources, helps to piece together a complex history.

59. Sources differ about details, but they pretty well agree on these generalizations. See Kennedy, Kovacic, and Rea, "Solid Rocket History," p. 13; Neal, *Ace in the Hole*, p. 85; my intvws. with Adelman, 23 February 2000, and Hall 13 November 1998; Hall, "USAF Engineer in Wonderland," p. 61. For the change in WDD's name, Robert C. Anderson, "Minuteman to MX: The ICBM Evolution," *TRW/DSSG/Quest* (Autumn 1979), p. 34, "SAMSO: A Look Back," reprinted from *Air Force Magazine*, August 1979.

60. Kennedy, Kovacic, and Rea, "Solid Rocket History," pp. 13–14; Neal, *Ace in the Hole*, pp. 85–86; Anderson, "Minuteman to MX," pp. 33–36. For the proposal to make Minuteman mobile, "Mobility Designed into Minuteman," *Aviation Week, including Space Technology*, 3 August 1959, pp. 93–94, seen in National Air and Space Museum file OM-720001-03, Minuteman I Missile. It is not strictly relevant to the arguments advanced in support of Minuteman, but of course Minuteman gave up range in exchange for its smaller size and more efficient operation. As of 1962, one set of statistics showed Atlas with a range of 9,000 miles. Titan I had a range of 6,300 miles, and Titan II, one of 9,000 miles. These figures compared with a range of only slightly over 5,500 for Minuteman I, but it was still able to carry out its intended mission. For these statistics, see Stanley M. Ulanoff, *Illustrated Guide to U.S. Missiles and Rockets* (Garden City, N.Y.: Doubleday, 1962), pp. 42–44.

61. Neal, *Ace in the Hole*, pp. 87–91; Hall, "USAF Engineer in Wonderland," pp. 61–62; oral history intvw. of Hall by Jacob Neufeld, 11 July 1989, pp. 20–21; my intvw. with Adelman; Memorandum for Record, Col. Ray E. Soper, Chief, Ballistic Division, Office of the Assistant Chief of Staff for Guided Missiles, subj: Secretary of Defense Review of Air Force Ballistic Missile Proposals, 10 February 1958, in Ethel M. DeHaven, "Aerospace—The Evolution of USAF Weapons Acquisition Policy, 1945–1961," vol. 3, June 1962 (AFSC Historical Publications Series 62-24-8), document 212, kindly provided by the Air Force Historical Research Agency. As indicated in the narrative, these sources disagree about details but agree in the overall picture they present. Neal places this meeting in October 1957, but Soper's memo gives the 8 February date.

62. Neal, *Ace in the Hole*, pp. 92–100; Hall, "USAF Engineer in Wonderland," pp. 62–63; intvw. of Hall by Neufeld, pp. 21–23; my intvw. with Adelman. Neal does not report Ramo-Wooldridge's opposition, but his account is otherwise generally consistent with those of Adelman and Hall, who do emphasize the opposition. Adelman's account concurs with Neal's in pointing to Schriever's role in funding subsequent to the initial $50 million, which Hall claims as being the result of his advocacy. Likely, there is some truth in both accounts. In his letter to me of 15 July 2000, Hall said that Adelman was not involved in this sequence of events and that McElroy granted the $50 million in the February briefing but that TRW "continued to sabotage the Minuteman program. . . . At this period, I had no support from any part of TRW. The effect of this constant noisy opposition was to have Mr. [William] Holaday, the missile czar, terminate the program. It took a great deal of effort on my part to have it resuscitated." Adelman remembers being involved in advocating Minuteman.

63. Hall, "USAF Engineer in Wonderland," p. 63; my intvws. with Hall and Adelman; Hunley, "Evolution," pp. 30–31; Anderson, "Minuteman to MX," p. 49, which gives Hall's tenure as first director of the Minuteman program as 1957–58. For his contributions to missile development, Hall was inducted into the Air Force Space and Missile Hall of Fame in 1999. See Timothy Hoffman, "Three Inducted into Air Force Space and Missile Hall of Fame," *Space Observer*, 8 October 1999, p. 3.

64. Anderson, "Minuteman to MX," pp. 34, 49; Kennedy, Kovacic, and Rea, "Solid Rocket History," p. 10; Neal, *Ace in the Hole*, p. 123.

65. "Launch on December 63rd: Gen. Sam Phillips Recounts the Emergence of Solids in the Nation's Principal Deterrent," *Astronautics & Aeronautics* 10 (October 1972): 62; Klager, "Early Polaris and Minuteman," pp. 14–15; Kennedy, Kovacic, and Rea, "Solid Rocket History," pp. 10, 15; Frank G. McGuire, "Minuteman Third-Stage Award Nears," *Missiles and Rockets*, 6 June 1960, pp. 42–43; "Minuteman 3rd Stage Is Big Advance," *Missiles and Rockets*, 17 October 1960, pp. 36–37; William Leavitt, "Minuteman—Ten Years of Solid Performance," *Air Force Magazine* 54 (March 1971): 26.

66. Kennedy, Kovacic, and Rea, "Solid Rocket History," pp. 14, 17; Klager, "Early Polaris and Minuteman," pp. 17–18; Neal, *Ace in the Hole*, pp. 136–37. Each of these sources provides part of the explanation in the narrative, but unfortunately, none of them address the important question of exactly how the solution was arrived at and who among the participants provided it. Moreover, the specifics of the solutions come from Klager, who described them for the second stage. Presumably, however, they also applied to stage 1. This is suggested by Kennedy, Kovacic, and Rea, who on p. 17 state with regard to the throat inserts, "The basic failure mechanisms of the early nozzle designs were identified and the need for tungsten throat inserts and nozzle entrance flow straightening was established." They make no mention of specific stages in this regard.

67. Kennedy, Kovacic, and Rea, "Solid Rocket History," p. 15; Neal, *Ace in the Hole*, pp. 137–38.

68. Phillips, "Launch on December 63rd," p. 62.

69. Ibid.; Neal, Ace in the Hole, pp. 147–51, 167–68; Leavitt, "Minuteman," p. 26; U.S.

Air Force, Astronautics Laboratory, fact sheet "Air Force Astronautics Laboratory," February 1988, pp. 2–3, on the rocket lab. On the early critics, see esp. Neufeld's intvw. with Colonel Hall, p. 22.

70. Phillips, "Launch on December 63rd," p. 62; Leavitt, "Minuteman," p. 26.

71. USAF ICBM Program Office, "Minuteman Weapon System," pp. 14, 17.

72. Sutton, "Polymers to Propellants," p. 13; E. S. Sutton, "From Polysulfides to CTPB Binders—A Major Transition in Solid Propellant Binder Chemistry," Paper 84-1236, presented at the AIAA/SAE/ASME 20th Joint Propulsion Conference, Cincinnati, Ohio, 11–13 June 1984, pp. 6–7, quotation from p. 7; E. J. Mastrolia and K. Klager, "Solid Propellants Based on Polybutadiene Binders," reprinted from Advances in Chemistry Series, Number 88: *Propellants, Manufacture, Hazards and Testing* (American Chemical Society, 1969), pp. 122–23; Adolf E. Oberth, *Principles of Solid Propellant Development* (Laurel, Md.: Chemical Propulsion Information Agency Publication 469, 1987), p. 2–8. My thanks to Dr. Klager and the CPIA for providing me copies of the last two publications.

73. Sutton, "Polysulfides to CTPB," p. 6; Irving Stone, "Minuteman Propulsion—Part I: Minuteman ICBM Solid Motor Stages Enter Production Phase," *Aviation Week & Space Technology,* 27 August 1962, p. 57. Details such as these are not available for all the motors and stages discussed in this chapter. I provide them where they are available to show the really enormous number of firms and organizations that contributed to the development of solid-propellant rocket technology.

74. Sutton, "Polysulfides to CTPB," pp. 7–8; Stone, "Minuteman Propulsion—Part I," p. 57; "Propulsion Characteristics Summary, TU-122, Wing I," 1 February 1963, extract from the "Gray Books," generously provided by Bill Elliott of the AFMC History Office; Thiokol Chemical Corporation, *Rocket Propulsion Data,* 3d ed. (Bristol, Pa.: n.p., 1961), entry for TU-122.

75. Stone, "Minuteman Propulsion—Part I," pp. 54–62; USAF "Propulsion Characteristics Summary, TU-122, Wing I"; Thiokol, *Rocket Propulsion Data,* TU-122 and TX-12. The Thiokol figures differ somewhat from the USAF data for Minuteman, perhaps because the USAF data were from 1962–63 instead of 1961, hence my use of the USAF data.

76. Klager, "Early Polaris and Minuteman," p. 16; Irving Stone, "Aerojet Second Stage Must Withstand Heaviest Stresses," *Aviation Week & Space Technology,* 3 September 1962, p. 71; "Propulsion Characteristics Summary, M-56, Wing I" extract from the USAF Gray Books, 3 February 1964; CPIA, *Rocket Motor Manual,* vol. 1, unit 362.

77. Klager, "Early Polaris and Minuteman," p. 15; Stone, "Aerojet Second Stage," pp. 68–69; "Propulsion Characteristics Summary, M-56, Wing I." Stone gives a different length, but I have followed the last source, which is an official Air Force document.

78. See note above. All three sources provided some different details, but the common information on these subjects was in agreement to a remarkable degree.

79. Stone, "Aerojet Second Stage," p. 73. Systems similar to PRIMISCO no doubt governed the production of the other three stages as well as the management of the entire missile.

80. Moore, "Solid Rocket Development," pp. 5–7; Hunley, "Evolution," pp. 27–28 and the sources cited there, including Davis Dyer and David B. Sicilia, *Labors of a Modern Hercules: The Evolution of a Chemical Company* (Boston: Harvard Business School Press, 1990), pp. 2, 9, 257–58, and my telephonic intvw. with Daniel W. Dembrow; comments of R.H. Woodward Waesche at the session on solid-propulsion history, 35th AIAA/ASME/SAE/ASEE Joint Propulsion Conference & Exhibit, 20–23 June 1999, Los Angeles; letter Edward W. Price to me, 28 April 2000; Herty and Warren, "U.S. Double-Base Solid Propellant Tactical Rockets," generously provided to me by Ed Price. In his comments on a draft of this chapter, Price wrote, "I believe the first CMDB propellants may have been made at ABL by mixing the dry ingredients, pouring the mixture into the mold, and then adding the solvent-plasticizer. Slurry mixing [as this process evidently was called] was adopted by everyone because it was

more amenable to large-scale motor production and gave latitude to get better mechanical properties."

81. C.B. Henderson, with comments by Rumbel incorporated, "The Development of Composite-Modified Double Base Propellants at the Atlantic Research Corporation," 30 September 1998, encl. in letter, Henderson to me, 8 October 1998.

82. Hunley, "Evolution," p. 28; Dyer and Sicilia, *Modern Hercules*, pp. 9, 318–20; Constance McLaughlin Green and Milton Lomask, *Vanguard: A History* (Washington, D.C.: Smithsonian Institution Press, 1971), pp. 228, 254, 287; Brian A. Wilson, "The History of Composite Motor Case Design," AIAA paper 93–1782, presented at the 29th Joint Propulsion Conference and Exhibit, 28–30 June 1993 in Monterey, Calif., no pagination (a series of slides rather than an actual paper); "Lightweight Pressure Vessels," portion of a technical memoir by H.W. Ritchey, c. 1980, kindly provided to me by Ernie Sutton.

83. Hunley, "Evolution," p. 28; Dyer and Sicilia, *Modern Hercules*, p. 322; CPIA, *Rocket Motor Manual*, unit 410, Polaris Model A2 Stage 2, pp. 1–2.

84. Hunley, "Evolution," pp. 27–28; Dyer and Sicilia, *Modern Hercules*, p. 322; CPIA, *Rocket Motor Manual*, unit 410, Polaris Model A2 Stage 2, pp. 1–3; Irving Stone, "Hercules Stage 3 Uses Glass Fiber Case," *Aviation Week & Space Technology*, 10 September 1962, pp. 162–65; "The Minuteman: The Mobilization Programme for a Thousand Minutemen," *Interavia* 3 (1961): 310; Oberth, *Solid Propellant Development*, p. 2–13; "Propulsion Characteristics Summary, 65–DS-15,520," 29 September 1961. I have used the length and diameter from the last source, which differs slightly from the CPIA figures. For further details on the fiberglass case, see "Minuteman 3rd Stage Is Big Advance," *Missiles and Rockets*, 17 October 1960, pp. 36–37.

85. "Propulsion Characteristics Summary, 65–DS-15,520," 29 September 1961; CPIA, *Rocket Motor Manual*, unit 410, Polaris Model A2 Stage 2, pp. 1–3; Stone, "Hercules Stage 3," p. 167.

86. Anderson, "Minuteman to MX," p. 39; J. M. Wuerth, "The Evolution of Minuteman Guidance and Control," *Journal of the Institute of Navigation* 23 (Spring 1976): 64.

87. "Propulsion Characteristics Summary, 65–DS-15,520," 29 September 1961; "Propulsion Characteristics Summary (M-56AJ-3, Wing I)," 10 April 1962; "Propulsion Characteristics Summary (TU-122, Wing I)," 20 January 1962; Ogden Air Logistics Center, "Minuteman Weapon System: History and Description," OO-ALC/MMG (Hill AFB, Utah, August 1990), pp. 27, 33.

88. "Propulsion Characteristics Summary (M-56AJ-3, Wing I)," 10 April 1962, and "Propulsion Characteristics Summary (M-56A1, Wing II)," 9 November 1962; Kennedy, Kovacic, and Rea, "Solid Rocket History," p. 16; Stone, "Aerojet Second Stage," p. 68; Klager, "Early Polaris and Minuteman," p. 16; Anderson, "Minuteman to MX," p. 43. I have taken the weight reduction from the first two sources rather than Stone since the former are official Air Force documents and likely to be more accurate. Stone placed the weight reduction for the case at 220 pounds and that for the nozzles at 25 pounds for a total of 245 pounds. The Propellant Characteristics Summaries show that the mass fraction improved from 0.865 to 0.897. Klager talked about the shift from steel to titanium as occurring in Minuteman II, but he apparently confused Wing II and Minuteman II. The other sources all indicate that the change occurred in Wing II.

89. Kennedy, Kovacic, and Rea, "Solid Rocket History," pp. 17–18; "Propulsion Characteristics Summary (M-56A1, Wing II)," 9 November 1962, and "Propulsion Characteristics Summary (SRJ19–AJ-1)," 22 January 1966.

90. Sparks and Friedlander, "Fifty Years," p. 8; Sutton, "Polysulfides to CTPB," p. 8.

91. Sparks and Friedlander, "Fifty Years," p. 8.

92. Gordon et al., *Aerojet*, p. IV-84.

93. Kennedy, Kovacic, and Rea, "Solid Rocket History," p. 18.

94. Ibid.; "Propulsion Characteristics Summary (M-56A1, Wing II)," 9 November 1962, and "Propulsion Characteristics Summary (SRJ19–AJ-1)," 22 January 1966; Hunley, "Evolution," p. 29 and the sources cited there.

95. J.D. Hunley, "The History of Solid-Propellant Rocketry: What We Do and Do Not Know," AIAA paper 99-2925, presented at the 35th AIAA/ASME/SAE/ASEE Joint Propulsion Conference and Exhibit in Los Angeles, 20–24 June 1999, p. 7; Sutton, "Polysulfides to CTPB," p. 8; Sutton, "Polymers to Propellants," p. 14.

96. "Propulsion Characteristics Summary (M-56A1, Wing II)," 9 November 1962, and "Propulsion Characteristics Summary (SRJ19–AJ-1)," 22 January 1966; Klager, "Early Polaris and Minuteman," p. 18; Kennedy, Kovacic, and Rea, "Solid Rocket History," p. 18; Fuhrman, "Polaris to Trident," p. 18; "Minuteman Weapon System," August 1990, p. 50. For the early contributions of NOTS, see Herty and Warren, "U.S. Double-Base Solid Propellant Tactical Rockets," pp. 6, 22–25 (my pagination). Notes by Ed Price in the copy he sent me of this paper indicate that the Naval Ordnance Test Station (where Price worked) made even more contributions than the two authors recognized.

97. MacKenzie, *Inventing Accuracy,* pp. 206–13; Wuerth, "Minuteman Guidance and Control," pp. 64–65, 67–68, 70, 74–75; Anderson, "Minuteman to MX," p. 44; Barry Miller, "Microcircuits Boost Minuteman Capability," *Aviation Week & Space Technology,* 28 October 1963, esp. p. 77.

98. Hunley, "Evolution," p. 32 and the sources cited there; comments of Edward W. Price on a draft of this chapter, 4 August 2000.

99. "Minuteman Weapon System," August 1990, p. 54; Kennedy, Kovacic, and Rea, "Solid Rocket History," p. 20; Wilson, "Composite Motor Case Design"; *Aerojet: The Creative Company,* pp. IV-104 to IV-105; "Propulsion Characteristics Summary (SRJ19–AJ-1)," 22 January 1966; "Propulsion Characteristics Summary (M57A1," 4 January 1966; "Propulsion Characteristics Summary (SR73–AJ-1), 15 April 1969; "Stage III for USAF's Minuteman III," *Aerojet General Booster,* September 1967; CPIA, *Rocket Motor Manual,* unit 457. The last source gives different percentages for PBCT (as it calls CTPB) and other propellant ingredients, but I have followed the Air Force's Propulsion Characteristics Summary in writing the narrative. A submerged nozzle is one that, instead of extending from the rear of the combustion chamber, is partly or wholly embedded in it. It displaces a small amount of the propellant but also shortens the missile or rocket. This enables a missile to fit in a smaller silo or in the confines of a submarine. For a launch vehicle, it shortens the gantry supporting the vehicle for launch and reduces the length of the ground elevator, wiring, and so on, thus reducing costs. On the debit side of the ledger, usually a submerged nozzle requires insulation on both sides of the submerged portion, adding to weight. Also, there is a performance loss because of entrapment of aluminum oxide in the nozzle cavity. I am indebted to Wilbur Andrepont for explaining these intricacies to me in e-mails of 9 August 1999 and 22 July 2000.

100. "Minuteman Weapon System," August 1990, p. 54; Kennedy, Kovacic, and Rea, "Solid Rocket History," p. 20; CPIA, *Rocket Motor Manual,* unit 412.

101. The three paragraphs above are based on a large variety of sources, especially Robert Lindsey, "UTC Chief Sees Tighter Rocket Market," *Missiles and Rockets,* 18 April 1966, p. 22; "Brainpower First in United's Space Venture," *Business Week,* 5 March 1960, pp. 138–44; "Highlights from the History of United Technologies" (January 1988), pp. 35–42, and a narrative, biographic materials, and chronology generously provided by CSD librarian Karen Schaffer; my telephonic intvws. with Barnet Adelman, 20 November 1996 and 23 February 2000; my telephonic intvw. with David Altman, 22 December 1999; Michael H. Gorn, *Harnessing the Genie: Science and Technology Forecasting for the Air Force, 1944–1986* (Washington, D.C.: Office of Air Force History, 1988), pp. 48–49, 89–93; and Michael H. Gorn, *The Universal Man: Theodore von Kármán's Life in Aeronautics* (Washington, D.C.: Smithsonian Institution Press, 1992), pp. 90–101, 136–37, 150–54.

102. Gordon et al., *Aerojet*, p. IV-38; Karl Klager, "Segmented Rocket Demonstration: Historical Development Prior to Their Use as Space Boosters," in J.D. Hunley, ed., *History of Rocketry and Astronautics: Proceedings of the Twenty-Fifth History Symposium of the International Academy of Astronautics* (San Diego: Univelt, 1997), pp. 159–70, with a diagram of the joint on p. 168; Wilbur C. Andrepont and Rafael M. Felix, "The History of Large Solid Rocket Motor Development in the United States," AIAA paper 94-3057, presented at the 30th AIAA/ASME/SAE/ASEE Joint Propulsion Conference, 27–29 June 1994, p. 2; comments of Ross Felix on a draft of this chapter, 22 August 2000.

103. Above nine paragraphs based on Klager, "Segmented Rocket Demonstration," pp. 163–87; Gordon et al., *Aerojet*, pp. IV-38 to IV-39; Andrepont and Felix, "Large Solid Rocket Motor Development," pp. 2, 19. The first and last of these sources differ in the dates they give for the Aerojet tests. I have followed Klager. Andrepont and Felix give the date of the third UTC test as 1 December 1981, but that is an evident misprint for 9 December 1961, which they give as the date on p. 3.

104. Above three paragraphs based principally on Andrepont and Felix, "Large Solid Rocket Motor Development," pp. 3, 19. Where p. 19 differs from p. 3, I have followed p. 3.

105. Hunley, "Evolution," p. 30, and the sources cited there, esp. G.R. Richards and J.W. Powell, "Titan 3 and 4 Launch Vehicles," *Journal of the British Interplanetary Society* 46 (1993): passim.

106. Hunley, "Evolution," pp. 30–31 and the sources cited there, esp. CPIA, *Rocket Motor Manual*, unit 559, Titan III and Titan IIIC Zero Stage; my intvw. with Herman P. Weyland, Paul G. Willoughby, Stan Backland, and J.G. Hill at Chemical Systems Division, 18 November 1996, information about the toughness of UTC's PBAN from senior research propellant chemist Weyland; my 1996 intvw. with Adelman; e-mail comments to me by Bernard Ross Felix, formerly vice president of engineering and technology, CSD, 8 January 1997, telephonic intvw. with him 10 February 1997, and his comments on the draft of this chapter, 22 and 23 August 2000. As he explained there, the term "zero stage" permitted the Martin Company to retain the nomenclature of stage 1 and stage 2 on its Titan IIIB liquid-propellant stages. He also explained that some of the shorthand in the CPIA description of the Titan SRM nozzle materials (p. 2 of unit 559) is not clear, stating that the Titan throat was similar to the shuttle SRB throat and that "carbon phenolic" was the simplified description. In both the shuttle and the Titan solid rocket motors, "phenolic resin glues together layers of fabric that have either been graphitized or carbonized in a furnace. This avoids the thermal stress problem that causes cracking on large-diameter monolithic graphite throats." The account in the narrative provides a greatly simplified overview of Titan SRM development. Many further details are available in Richards and Powell, "Titan 3 and 4."

107. Andrepont and Felix, "Large Solid Rocket Motor Development," pp. 4–6; Defense Metals Information Center, Battelle Memorial Institute, "Report on the Third Maraging Steel Project Review," Memorandum 181 (Columbus: DMIC, 1963), esp. pp. 1–3, A-17, A-44.

108. Above four paragraphs based on Andrepont and Felix, "Large Solid Rocket Motor Development," pp. 6, 9–10, 20; Sutton, "Polymers to Propellants," p. 17.

109. Andrepont and Felix, "Large Solid Rocket Motor Development," pp. 10–11, 20.

110. Ibid., pp. 11–14, 17, 20; Gordon et al., *Aerojet*, pp. IV-42 and IV-43; Sutton, "Polymers to Propellants," p. 17.

111. Linda Neuman Ezell, *NASA Historical Data Book*, vol. 3, *Programs and Projects 1969–1978*, NASA SP-4012 (Washington, D.C.: NASA, 1988), pp. 48–49; Dennis R. Jenkins, *Space Shuttle, The History of Developing the National Space Transportation System: The Beginning through STS-75* (Indian Harbour Beach, Fla.: Dennis R. Jenkins, 1996), pp. 132, 137. For the relationship of United Space Boosters to UTC, or Chemical Systems Division of United Technologies Corporation, as it became after 1975, "UTC Prepares for Shuttle Booster Role," *Aviation Week & Space Technology*, 3 January 1977, p. 15.

112. Jenkins, *Space Shuttle*, pp. 245–47, 277–79; Hunley, "Evolution," pp. 31 and 37 plus the sources cited there, especially my 1996 intvw. with Adelman; Andrepont and Felix, "Large Solid Rocket Motor Development," p. 14; CPIA, *Rocket Motor Manual*, unit 556; T.A. Heppenheimer, draft "History of the Space Shuttle," vol. 2, "Development of the Shuttle, 1972–1981," pp. 210–11. Heppenheimer and Jenkins differ as to the number of pins in the joint, Jenkins saying 177, which Andrepont and Felix also support, and Heppenheimer, 180. Jenkins also gives the overall length of the SRB as 149.16 feet, which may include elements not included in the CPIA manual, whose length I have followed.

113. Heppenheimer, "Development of the Shuttle," pp. 213–15; Andrepont and Felix, "Large Solid Rocket Motor Development," pp. 14, 21. Heppenheimer says the deflection was 7.1 degrees, which in any event is higher than the 6 degrees afforded the Titan IV solid rocket motor by its nitrogen tetroxide system. But Jenkins, *Space Shuttle*, p. 249, agrees with Andrepont and Felix on the 8 degrees.

114. CPIA, *Rocket Motor Manual*, units 556 and 613; Jenkins, *Space Shuttle*, pp. 245–50; Heppenheimer, "Development of the Shuttle," pp. 207, 212; Andrepont and Felix, "Large Solid Rocket Motor Development," pp. 14, 21; Hunley, "Evolution," pp. 30–31, 37–38 and the sources cited there; *Aeronautics and Space Report of the President*, Fiscal Year 1998 Activities (1999), Appendix C; *Aeronautics and Space Report of the President*, Fiscal Year 1995 Activities (1996), Appendix B-3.

115. Jenkins, *Space Shuttle*, pp. 251–54.

116. Andrepont and Felix, "Large Solid Rocket Motor Development," p. 15; William B. Scott and Michael A. Dornheim, "Upgraded Titan 4 Solid Rocket Motor Destroyed During First Full-Scale Test," *Aviation Week & Space Technology*, 8 April 1991, p. 24; CPIA, *Rocket Motor Manual*, unit 627, Titan IV SRMU, July 1996, generously provided by Tom Moore along with the updated unit 629 for the Titan IV SRM, July 1996. For the new name of Phillips Lab, Jay Levine, "Director Envisions Big Future for Labs," *Antelope Valley Press*, 1 October 1997, p. A1.

117. Andrepont and Felix, "Large Solid Rocket Motor," pp. 15–16, 21; CPIA, *Rocket Motor Manual*, unit 627, Titan IV SRMU, July 1996, and ibid., unit No. 629, "Titan IV SRM," July 1996.

118. Andrepont and Felix, "Large Solid Rocket Motor," p. 16; CPIA, *Rocket Motor Manual*, unit 627, Titan IV SRMU, July 1996; and ibid., unit 629, Titan IV SRM, July 1996; Roger Guillemette, "New & Improved Titan Launch Clears Path for Cassini," *Quest: The Magazine of Spaceflight* 5, no. 4 (1996): 38; Bruce A. Smith, "Titan 4 Officials Focus on Burn-Through," *Aviation Week & Space Technology*, 23 August 1993, p. 26; Scott and Dornheim, "Upgraded Titan 4 Solid Rocket Motor Destroyed," p. 23.

119. Guillemette, "New & Improved Titan," p. 38; "First Flight of Titan 4B Marks Major USAF Advance," *Aviation Week & Space Technology*, 3 March 1997, p. 29.

120. William Harwood, "New Titan Rocket Boosters Glean Near-Perfect Results," *Space News*, 3–9 November 1997, p. 6; *Aeronautics and Space Report of the President*, FY 1998, p. 69; "USAF Reports Cause of Titan IV Failure; Fleet Remains Grounded," *Aerospace Daily*, 19 January 1999, item 12; "Titan IVB Lifts Off with Defense Support Program Spacecraft," *Aerospace Daily*, 12 April 1999, item 12; "Titan IV Launches Classified Payload," *Aerospace Daily*, 25 May 1999, item 4.

121. "USAF Ready for First Launch of Satellites by Refurbished Minuteman," *Aerospace Daily*, 2 December 1999, item 11; Lynn Stodghill, Philip Newton, L. B. Mobley, "Use of Excess ICBM Assets to Support U.S. Space Launch Requirements," AIAA paper JPC 99-2513, presented at the 35th AIAA/ASME/SAE/ASEE Joint Propulsion Conference, 20–24 June 1999.

122. Steven J. Isakowitz, *International Reference Guide to Space Launch Systems*, 2d ed. (Washington, D.C.: AIAA, 1995), pp. 256–62; *Aeronautics and Space Report of the President*,

Fiscal Year 1995 Activities (1996), p. 15, Fiscal Year 1996 Activities (1997), pp. 15, 84, 86. Orbital Sciences also had a four-stage solid-propellant launch vehicle, the Taurus, with stages 2 through 4 derived from Pegasus. There has not been space or time to include it and other solid-propellant rockets, such as the solid, strap-on boosters for a variety of the Delta launch vehicles, in this chapter.

123. "USAF Readies for First Minuteman Space Launch Mission," *Aerospace Daily,* 12 October 1999, item 1; "USAF Ready for First Launch of Satellites by Refurbished Minuteman," *Aerospace Daily,* 2 December 1999, item 11; "Minuteman II Space Launch Postponed for Destacking," *Aerospace Daily,* 8 December 1999, item 11; "University Satellites Orbited by Minuteman II with Pegasus Stages," *Aerospace Daily,* 31 January 2000, item 17. The delay of the launch occurred because two critical avionics boxes failed preflight tests. The *Aerospace Daily* articles do not call the hybrid launch vehicle the Minotaur; nor do they spell out all the acronyms for the orbiting spacecraft.

124. Had there been room in this chapter to discuss in some detail the efforts to understand and reduce combustion instabilities, the examples of interservice cooperation would have been magnified. Identification of unwanted combustion instabilities began as early as 1941. Efforts to cope with the problem proceeded from trial-and-error by resident "magicians" in each company or organization to sophisticated laboratory testing and increasing understanding of the extraordinarily complex phenomenon. With the Polaris program, a realization of the need to deal with the problem led to a Tri-Service Committee to improve research through the Office of Naval Research, the Air Force's Office of Scientific Research, and the Army's Research Office, with supplemental funding by the Advanced Research Projects Agency. Frank T. McClure of the Applied Physics Laboratory at Johns Hopkins University was responsible for improving communications about research in this area, and a panel he organized on combustion instability led to the present JANNAF Combustion Subcommittee which has promoted exchange of research results. As ARPA funding decreased, NASA kept funding strong until 1968. The result was a great deal of progress but "while qualitative understanding of the phenomenon is now good, the goal of quantitative design for stability still escapes us," according to one of the pioneers in the field, Edward W. Price. To achieve high performance, motor designers tend to make motors that are close to or over stability limits. And as of yet, "present knowledge has not been reduced to quantitative engineering design principles, and may never be, because of the complexity of the phenomenon." Despite this, there has been a great deal of cooperation across institutional, company, and corporate lines to reach the level of understanding that now exists. The literature on the subject is immense, some of it being listed in Hunley, "Evolution," esp. n. 87, and more in E.W. Price, "Solid Rocket Combustion Instability—An American Historical Account," in *Nonsteady Burning and Combustion Stability of Solid Propellants,* ed. Luigi De Luca, Edward W. Price, and Martin Summerfield, vol. 143 of *Progress in Astronautics and Aeronautics* (Washington, D.C.: AIAA, 1992), pp. 1–16, on which this footnote is partially based. For the quotations, letter, Ed Price to me, 12 October 1999, which provides the other principal source of the remarks in this note.

125. A further area of complexity lay in the fact that a lot of the cutting-edge technology developed independently of specific missile programs through research and development funds granted to companies and other funding of government laboratories. A further source was funds to universities by the Office of Naval Research and the Air Force's Office of Scientific Research. I have found virtually no information about these matters in the extensive literature I have consulted, so I rely for the information in this footnote on comments by Ed Price on a draft of this chapter, 4 August 2000.

7

The Biggest of Them All

Reconsidering the Saturn V [1]

Ray A. Williamson

Introduction

On 25 May 1961 President John F. Kennedy announced to the nation a goal of sending an American safely to and from the Moon before the end of the decade. This decision involved much study and review prior to making it public, and tremendous expenditure and effort to make it a reality by 1969.[2] Only the building of the Panama Canal rivaled the Apollo program's size as the largest nonmilitary technological endeavor ever undertaken by the United States; only the Manhattan Project was comparable in a wartime setting. The human spaceflight imperative was a direct outgrowth of it; Projects Mercury (at least in its latter stages), Gemini, and Apollo were each designed to execute it. It was finally successfully accomplished on 20 July 1969, when Apollo 11 astronaut Neil Armstrong left the Lunar Module and set foot on the surface of the Moon.

Apollo was a remedial action ministering to a variety of political and emotional needs floating in the ether of world opinion, and it addressed these problems very well. It was a worthwhile action if measured only as a reaction to a cold war crisis with the Soviet Union. In announcing Project Apollo, Kennedy put the world on notice that the United States would not take a back seat to its superpower rival. John Logsdon commented, "By entering the race with such a visible and dramatic commitment, the United States effectively undercut Soviet space spectaculars without doing much except announcing its intention to join the contest."[3]

It also gave the United States an opportunity to shine. The lunar landing was so far beyond the capabilities of either the United States or the Soviet Union in 1961 that the early lead in space activities taken by the Soviets would not predetermine the outcome. The United States had a reasonable chance of overtaking the Soviet Union in space activities and recovering a measure of lost status.

In the end a unique confluence of political necessity, personal commitment and activism, scientific and technological ability, economic prosperity, and public mood made possible the 1961 decision to carry out a forward-looking lunar landing program. In essence, a complex web or system of ties between various people, institutions, and interests allowed the Apollo decision.[4] It then fell to NASA and other organizations of the federal government to accomplish the task set out in a few short paragraphs by President Kennedy.

Origins of Saturn

Meeting President Kennedy's 1961 challenge to put people on the Moon before 1970 required much larger launch vehicles than were then available; in many ways, the race to the Moon was a rocket-building competition. Because planning for such large vehicles had been begun by the Wernher von Braun team in Huntsville, Alabama, and others even before NASA was officially established in 1958, the agency was able to respond quickly. Among other things, NASA sped up work on technologies that led to the Saturn I and to the huge Saturn V, which in its final form was capable of lifting 260,000 pounds to low-Earth orbit (LEO). Although the roots of the design of the Saturn V ultimately reach back to the V-2, the Saturn evolved along a different development path from the Redstone, Atlas, Titan, Thor/Delta, and other launchers that were originally designed as missiles to carry nuclear warheads efficiently. The Saturn family was the first designed as pure space boosters.

Well before Kennedy's speech to Congress, von Braun's team at the Army Ballistic Missile Agency (ABMA) had begun to consider building a large, multi-stage rocket capable of launching heavy objects into space.[5] Von Braun and many of his engineering team had the Moon and Mars as their ultimate goal, but also had in mind an orbiting space station.[6] U.S. officials had been astonished by the lift capacity of the initial Soviet rockets. Although U.S. intelligence had known that the Soviets were building rockets based on the V-2, the United States was unprepared for the scope of this effort. The Eisenhower administration decided that the United States might need a much larger vehicle than was available, one capable of launching large military payloads and perhaps humans and the gear to support them. Building and testing a successful high-power engine was the most difficult of the many tasks planners faced in developing such a vehicle. Hence, engineers began to tackle the difficult problem of providing the propulsion to propel such a large payload into space.

By late 1957, they had settled on a launch design that would employ a first stage propelled by a cluster of eight powerful engines based on the S-3D engine from the Jupiter IRBM. In 15 August 1958, the newly created Advanced

Research Projects Agency (ARPA), which was organizing the U.S. military space effort, issued orders to begin work on a new large launcher.[7] Increasing the S-3D's thrust by 14 percent made it possible to achieve 1.5 million pounds of thrust in the cluster of eight engines. The engine was named the H-1; the launcher was tentatively named Juno V. Team members adopted a clustered approach out of necessity—building a brand-new, high-thrust engine would have been too expensive.[8] ARPA officials were forcing the von Braun team to live on low budgets and encouraging it to use off-the-shelf hardware wherever possible. As a result, the designers of the Juno V became quite inventive.[9] Although engine clusters raise many technological challenges, by meeting them at this early stage the team was able to provide a firm base for the development of later engine clusters.[10] ARPA conceived of the Juno V as a static test vehicle, but von Braun's team had clearly intended that it serve as the basis for a new launcher, which von Braun and his associates called "Saturn."[11] Shortly after ARPA gave ABMA the green light to proceed with Saturn, NASA came into being officially, and the issue of transferring ABMA to NASA began to be discussed in earnest. On 2 November 1959 President Eisenhower approved that transfer.

By late December 1959, NASA and the Department of Defense (DoD) had already made many of the initial technology decisions that would lead first to the Saturn IB launch vehicle, and then to the huge Saturn V. NASA, working with DoD, organized the Saturn Vehicle Team, chaired by Abe Silverstein of NASA. The Silverstein Committee made three important technological choices that set the stage for later Saturn developments. They decided (1) to use liquid hydrogen (LH2) as the fuel for the upper stages of the Saturn booster, (2) to develop a series of multistage rockets, and (3) to follow an evolutionary path for growth in which each succeeding vehicle used the proven stages of the preceding one. The Silverstein Committee saw three primary functions for the Saturn family: (1) lunar and deep-space missions with escape payload of 9,900 pounds, (2) geostationary orbit payloads of 4,950 pounds, and (3) missions carrying humans into low-Earth orbit in the Dyna-Soar program, an Air Force human space-flight effort. These choices, while they introduced some serious technical hurdles, were the backbone of Saturn's ultimate success as a launch vehicle.

The decision to use high energy LH2 as a fuel was the most controversial of the three decisions made in the earliest period of the Saturn program. It was the also the crucial one in allowing the program to develop efficient boosters. Just after the turn of the century, the Russian Konstantin E. Tsiolkovskiy and the American Robert H. Goddard had determined that using liquid hydrogen as a fuel in a liquid oxygen (LOX) environment would provide superior specific impulse (Isp).[12] In 1923 Hermann Oberth even suggested that the LOX-hydrogen combination would be especially appropriate for upper stages of rockets.[13] Yet liquid hydrogen, which requires cooling to -423 degrees F, is hard to handle and causes imbrittlement of many metals. Nevertheless, with von Braun in concurrence, Silverstein was able to convince the other committee members to accept LH2 as a fuel, despite its handling problems. As Sloop has noted, "It was a very bold and crucial decision to stake the success of the entire manned space program on a relatively new high-energy fuel, but subsequent developments proved it to be a sound decision and a key one in the success of the Saturn V and the Apollo missions."[14]

The Centaur upper stage currently in development at NASA in the late 1950s used two hydrogen-fueled Pratt & Whitney RL-10 engines that each developed a thrust of 15,000 lbf. By clustering these engines in a group of six, NASA planned to build a powerful second stage for the Saturn I, called the S-IV. It was to be the first major Saturn stage to be built under contract by industry, rather than developed within ABMA.

The decision to award the S-IV contract to the Douglas Aircraft Company illustrates the importance of subjective factors in NASA's choice of contractors. Two companies—Convair and Douglas—placed well above the other nine that submitted proposals. In choosing between them, NASA officials considered not only technical competence, but their judgment of the firms' ability to manage a large, complex contract and the firms' business acumen. Convair, which was developing the Centaur upper stage, placed slightly higher on technical competence but lower in the latter two categories. NASA Administrator T. Keith Glennan believed that Douglas Aircraft's proposal was more imaginative. He was also concerned that giving the S-IV contract to Convair would inadvertently create a monopoly in the development of cryogenic upper stages.[15] NASA officials were well aware of the need to develop a broad, competitive contractor base from which to choose, especially in building systems that required the development of new, untried technologies. As a result, NASA announced the choice of Douglas Aircraft Company on 26 May 1960. The closeness of the decision, and the subjective reasons for the selection of Douglas, caused some concern within Congress, which directed the General Accounting Office (GAO) to investigate. The GAO report, however, generally sustained NASA's decision.[16]

Modern launch systems consist of hundreds of interacting systems, each of which is itself composed of thousands of smaller subsystems and parts. Designing and successfully launching moderate and small-sized launchers is a major challenge. For systems the size and complexity of the Saturn I, its descendent the Saturn IB, and Saturn V Moon rocket, the task seemed daunting. Building the Saturn vehicles forced NASA and the aerospace industry to solve myriad practical problems, including the handling of large structures, flawless welding, and the testing and tracking of millions of components. It also required the development of new manufacturing methods. For example, Douglas Aircraft and NASA had to overcome a panoply of obstacles in order to manufacture the S-IV to a standard sufficient to carry people reliably and safely to space. In order to build a rocket stage of requisite size and strength, designers decided to carry the two propellants in only two tanks, one above the other, and to give them a common bulkhead. The size of the S-IV and the decision to use large propellant tanks brought its own production problems. The tanks' welded seams needed to be flawless. New machinery needed to be developed in order to handle the large tanks. New fabrication methods had to be invented in order to create the common bulkhead. In addition, Douglas also had to build special facilities to handle components the size of the tanks. Historian Roger Bilstein has commented that the development of Saturn hardware "frequently came down to a question of cut-and-try."[17] This approach, of course, made it extremely difficult to estimate the development cost of any of the launchers.

Building Saturn I and IB

Saturn I was in essence a research and development project, designed to gather data and experience with large launch vehicles. NASA made the first flight to test its first stage on 27 October 1961, carrying only a dummy S-IV stage. The first flight of an operating S-IV second stage was made on 29 January 1964. On 30 July 1965, the Saturn I made its last flight, having prepared the way for the more powerful Saturn IB. During its ten flights, the Saturn I had been used in a variety of engineering experiments in low-Earth orbit and had given NASA's engineers valuable insights into the complexities of building and launching a large cryogenic rocket.

The launch requirements considered by the Silverstein Committee demanded an even larger propulsion stage than the S-IV, and instead of uprating the RL-10, or somehow adding more of them to the cluster, the committee began to look toward a much larger, more powerful single engine that would generate 200,000 lbf. On 1 June 1960, a source evaluation board chose the Rocketdyne Division of North American Aviation to build a high-thrust cryogenic rocket engine called the J-2. Marshall Space Flight Center (MSFC) developed the concept and monitored the contractor's work, while Rocketdyne attempted to bend metal around MSFC's ideas.

From the beginning, the Saturn IB was designed to carry humans. Hence, the final engine contract, which was awarded to Rocketdyne in September 1960, contained an important phrase: "To insure maximum safety for manned flight."[18] In other words, although reliability had been an important ingredient of earlier designs, for the first time a contract specified that a rocket engine was to be designed with human safety as part of the initial specifications.[19] Because the Saturn IB would carry humans, each stage of the design and manufacturing process was closely scrutinized for high reliability, and each part was tested individually as well as in concert with other parts. Rocketdyne engineers faced serious problems finding appropriate metals and other materials that would work properly in a liquid hydrogen environment. They also had to trace down every leak in great detail, for a small amount of gaseous

Table 1 Saturn I Flights			
Number	Mission Date	Payload	Objectives/ Accomplishments
SA-1	27 October 1961	Dummy	Saturn I launch vehicle research and development. Tested S-I stage propulsion, verified structure and aerodynamics.
SA-2	25 April 1962	Water (95 tons)	Launch vehicle test and water ballast payload/experiment. Observed water dispersion at high altitude, by releasing 22,900 gallons.
SA-3	16 November 1962	Water (95 tons)	Launch vehicle test and water ballast payload/experiment. Observed water dispersion at high altitude, by releasing 22,900 gallons.
SA-4	28 March 1963	Dummy	Tested premature engine cutoff capabilities.
SA-5	29 January 1964	Dummy	First flight test of S-IV second stage.
SA-6	28 May 1964	BP-13	Carried first boilerplate (BP) Apollo Command and Service Modules; also carried live launch escape system. Insertion into orbit following premature cutoff of one first stage engine.
SA-7	18 September 1964	BP-15	Similar testing as with SA-6; also tested launch escape jettison.
SA-8	25 May 1965	Pegasus 11 BP-26	Carried boilerplate Apollo space craft and Pegasus satellite that conducted operational meteoroid experiment near Earth environment. First CCSD-built S-1 stage.
SA-9	16 February 1965	Pegasus 1 BP-16	Carried boilerplate Apollo space craft and Pegasus satellite that conducted operational meteoroid experiment near Earth environment.
SA-10	30 July 1965	Pegasus III BP-9	Carried boilerplate Apollo space craft and Pegasus satellite that conducted operational meteoroid experiment near Earth environment.

hydrogen in the wrong place could lead to a devastating explosion. After pursuing a number of intermediate short-duration tests for some nine months previous, Rocketdyne successfully ran the first model of the J-2 in a 250-second test on 4 October 1962.[20]

Image 7-1: A wind tunnel test of the Apollo/Saturn IB's aerodynamic integrity at the Langley Research Center, Hampton, Virginia, 2 March 1963. (NASA Photo no. L-1963-01637)

Image 7-2: The first stages of the Saturn IB *(left to right):* S-IB-7, S-IB-9, S-IB-5, and S-IB-6, in the final assembly area of Michoud Assembly Facility (MAF) in 1967. (NASA Photo no. MSFC 6760550)

When this contract was let in 1960, NASA had not yet decided which vehicle would use the powerful engine. Outside of NASA, there was relatively little interest in pursuing a program that would require the lift capacity of an upper stage that used the J-2 rocket. However, President Kennedy's May 1961 decision to "shoot the Moon" dramatically changed the situation. By July 1962, NASA settled on proceeding with the uprated S-IV, called the S-IVB, which it planned to use as the second stage of the Saturn IB; the stage would be powered by a single J-2 engine. The Saturn IB would loft an Apollo spacecraft to low-Earth orbit as part of the sequence of tests that would lead to a landing on the Moon. This powerful launcher, capable of placing 41,000 pounds into an orbit 110 miles above the earth, had an important role in the execution of the Apollo program.

Saturn IB made its first flight two years after the first flight of Saturn I, on 26 February 1966, using the S-IVB second stage. On 11 October 1968, it carried Apollo 7 into orbit for a ten-day, twenty-hour flight, during which

Image 7-3: The Saturn IB launch vehicle lifting off from Launch Complex 39B at the Kennedy Space Center, Florida, on 16 November 1973. It carries Skylab 4 astronauts Gerald P. Carr, Edward G. Gibson, and William R. Pogue for the third and final mission to the orbiting Skylab. (NASA Photo no. MSFC 6760550)

Table 2
Saturn IB Test and Operational Flights

Number	Vehicle	Launch Mission Date	Payload	Objectives/Accomplishments
AS-201 (1A)	S-IB SA-201	26 February 1966	CSM-009	Unpiloted test of Command Module heat shield. Demonstrated structural integrity and compatibility of launch vehicle and confirm launch loads. Demonstrated separation of first and second stages of Saturn, LES and boost protective cover from CSM. CSM from instrument unit/spacecraft/lunar module (LM) adapter, and CM from SM. Verified operations of Saturn propulsion, guidance and control, and electrical subsystems. Verified operation of spacecraft subsystems and adequacy of heat shield for reentry from low Earth orbit.
AS-203 (2)	S-IB SA-203	5 July 1966	Nose Cone, LH2 in S-IVB	Unpiloted test of Saturn S-IVB stage. Evaluated performance on S-IVB instrument unit stage under orbital conditions and obtained flight information on venting and chill-down systems, fluid dynamics and heat transfer of propellant tanks; attitude and thermal control system, launch vehicle guidance, and checkout in orbit
AS-202 (3)	S-IB SA-202	25 August 1966	Spacecraft-011	Unpiloted test of rocket's reliability. Demonstrated structural integrity and compatibility of launch vehicle and confirm ed launch loads. Demonstrated separation of first and second stages of Saturn, LES and boost protective cover from CSM. CSM from instrument unit/ spacecraft/lunar module (LM) adapter, and CM from SM. Verified operations of Saturn propulsion, guidance and control, and electrical subsystems. Verified operation of spacecraft subsystems and adequacy of heat shield for reentry from low-earth orbit. Evaluated emergency detection system in open-loop configuration. Evaluated heat shield ablator at high reentry rates. Demonstrated operation of mission support facilities.

Number	Vehicle	Launch Mission Date	Payload	Objectives/Accomplishments
Apollo 1	S-IB SA-204	Cancelled	Spacecraft-012	First planned piloted mission; crew killed in 27 January 1967 fire during rehearsal.
Apollo 5	S-IB SA-204	22 January 1968	LM-1 and nose cone	Unpiloted test of Lunar Module. Verified operation of Lunar Module ascent and descent propulsion systems. Evaluated Lunar Module staging. Evaluated S-IVB instrument unit performance.
Apollo 7	S-IB SA-205	11–22 October 1968	Apollo CSM-101	First piloted Apollo flight; practiced rendezvous in Earth orbit. The primary objectives for the Apollo 7 engineering test flight were simple: "Demonstrate CSM/crew performance; demonstrate crew/space vehicle/mission support facilities performance during a manned CSM mission; demonstrate CSM rendezvous capability."
Skylab 2	S-IB SA-206	25 May– 22 June 1973	Apollo CSM-116	The crew rendezvoused with Skylab on the fifth orbit. After making substantial repairs, including deployment of a parasol sunshade that cooled the inside temperatures to 23.8 degrees C (75 degrees F), by June 4 the workshop was in full operation. In orbit the crew conducted solar astronomy and Earth resources experiments, medical studies, and five student experiments; 404 orbits and 392 experiment hours were completed; three EVAs totaled 6 hours, 20 minutes. Crew: Charles C. Conrad Jr., Paul J. Weitz, Joseph P. Kerwin.
Skylab 3	S-IB SA-207	28 July– 25 September 1973	Apollo CSM-117	Continued maintenance of the space station and extensive scientific and medical experiments. Completed 858 Earth orbits and 1,081 hours of solar and Earth experiments; three EVAs totaled 13 hours, 43 minutes. Crew: Alan L. Bean, Jack R. Lousma, Owen K. Garriott.

Table 2 (cont'd)
Saturn IB Test and Operational Flights (cont'd)

Number	Vehicle	Launch Mission Date	Payload	Objectives/Accomplishments
Skylab 4	S-IB SA-208	16 November 1973–8 February 1974	Apollo CSM-118	Last of the Skylab missions; included observation of the Comet Kohoutek among numerous experiments. Completed 1,214 Earth orbits and four EVAs totaling 22 hours, 13 minutes. Crew: Gerald P. Carr, William R. Pogue, Edward G. Gibson.
Apollo-Soyuz Test Project	S-IB SA-210	15–24 July 1975	Apollo CSM-111	The Soyuz was launched just over seven hours prior to the launch of the Apollo CSM. Apollo then maneuvered to rendezvous and docking 52 hours after the Soyuz launch. The Apollo and Soyuz crews conducted a variety of experiments over a two-day period. After separation, Apollo remained in space an additional 6 days. Soyuz returned to Earth approximately 43 hours after separation. Crew: Thomas P. Stafford, Vance D. Brand, Donald K. Slayton.

astronauts tested the hardware necessary to travel to and from the Moon. That was also the last flight for the Saturn IB supporting what is usually referred to as Project Apollo, the NASA effort to land Americans on the Moon. It not only carried the first Apollo spacecraft into orbit during the test phases of the Apollo program of the mid-1960s but also served to ferry astronauts to Skylab in the 1970s and was the launch vehicle used in the Apollo-Soyuz mission of 1975.[21]

The Mighty Saturn V

The flights of the earlier Saturn rockets were followed in 1967 by the first Saturn V test flights and in late 1968 by the first Saturn V mission to carry astronauts, when the Apollo 8 mission launched Frank Borman, Jim Lovell, and Bill Anders into space for the first human flight around the Moon.[22]

The Saturn V dwarfed even the powerful Saturn I and Saturn IB launchers. Standing 363 feet high and weighing 500,000 pounds unfueled, the Saturn V was capable of launching over 200,000 pounds into low-Earth orbit. It was built to place three humans on the Moon and allow them to take sufficient fuel and equipment to return to Earth. The Saturn V had three stages: the S-IC first stage, powered by five F-1 LOX/kerosene engines; the S-II second stage, propelled by five J-2 LOX/LH2 engines; and the S-IVB third stage, with one J-2 engine.

Rocketdyne, which NASA later chose to build the J-2 engine, had in January 1959 received the contract to build the F-1.[23] The contract for the giant power plant, which would employ RP kerosene fuel and LOX, stipulated that it should develop 1,500,000 lbf, nearly four times the thrust of the Navaho missile engine from which it was derived.[24] Rocketdyne's experience in the Navaho program made it the logical candidate for the task. In awarding this contract, NASA was betting that it could bypass the more common evolutionary approach to engine development and make a revolutionary jump to this enormous engine.

At the time of this decision, the United States had no program to attempt a Moon landing. NASA did not make a final decision about the configuration of the first stage of the Saturn Moon rocket until 10 January 1962, after it had already chosen Boeing Aircraft Company to build it. That choice was based on the recognition that a booster with five F-1 engines in its first stage might be able to accomplish the lunar landing mission in a single launch, if NASA were to adopt the controversial lunar orbital rendezvous approach to a lunar landing. With this decision, the large booster was named Saturn V.[25]

As planning for the lunar landing mission had proceeded at an intense pace following Kennedy's May 1961 speech, NASA officials gave serious consideration to the need for a booster even larger than that which became the Saturn V; that booster, named Nova, would use eight F-1 engines in its first

Figure 1

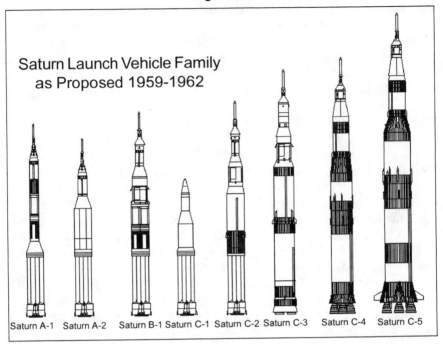

Saturn Launch Vehicle Family
as Proposed 1959-1962

Saturn A-1   Saturn A-2   Saturn B-1   Saturn C-1   Saturn C-2   Saturn C-3   Saturn C-4   Saturn C-5

stage. The difficulty of a rapid evolution to such a gigantic rocket was one of the reasons that mission planners in the second half of 1961 began to converge on some kind of rendezvous approach to accomplishing the lunar mission; adopting such an approach would allow the use of one or several Saturn rockets. A rocket the size of Nova, and all of the very large ground facilities required to launch it, would not be needed. Although studies of Nova continued until 1963, development of the vehicle was never initiated.

Although the difficulties of working with LOX/LH2 on the J-2 engine created novel complications for rocket engine designers, the sheer size and thrust of the LOX/RP F-1 engine also presented formidable challenges. David E. Aldrich, the manager of MSFC's Engine Program Office, acknowledged that "the development of the F-1 engine, while attempting to stay within the state of the art, did, by size alone, require major facilities, test equipment, and other accomplishments which had not been attempted prior to F-1 development."[26] The F-1 was a gimbaled engine whose bell-shaped expansion nozzle was regeneratively cooled by liquid oxygen. Although nearly every subsystem brought its own technological hurdles and special challenges, the F-1 injector, which controlled the flow and pattern of both fuel and oxygen into the thrust chamber, turned out to be the stiffest challenge of all. The injector forced fuel through thirty-seven hundred orifices into the combustion chamber to meet oxygen that entered from twenty-six hundred additional openings. The injector had to endure greater heat and pressure than in any previous engine.

Unfortunately, it proved impossible simply to scale up previous designs to the required size. Initial tests with early models of the F-1 injector led to unacceptable combustion instability that could not be stopped short of cutting off the flow of fuel.[27] New designs, based on tests with scale models and the use of high-speed photography in a specially designed test chamber looked promising, but when scaled up to F-1 size and tested, they also failed. On 28 June 1962 one of these tests resulted in the loss of an F-1 test engine. By early 1963, NASA Associate Administrator Robert Seamans was quite concerned; he told the head of the Apollo program, Brainerd Holmes, that "as you and Wernher [von Braun] know, I feel that F-1 instability may seriously delay the MLLP [manned lunar landing program] and consequently is the technological pacing item."[28]

Eventually, after empirical "cut-and-try" redesign and exhaustive testing, coupled with an intensive theoretical attack on injector combustion instability, Rocketdyne and NASA engineers developed an injector that would pass muster. Rocketdyne, working closely with NASA and university researchers, came up with a flight-rated model by January 1965. Still, despite the satisfaction of having developed a working engine, the Rocketdyne engineers noted that "the causes of such instability are still not completely understood."[29]

The decision to build a cryogenic second stage for the launch vehicle that became the Saturn V was also rooted in the Silverstein Committee report of 1959. Committee members knew that an extremely powerful second stage would be needed to launch humans to the Moon. Soon after the committee issued its report, NASA designers had begun to define the general outlines of the S-II stage that would eventually become the second stage of the Saturn V. Before he left NASA in January 1961, Administrator T. Keith Glennan wrote, "The Saturn program is left in mid-stream if the S-II stage is not developed and phased in as the second stage of the C-2 launch vehicle."[30] (The C-2 was a proposed Saturn version using two F-1 engines in its first stage.)

Soon after James E. Webb was sworn in as the new NASA Administrator in February 1961, Marshall Space Flight Center started contractor selection. The S-II, at that point to be propelled by four J-2 engines, was to be the largest rocket project to be given to U.S. industry. Although thirty interested contractors attended the first meeting, only seven submitted bids,[31] of which four survived source evaluation board scrutiny. Those remaining, Aerojet, Convair, Douglas, and North American Aviation, faced the difficult task of attempting to bid on a contract that was still largely undefined. NASA had not yet decided the size of the S-II stage, nor had it settled on any of the myriad other specifications of the project. Yet NASA was under considerable pressure to get the design process under way, and it needed to choose a contractor as soon as possible.[32] NASA was attempting to build a team, and it needed to select a contractor that would manage the construction well. "I wish to emphasize at this point that the important product that NASA will buy in this procurement is the efficient management of a stage system."[33]

NASA chose North American Aviation for the job on 11 September 1963, a decision that raised some eyebrows outside of NASA. North American already had the contract for building the Apollo capsule and was not considered to be a major player in the launch vehicle area. However, North American had built the highly successful X-15 research airplane, and because NASA emphasized the importance of a strong management team, it selected North American largely on that basis.

The relationships between NASA and North American on the S-II contract provide important insights into the development of NASA as an institution. Although NASA was determined here, as in other contracts, to use the intellectual capacity and manufacturing experience of American industry, because NASA retained its own cadre of engineers and other specialists, it often found itself second-guessing North American. In an area where both NASA and the contractor were "pushing the envelope" of the state of the art, misunderstandings and disagreements over the best way to proceed inevitably arose, which led to tensions between individuals and sometimes whole departments in the two organizations. Additionally, North American's approach was one of an aircraft company, used to building high-performance flying machines. By contrast, MSFC, NASA's lead center for the Saturn V, had little experience in aircraft procedures but a lot of experience in building hefty rockets that looked more like boilers than aircraft. MSFC, for example, had designed and built the first stage of the Saturn I. Added to these tensions was the fact that the U.S. Navy had oversight over construction of new government facilities for building the S-II. Coordination among the three entities was not always smooth.

In many respects, the final S-II stage resembled an aircraft component more than a rocket. It was much more efficient than the first stage, not only because it carried more efficient rocket engines but also because it was lighter for a given strength factor. During the design and manufacturing process in the early 1960s, NASA continually asked North American to shed pounds on the S-II, as the Apollo payload relentlessly gained weight. In order to add one additional pound of payload delivered to orbit, NASA had to cut 30 pounds from the S-IC first stage, nearly 11 pounds from the S-II second stage, or only 1 pound from the S-IVB. Thus, it would have been most effective to remove weight from the third, S-IVB stage. However, the S-IVB was already in production at McDonnell Aircraft by the time NASA began to experience the most severe weight problems. Taking 30 pounds from the S-IC for each pound of overweight in the Apollo stage was also not very feasible. The burden fell on the S-II stage, and on the NASA–North American Aviation team.

The need to find innovative ways in which to shave weight from the stage, and the requirements for innovative manufacturing processes, took their toll on both MFSC and North American Aviation. By 29 September 1965, when an S-II test article failed catastrophically during a test designed to simulate forces that the S-II would experience at the end of the first stage burn, concern within NASA over North American's management reached major proportions. Apollo Program manager Gen. Samuel C. Phillips of NASA Headquarters was appointed head of a so-called Tiger Team to investigate the problems. The Tiger Team effort served two purposes—it helped NASA investigate the problems and recommend solutions, and it put North American on notice that NASA considered the perceived problems extremely serious. The resulting report, sent to Lee Atwood, president of North American Aviation, on 19 December 1965, was extremely critical of North American's management of the S-II and also on its handling of spacecraft development. It was a wake-up call. By this time, problems with the S-II threatened to hold up the first launch of the Saturn V.

Stung by the criticism, the company responded to this crisis by reorganizing its management team and rethinking how it organized the work on the S-II. Among other things, it brought in new top managers and improved the sharing of information on the progress and problems experienced by the company's engineering teams. Still, despite making significant progress,[34] North American continued to experience problems. On 28 May 1966, a second S-II stage (S-II-T) failed during a pressure check of the LH2 tank. The loss once again indicated poor management control. The tank exploded as technicians filled it with helium during a test for leaks. Unfortunately, they were unaware that other technicians had previously disconnected the pressure sensors and relief switches that would have prevented an explosion. The accident injured five men. To add to the problem, the team investigating the S-II-T failure found tiny cracks and other problems in the test article. Nevertheless, North American continued to make progress, and by 9 January 1967, Phillips could report that the company had markedly improved its management and test procedures.[35]

Unfortunately, on 27 January 1967, a deadly fire broke out in the Apollo Command Module during a test with crew aboard, killing three astronauts. Because North American also had responsibility for building the Command Module, NASA asked for and got further reorganization of the company's top management. Harrison Storms, president of the Space and Information Systems Division, and the highest official directly overseeing Command Module and S-II construction, was replaced and moved to a slot as corporate vice president.[36] NASA continued to follow closely the development of the S-II. On 9 November 1967, after numerous delays caused not only by problems with the S-II stage but also by other aspects of the Saturn V development, the SA-501 launch vehicle successfully carried the crewless Apollo 4 capsule into orbit for tests.

The Apollo 4 flight was the first "all-up" test of the Saturn V launch vehicle. For NASA, it was also a major risk, because not only had the stages never been launched together, but neither the S-IC nor the S-II had flown at all. The S-IVB stage, the Command Module, and the instrument unit that provided inertial guidance and avionics to the vehicle had been tested on Saturn IB flights. As one writer put it, "The all-up concept is, in essence, a calculated gamble, a leap-frogging philosophy which advocates compression of a number of lunar landing preliminaries into one flight. It balances the uncertainties of a number of first-time operations against a 'confidence factor' based on the degree of the equipment reliability achieved through the most exhaustive ground-test program in aerospace history."[37]

This all-up test was the result of an earlier decision by George E. Mueller, Associate Administrator of the Office of Manned Space Flight. In September 1963, when Mueller took his post, NASA was beginning to feel the enormity of meeting Kennedy's deadline for reaching the Moon. It was also experiencing the first hint of a shrinking yearly budget. President Johnson was under considerable pressure to keep NASA's 1964 budget under $5 billion, versus a $5.75 billion budget request to President Kennedy earlier in the year. Mueller notified the directors of the Manned Spacecraft Center, Houston; Launch Operations Center,[38] Cocoa Beach; and Marshall Space Flight Center, Huntsville that the first Saturn IB flight and the first Saturn V flight would be made with all stages operating. Both should also carry complete spacecraft.

Although Mueller's directive caused considerable debate among the highly conservative staff at the centers, particularly at Marshall, eventually even the Marshall team came around.[39] Nevertheless, everyone recognized the risks, and it was with considerable trepidation that the launch team prepared the first Saturn V for launch. Yet at 7:00 A.M. Eastern Standard Time, 9 November 1967, SA-501 lifted off the pad, carrying the Apollo 4 Command Module and performing nearly flawlessly. The risk had paid off. With one exception, the remainder of the Saturn V launches were also highly successful. In all, NASA launched thirteen Saturn Vs between November 1967 and May 1973.

The one troublesome launch was SA-502, or Apollo 6. NASA had planned to fly both it and SA-503 without a crew. However, on 16 November, in light of the success of SA-501, Gen. Samuel C. Phillips decided that tests were going so well that if the flight of Apollo 6 proved successful, NASA would proceed directly to human flights with AS-503.[40] As it turned out, Phillips's optimism was short-lived. AS-502 lifted off from Launch Complex 39 on 4 April 1968. All went well until about 125 seconds into the flight, near the end of first stage burn, when the launcher began to experience strong longitudinal oscillations that created a "pogo effect" for nearly 10 seconds. Despite the pogo, separation and ignition of the second stage occurred normally, but after 4.5 minutes of operation, its number two engine shut down, followed a second later by shutdown of the number three engine. The instrument unit, which performed the vehicle's guidance and control, compensated by steering the rocket into a new trajectory and causing the three remaining J-2 engines to fire longer than planned. Following a normal third-stage firing and shutdown of its single J-2 engine, Apollo 6 coasted into Earth orbit. After waiting two orbits, NASA flight controllers signaled the J-2 third stage engine to restart in order to complete as much of the flight plans as possible. Despite many attempts, it failed to function. Flight controllers finally gave up and managed to separate the Command Service Module (CSM) and Command Module from the third stage. They finally resorted to using the smaller engine on the CSM to position the Command Module for a successful reentry.

The flight was successful in proving that even with two-second stage J-2 engines out, the Command Module could still reach orbit and return safely. The pogo phenomenon had been experienced on Gemini-Titan and other launches, but not with such intensity. Although the vibrations were apparently not severe enough to harm the vehicle or the astronauts directly, it would have caused extra stress to the astronauts, and NASA officials decided they should not risk the possibility of stronger vibrations with people aboard. Intensive investigation by a specially constituted pogo task force composed of representatives of NASA, industry, and the universities disclosed that the F-1's thrust chamber and combustion chamber vibrated at about 5.5 hertz during burning. The vehicle as a whole vibrated with a variable frequency. When the vehicle vibrations reached 5.5 hertz, the two effects combined to produce the pogo effect. The pogo team was able to devise a repair that involved "de-tuning" the F-1 engine to change its frequency of vibration.[41]

The J-2 problem was much more serious, in part because NASA had no idea what might have gone wrong on the two engines. Fortunately, the second stage was extremely well instrumented; one of the thermocouples showed a temperature drop about 70 seconds into second stage burn, indicating a leak of cold gas. Then, just before engine shutdown, another thermocouple registered a suddenly higher temperature, suggesting that there had been an eruption of hot gas, probably from the igniter fuel line. With these data in hand, NASA and Rocketdyne engineers began to perform extensive tests on the J-2 fuel lines. At sea-level temperatures and pressures they could not reproduce the failure. But by pumping liquid hydrogen through eight separate lines in a vacuum chamber, thereby simulating operational conditions, they were able to cause every one of them to fail about 100 seconds into the test. Once the engineers had reproduced the failure in the laboratory, they were then able to devise a suitable repair, and the Apollo program was back on track.[42]

Saturn Launch Operations

The relatively mundane tasks of assembling all of the launch vehicle's parts and preparing the vehicle for liftoff are easily overlooked when examining the

Table 3
Saturn V Test and Operational Missions

Number	Launch Vehicle	Mission Date	Payload	Objectives/Accomplishments
Apollo 4	Saturn V SA-501	9 November 1967	Apollo CSM-017 (Block 1), LTA-10R (Lunar Module Test Article)	First flight of the Saturn V; first "All-Up" testing flight demonstrated capabilities of launch vehicle and CM heat shield at lunar reentry velocities.
Apollo 6	Saturn V SA-502	4 April 1968	Apollo CM-020, SM-014 (SM-020 destroyed by tank explosion), LTA-2R	Final "man-rating" (last non-crewed flight).
Apollo 8	Saturn V SA-503	21–27 December 1968	Apollo CSM-103, LTA-B	First launch of piloted Saturn V mission and first piloted circumlunar mission. Crew: Frank Borman (CDR), Jim Lovell (CMP), Bill Anders (PLT).
Apollo 9	Saturn V SA-504	3–13 March 1969	Apollo CSM-104 (Gumdrop), LM-3 (Spider)	All lunar landing hardware tested on this mission, LM given a thorough shakedown testing all systems. Crew: Jim McDivitt (CDR), Dave Scott (CMP), Rusty Schweickart (LMP).
Apollo 10	Saturn V SA-505	18–26 May 1969	Apollo CSM-106 (Charlie Brown)/ LM-4 (Snoopy)	Tested LM down to 50,000 feet above the lunar surface, abort modes tested. Crew: Tom Stafford (CDR), John Young (CMP), Gene Cernan (LMP).
Apollo 11	Saturn V SA-506	16–24 July 1969	Apollo CSM-107 (Columbia), LM-5 (Eagle)	First lunar landing (Sea of Tranquility). 46 pounds of sample retrieved during 2.5 hour EVA. Duration of Lunar visit: 21.6 hours. Neil Armstrong (CDR), Mike Collins (CMP), Buzz Aldrin (LMP).
Apollo 12	Saturn V SA-507	14–24 November 1969	Apollo CSM-108 (Yankee Clipper), LM-6 (Intrepid)	Second lunar landing (Ocean of Storms). Saturn V struck by lightning soon after liftoff, knocks spacecraft systems out temporarily. Once in orbit all systems check out as unaffected. EVAs totaling almost 8 hours with 75 pounds of lunar material collected. The Surveyor 3 lander was

Number	Launch Vehicle	Mission Date	Payload	Objectives/Accomplishments
Apollo 13	Saturn V SA-508	11–17 April 1970	Apollo CSM-109 (Odyssey), LM-7 (Aquarius)	examined and samples retrieved. Crew: Pete Conrad (CDR), Dick Gordon (CMP), Al Bean (LMP). Normal launch except for a shutdown of the center J-2 engine on the S-II. Explosion of an oxygen tank while en route to the Moon cripples the CSM; the LM is used as a lifeboat and returns the crew to Earth. Mission aborted. Crew: Jim Lovell (CDR), Ken Mattingly (CMP), Fred Haise (LMP).
Apollo 14	Saturn V SA-509	31 January – 9 February 1971	Apollo CSM-110 (Kitty Hawk), LM-8 (Antares)	Third lunar landing (Fra Mauro). Two EVA's totaling 9.1 hours with 94 pounds of lunar material collected. Total surface stay: 33.5 hours. Crew: Alan Shepard (CDR), Stu Roosa (CMP), Ed Mitchell (LMP).
Apollo 15	Saturn V SA-510	26 July–7 August 1971	Apollo CSM-112 (Endeavour), LM-10 (Falcon)	Fourth lunar landing (Hadley-Appenine region). First flight to use Lunar Rover (LRV), total distance traveled: 17 miles, 169 pounds of lunar material collected during 3 EVAs totaling 18.5 hours. Crew: Dave Scott (CDR), Al Worden (CMP), Jim Irwin (LMP).
Apollo 16	Saturn V SA-511	16–27 April 1972	Apollo CSM-113 (Casper), LM-11 (Orion)	Fifth lunar landing (Descartes Highlands). Lunar Rover expeditions travel 16.7 miles over three EVAs totaling 20.2 hours with 213 pounds of Lunar material collected. Crew: John Young (CDR), Tom Mattingly (CMP), Charlie Duke (LMP).
Apollo 17	Saturn V SA-512	17–19 December 1972	Apollo CSM-114 (America), LM-12 (Challenger)	Sixth and final lunar landing (Taurus-Littrow). EVAs totaling 22.1 hours with 22 miles traveled in the Lunar Rover. 258 pounds of lunar material collected. Crew: Gene Cernan (CDR), Ron Evans (CMP), Jack Schmitt (LMP).
Skylab 1	Saturn V SA-513	14 May 1973	Skylab Orbital Workshop built from S-IVB-212	Last flight of Saturn V. Skylab launch successful except for premature deployment of Micrometeroid shield on station exterior. Shield was ripped off station by aerodynamic forces, taking Solar Array Wing No. 1 with it. After repair by astronauts, three crews successfully lived and worked aboard orbital workshop.

development of large, powerful launch vehicles. However, a well-organized manufacturing and logistics chain, and smooth-running launch operations, are absolutely crucial to a successful launch. The manufacturing, assembly, preparation, and launch of the completed Saturn V constituted an engineering and organizational marvel. It required huge machines for handling the Saturn V's three stages, as well as barges, specially modified aircraft, and trucks for transporting them to the launch site at Cape Canaveral. It involved a logistics chain that stretched across the United States, fed by major manufacturing sites along the East, West, and Gulf Coasts.

Marshall Space Flight Center had responsibility for construction of the launch vehicles; the Manned Spacecraft Center in Houston was responsible for the spacecraft and for mission control once liftoff had occurred; and the Launch Operations Center in Florida had charge of ensuring a safe, successful launch. Merritt Island near Cape Canaveral was chosen for the Launch Operations Center in part because it offered ready access by barge from manufacturing sites on the West and Gulf Coasts. Its location on the Atlantic Ocean simplified the safety precautions during launch. In addition, the Air Force already maintained a launch facility immediately southeast of NASA's launch range. After some rather tense negotiations, agreement was reached on sharing responsibilities for range communications and other facilities between NASA and the Air Force.

Early in the Saturn program, officials realized they would need massive facilities in which to erect the massive Saturn V and prepare it for launch.[43] After some discussion regarding whether to construct the vehicle and mate it with its payload on the launch pad, NASA engineers decided that the most efficient operation would result from keeping vehicle construction and payload integration separate from launch operations. Hence, they conceived of a large enclosed structure capable of holding the entire vehicle and its payload: the Vertical Assembly Building (VAB). The building covers eight acres, and its high bays stand 441 feet high. NASA officials anticipated a high launch rate and designed the VAB to accommodate four fully assembled Saturn V launch vehicles in order to minimize their time on the launch pad.[44] Each Saturn V was erected on a massive device called a mobile launcher, which supported the launcher from the initial assembly through the launch.

For safety purposes, the VAB had to be located far away from the launch pad. Originally NASA had explored the possibility of building a shallow canal between the VAB and the launch pads and floating the assembled launcher to the pad on a barge. However, tests at the Navy's David Taylor Basin near Washington, D.C., soon showed that the mobile launcher's huge gantry and the launcher would act like an enormous sail, making steering such a mammoth contraption impossible. After considering a rail line and rejecting it because of the enormous forces the rails and their bedding would have had to sustain, NASA settled on a large tractor built by the Marion Power Shovel Company, which had built similar tractors for strip-mining coal.[45] After the launcher

Image 7-4: The Apollo 11 Saturn V launch vehicle lifts off with astronauts Neil A. Armstrong, Michael Collins, and Edwin E. Aldrin Jr. at 9:32 A.M. Eastern Daylight Time on 16 July 1969, from Kennedy Space Center's Launch Complex 39A. (NASA Photo no. 69PC-0447)

was assembled in the VAB, a huge crawler-transporter lifted the assembled vehicle spacecraft atop its mobile launcher, which together weighed nearly 12 million pounds, and slowly crawled the 3.5 miles to the launch pad.[46]

This system was used thirteen times to bring a Saturn V and its payload to the launch pad. There were two test launches carrying Apollo spacecraft without crews. The first launch with a crew aboard was the Apollo 8 mission to lunar orbit in December 1968; this was followed by the Apollo 9 Earth-orbital test of the Lunar Module in February 1969 and the Apollo 10 "dress rehearsal" in May. Then there were seven launches of crews to the Moon, beginning of course with the July 1969 Apollo 11 mission and ending, prematurely in terms of the original plans, with the Apollo 17 mission in December 1972. A final launch in May 1973 carried the Skylab space station, which was in fact a modified S-IVB Saturn V upper stage, to Earth orbit.

Legacies of Saturn

There are several legacies of Saturn technology to the U.S. and the world. These revolve around three key areas: space hardware, nonspace applications of space-related technology, and the management of large-scale technological systems. First, in the area of space hardware, building the Saturn V may go down as one of the most complex efforts ever in the history of the United States. To meet the challenge, the space agency's annual budget increased from $500 million in 1960 to a high point of $5.2 billion in 1965. Between 1959 and 1973 NASA spent $23.6 billion on human space flight, exclusive of infrastructure and support, of which more than $20 billion was for Apollo, a figure over $150 billion in 1994 dollars when accounting for inflation. The NASA funding level represented 5.3 percent of the federal budget in 1965, more than half of which went directly to Apollo. A comparable percentage of the more than $1.2 trillion federal budget in 1994 would have equaled more than $69 billion for NASA, whereas the agency's actual budget then stood at less than $14 billion.[47]

Although Kennedy's decision to sprint to the Moon opened the funding floodgates to accomplish Apollo on a crash basis, the schedule also dictated that decisions be made that were probably not in the long-term best interests of the orderly exploration of space. Two major technological areas where this was true are readily apparent. First, it dictated that Apollo be accomplished using as many "off-the-shelf" technologies as possible. The result was that many of the systems were predicated on 1950s or very early 1960s technology. The engineering that went into the program was largely obsolete by the time of the landings.

Second, in the earliest years of the agency a planning committee had formulated a series of activities that the civilian space agency would faithfully pursue in the decades that followed for humans to explore space. These included in this order: (1) Earth-orbiting spacecraft, (2) a laboratory in space, (3) expeditions to the Moon (beginning with robots), (4) robots to Venus and Mars, and (5) expeditions to the nearby planets. "The ultimate objective of space exploration," the committee agreed, "is manned travel to and from other planets."[48] The plan set out a program of activities with a piloted landing on the Moon as something that would not happen until the well into the 1970s.[49] A centerpiece of the plan was an orbital laboratory—a space station—that would serve as the pioneer outpost on the space frontier, offering a haven from the rigors of the "wilderness" and a jumping off point for forays into the unknown.[50]

The tight schedule for Apollo changed that plan. One of the critical management decisions made by NASA to meet the crash schedule was the lunar orbital rendezvous mode of going to the Moon, and this 1962 decision bypassed the need to construct an Earth-orbiting space station as a jumping off point into deep space even as it sped up the accomplishment of a landing before the end of the decade. Wernher von Braun led the opposition to this mode in part because it provided a logical rationale for the space station that he had been advocating at least since 1952. Such a station, of course, would have been a logical technological legacy to the lunar landing project, useful for follow-on exploration missions. But the expediency of the lunar schedule dictated the station's deferral, and NASA is still wrestling with it.[51]

The technological legacy of Saturn hardware was not all bad news. Many of the thousands of components developed as a part of the project found usage either directly or in a more advanced form in later space projects. Heavy-lift launch vehicle technology, the most critical component of the project, went directly into the Space Shuttle and into other systems that followed. Alloys and systems developed for Saturn for all types of functions found use. And the follow-on Apollo Applications Program, designed by NASA to employ Saturn technology, had at its center a relatively small orbital space platform, called Skylab, that could be tended by astronauts. It would be, NASA officials hoped, the precursor of a real space station. It made extensive use of Saturn and Apollo equipment, an idea that had been germinating within NASA since 1963, by using a reconfigured and habitable third stage of the Saturn V rocket as the basic component of the orbital station. Carried out on a shoestring, in large measure because of the use of equipment developed and built with Project Apollo funding, the direct Skylab expenditure cost less than $3 billion. The orbital workshop was launched on 14 May 1973, the last use of the giant Saturn V launch vehicle, and it was the location of three extended space flights that yielded valuable scientific data about the effects of space flight on humans.[52]

Much has been made over the years of what NASA calls "spinoffs," commercial products that had at least some of their origins as a result of space-flight-related research. Most years the agency puts out a book describing some of the most spectacular, and they range from laser angioplasty to body imaging for medical diagnostics to imaging and data analysis technology. Spinoffs have not only been Tang and Teflon, neither of which were actually developed for the space program.[53] NASA has spent a lot of time and trouble trying to track these benefits of the space program in an effort to justify its existence, and the NASA History Office has more than five linear feet of documentation relative to the subject. With the caveat that technology transfer is an exceptionally complex subject that is almost impossible to track properly, these various studies show much about the prospect of technological lagniappe from the U.S. effort to get to the Moon.

Whether good or bad, no amount of cost-benefit analysis, which the spinoff argument essentially makes, can sustain NASA's historic level of funding. More useful, one may assert, is a counterfactual question. How would life today be different if there were no space program? There can be no fully satisfactory answer to that question. One person's vision is another's belly-laugh. But perhaps we can begin with the elimination of the microchip. Whether our life would be significantly different is problematic, but I think many of the high-technology capabilities we enjoy—starting with biomedical diagnostics and related technologies and ending with telecommunications breakthroughs—might well have followed different courses and perhaps have lagged beyond their present breakneck pace as a result. Some of us might well think that a positive development, though one may doubt most would want to go back to typewriters, problematic global communication, and so on. The point, of course, is that the past did not have to develop in the way that it did, and that I believe there is evidence to suggest that Saturn technology and the larger space program pushed technological development in certain paths that might have not been followed otherwise, both for good and ill.[54]

Finally, if there was a truly outstanding legacy to Saturn and the larger Apollo program it was the knowledge gained about the management of large-scale technological systems. Project Apollo was a triumph of management in meeting enormously difficult systems engineering, technological, and organizational integration requirements. James E. Webb, NASA Administrator at the height of the program between 1961 and 1968, always contended that Apollo was much more a management exercise than anything else, and that the technological challenge, while sophisticated and impressive, was largely within grasp at the time of the 1961 decision.[55] More difficult was ensuring that those technological skills were properly managed and used.

Webb's contention was confirmed in spades by the success of Apollo. NASA leaders had to acquire and organize unprecedented resources to accomplish the task at hand. To do so they employed a "program management" concept that centralized authority over design, engineering, procurement, testing, construction, manufacturing, spare parts, logistics, training, and operations.[56]

One of the fundamental tenets of the program management concept was that three critical factors—cost, schedule, and reliability—were interrelated

Image 7-5: This image demonstrates the size of the Saturn V, which stood 363 feet high. Note the workmen at the lower right of the photograph. This was the Apollo 12/Saturn V space vehicle from the VAB's High Bay 3 at the start of the 3.5 mile rollout to Launch Complex 39A on 8 September 1969. (NASA Photo no. 69PC-0529)

and had to be managed as a group. Many also recognized these factors' constancy; if program managers held cost to a specific level, then one of the other two factors, or both of them to a somewhat lesser degree, would be adversely affected. This held true for the Apollo program. The schedule, dictated by the president, was firm. Since humans were involved in the flights, and since the president had directed that the lunar landing be conducted safely, the program managers placed a heavy emphasis on reliability. Accordingly, Apollo used redundant systems extensively so that failures would be both predictable and minor in result. The significance of both of these factors forced the third factor, cost, much higher than might have been the case with a more leisurely lunar program such as had been conceptualized in the latter 1950s. As it was, this was the price paid for success under the Kennedy mandate, and program managers made conscious decisions based on knowledge of these factors.[57]

The program management concept was recognized as a critical component of Project Apollo's success in November 1968, when *Science* magazine, the publication of the American Association for the Advancement of Science, observed, "In terms of numbers of dollars or of men, NASA has not been our largest national undertaking, but in terms of complexity, rate of growth, and technological sophistication it has been unique. . . . It may turn out that [the space program's] most valuable spin-off of all will be human rather than technological: better knowledge of how to plan, coordinate, and monitor the multitudinous and varied activities of the organizations required to accomplish great social undertakings."[58] Understanding the management of complex structures for the successful completion of a multifarious task was an important outgrowth of the effort.

This management of complex structures for Saturn involved more than 500 hundred contractors working on both large and small aspects of the effort. For example, the prime contracts awarded to industry for the principal components of just the Saturn V included The Boeing Company for the S-IC, first stage; North American Aviation for the S-II, second stage; the Douglas Aircraft Corporation for the S-IVB, third stage; the Rocketdyne Division of North American Aviation for the J-2 and F-1 engines; and International Business Machines (IBM) for the Saturn instruments. These prime contractors, with more than 250 subcontractors, provided millions of parts and components for use in the Saturn launch vehicle, all meeting exacting specifications for performance and reliability. The total cost expended on development of the Saturn launch vehicle was massive, amounting to $9.3 billion. So huge was the overall Apollo endeavor that NASA's procurement actions rose from roughly forty-four thousand in 1960 to almost three hundred thousand by 1965.[59]

Getting all of the personnel elements to work together challenged the program managers, regardless of whether they were civil service, industry, or university personnel. There were various communities within NASA that differed over priorities and competed for resources. The two most identifiable groups were the engineers and the scientists. Engineers usually worked in teams to build hardware that could carry out the missions necessary to a successful Moon landing by the end of the decade. Their primary goal involved building vehicles that would function reliably within the fiscal resources allocated to Apollo. However, space scientists engaged in pure research and were more concerned with designing experiments that would expand scientific knowledge about the Moon. They also tended to be individualists, unaccustomed to regimentation and unwilling to concede gladly the direction of projects to outside entities. The two groups contended with each other over a great variety of issues associated with Apollo. For instance, the scientists disliked having to configure payloads so that they could meet time, money, or launch vehicle constraints. The engineers, likewise, resented changes to scientific packages added after project definition because these threw their hardware efforts out of kilter. Both had valid complaints and had to maintain an uneasy cooperation to accomplish Project Apollo.

The scientific and engineering communities within NASA, additionally, were not monolithic, and differences among them thrived. Add to these groups representatives from industry, universities, and research facilities, and competition on all levels to further their own scientific and technical areas was the result. The NASA leadership generally viewed this pluralism as a positive force within the space program, for it ensured that all sides aired their views and emphasized the honing of positions to a fine edge. Competition, most people concluded, made for a more precise and viable space exploration effort. There were winners and losers in this strife, however, and sometimes ill will was harbored for years. Moreover, if the conflict became too great and spilled into areas where it was misunderstood, it could be devastating to the conduct of the lunar program. The head of the program worked hard to keep these factors balanced and to promote order so that NASA could accomplish the landing within the time constraints of the president's directive.[60]

Conclusion

With the exception of its second flight, the Saturn V performed almost flawlessly in each of its missions. The Saturn launch system was truly a triumph of U.S. organizational, management, and technological capabilities. However, as the United States shifted priorities away from large-scale space exploration after achieving the goal of a successful lunar landing, there were no new missions requiring the booster's power. NASA's original order was for fifteen Saturn V rockets. As early as 1968, NASA faced the issue of whether it would need more Saturn Vs and decided to wait before ordering the long-lead-time components involved. Further, it became clear in 1970 that there would no early approval of a large space station. As NASA in 1971 and 1972 struggled with getting approval to develop a new space transportation system, the Space

Shuttle, NASA officials reluctantly decided that they had no choice but to give up hopes of preserving the two remaining Saturn V boosters for future use and of maintaining production capabilities for additional vehicles.

Thus the two remaining Saturn V rockets became museum pieces, reminders of a time when the United States pioneered the space frontier beyond Earth orbit. They may be seen today at NASA's Kennedy Space Center and Johnson Space Center. A full-scale mockup may be seen at the Marshall Space Flight Center. Those who actually saw them in use, not as they exist today, were indeed fortunate.

Notes

1. This chapter is based on Ray A. Williamson, "Access to Space: Steps to the Saturn V," in John M. Logsdon, gen. ed., *Exploring the Unknown: Selected Documents in the History of the U.S. Civil Space Program*, vol. 4, *Accessing Space*, NASA SP-4407 (Washington, D.C.: NASA, 1999), pp. 1–31. There are several excellent accounts of the building of the Saturn. These include Roger E. Bilstein, *Stages to Saturn: A Technological History of the Apollo/Saturn Launch Vehicle*, NASA SP-4206 (Washington, D.C.: NASA, 1980), and Charles Murray and Catherine Bly Cox, *Apollo: The Race to the Moon* (New York: Simon & Schuster, 1989). This chapter does not seek to be comprehensive in discussing the development of Saturn but would refer readers to these more detailed accounts.

2. The classic work on this subject is John M. Logsdon, *The Decision to Go to the Moon: Project Apollo and the National Interest* (Cambridge, Mass.: MIT Press, 1970).

3. John M. Logsdon, "An Apollo Perspective," *Astronautics & Aeronautics*, December 1979, pp. 112–17, quote from p. 115.

4. John Law, "Technology and Heterogeneous Engineering: The Case of Portuguese Expansion," pp. 111–34; and Donald MacKenzie, "Missile Accuracy: A Case Study in the Social Processes of Technological Change," pp. 195–222, both in Wiebe E. Bijker, Thomas P. Hughes, and Trevor J. Pinch, eds., *The Social Construction of Technological Systems: New Directions in the Sociology and History of Technology* (Cambridge, Mass.: MIT Press, 1987).

5. Development Operations Division, Army Ballistic Missile Agency, "Proposal: A National Integrated Missile and Space Vehicle Development Program," 10 December 1957, Dwight D. Eisenhower Library, Abilene, Kans.

6. Wernher von Braun, "Crossing the Last Frontier," *Colliers*, 22 March 1952, pp. 23–29, 72–73. Reprinted in John M. Logsdon, gen. ed., *Exploring the Unknown: Selected Documents in the History of the U.S. Civil Space Program*, vol. 1, *Organizing for Exploration*, NASA SP-4407 (Washington, D.C.: NASA, 1995), pp. 179–88.

7. ARPA Order Number 14–59, 15 August 1958, NASA Historical Reference Collection, NASA History Division, NASA Headquarters, Washington, D.C. (hereafter cited as NASA Historical Reference Collection).

8. As it was, the ABMA team ran into difficulties uprating the S-3D engine from 165,000-lbf pounds of thrust to 188,000-lbf because the more powerful engine developed a combustion instability that threatened to destroy the engine. This led to a costly redesign.

9. "The dire need made us more inventive, and we bundled the containers to be loaded with propellants." Willy Mrazek, quoted in Bilstein, *Stages to Saturn*, p. 30.

10. ABMA engineers also clustered the propellant tanks by using eight Redstone tanks, which alternately held RP fuel (a form of kerosene) and LOX, surrounding a single large Jupiter tank in the center that carried RP fuel. See Bilstein, *Stages to Saturn*, p. 82.

11. In writing about it a few years later, von Braun noted that "Juno V was, in fact, an infant Saturn." See Wernher von Braun, "The Redstone, Jupiter, and Juno," in Eugene M. Emme, ed., *The History of Rocket Technology: Essays on Research, Development, and Utility* (Detroit: Wayne State University Press, 1964), p. 120.

12. Konstantin E. Tsiolkovskiy, *Collective Works of K.E. Tsiolkovskiy,* A.A. Blagonrovov, ed., vol. 2, *Reactive Flying Machines,* TTF-237 (Washington, D.C.: NASA, 1965), pp. 78–79; Robert H. Goddard, *The Papers of Robert H. Goddard,* 3 vols., Esther Goddard and G. Edward Pendray, eds. (New York: McGraw-Hill, 1970).

13. Hermann Oberth, *Rockets in Planetary Space,* TTF-9227 (Washington, D.C.: NASA, 1965).

14. John L. Sloop, "Technological Innovation for Success: Liquid Hydrogen Propulsion," in Frederick C. Durant, ed., *Between Sputnik and Shuttle: New Perspectives on American Astronautics* (San Diego: Univelt, 1985), pp. 225–39.

15. James E. Webb, who became the second NASA Administrator, noted in 1963 hearings before the House of Representatives that "one of the principal factors cited in the selection of the Douglas Aircraft Company was that the addition of the company would broaden the industrial base in the hydrogen technology field." 1963 NASA Authorization Hearings, U.S. House of Representatives, pt. 2, p. 825. On Webb's career, see W. Henry Lambright, *Powering Apollo: James E. Webb of NASA* (Baltimore: Johns Hopkins University Press, 1995).

16. Comptroller General of the United States to Overton Brooks, Chairman, Committee on Science and Astronautics, 22 June 1960, NASA Historical Reference Collection.

17. Roger Blistein, "The Saturn Launch Vehicle Family," in Richard P. Hallion and Tom D. Crouch, eds., *Apollo: Ten Years Since Tranquility Base* (Washington, D.C.: Smithsonian Institution Press, 1979), p. 117.

18. Bilstein, *Stages to Saturn,* p. 141.

19. One of the best ways to ensure safety to astronauts and their launch crew is to build a highly reliable vehicle. However, even a vehicle of high reliability may not have adequate safety margins for human flight if it fails catastrophically and does not provide some means to protect its passengers. Conversely, a vehicle meeting a lower reliability rating could, in principle, be safer for humans if it incorporated sufficient means to ensure the crew's ability to survive a failure.

20. Blistein, *Stages to Saturn,* p. 143.

21. See W. David Compton and Charles D. Benson, *Living and Working in Space: A History of Skylab,* NASA SP-4208 (Washington, D.C.: NASA, 1983); Edward Clinton Ezell and Linda Neuman Ezell, *The Partnership: A History of the Apollo-Soyuz Test Project,* NASA SP-4209 (Washington, D.C.: NASA, 1978).

22. Robert Zimmerman, *Genesis: The Story of Apollo 8* (New York: Four Walls Eight Windows, 1998).

23. The F-1 design had its origins in earlier studies at the Army Ballistic Missile Agency. See David E. Aldrich, "The F-1 Engine," *Astronautics,* February 1962, p. 40.

24. The Navaho engine developed 415,000 lbf. The original Air Force goal had been an engine of 1,000,000-lbf, but the company was not able to reach that goal until March 1959, when it fired up a "boilerplate" thrust chamber and injector that achieved it. By then, the program had been transferred to NASA. Bilstein, *Stages to Saturn,* p. 106.

25. NASA chose Boeing in December 1961.

26. Aldrich, "F-1 Engine," p.40.

27. A.O. Tischler, Chief, Liquid Fuel Rocket Engines, NASA, to David Aldrich, Program Engineer, Rocketdyne, 29 July 1959, NASA Historical Reference Collection.

28. Memorandum from R.C. Seamans Jr. to Brainerd Holmes, 23 January 1963, NASA Historical Reference Collection.

29. William Brennan, "Milestones in Cryogenic Liquid Propellant Rocket Engines," AIAA Paper 67–978, October 1967, p. 9, quoted in Bilstein, *Stages to Saturn*, p. 116.

30. T. Keith Glennan, "Memorandum for the Administrator," 19 January 1961, NASA Historical Reference Collection.

31. Aerojet General Corporation, Chrysler Corporation, Douglas Aircraft Corporation, Lockeed Aircraft Corporation, Martin Company, and North American Aviation.

32. The Marshall project team felt considerable pressure to move as quickly as possible on the design of this crucial stage. "You can see that we have a whole lot of doubt in what we say here, and there are a lot of conflicting problems. We are presently trying to resolve them. We could have asked you not to come here today and could have taken, say, six weeks time to resolve these problems internally, in which case we would have lost six weeks on the S-II contract." Marshall spokesman, MSFC, "Minutes of the Phase II Pre-Proposal Conference for Stage S-II Procurement on June 21, 1961," JSC historical files, Houston, Tex.

33. Wilbur Davis, MSFC Procurement and Contracts Office, quoted in ibid.

34. George Mueller commented to Lee Atwood, "Your recent efforts to improve the stage schedule position have been most gratifying and I am confident that there will be continuing improvement." George E. Mueller to Lee Atwood, 23 February 1966, NASA Historical Reference Collection.

35. Samuel Phillips to Associate Administrator, "S-II-T Failure Corrective Action," 9 January 1967, NASA Historical Reference Collection.

36. For a journalistic and rather biased (in favor of Storms) account of this relationship and its development, see Mike Gray, *Angle of Attack* (New York: W.W. Norton, 1992).

37. James J. Haggerty, "Apollo 4: Proof Positive," *Aerospace* (Winter 1967): 3. Quoted in Blistein, *Stages to Saturn*, p. 348.

38. Later renamed the NASA John F. Kennedy Space Center.

39. Blistein, *Stages to Saturn*, pp. 348–51.

40. General Sam Phillips to NASA centers, teletype, 15 November 1967, NASA Historical Reference Collection.

41. Wernher von Braun, "The Detective Story Behind Our First Manned Saturn V Shoot," *Popular Science*, November 1968.

42. Blistein, *Stages to Saturn*, pp. 360–62.

43. The story of these facilities are related in Charles D. Benson and William Barnaby Faherty, *Moonport: A History of Apollo Launch Facilities and Operations*, NASA SP-4204 (Washington, D.C.: NASA, 1978).

44. However, ultimately it was equipped to handle only three.

45. The VAB, Launch Pads 39A and 39B, the mobile launchers, and the crawler-transporters are still in use in the Space Shuttle program, although they have been modified to accommodate the rather different shape, size, and load requirements of the shuttle system. Also, the VAB is now the *Vehicle* Assembly Building, versus the Apollo-era *Vertical* Assembly Building.

46. Walter Flint, "Operational Support for Apollo," in Allan A. Needell, ed., *Apollo: Ten Years Since Tranquility Base* (Washington, D.C.: Smithsonian Institution Press, 1979), pp. 109–14.

47. *Aeronautics and Space Report of the President, 1988 Activities*, annual report (Washington, D.C.: NASA, 1990), p. 185; Linda Neuman Ezell, *NASA Historical Data Book*, vol. 2, *Programs and Projects, 1958–1968*, NASA SP-4012 (Washington, D.C.: NASA, 1988), pp. 122–23.

48. NASA, "Minutes of Meeting of Research Steering Committee on Manned Space Flight," 25–26 May 1959, p. 2, NASA Historical Reference Collection.

49. NASA Office of Program Planning and Evaluation, "The Long Range Plan of the

National Aeronautics and Space Administration," 16 December 1959, p. 28, NASA Historical Reference Collection.

50. See Howard E. McCurdy, *The Space Station Decision: Incremental Politics and Technological Choice* (Baltimore: Johns Hopkins University Press, 1990).

51. This story has been told in John M. Logsdon, "Selecting the Way to the Moon: The Choice of the Lunar Orbital Rendezvous Mode," *Aerospace Historian* 18 (Summer 1971): 63–70; Courtney G. Brooks, James M. Grimwood, and Loyd S. Swenson Jr., *Chariots for Apollo: A History of Manned Lunar Spacecraft*, NASA SP-4205 (Washington, D.C.: NASA, 1979), pp. 61–86; Bilstein, *Stages to Saturn*, pp. 57–68. See also "Concluding Remarks by Dr. Wernher von Braun about Mode Selection given to Dr. Joseph F. Shea, Deputy Director (Systems), Office of Manned Space Flight," 7 June 1962, NASA Historical Reference Collection; John C. Houbolt, "Lunar Rendezvous," *International Science and Technology* 14 (February 1963): 62–65.

52. On Skylab, and the use of the Saturn launcher for it, see Compton and Benson, *Living and Working in Space*.

53. See Sarah L. Gall and Joseph T. Pramberger, *NASA Spinoffs: 30 Year Commemorative Edition* (Washington, D.C.: NASA, 1992).

54. See Frederick I. Ordway III, Carsbie C. Adams, and Mitchell R. Sharpe, *Dividends from Space* (New York: Thomas Y. Crowell, 1971).

55. See James E. Webb, *Space Age Management: The Large Scale Approach* (New York: McGraw-Hill, 1969).

56. Albert F. Siepert, memorandum to James E. Webb, 8 February 1963, NASA Historical Reference Collection; Sarah M. Turner, "Sam Phillips: One Who Led Us to the Moon," *NASA Activities* 21 (May/June 1990): 18–19.

57. Aaron Cohen, "Project Management: JSC's Heritage and Challenge," *Issues in NASA Program and Project Management*, NASA SP-6101 (Washington, D.C.: NASA, 1989), pp. 7–16; C. Thomas Newman, "Controlling Resources in the Apollo Program," *Issues in NASA Program and Project Management*, NASA SP-6101 (Washington, D.C.: NASA, 1989), pp. 23–26; Eberhard Rees, "Project and Systems Management in the Apollo Program," *Issues in NASA Program and Project Management*, NASA SP-6101 (02) (Washington, D.C.: NASA, 1989), pp. 24–34.

58. Dael Wolfe, Executive Officer, American Association for the Advancement of Science, editorial for *Science*, 15 November 1968.

59. Bilstein, *Stages to Saturn*, passim, and appendix E.

60. McCurdy, *Inside NASA*, pp. 11–98.

– 8 –

Taming Liquid Hydrogen

The Centaur Saga

Virginia P. Dawson

To the Huntsville people the methods and philosophies of the Atlas and Centaur
were as mysterious as the dark side of the Moon.

Deane Davis, General Dynamics

Introduction

In May 1962 a Centaur upper stage, mated to an Atlas rocket, exploded 54
seconds after launch. Centaur's tank—filled with liquid hydrogen fuel—split
open when its insulation came loose and tore a hole through its paper-thin
wall, immediately engulfing the rocket in a huge fireball. This disaster
prompted the Subcommittee on Space Sciences, headed by Congressman Joseph E. Karth of Minnesota, to call for a probe. Centaur had been pitched to
Congress as crucial to the space program. Two years earlier Abe Silverstein,
Associate Administrator for Space Flight Development Programs, had declared
during the 1960 NASA budget hearings that Centaur was "the kind of thing
upon which our whole future technology, I think, rests—that is, the development of an early capability with these high-energy propellants."[1] Based on
NASA's strong support of Centaur development, Congress had authorized increased funding in June 1960 and amended the contract with General Dynamics to require not six vehicles, but ten—"this being considered an absolute
minimum to prove out a vehicle design upon which so much of the national
space program was beginning to depend."[2] Two years later, when Centaur exploded, this confidence turned to bitter disappointment. Homer E. Newell,
Director of Space Sciences for NASA, declared to Congress in characteristic
understatement, "Taming liquid hydrogen to the point where expensive op-

erational space missions can be committed to it has turned out to be more difficult than anyone supposed at the outset."[3]

Developing liquid hydrogen as a rocket fuel involved mind-boggling technical challenges. Because liquid hydrogen is cryogenic, it can only be converted from a gas to a liquid at extremely low temperatures. Liquid hydrogen must be stored at -423 degrees Fahrenheit. To keep from evaporating, or "boiling off," tanks must be carefully insulated from all sources of heat, such as rocket engine exhaust, air friction during flight through the atmosphere, and the radiant heat of the Sun. When liquid hydrogen absorbs heat, it expands rapidly, making venting essential to reduce gas pressure. These daunting problems, however, are offset by liquid hydrogen's definite advantages. Its high specific impulse, or efficiency in producing thrust in relation to weight, makes liquid hydrogen among the lightest and most powerful propellants available to designers. A rocket fueled by liquid hydrogen (with liquid oxygen as the oxidizer) can loft approximately 40 percent more payload per pound of liftoff weight than conventional kerosene-based fuels. This was definitely a prize worth the gamble, especially since U.S. launch vehicle capability at this time was still well behind the Soviet Union.[4]

Rocket pioneers Goddard, Oberth, and von Braun all had considerable experience using liquid oxygen (which is also cryogenic) as an oxidizer, but they apparently never experimented with liquid hydrogen. The tragic explosion of the *Hindenberg* dirigible in the 1930s probably reinforced their apprehension over hydrogen's volatility. After World War II advances in the technology of the handling, storage, and firing of liquid hydrogen gave engineers hope that the fuel would ultimately prove reliable.[5]

Congressional hearings do not often reveal the drama, the differences in engineering style, or the rivalries that exist between people and the institutions they represent. However, implicit in the 1962 Centaur hearings is the tension between two entirely different philosophies of design and approach to the management of large-scale technology projects. This chapter will trace the evolution of the Centaur design and suggest why NASA chose to continue to support the development of this rocket. The 1962 Centaur hearings marked a turning point in Centaur's fortunes, because after this point NASA never wavered in its belief that Centaur would become operational. In its decision to continue to back Centaur development, Headquarters asserted its authority over Wernher von Braun, Director of the Marshall Space Flight Center in Huntsville, Alabama, who had privately urged its cancellation.

Von Braun was used to designing, building, and testing a rocket prototype before turning it over to an industry contractor for closely supervised production.[6] The "everything-under-one-roof approach" had prevailed at Peenemünde and dovetailed with the U.S. Army arsenal tradition of weapons development. In contrast, the U.S. Air Force, lacking extensive in-house expertise, allowed its industry contractors considerable autonomy. Centaur was developed by the Convair/Astronautics Division of General Dynamics, located in San Diego, California. After the first Centaur blew up, General Dynamics had to defend both the choice of liquid hydrogen fuel and Centaur's basic design. The innovative "egg-shell" design the company had pioneered for Atlas was radically different from the large, heavy structures favored by the von Braun team. John Sloop, an engineer-turned-historian to whom this chapter is greatly indebted, has revealingly described the differences in approach to rocket design in his study *Liquid Hydrogen as a Propulsion Fuel, 1945–1959:* "The design of the V-2, Redstone, Jupiter, and Saturn all reflect the conservatism of von Braun and his team. They looked askance at such lightweight structural innovations as [Karel J.] Bossart's thin-wall, pressurized tanks for the Atlas ICBM, which they jokingly referred to as 'blimp' or 'inflated competition.' . . . This conservative design philosophy mitigated against the use of liquid hydrogen which, more than conventional fuels, depended upon very light structures to help offset the handicap of low density."[7] Vincent L. Johnson, the new program manager for Centaur at Headquarters, and Homer Newell believed that the United States space program would benefit from General Dynamics' more innovative approach to rocket design and development. Certainly, the policy of NASA Administrator T. Keith Glennan favored development of new technology by industry over building up a large in-house expertise.[8] Webb, who succeeded Glennan in January 1961, had more confidence in large government-directed technology enterprises but also believed that NASA needed a range of launch vehicles for different missions. The Centaur decision represented the refusal of Headquarters to be strong-armed into limiting NASA's launch vehicle options.[9]

The Balloon Structure

The structure of both the Atlas and Centaur rockets is a thin monocoque steel shell. This foil-like shell holds its shape when pressurized, like a balloon. The skin of the rocket has the dual function of serving both as the external wall and as the walls of the fuel and oxidizer tanks. The tanks are separated by a common bulkhead, and there is no internal bracing and no insulation surrounding the propellants. The result is a dramatic saving in weight, always a critical factor in any successful rocket design.

Karel J. "Charlie" Bossart designed the Atlas during the early postwar years at Convair in San Diego, California. Deane Davis, who served as manager of the Centaur Program Office at General Dynamics in the mid-1960s, described the father of the Atlas as a "shy, quiet, lovable self-effacing man who was quite tolerant of the criticisms pointed at the innovations inherent in the Atlas Design, particularly his pride and joy: its balloon structure."[10] An engineer with a background in structures, Bossart hit upon the weight-saving idea of using pressure-stabilized tanks. This was not an entirely new idea. Hermann

Oberth in Germany had discussed it in his writings. After World War II, the stabilized tank was suggested in designs for two U.S. Navy satellites. Bossart arrived at the idea independently, however, and was the first to make it work.[11]

Charlie Bossart's background was as unconventional as his design for Atlas. Born in Antwerp, Belgium, he graduated from the Free University of Brussels in 1925 with a degree in mining engineering. During his last year there, an optional course in aeronautical engineering fired his ambition to study aeronautical engineering at MIT. He earned a master's degree in 1927, returned to Belgium for military service, then emigrated to the United States in 1930. He took a job in aircraft structures first at General Aviation Corporation and then at United Aircraft and Transport Corporation. In 1941 he moved to Downey, California, to work for Vultee Aircraft (the company became Consolidated Vultee, or Convair, in 1943), and later to San Diego when the program moved there during full-scale development.

At Convair, Bossart became project engineer for one of the nation's first ballistic missile development projects, the MX-774. This project was funded by the Air Technical Service Command. Like all early postwar missile projects, Bossart's group started by making extensive studies of the V-2. The V-2 had a stiffened, or "rings and stringer," tank design. Metal bands or rings, connected by horizontal stringers, reinforced the walls of the rocket to provide rigidity and an extra margin of safety. Bossart thought rings and stringers unnecessary. His much lighter design dispensed with internal buttressing in favor of a pressure-stabilized tank. In an interview he noted how his background in aircraft structures influenced his novel approach:

> You see most of my previous experience was in structures. So the first thing I did was to decide what kind of a structure we're going to use for this missile. Well, we knew that we had to have a certain pressure in the tank to maintain the required net positive suction head for the pump. So the first question was, how thin does the skin have to be to resist that pressure? With the skin thickness thus arrived at and assuming the tank to be under pressure, how much stiffening would you have to add to make this thing capable of taking the compression and the bending anticipated in flight? And lo-and-behold, we didn't need to add any.[12]

Though the balloon structure became controversial after it was adopted by Dr. Krafft A. Ehricke for Centaur, in the early postwar period, lack of funding, not objections to the design, slowed Atlas MX-774 development. The Air Force canceled the project in July 1947, but Bossart was permitted to flight test three vehicles that had already been completed. The partial success of the tests convinced Convair to allow Bossart and his team to keep working on the fuselage and flight-control system on a part-time basis, rather than give up what appeared a promising lead in missile technology. In 1951 Convair was awarded another study contract for development of the missile, renamed Project MX-1593.[13]

Atlas development proceeded in fits and starts until 1953, when it won the backing of two influential Air Force officials, Trevor Gardner, Assistant Secretary of the Air Force for Research and Development, and Brig. Gen. Bernard Schriever. The story of development of the Atlas missile from the management point of view has received deserving historical attention.[14] Gardner and Schriever were convinced that the development of a powerful and light thermonuclear warhead now made an intercontinental ballistic missile (ICBM) technically feasible. Concerned that the United States was in a race with the Soviet Union in ICBM development, they advocated setting up a crash program comparable to the Manhattan Project. A blue ribbon committee of scientific, industrial, and academic leaders, known informally as the "Teapot Committee," reviewed Atlas development. The committee endorsed the Atlas missile but apparently did not believe General Dynamics capable of managing a project of the magnitude envisioned. The Ramo-Wooldridge Corporation took over technical direction of the program, reporting directly to Bernard Schriever, head of the Western Development Division for the Air Force. Bossart recalled that he received "considerable pressure" from Ramo-Wooldridge to discard the pressure-stabilized tank design. Its light weight was the deciding factor in retaining it. Bossart admitted the bulkhead would collapse unless the tanks were pressurized correctly, but development was too far along to change to a reinforced structure.[15] Though the Atlas missile was considered an outstanding success when it became operational in 1960, within two years it was slated for obsolescence—to be supplanted by the smaller, more efficient Minuteman solid-fuel missile and the much larger Titan II. However, when mated with an upper stage like Agena, and later, with the far more powerful Centaur, it would provide NASA with a superior rocket for planetary exploration. Development of Atlas/Centaur into a dependable launch vehicle, however, would be neither inexpensive nor easy.

Ehricke and Centaur

Krafft Ehricke had dreamed of space travel since his early teens. Like many of the early German rocket pioneers, the famous 1929 Fritz Lang film *Frau im Mond (Woman in the Moon)* inspired Ehricke as a youth. "I felt, 'My God, it really must be possible to get to the moon,' which for an 11-year-old boy is a kind of revelation," he recalled. After seeing the film many times, he began to study the early rocket pioneers on his own. He found the mathematics of Oberth's *Wege zur Raumschiffahrt (Ways to Space-flight)* almost impenetrable, but his study whetted his appetite for aeronautical engineering. He entered the Technical University in Berlin. During his student years Ehricke took part in the German Rocket Society's experiments, but he recalled he and his colleagues avoided using liquid hydrogen/liquid oxygen as a rocket fuel because they feared the mixture might explode prematurely. "There was so little money available we simply couldn't afford this sort of thing," he said. His graduate studies in nuclear physics at Humboldt University under Werner Heisenberg were interrupted by the war. He served at the Russian front before being recruited for Peenemünde. There Ehricke came under the spell of Walter Theil, von Braun's expert in rocket engine development. Theil inspired Ehricke to think about the development of "multi-million pound rocket engines," presumably for space exploration. Learning of Heisenberg's and Robert Pohl's success in building a nuclear reactor, Ehricke and Theil dreamed of nuclear rocket propulsion with hydrogen as the working fluid.[16]

Project Paperclip brought Ehricke to the United States after the war, where he worked with von Braun at White Sands, New Mexico, and later at the Army Ballistic Missile Agency in Huntsville. Ehricke became better acquainted with liquid hydrogen after Von Braun asked him to critique a report published in 1947 by Jet Propulsion Laboratory (JPL) rocket researcher Richard Canright. At this time there was little hard data to guide rocket designers in the selection of a fuel. Exhaust velocity was usually the main criterion, but Canright argued that propellant density must also be considered, since propellant density can affect vehicle size and weight. His report compared the relative importance of propellant density versus exhaust velocity in assessing rocket performance and concluded that exhaust velocity was the more important of the two. Liquid hydrogen and hydrazine garnered high praise in his ranking of rocket fuels and he urged that more research be focused on high-energy liquid propellants. Though Ehricke found the report suggestive, von Braun apparently was unimpressed.[17]

Anxious to leave Huntsville, Ehricke took a job at Bell Aircraft under Walter Dornberger in 1952, where he worked briefly on the Agena upper stage. He moved on to Convair in 1954, the same year that Convair became the Convair/Astronautics Division of General Dynamics.[18] In early 1956 Ehricke, working under Bossart as a design specialist, was caught up in the white-hot heat of Atlas development. Always a visionary, he began to play with the idea of using the Atlas, mated to a liquid hydrogen upper stage, to launch a satellite. He recalled, "Going back to the old ideas of Oberth, I said, 'I have a relatively dense first stage. And it so happens that the second stage, because it is less dense, fits just beautifully on the first stage. Now all we have to do is remove the neck from Atlas and make it cylindrical all the way and we have a 10-foot diameter base. For a second stage, that's just beautiful."[19]

During design studies, Ehricke investigated a number of propellant combinations, including fluorine/hydrogen, fluorine/hydrazine, ozone/methane, and oxygen/hydrogen. Oxygen/hydrogen was the clear winner because the technology for using oxygen as the oxidizer already existed. What made liquid hydrogen attractive was its high specific impulse "near the upper limit obtainable with chemical fuels."[20] Ehricke tried to see technical choice within the context of future rocket development. "I felt one of the advantages of using oxygen-hydrogen was that it paved the way for the nuclear whereas fluorine wouldn't," he told John Sloop in 1974.[21]

Ehricke's original design bore little affinity to Atlas. He envisioned a rocket with two tanks, one filled with hydrogen, the other with oxygen. Each of these tanks had inner tanks filled with their respective cryogenic liquids. These inner tanks were intended to supply fuel for restarting the engines in orbit after a coast period. They obviously added weight and made the design very complicated.[22] Ehricke proposed a pressure-fed engine system to avoid the arduous development of a hydrogen pump.[23] The proposed rocket would be powered by four 7,500-pound-thrust oxygen/hydrogen engines designed by the Rocketdyne Division of North American Aviation.

Ehricke's first Centaur proposal, "A Satellite and Space Development Program," was submitted to the Air Force in December 1957, only two months after the launch of *Sputnik I*. General Dynamics asked for $15 million to begin work on the 30,000-pound upper stage. Though the Air Force did not accept his first proposal, Ehricke received enough encouragement to continue to develop the idea. Optimistic that the new national space agency would find his ideas attractive, he called on Abe Silverstein at NACA Headquarters in Washington in June 1958. Though Silverstein shared Ehricke's enthusiasm for liquid hydrogen, NASA had not yet received funding. He suggested that the Advanced Research Projects Agency (ARPA), the new agency responsible for the military's initiatives in space, might be interested.[24]

A Bold New Design

When Ehricke submitted his Centaur proposal to officials at ARPA in June 1958, he had no idea that Pratt & Whitney was in the process of developing a liquid hydrogen rocket engine. ARPA suggested he submit a new design based on a strikingly innovative engine Pratt & Whitney had just designed for a hydrogen-powered aircraft. Kenneth E. Newton, vice president and program manager of launch vehicle programs, recalled the excitement the RL10 engine generated at General Dynamics: "I remember the first time we got a set of drawings of the Pratt-Whitney engine. We were so enthralled with their simplicity."[25]

Pratt & Whitney's engine was built for a highly classified high-altitude spy plane called Project Sun Tan.[26] The engine's two-stage centrifugal pump and heat exchanger were its most advanced features. As hydrogen flowed through tubes into the combustion chamber walls on the way to the turbine it picked up heat and expanded to provide energy to drive the turbine and turbopumps. At the same time the cryogenic hydrogen cooled the thrust chamber walls. This process called "regenerative cooling," or a "bootstrap cycle," was the key to the efficiency and reliability of the engine—"a simple, elegant

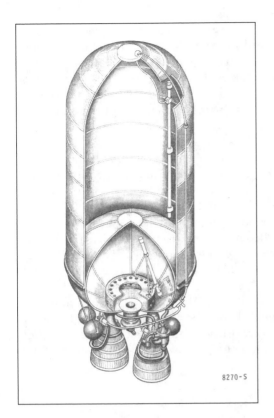

8270-S

Image 8-1: This drawing of a Centaur rocket shows the liquid hydrogen and liquid oxygen tanks separated by an intermediate bulkhead. LH2 was in the upper part, while LO2 was in the lower portion of the rocket, separated by a double-walled intermediate bulkhead. (NASA Photo no. 8270-S)

system that works with the unique properties of liquid hydrogen rather than battling against them."[27]

Inspired by the elegance of the RL10's regenerative cooling, Ehricke proposed a bold new design for Centaur. To capitalize on the low density and extreme cold of liquid hydrogen, he designed a single, or "integral," tank, separated by a double-walled intermediate bulkhead. The bulkhead consisted of two thin metal skins filled with a quarter inch layer of fiberglass insulation. When filled with liquid oxygen and liquid hydrogen, the extreme cold of the liquid hydrogen produced a vacuum within the hollow bulkhead. This thermal barrier prevented heat transfer between the liquid oxygen and the much colder liquid hydrogen. During his testimony at the 1962 Centaur failure investigation, Ehricke explained the relationship between the double bulkhead design and bootstrap cycle of the RL10 he so admired. "Using this design, we achieve, a very high mass ratio for a vehicle of this size with low-density propellants. The membrane bulkhead design uses cryogenic vacuum formation as the logical utilization of an existing extreme temperature environment; just like Pratt Whitney's bootstrap method of engine start [i.e., regenerative cooling] was an ingenious and logical utilization of the special physical characteristics of liquid hydrogen."[28] Deane Davis recalled that the double-walled bulkhead was literally created on a tablecloth and turned out to be the most fundamental part of the design. "We had that integral [double-walled] bulkhead before we even knew how many engines we had on the thing or how long the thing was going to be," he said. "That was the kind of invention that sprung the whole thing."[29]

Though the bulkhead design drew on one of the essential features of Atlas, Ehricke obviously took this for granted, since he did not mention his debt to Atlas in his testimony or interviews. The Atlas bulkhead was not double-walled since insulation between the fuel and oxidizer was not required. The double-walled integral bulkhead would be not only the most innovative feature of Centaur, but also the most technically demanding, as General Dynamics engineers would later discover. The new "Proposal for a Mars Probe," submitted August 1958, featured a 30,000-pound rocket, 10 feet in diameter, with two Pratt & Whitney pump-fed engines, with 15,000 lbf each. The new Centaur had a pressure-stabilized tank like Atlas with aerodynamic insulation and nose fairing, which would be jettisoned after the rocket left the atmosphere. In November, ARPA accepted Ehricke's proposal. Air Force Lt. Col. John Seaberg, previously a champion of liquid hydrogen–fueled aircraft, became the USAF project manager. Now Centaur's troubles would begin.

Marshall's "Unruly Wards"

Originally, the goals of the Centaur were modest. Considered a bridge between the underpowered Atlas/Agena and much larger future boosters like Saturn, Centaur was intended merely to prove the feasibility of a liquid hydrogen/liquid oxygen rocket. Because it was so experimental, no mission was assigned to Centaur. Development was proposed to take twenty-five months with the first flight to occur in January 1961. Cost was projected at $36 million for six "flight articles." The guidance system, test facilities, and launch complex were not included in the original bare bones contract. To keep down costs, General Dynamics was directed to use Atlas hardware, if possible, rather than design from scratch.[30]

When ARPA assigned Advent to Centaur in December 1958, expectations began to rise. Advent was to be a twenty-four-hour military communications satellite with the ambitious requirement of a long coast and the necessity of restarting the engine twice in space. The design for Centaur was frozen about the time the program was transferred to NASA, 1 July 1959. Seaberg managed Centaur from the Air Research and Development Command (ARDC) offices in Washington, reporting to Milton Rosen at NASA. However, the original Air Force contracts remained in effect.

Development of Centaur during this period, Ehricke recalled, was extremely difficult. Rosen, not a particular fan of Centaur, insisted General Dynamics take over development of the small conventional Vega rocket. Ehricke, now in charge of both projects, watched in dismay as Vega siphoned off personnel and funds needed for the more technically challenging Centaur. That August the Centaur program office was transferred to the Air Force Ballistic Missile Development complex in Los Angeles and a Centaur project team was set up. Program cost increased to $42 million when the guidance system by Minneapolis Honeywell was added. Construction of additional test facilities was also begun at this time. Rosen was succeeded at NASA Headquarters by Cmdr. William Schubert in January 1960. At the same time, a Centaur Project Technical Team was set up at Headquarters under Col. Donald H. Heaton.

After the Vega program was canceled in December, more important missions were assigned to Centaur. They included lunar and the planetary exploration programs, Surveyor and Mariner. Surveyor's mission was to prove the feasibility of a soft landing on the Moon, paving the way for landing the Apollo astronauts within the decade. Centaur program costs rose to $63 million after General Dynamics agreed to build four more vehicles, in addition to the six already on the drawing boards.[31]

In July 1960 Centaur was assigned to the Marshall Space Flight Center, along with Thor-Agena and Atlas-Agena. Because the giant Saturn vehicle was always a top priority at Marshall, Marshall paid little attention to Centaur. With the first launch projected for June 1961, construction began on the Centaur launch complex (Pad 36) at the Army Atlantic Missile Range at Cape Canaveral. However, as the costs and visibility of Centaur increased, technical and management problems at General Dynamics threatened to scuttle the program. Probably at Headquarters' urging, Marshall took steps to assert more direct control over the program.

On 1 January 1962 the Air Force contracts with General Dynamics were converted to NASA contracts and a program office was set up in Huntsville to provide more oversight of the industry contractor. Hans Hueter was named director of Marshall's Light and Medium Launch Vehicles Office with Frances Evans to take over Centaur as project manager.

At General Dynamics management was also strengthened. Krafft Ehricke was replaced by Grant Hansen, a vice president who reported directly to James R. Dempsey, president of General Dynamics/Astronautics. Though relatively new to General Dynamics, Hansen had worked for Douglas Aircraft between 1948 and 1960 in missile and space systems development. Hansen's management style was tough and effective. Of the management change, he said, "And then finally come sort of a showdown and, as Dempsey explained it to me, they felt that Krafft was a tremendously imaginative, creative idea man, a hell of a good engineer, but that he wasn't enough of a S.O.B. to manage a program like this. He had functional department people who would tell him to

Image 8-2: A Centaur rocket is suspended from an overhead crane at its production dock at General Dynamics/Astronautics in San Diego, in 1964, to begin a journey to Cape Canaveral for eventual launch. (NASA Photo)

get lost, and he would be willing to do just that, and they needed somebody who wouldn't."[32] NASA insisted that General Dynamics shift from a matrix organization to a project form of management. This involved physically locating the entire program in its own space, instead of assigning different phases of the work on a functional basis to different managers throughout the company. Under the new plan all managers reported directly to Hansen as Centaur Program Director. General Dynamics increased the number of employees specifically assigned to Centaur from a handful to about one thousand. The shift to project management guaranteed that Centaur would no longer play second fiddle to Atlas in terms of priority within the company. To assist Hansen in implementing the new management structure, the company contracted with a consulting firm, Robert Heller and Associates. Deane Davis, who had previously functioned as Ehricke's right-hand man, now became Deputy Program Director for Technical Control.

Von Braun probably expected Hans Hueter to impose on General Dynamics the kind of discipline that the Germans favored. But development was too far along. Deane Davis recalled that not much love was lost between the Germans and their "unruly wards at Astronautics." Hueter may have been one of the few that they respected, but he had the "seemingly impossible task," from Davis's point of view, of attempting reconcile "two diametrically opposed design and administrative concepts into a successful whole." Bossart, in particular, was acutely aware of "the vast difference in basic structural philosophy between himself and the Huntsville group; his light egg-shell concept of structure *vs* the German's traditional 'bridge design,' as he called it."[33]

Davis recounted an anecdote to illustrate this difference.[34] Although the exact date of this incident is not clear, it must have occurred some time between January 1962 and early May, when the first Centaur failed. As Marshall tried to assert its authority during this period, the General Dynamics plant in San Diego was suddenly "deluged with Huntsville personnel attempting to understand their new creature." Apparently, Bossart kept a low profile during these visits, but a confrontation occurred, despite the best efforts of Davis and others. After the Germans took the plant tour and visited the new Sycamore Canyon hot-firing test facility, they gathered with their counterparts at General Dynamics for a late afternoon briefing. Trouble began when Bossart, seated next to Willie Mrazek, von Braun's structural section chief, began to argue in loud whispers over the question of Centaur's structure. "Charlie was trying to explain its merits to a disbelieving Mrazek and from experience," Davis recounted. "I knew that the only solution was to get those two gentlemen separated from the briefings so they could have at it."

Bossart led Mrazek out into the factory yard where a rocket tank stood gleaming in the sunlight. Mrazek asked, "What's inside it?" to which Bossart responded, "Nitrogen." (Nitrogen is used to maintain pressure in the tank before the rocket is filled with propellants. Without pressurization, the rocket's thin skin would wrinkle and collapse.) Mrazek, an advocate of solid, reinforced rocket walls, could not understand how nitrogen, kept at the relatively low pressure of eight to ten pounds per square inch, was sufficient to keep the tank rigid.

To prove the strength of the balloon structure, Bossart invited Mrazek to take a hammer and give the tank a whack. Failing to put even the slightest dent in the tank, he tried again, this time giving the side of the tank a such a glancing blow that the hammer flew out of his hand, knocking his glasses off, but again leaving the surface unscathed. Though Bossart may have demonstrated the strength of the structure of the tank, this incident did not endear General Dynamics to Mrazek nor win the von Braun group's faith in Centaur's reliability. But it was not only differences in design philosophy but also approach to management that ultimately doomed the relationship between Marshall and General Dynamics. Davis explained: "To the Huntsville people the methods and philosophies of the Atlas and Centaur were as mysterious as the dark side of the Moon. On the Astronautics side, still riding the crest of their brilliant success with the Atlas, and with no reason to believe that the Centaur would prove to be any different, the new Centaur people resented what they felt was undue interference in their established manner of creating, producing, and operating their product. After all, they said repeatedly, the Atlas and Centaur were their inventions."[35]

Centaur's Troubles

The team at General Dynamics began to encounter serious problems with Centaur as early as 1960. To solve these problems the company initiated an ambitious test program. The company built a flow test facility at Point Loma and a static test facility at Sycamore Canyon. It also contracted for testing at Langley and Lewis Research Centers. By 1961 program cost had reached $100 million.

Testing revealed many complex problems with liquid hydrogen. For example, General Dynamics engineers discovered an unacceptably high heat transfer rate across the bulkhead separating the two cryogenic liquids. Hydrogen was leaking through minute holes in the welds. These holes were less than one-ten-thousandth of an inch, but they destroyed the vacuum necessary to prevent heat transfer between the oxygen and far colder hydrogen tank. When called to testify to the Congress in May 1962 Ehricke admitted, "We have definitely in this particular case, if I might say so, gambled at a very low weight and found that we have to correct ourselves. We have now under fabrication bulkheads which have greater wall thickness at the points of welding so that we are confident that we can overcome this problem."[36]

Another perplexing problem was why, after filling the tank with liquid hydrogen, the metal of the lower bulkhead would suddenly wrinkle. It was discovered that the extreme cold of the cryogenic fuel caused "cryoshock." To prevent cryoshock, the bulkhead had to be gradually cooled down. With this knowledge, the launch team at Cape Canaveral developed carefully orchestrated tanking procedures.[37]

An important question centered on how liquid hydrogen behaved in the weightless, or zero gravity, environment of outer space. The Advent satellite's intended twenty-four-hour orbit had the ambitious requirement of restarting the engine twice. The first coast period or "parking orbit" was slated to last almost an hour, the second slightly more than five hours. During both coast periods the longitudinal axis of the vehicle would be pointed toward the Sun, creating the danger of liquid hydrogen boiling off.

There were many unknowns involved in restarting an engine in space. A major issue was how to make sure there was adequate liquid hydrogen at the aft end to ignite the engines. Would the liquid hydrogen in a partially filled tank continue to wet the walls of the tank? Would pressure buildup make venting necessary? Could enough pressure be maintained to prevent bulkhead collapse? To gather data, NASA and General Dynamics initiated an ambitious zero-gravity test program using specially equipped aircraft. Tests were carried out by NASA Lewis Research Center in Cleveland, Ohio, and at Wright-Patterson Air Force Base, near Dayton, Ohio. The usual method for an airplane to get several seconds of zero-gravity test time is to pull up suddenly out of a full-power dive at approximately 20,000 feet. Experiments suspended in the bomb bay were filmed. Later a series of Aerobee rockets carrying dewars of liquid hydrogen would be used for research into these questions.[38]

Another unknown was the effect of the cryogenic fuel on the insulation of the tank, and how to attach it to the vehicle so that it could be jettisoned in flight. General Dynamics chose an insulation consisting of a foam material covered by very thin fiberglass. It was held to the rocket by metal bands. At approximately Mach 10 (when the vehicle left the sensible atmosphere), the explosive bolts on the bands were supposed to fire, allowing the insulation to fall away from the vehicle.

NASA was dubious that General Dynamics had grasped the unique problems of insulating a liquid hydrogen tank. A memo in March 1961 from Lewis Research Center complained of the need to "cajole" aerodynamic test data out of General Dynamics and noted during a visit to San Diego that the contractor seemed not to understand the need for purging with helium gas between the insulation and the skin of the rocket to prevent air from contaminating the insulation. The memo concluded, "NASA is worried about the aerodynamic flutter of the panels and the likelihood of developing leaks in the sealed panels while in atmospheric flight, thus setting up a pumping cycle of air going in beneath the panels and liquid air running out. These worries probably will not be resolved until the first full-scale Centaur flight. Nothing being done now gives any real basis for optimism."[39] This memo was prophetic. When the first Centaur did fly, problems with this insulation caused the rocket to fail.

RL10 Woes

Testing the RL10 engine at Pratt & Whitney went smoothly at first. By July 1959 it had accumulated 230 successful firings in a horizontal test stand. It started, stopped, and restarted with surprising reliability. Testing the engine in a vertical stand was initiated in October 1960 with forty nonfiring cold flows and one successful dual engine firing of 5 seconds duration. However, during the second hot firing in the vertical stand, the engine exploded. Investigation suggested that this was a simple operational problem. Pratt & Whitney fixed the damaged test stand and resumed testing in January 1961. However, when the engine exploded again during an important demonstration to Air Force and NASA officials, Pratt & Whitney engineers looked more deeply into the cause. Through testing, they quickly discovered that in the vertical position, gravity acting on the flow of oxygen prevented proper ignition. By installing a separate oxygen feed to the igniter the problem was solved.[40]

At this time Pratt & Whitney developed a new test stand (E-5) to simulate Centaur propulsion system firings. The stand could accommodate two RL10 engines, propellant supply ducts, pneumatic system, propellant boost pumps, and boost-pump hydrogen peroxide gas generator drive systems. On 24 April 1961 two RL10 engines successfully completed their first full-duration test on the test stand. By May 1962 the RL10 engine had completed over seven hundred hot firings.[41]

When called to testify at the Centaur hearings in May 1962, Bruce Torel, the RL10 engine's program manager, asserted proudly that the development problems of the "free world's first liquid hydrogen rocket engine" had been solved.[42] He was correct. However, Pratt & Whitney would have to wait several years for General Dynamics to solve Centaur's many problems before the RL10 would fly.

Deciding the Fate of Centaur

On the first day of the 1962 hearings, members of the committee grilled those responsible for Centaur's failure. Both the General Dynamics spokesmen and their NASA Marshall managers knew that this was by no means a routine investigation. It was clear the fate of the Centaur program hung in the balance. Von Braun accepted responsibility. He thought the government had not "penetrated" the program sufficiently. Because Marshall had taken over management of the program from the Air Force after many of the important design decisions had already been made, it had difficulty asserting control over the industry contractor. "The only excuse one can have for it is that it started out as a little exploratory program and grew and grew and grew into a major program, and it wasn't intended that way from the outset."[43] Von Braun concluded that greater in-house competence was the key to proper supervision: "I think the general lesson we have learned from Centaur management is probably that one should never underestimate the magnitude of a program where so many new and unproven ideas are tried out and I think we will always get in difficulties, as a Government agency, unless we build up a competence in the Government that we can really stay on top of the problems right from the outset and learn how to identify potential problem areas before we have explosions and fires and setbacks."[44] Industry would not have the same free hand when it came to Saturn development, he assured the Congress. "We are really making an all-out effort to stick together with the contractor before major design decisions are even made and have our men argue with his men as to whether this is really the way to go."[45]

Von Braun also expressed his reservations with regard to the basic design

Image 8-3: Centaur rocket installation in Propulsion Systems Laboratory No. 1 on 17 April 1962. The RL10 was developed by Pratt & Whitney in the late 1950s and tested at the Lewis Research Center (now known as the John H. Glenn Research Center at Lewis Field). This engine was the propulsion system for NASA's upper-stage Centaur rocket and was significant for being the first to use liquid hydrogen and liquid oxygen as propellants. The Centaur suffered a number of early failures but later proved to be a very successful upper stage for numerous commercial, NASA, and military payloads. (NASA Photo no. C1962-60072)

of Centaur. He called the approach that General Dynamics had taken, imaginative, but risky. "In order to save a few pounds, they have elected to use some rather, shall we say marginal solutions where you are bound to buy a few headaches before you get it over with. Ultimately when you are successful you have a real advanced solution." When pressed for an example, von Braun referred to Centaur's pressurized tank: "This is certainly a very great advance in the art, but looking backward now I would say we certainly have bought a lot of trouble by switching to that method. It is a great weightsaver but it is also a continuous pain in the neck."[46]

Ehricke and Hansen defended the technical decisions that had determined the rocket's structure and fuel choice several days later. Asked about the difference in design philosophy between General Dynamics and Huntsville, Hansen admitted that Huntsville's was more conservative. In contrast to the 35 percent margin of safety generally favored by the von Braun team, General Dynamics followed the Air Force design practice of allowing a 25 percent margin. He explained, "We are inclined, I think, to be willing to take a little bit more of a design gamble to achieve a significant improvement, whereas I think they build somewhat more conservatively."[47]

Ehricke tackled the issue of the balloon structure. He pointed out that it was also used on Atlas, a rocket deemed reliable enough to send Mercury astronaut John Glenn into orbit around the earth. In Centaur the balloon structure optimized the unique characteristics of liquid hydrogen—its low density and extreme cold. Ehricke also strongly defended the use of liquid hydrogen fuel. "Hydrogen itself has turned out to be less of a culprit than many thought initially," he said. "Hydrogen, like all chemical fluids behaves fine if you know its little 'idiosyncrasies' and treat it correctly. But you have to go through a development program such as ours first."[48] If the nation wanted to fly missions to the planets and beyond, NASA's launch vehicle program needed upper stages powered by high performance fuels. In addition to pioneering a new propulsion technology, Centaur was breaking new ground in other technical areas, such as restarting the vehicle after an extended coast in zero gravity.[49] In Ehricke's view, lack of a DX-priority rating, the highest classification in terms of national security, was holding back Centaur development. "Like other space programs, Centaur was to grow as it went along," he said. "Yet, unlike Mercury and Saturn, Centaur could never attain DX-priority although its contributions are as much a cornerstone of our future space capability as those of a manned space capsule and a high thrust booster."[50]

Between the May hearings and October 1962 NASA Headquarters reached the decision to transfer management of Centaur from Marshall to Lewis Research Center. Several facts related to this decision are clear. First, both von Braun and Brian Sparks, Deputy Director of the Jet Propulsion Laboratory advocated canceling Centaur. On 13 September Sparks wrote a strongly worded letter to Homer Newell "regarding the deplorable situation the current Centaur program has created in the lunar and planetary efforts assigned to this Laboratory." He recommended "immediate replacement of the Atlas/Centaur with the SaturnC-1/Agena D." On 20 September von Braun "clearly, strongly, and unequivocally" also recommended this solution to the Centaur dilemma. In his letter of 21 September Brian Sparks again emphasized the need to replace Atlas/Centaur with Saturn C-1/Agena D. He wrote, "I feel compelled to emphasize to you the necessity for gearing our national Lunar and Planetary programs to be strongly competitive with the Soviet Union. Timeliness and weight lifting capability are now well established advantages of the Soviet programs. There is indeed no real comfort to be found in the failures to date in the doggedly determined effort of the Soviets to achieve their lunar and planetary objectives. Our international position and our national pride can be significantly enhanced by the earliest possible exploitation of the C-1/Agena D."[51]

It is also clear that NASA Headquarters had no intention of honoring this request. Indeed, the decision had already been made and the transfer to Lewis Research Center in progress. Abe Silverstein, who had played a strong role as advocate of liquid hydrogen at Headquarters as Director of Space Flight Programs, had recently returned to the Center as Director.[52] He agreed to take over management of the project. Bruce Lundin, later in charge of launch vehicle operations at Lewis, recalled that when Vince Johnson called the two of them to Headquarters to consider the taking over the troubled program, they were not enthusiastic: "Abe said, 'Well someone's got to do it,' and I said, 'Well, we'll do our best.'"[53] Shortly after the transfer, the decision was replayed in the press. *Washington Evening Star* reporter William Hines wrote, "In its failure to meet deadlines, specifications and cost estimates, Centaur has poorly qualified for the title of Space Age Turkey No. 1." He called the transfer "an action that smacked of desperation" and scoffed at the idea of granting DX-priority to Centaur.[54]

Three years later Centaur had yet to prove itself. During the 1966 NASA budget authorization hearings, Representative Joseph Karth and others were impatient with the results. They were unhappy about the enormous costs of the program—now about $600 or $700 million. They thought the price of the rocket reflected NASA's dependency on General Dynamics as a sole source supplier. Implicit in their line of questioning was the issue of Centaur's unique design—the thin-skinned balloon structure and the intermediate bulkhead, in particular. Karth pointed out that liquid hydrogen could no longer be considered the culprit, since the Saturn V had high-energy upper stages.[55] (These stages had the solid construction that von Braun favored.) This criticism of Centaur by Congress betrays a lack of understanding and appreciation of Centaur's real significance for it is doubtful that hydrogen-fueled upper stages for Saturn would have been developed without the pioneering work on the fuel by General Dynamics and Lewis Research Center.

What remains to be explained is why NASA Headquarters was unwilling

Image 8-4: Viking 1 was launched by a Titan-Centaur rocket from Complex 41 at Cape Canaveral Air Force Station on 20 August 1975 to begin a half-billion mile, eleven-month journey through space to explore Mars. The 4-ton spacecraft went into orbit around the red planet in mid-1976. (NASA Photo no. 75PC-0443)

to permit the cancellation of Centaur in 1962 after the failure of the first launch. According to Erasmus Kloman, a management consultant who analyzed the management of the Surveyor program in the 1970s, this decision was intended as a "rebuke" to Marshall and the JPL and an assertion of the authority of Headquarters to determine the nation's launch vehicle needs. "Headquarters, after carefully reviewing the situation, confirmed its position that the Centaur concept was both technically feasible and essential to the launch vehicle pro-

Image 8-5: A Centaur upper stage is being prepared for launch in 1998. (NASA Photo)

gram for the space effort," he recalled. "It thus rejected the recommendation of senior management at MSFC and JPL. Responsibility for Centaur was transferred abruptly then to Lewis Research Center. This was interpreted as a rebuke to MSFC and a signal to the other centers that they could not back out of major development commitments assigned by Headquarters."[56] Headquarters assumed the burden of the high cost of development and risked the ire of Congress when things went wrong—as they continued to do through the 1960s—because it did not want to limit the country's launch vehicle options. In hindsight, that risk appears justified. Today liquid hydrogen is one of the hallmarks of American rocket propulsion technology. It is the main energy source for boosting the Space Shuttle into space. Centaur, now managed by Lockheed Martin, has proved a mainstay of America's commercial rocket fleet, though the delicate balloon structure—Charlie Bossart's pride and joy—has never been used in the design of other NASA rockets. Centaur is the legacy of two brilliant European visionaries—Bossart and Ehricke—who proposed, and saw realized, a rocket design that defied its harshest critics and outlived von Braun's mammoth and sturdy Saturn V.

Notes

1. *1960 NASA Authorization, Hearings before the Committee on Science and Astronautics*, U.S. House of Representatives, 86th Cong., H.R. 6512, April and May 1959, p. 391.
2. *House Subcommittee on Spaces Sciences of the Committee on Science and Astronautics*, U.S. House of Representatives, 87th Cong., 15 and 18 May 1962, p. 7 (hereafter cited as Centaur Program, 1962 Hearings.)
3. Centaur Program, 1962 Hearings, p. 11.
4. Glennan said in his diary, p. 23, "In truth, we lacked a rocket-powered launch vehicle that could come anywhere near the one possessed by the Soviets. And it would take years to achieve such a system, no matter how much money we spent." See J.D. Hunley, ed., *The Birth of NASA: The Diary of T. Keith Glennan*, NASA SP-4105 (Washington, D.C.: NASA, 1993).
5. See John Sloop's masterful *Liquid Hydrogen as a Propulsion Fuel, 1945–1959*, NASA SP-4404 (Washington, D.C.: NASA, 1978). Sloop's papers in the NASA Historical Reference Collection, Washington, D.C., are also a valuable source.
6. On the arsenal approach, see Michael J. Neufeld, *The Rocket and the Reich: Peenemünde and the Coming of the Ballistic Missile Era* (New York: Free Press, 1995), p. 108.
7. Sloop, *Liquid Hydrogen*, p. 208. See also G.R. Richards and Joel W. Powell, "The Centaur Vehicle," *Journal of the British Interplanetary Society* 42 (1989): 99–120.
8. Hunley, *Birth of NASA*, p. 5.
9. Erasmus Kloman, *Unmanned Space Project Management: Suveyor and Lunar Orbiter*, NASA SP-4901 (Washington, D.C.: NASA, 1972), p. 32. See also unpublished draft, chap. 2, pp. 62–64, NASA Marshall Space Flight Center Archives, Huntsville, Alabama.
10. Deane Davis, "Seeing Is Believing," *Spaceflight* 25 (5 May 1983): 196.
11. Sloop, *Liquid Hydrogen*, pp. 41, 44, 176–77.
12. Interview with K.J. Bossart by John Sloop, 27 April 1974, NASA Historical Reference Collection, Washington, D.C. For a superb article on Bossart and the balloon tank, see Richard E. Martin, "The Atlas and Centaur 'Steel Balloon' Tanks: A Legacy of Karel Bossart,"

reprint of General Dynamics Corporation 40th International Astronautical Congress paper, IAA-89-738, 1989.

13. On Project MX-774, see Jacob Neufeld, *The Development of Ballistic Missiles in the United States Air Force, 1945–1960* (Washington, D.C.: Office of Air Force History, 1990), pp. 44–50.

14. See John Clayton Lonnquest, "The Face of Atlas: General Bernard Schriever and the Development of the Atlas Intercontinental Ballistic Missile, 1953–1960," Ph.D. diss., Duke University, 1996. Also Davis Dyer, "Necessity Is the Mother of Convention: Developing the ICBM, 1954–1958," *Business and Economic History* 22 (1993): 194–209, and Davis Dyer, *TRW: Pioneering Technology and Innovation since 1900* (Boston: Harvard Business School Press, 1998), pp. 167–94.

15. Interview with K.J. Bossart by John Sloop, 27 April 1974.

16. Interview with Krafft Ehricke by John Sloop, 26 April 1974. See also Sloop's *Liquid Hydrogen as a Propulsion Fuel: 1945–1959*, pp. 191–95.

17. John Sloop, *Liquid Hydrogen*, pp. 47–48. Also interview with Krafft Ehricke by John Sloop, 26 April 1974.

18. Interview with Krafft Ehricke by John Sloop, 26 April 1974.

19. Interview with Krafft Ehricke by John Sloop, 26 April 1974.

20. Centaur Program, 1962 Hearings, p. 64.

21. Interview with Krafft Ehricke by John Sloop, 26 April 1974.

22. Statement by Deane Davis in interview by John Sloop with Convair Aerospace Division of General Dynamics, 29 April 1974, NASA Historical Reference Collection. The original design has not been located.

23. *Centaur Technical Handbook and Log (U)*, General Dynamics, 12 March 1963, "Old RL10 Records," DEB Vault, NASA Glenn, Cleveland, Ohio.

24. Sloop, *Liquid Hydrogen*, pp. 191–97; Virginia Dawson, *Engines and Innovation: Lewis Laboratory and American Propulsion Technology*, NASA SP-4306 (Washington, D.C.: NASA, 1991), p. 168.

25. Interview by John Sloop with Convair Aerospace Division of General Dynamics, 29 April 1974.

26. On the Sun Tan project, see Sloop, *Liquid Hydrogen*, pp. 154–66. Although the Sun Tan project was canceled in 1958, Lt. Col. John Seaberg convinced the Advanced Research Projects Agency to fund a project known simply as the high energy upper stage. Thus, when Ehricke submitted his proposal to ARPA, he found the agency receptive.

27. Joel E. Tucker, "The History of the RL 10 Upper-Stage rocket Engine, 1956–1980," in *History of Liquid Rocket Engine Development in the United States, 1955–1980*, ed. Stephen E. Doyle, AAS History Series (San Diego: Univelt, 1992), 13:123–51. See also Dick Mulready, *Advanced Engine Development at Pratt & Whitney* (Warrendale, Pa.: Society of Automotive Engineers, 2001).

28. Centaur Program, 1962 Hearings, p. 67.

29. Interview by John Sloop with Convair Aerospace Division of General Dynamics, 29 April 1974.

30. Centaur Program, 1962 Hearings, pp. 66–67.

31. Ibid., p. 7.

32. Interview by John Sloop with Convair Aerospace Division of General Dynamics, 29 April 1974.

33. Davis, "Seeing Is Believing," p. 196.

34. Ibid., pp. 196–98.

35. Ibid.

36. Centaur Program, 1962 Hearings, p. 97.

37. Interview by John Sloop with Convair Aerospace Division of General Dynamics, 29 April 1974.

38. See "Proposed Appendix to Centaur Review Team Report, NASA Zero-Gravity Research Program Pertaining to Centaur," 8 April 1960, Box 254, NASA Glenn Research Center Records, NARA, Chicago.

39. "Memo Vernon H. Gray to Acting Director," 17 March 1961, NASA Glenn Research Center Records, NARA 254, Chicago.

40. Joel E. Tucker, "The History of the RL 10 Upper-Stage rocket Engine, 1956–1980," in Doyle, *History of Liquid Rocket Engine Development*, pp. 123–51.

41. Ibid.

42. Centaur Program, 1962 Hearings, pp. 111 ff.

43. Ibid., p. 59.

44. Ibid.

45. Ibid.

46. Ibid., pp. 58–59.

47. Ibid., p. 95.

48. Ibid., p. 69.

49. Ibid., pp. 67–68.

50. Ibid., p. 68.

51. Letter Brian Sparks to Homer Newell, 21 September 1962, Surveyor files, NASA Historical Reference Collection.

52. On the politics concerning this demotion, see Dawson, *Engines and Innovation*, pp. 169–70.

53. Interview with Bruce Lundin by Virginia Dawson, NASA Glenn, 7 March 2000.

54. *Washington Evening Star*, 15 October 1962, Centaur files, NASA Historical Reference Collection.

55. *1966 NASA Authorization Hearings*, Subcommittee on Space Science and Applications, H.R. 3730, March 1965, pt. 3, pp. 1139–47.

56. Kloman, *Unmanned Space Project Management*, p. 32.

– 9 –

Broken in Midstride

Space Shuttle as a Launch Vehicle

Dennis R. Jenkins

The Beginning

Perhaps more so than most ideas, the Space Shuttle was the result of a truly grandiose plan. The United States was riding high on the successes of the early human space flights, particularly the Apollo Moon program, and visions of "space stations" and "space shuttles" were firmly implanted in the minds of both engineers and the public. Arthur C. Clarke and Stanley Kubrick furthered these ideas to the music of the Blue Danube in the movie *2001: A Space Odyssey*. The future was bright.

Surprisingly, launch vehicles had evolved remarkably little in the twenty years between the launch of the first Jupiter C and the development of the Space Shuttle. In fact, most of the launch vehicles in use during 1970 could directly trace their roots back to the first generation of military missiles developed in the late 1950s and early 1960s—Delta was a development of the Thor IRBM, Atlas was a variant of the Atlas ICBM, and the Titan III owed much to the Titan ICBM. Only NASA's Saturn family had been developed from the ground up as space launch vehicles instead of military missiles, and they were already out of production.

Granted, each of the boosters had improved dramatically over the years since they had first been developed, but their heritage was clearly evident in how they were manufactured, processed, and launched. All of these boosters were "expendable"—meaning that the entire launch vehicle either followed the payload into space or dropped into the ocean. It was an expensive way to do business, or so it seemed.

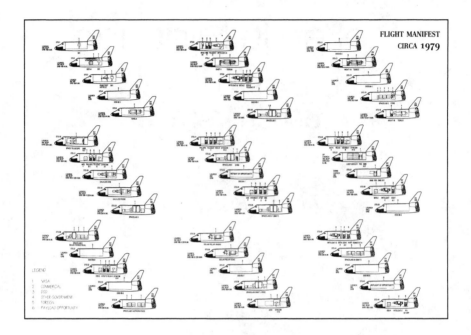

Image 9-1: Proposed Space Shuttle flight manifest, 1979. (Dennis R. Jenkins Collection)

An Idea Emerges

Many people likened the idea of using expendable launch vehicles to throwing a car or airplane away after every trip. Engineers, always looking for a more elegant answer to every problem, determined that if the launch vehicle could be made reusable, the cost of getting to space would necessarily come down. The logic appeared sound on the surface—only later would it become evident that designing a reusable spacecraft presented some enormous problems.

With an idea in hand, the U.S. Air Force tasked the aerospace companies with coming up with the quickest way to develop a reusable vehicle as part of a study under SR-89774. This study began in 1957 and similar studies continued through 1965. Contracts were issued to several contractors to determine the feasibility of a reusable space launch vehicle (RSLV)—each focusing on a different existing or proposed booster. In all cases, the contractor was to examine the minimal modifications needed to permit winged, fly-back recovery with a horizontal landing.[1]

Boeing concentrated on the Saturn family of boosters being developed by NASA. Interestingly, similar work was also accomplished by Boeing for NASA under contract NAS8-5036 through the end of 1962. Martin and North American also investigated reusable Saturns as part of the NASA studies. The Boeing study concluded that adding wings to the first stage of the Saturn C-5 was feasible and could result in a fully reusable fly-back first stage by 1970. The concept was considered economically feasible given a launch rate of twenty-four per year, but resulted in a 22 percent reduction in payload capacity to low-Earth orbit. One interesting item from the later NASA-sponsored studies is that Boeing situated the proposed KSC runway for the booster in roughly the same location as the shuttle landing facility would be built a decade later.[2]

The Atlas booster was the subject of studies at Convair. Although initiated as part of SR-89774 in 1957 these efforts continued through at least the end of 1965 as part of ongoing NASA space station resupply and Air Force recoverable booster studies. Martin Marietta was asked to investigate modifications to the Titan II booster, and over twenty thousand man-hours were invested in a configuration that ultimately included a large set of aft-mounted wings, forward canards, and a pair of turbojet engines in nacelles mounted about midbody.[3]

These studies all concluded that reusable boosters could substantially reduce the cost of putting payloads into orbit, assuming a high (some predictions were as high as 50 per year) launch rate. This was encouraging since some factions within the Air Force wanted a permanent manned presence in space, and others wanted to use space as a rapid way to get supplies from one point on the Earth to another (antipodal delivery). The budding national intelligence community was also interested in reducing the cost of launching future spy satellites. But adapting an existing booster to be reusable was not considered optimal since the construction and technology still represented the old way of doing business.

There had, of course, been earlier studies and proposals for reusable vehicles. As early as 1928 Eugen Sänger in Germany had proposed a rail-launched antipodal aircraft dubbed Silverbird. Wernher von Braun had proposed a winged "ferry rocket" during the early 1950s as part of his ambitious plan to go to Mars. On the Air Force side, the abortive Aerospaceplane studies during the late 1950s and the X-20 Dyna-Soar during the early 1960s also represented reusable spacecraft. But with the exception of the Dyna-Soar, none of these represented anything more than interesting glimpses into the future, and Dyna-Soar was canceled before it flew.[4]

One of the earliest (1961–65) serious proposals for a reusable spacecraft can be found in a Martin Marietta two-stage-to-orbit (TSTO) concept called Astrorocket. With a gross liftoff weight (GLOW) approaching 2,500,000 million pounds, a large delta-winged booster with a high mounted wing carried a small orbiter with a low mounted delta-wing to a staging velocity of 8,800 feet per second (fps). The three-man orbiter could remain in orbit for up to two weeks, while the fly-back booster used large turbojet engines to return directly to its landing site. The integrated vehicle was to be launched vertically from the Cape Canaveral Air Force Station (CCAFS) in Florida, and both stages returned to horizontal landings at the Cape.[5]

Astrorocket reflected the consensus in the technical community that any advanced launch vehicle should be fully reusable and be more of an "aircraft" than previous boosters. The general configuration of Astrorocket was similar in many respects to the multitude of designs that would follow it during Air Force and NASA studies on the way toward developing a space shuttle. But although it gave a fuzzy look into the future, Astrorocket was never seriously considered for further development since nobody believed that the available technology could actually build the vehicle at a reasonable expense.

During the early 1960s, NASA's Manned Spacecraft Center (MSC)[6] and Marshall Space Flight Center (MSFC) joined together in several joint study efforts on vehicles that greatly resembled space shuttles. Seven large aerospace companies participated in these studies, including Boeing, Convair, Douglas, Lockheed, Martin Marietta, McDonnell, and North American Rockwell. The studies did not try to actually design an operational vehicle but focused on evaluating whether the state-of-the-art existed to *attempt* development of a fully reusable spacecraft. The outcome was encouraging enough that the NASA ad hoc Committee on Hypersonic Lifting Vehicles endorsed the development of a TSTO "shuttle-craft" in June 1964.[7]

Even as Apollo flew toward the Moon, teams at each of the NASA space centers and all of the major aerospace contractors began investigating concepts for huge fully reusable two-stage vehicles designed to dramatically lower the cost of getting into space. The idea was that if you did not "throw anything away," like all previous launch vehicles, then the vehicle could be operated more like an airliner, with the attendant reduction in costs—perhaps as low as $100 per pound. This compared to about $1,000 per pound for the expendable vehicles then in use. The promise was to achieve routine access to space, and launch rate estimates of 50–100 missions per year were common; more fantastic ones were in the 200–300 range.

In late 1968, the Air Force Flight Dynamics Laboratory (AFFDL) contracted with Lockheed and McDonnell Douglas to investigate various space shuttle concepts. This study included the first applications of the "stage-and-a-half" concept that had first been advocated during a 1965 Air Force reusable launch vehicle (RLV) study. With this concept, fuel was moved into expendable drop tanks, allowing a much smaller and lighter orbiter. This began the Integral Launch and Reentry Vehicle (ILRV) System effort, a phrase that would be discontinued by the Air Force after NASA began the Phase A Space Shuttle studies under the same terminology.[8]

On 30 October 1968, the Manned Spacecraft Center and the Marshall Space Flight Center issued a joint request for proposals (RFP) for an eight-month study of an Integral Launch and Reentry Vehicle (ILRV). Both centers emphasized the fact that this RFP would not necessarily result in an actual

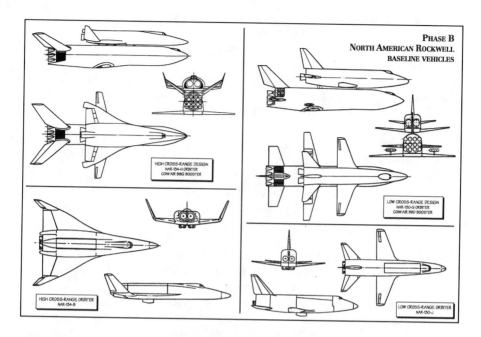

Image 9-2: North American Rockwell Phase B baseline vehicle design for a fully reusable two-stage rocket, 1967. (Dennis R. Jenkins Collection)

development contract being awarded and stated that the studies should concentrate on "economy and safety rather than optimized payload performance." All of the studies were to be based on the anticipated requirement to place between 5,000- and 50,000-pound payloads into low-Earth orbits between 115 and 345 miles, with 300 miles being the baseline for comparisons. A cross-range of 450 miles was required for safety considerations, although it was generally known that certain factions (namely, the Air Force) desired a greater capability. Five companies, the Convair Division of General Dynamics, Lockheed, McDonnell Douglas, Martin Marietta, and North American Rockwell expressed an interest in participating in the study, which was Phase A of the Space Shuttle design and development cycle.[9]

In NASA's new four-phase process, Phase A was labeled advanced studies, Phase B was project definition, Phase C was vehicle design, and Phase D was production and operations. As the Space Shuttle program progressed this changed slightly, with Phase A later being termed preliminary analysis instead of advanced studies, and Phases C and D being combined. The rationale behind this phased project planning (PPP) approach, which had been formally adopted by NASA in August 1968, was to foster competition throughout the development process, gradually reducing the number of competitors as the phases advanced. It was felt that this would leave only those contractors that could reasonably be expected to succeed to complete the task by the time Phase C (design) contracts were awarded, and that one ultimate winner would be selected for Phase D.

In February 1969, NASA awarded study contracts to General Dynamics/ Convair, Lockheed, McDonnell Douglas, and North American Rockwell for Phase A studies, although Martin Marietta also participated using company funds. The rationale for not accepting Martin's proposal is unclear but was most probably based on concerns over the proposed vehicle's unusual configuration (see fig. 003). In any case, NASA had always maintained that only four contractors would participate in Phase A, with two continuing onto Phase B. Langley monitored[10] the McDonnell Douglas contract, while MSFC monitored Lockheed and General Dynamics and MSC performed the same tasks for North American's efforts. Additionally, the Air Force's Space Division and The Aerospace Corporation functioned in oversight roles, examining NASA's activities and coordinating these studies with their own in-house efforts and the ongoing stage-and-a-half studies being conducted by the FDL.[11]

In the meantime, the Space Shuttle Task Group (SSTG) headed by Leroy E. Day[12] was evaluating potential uses for the vehicle that would eventually emerge from the development process. In July 1969 the SSTG issued its final report, concluding that the ILRV-class vehicles should be capable of performing six different types of missions:[13]

- Logistical support of a low-Earth orbiting space station, including ferrying personnel, supplies, and propellants to the space station, and experiment data and manufactured products from the station back to Earth
- Placement and retrieval of satellites into low-Earth orbit
- Propellant delivery to spacecraft in low-Earth orbits
- Low-Earth orbiter satellite servicing and maintenance
- Delivery of other payloads to low-Earth orbits
- Short (one to three days) duration human orbital research missions

The concept of a fully reusable space transportation system to make access to low-Earth orbit more routine and less expensive finally began gaining momentum during 1969 as part of NASA's ambitious post-Apollo plans. These plans were fueled by the astounding success of the first lunar landing in July, and rapidly grew to proportions unthought-of, even during the peak of the "space race." When the Vice President's Space Task Group (STG) report was released in September 1969, it had offered a choice of three long-range space plans:[14]

- An $8–10 billion per year program involving a manned Mars expedition, a manned space station in lunar orbit, and a fifty-person Earth-orbiting space station serviced by a reusable space shuttle

- An intermediate program, costing less than $8 billion per year, that deleted the lunar-orbiting space station but kept the other elements
- A relatively modest $5 billion per year program that would embrace an Earth-orbiting space station and a space shuttle

When President Nixon rejected all of these proposals, NASA was left with the task of identifying a project that could both gain enough political support for approval, and still be sizable enough to keep its development engineers and contractors usefully occupied during the 1970s. The space shuttle emerged as that project because it was the logical first step to implementing any future plans for a space station. It was also advertised as an "economical" way to continue sending man into space, a feature that was very important during the budget conscious period that saw the Vietnam War and a major economic recession.

According to the President's Science Advisory Committee (PSAC), twelve different launch systems could be replaced "with a STS [Space Transportation System] used jointly by both DoD [Department of Defense] and NASA as a national transportation system capability." The space shuttle would also offer "provision for national security contingencies by the ready availability of transportation to orbit on short notice, with sufficient maneuverability and cross-range capability for a variety of missions."[15]

Once it had decided that the Space Shuttle was its top priority program, NASA Headquarters unsuccessfully attempted to get the project approved by the White House during the fall of 1970. A new NASA Administrator, James C. Fletcher, was appointed in early 1971 and was determined to get the shuttle initiative approved that year. NASA calculated that in order to get White House and congressional approval to develop the new Space Transportation System they would have to significantly reduce the cost of access to space, and also get the DoD to commit to use the system for all of its launch needs once it became available. Thus, providing a vehicle that met all of the DoD requirements became a key element in the strategy for program approval. In addition, developing easy access to space was just the first step in NASA's ultimate goal, the development of a permanent manned space station.[16]

The *Only* Launch Vehicle

NASA knew that in order to get White House and congressional approval to develop the new Space Transportation System they would also have to get the Department of Defense to commit to use the system for all of its launch needs once it became available. Thus, providing a vehicle that met all of DoD's requirements became a key element in the strategy for program approval. In addition, developing easy access to space was just the first step in NASA's ultimate goal: the development of a permanent manned space station.

During 1970 the space shuttle concept was formally designated as the "Space Transportation System,"[17] and shortly thereafter Secretary of the Air Force Robert Seamans and Administrator Thomas O. Paine of NASA established the joint USAF-NASA Shuttle Coordination Board, also known as the STS Committee. The intended goal of the Space Transportation System was to provide "an economical capability for delivering payloads of men, equipment, supplies, and other spacecraft to and from space by reducing operating costs an order of magnitude below those of present systems." The committee would continue to review the shuttle program and recommend various changes in direction that would ensure the system would meet the operational needs of both DoD and NASA.

While meeting DoD requirements implied a large, high-performance space shuttle, NASA also had to meet its commitment to the Office of Management and Budget (OMB) to make access to space more economical. The Space Shuttle was the first NASA manned space program to be subjected to formal economic analysis and requirements. After an internal study by NASA failed to impress the OMB, contracts were issued in June 1970 to The Aerospace Corporation for estimating payload and launch vehicle costs, to the Lockheed Missiles and Space Company for analyzing the impacts of shuttle capabilities on reducing payload delivery costs, and to Mathematica for a comparison of total costs of carrying out likely future NASA, DoD, and commercial space missions using the shuttle and other launch vehicles. NASA Deputy Administrator George Low later commented that the Mathematica analysis' impact on the Space Shuttle decision was "influential and unfortunate." This was because although the analysis concluded the Space Shuttle could indeed significantly reduce the costs of launching payloads, the existence of an official study making such claims forced NASA to publicly maintain that the Shuttle was a good investment on economic grounds.[18]

It became clear, even during NASA's early in-house analyses, that any economic justification depended crucially on the Space Shuttle being the only U.S. launch vehicle during the 1980s and later. In particular, it was crucial for NASA to gain agreement from the national security community to use the Space Shuttle to launch all military and intelligence payloads, which were projected to be roughly one-third of all future space traffic. Thus DoD support was crucial on both political and economic grounds.

Accommodating all potential DoD missions required an orbiter that could handle payloads up to 60 feet long and could launch 40,000 pounds into polar orbit or over 65,000 pounds into a due-east orbit. Of the critical design parameters of the Space Shuttle, only the maximum payload width of 15 feet was based primarily on a potential NASA requirement, this being a projection that any future space station would be built using modules of that diameter. But even though DoD requirements drove several important aspects of the Space Shuttle design, it was not clear how strong military interest really

was. Robert Seamans saw "no pressing need" for the shuttle but characterized it as "a capability the Air Force would like to have." Thus, although it
drove several significant design requirements, the Air Force was still not totally committed to the concept of a common space transportation system and
stated that it would continue to develop and purchase its own expendable
(Titan and Atlas) boosters.

The Department of Defense also did not want to contribute any substantial amount of money to the Space Shuttle development effort, other than
funds to build a launch complex at Vandenberg AFB. Even then, the Air Force
was not sure when the complex would be built, or how many DoD flights a
year would be launched from it. In the end, DoD also contributed the production facilities at Air Force Plan 42 in Palmdale, California, where the orbiters
were manufactured.

In 1971 the Air Force finally agreed not to develop any new expendable
boosters, although they would continue to purchase existing designs, at least
until shuttle proved itself in operation. Leaders of the national security establishment (DoD, CIA, NSA, etc.) communicated their support of the Space
Shuttle program both to the White House and to Congress later in 1971. In
reality, except for a few new-build Titan IIIs and some surplus Atlas ICBMs
that were being converted to launch vehicles, the production of expendable
boosters would essentially cease in the early 1980s. Since the government
essentially controlled all expendable launch vehicles, this commitment to transfer all launches to the Shuttle would also effect any commercial companies
desiring access to space. The decision to replace all launch systems with the
Space Shuttle would have a significant impact to everyone desiring access to
space after the *Challenger* accident in 1986 grounded the Space Shuttle fleet
for almost three years.

Phase A Conclusions

The Phase A studies, supported by more than two hundred man-years of engineering effort backed up by extensive wind tunnel testing, materials evaluations, and structural design, resulted in the evaluation of four basic baseline
vehicles. These included straight-wing, delta-wing, and stowed-wing/variable-
geometry-wing vehicles as well as lifting bodies. It was found that the powerful boost engines proposed by MSFC did not necessarily provide the propulsion
capabilities needed for the vehicles being investigated since the majority of
the contractors opted for a less powerful alternate engine design. This was
partially because of their lighter weight but also as a result of abort profile
studies that indicated that a larger number of less powerful engines were pref-
erable over a smaller number of more powerful engines during emergencies.[19]

The Phase A studies, combined with the Air Force ILRV studies, confirmed
that cross-range was the major sticking point when trying to consolidate the

Air Force requirements with the needs of NASA. The Air Force wanted the Space Shuttle to be able to land at its launch site after only one orbit. During this time the Earth had rotated approximately 1,265 miles, meaning the returning spacecraft had to be able to fly at least that distance during reentry. On the other hand, NASA simply wanted an opportunity to land back at the launch site only once every twenty-four hours, requiring a relatively modest 265 miles in cross-range capability. However, many emergency scenarios required at least 450 miles cross-range to enable a return-to-launch-site abort capability, leading to most designs providing at least that amount of cross-range.

The lifting-body concept was found to be the most poorly suited to space shuttle applications, primarily because the body shape did not lend itself to efficient packaging and installation of a large payload bay, propellant tanks, and major subsystems. The complex double curvature of the body resulted in a vehicle that would be difficult to fabricate, and further, the body could not easily be divided into subassemblies to simplify manufacture. Also, the lifting body's large base area yielded a relatively poor subsonic lift-to-drag ratio (L/D), resulting in a less attractive cruise capability.

The variable-geometry designs were found to have many attractive features, including low inert/burnout weight and the high hypersonic L/D needed to meet the maximum cross-range requirements. In addition, the stowed-wing approach permitted the wing to be optimized for the low-speed flight regime. Drawbacks included a high vehicle weight to projected planform area ratio, which would result in a higher average base temperature relative to either the straight- or delta-wing designs. In addition, significantly increased design and manufacturing complexity would result from the mechanisms required to operate the wing and transmit the flight loads to the primary structure. The maintenance required between flights was expected to be high, and insufficient data existed to reliably determine potential failure modes which were thought to be numerous.

Maxime A. Faget at MSC was not a believer in lifting reentry, despite his proposed design having wings—he still held to the idea of a high-drag blunt body. To operate his design as a blunt body, Faget proposed to reenter the atmosphere at an extremely high angle of attack with the broad lower surface facing the direction of flight. This would create a large shock wave that would carry most of the heat around the vehicle instead of into it. The vehicle would maintain this attitude until it got below 40,000 feet and about 200 mph, when the nose would come down and it began diving to pick up sufficient speed for level flight. The vehicle would then fly to the landing site, touching down at about 140 mph. Since the only "flying" was at very low speeds during the landing phase, the wing design could be selected solely on the basis of optimizing it for subsonic cruise and landing—hence the simple straight wing proposed by Faget, and used on the North American Phase A design and the MSC-designed DC-3. The design did have one major failing, at least in the eyes of the Air Force: since it did not "fly" during reentry, it had almost no cross-range capability.[20]

Faget had convinced many that his simple straight-wing concept would be more than adequate. But not everybody agreed. In particular, Charles Cosenza and Alfred Draper at the Air Force Flight Dynamics Laboratory (AFFDL) did not accept the idea of building a shuttle that would come in nose high, then dive to pick up flying speed. With its nose so high the vehicle would be in a classic stall, and the Air Force disliked both stalls and dives, regarding them as preludes to an out-of-control crash. Draper preferred to have the shuttle enter a glide while still at hypersonic speeds, thus maintaining much better control while still continuing to avoid much of the severe aerodynamic heating.[21]

But if the shuttle was going to glide across a broad Mach range, from hypersonic to subsonic, it would encounter another aerodynamic problem: a shift in the wing's center of lift. At supersonic speeds the center of lift is located about midway down the wing's chord (the distance from the leading to the trailing edge)—at subsonic speeds it moves much closer to the leading edge. Keeping an aircraft in balance requires the application of an aerodynamic force that can compensate for this shift. Another MSC design, the Blue Goose, accomplished this in an extreme manner by translating the entire wing fore-aft as the center of lift changed.

The Air Force had extensive experience with this phenomena and had found that a delta planform readily mitigated most of the problem. Draper proposed that both stages of any reusable shuttle should use delta wings instead of straight ones. Faget disagreed, pointing out that his design did not "fly" at any speed other than low subsonic—at other speeds it "fell" and was not subject to center-of-lift changes since it was not using lift at all.[22]

To achieve landing speeds low enough to satisfy Faget, a delta wing would need to be very large since deltas tend to land "hot" under the best of circumstances. A straight wing with a narrow chord, argued Faget, would be light and have relatively little area that needed a thermal protection system. To achieve the same landing speed, a delta would be large, add considerable weight, and greatly increase the area that required thermal protection.

Draper argued that delta wings had other advantages. Since it was relatively thick where it joins the fuselage, delta wings offered more room for landing gear and other systems that could be moved out of the fuselage. Its sharply swept leading edge produced less drag at supersonic speeds, and its center of lift changed slowly compared to a straight wing. But the delta offered one other advantage—one that became increasingly importantly as the military became more interested in using space shuttle. Compared to a straight wing, a delta produces considerably more lift at hypersonic speeds. This allowed a returning shuttle to achieve a substantial cross-range during reentry, a capability highly prized by the Air Force.

Not accepting Faget's position that his vehicle did not "fly" at high speeds,

Draper argued that any straight wing concept would require additional aerodynamic surfaces (called canards) on the forward fuselage to compensate for the shifting center of lift. Canards produce lift of their own, and this would cause the main wings to be mounted farther rearward, where they would probably disturb the center of gravity. It was a complicated problem, and one that would take at least another year to solve.

To help resolve the controversy, the two concepts baselined for study during Phase B included a Faget-style straight-wing low cross-range orbiter, and a Draper-supported delta-wing high cross-range orbiter. The straight-wing orbiter would be configured to provide design simplicity, low weight, decent handling, and good landing characteristics. The vehicle would be designed to enter at a high angle of attack to minimize heating and to facilitate the use of a heat shield constructed from materials available in the early 1970s. The delta-wing orbiter would be designed to provide the capability to trim over a wide angle-of-attack range, allowing initial reentry at a high alpha to minimize the severity of the reentry environment, then transitioning to a lower angle of attack to achieve a higher cross-range. Phase B would use these two government-provided concepts as departure points for continued evaluation and subsystem studies.

Phase B

NASA issued two Phase B contracts on 6 July 1970 for continued study of the Integral Launch and Reentry Vehicle. The contracts were based on an evaluation of the Phase A results, and the companies' responses to the Phase B Statement of Work (SOW) issued in February 1970. The major program-level requirements established for Phase B by the Shuttle System Specification, dated 1 June 1970, included[23]

- the shuttle shall be a fully reusable two-stage vehicle,
- the integrated vehicle shall be launched vertically, and landed horizontally,
- the Initial Operational Capability shall be in late 1977,
- the reference mission shall be a 310-mile 55-degree inclination circular orbit from a launch site located at 28.5 degrees north latitude (KSC),
- the payload bay shall have a clear volume 15 feet in diameter by 60 feet in length, with a reference mission capacity of 15,000 pounds,
- the booster and orbiter shall have a go-around capability during landing operations [basically implying the use of air-breathing engines],
- launch rates will vary from 25 to 75 per year,
- total turnaround time from landing to launch shall be less than two weeks, and

- a 43-hour turnaround capability shall be provided for rescue missions.

The two contractors selected to compete during Phase B included a team consisting of McDonnell Douglas and Martin Marietta, and North American Rockwell, who was joined by Convair as a risk-sharing subcontractor. This phase was to involve a more detailed analysis of mission profiles and a preliminary look at the vehicles needed to complete them.

Both contractors would start with the baseline vehicles that were selected as a result of the Phase A studies. These designs included a straight-wing low cross-range orbiter that was based heavily on the MSC-002 (DC-3) that Maxime Faget had designed at the Manned Spacecraft Center, although by this time it was becoming increasingly evident that the straight-wing design suffered some serious handicaps during reentry and hypersonic flight. The other design was a high cross-range orbiter that used a severely clipped delta wing, and was thought to be superior during the reentry phase but at this point had a very poor payload fraction (total payload versus total vehicle weight). NASA did not specify any details for the booster stage, leaving the configuration of these entirely to the competing development teams.

One of the first steps in the Phase B studies would be to "size" the vehicles—determining the boost/payload performance, the optimum staging velocity, the thrust-to-weight ratio at liftoff, the number of engines, and so on. Economics was continuing to play a major role in the design of the Space Shuttle, but the primary consideration was beginning to shift from providing the most economical long-term operational system to reducing the near-term development costs.[24]

Alternate Concepts

In the simplest form, there are two major ways to design a fully reusable launch vehicle. The first is known as "single-stage-to-orbit" (SSTO) and is defined as a single vehicle that takes off and recovers intact. It is both the booster and orbital vehicle. The second is a two-stage-to-orbit (TSTO) concept in which two completely reusable vehicles are used. In general, a large hypersonic vehicle carries a smaller orbital vehicle through the densest parts of the atmosphere, then launches ("stages") the orbiter to continue into space— think of the X-15 being launched by the NB-52 as an example. Both vehicles recover independently at their landing sites. There are, of course, many variations to this, such as the "trimese" concepts investigated by McDonnell Douglas that used two recoverable boosters and a reusable orbiter.[25]

One of the great drawbacks to fully reusable configurations, such as the abortive SSTO Aerospaceplane from the late 1950s,[26] had been the extremely high research, development, test, and evaluation (RDT&E) costs inherent in their development and procurement. Because of this trend, maneuvering orbital reentry vehicle designs had always tended to be small, as evidenced by Dyna-Soar during the early 1960s. Thus large concepts such as Aerospaceplane surprised nearly no one when they fell by the wayside.

In a 1968 AIAA paper, AFFDL's Alfred C. Draper and Charles Cosenza had argued convincingly that partially expendable concepts such as stage-and-a-half would make larger orbiters more attractive, freeing up internal volume for payload and thus not requiring the development of vehicles large enough to carry their engines, propellants, and payload internally within a fully reusable structure. The downside was that these designs continued to "throw away" some parts—usually the propellant tanks. Although an economic case could be made for these concepts, they were not considered "elegant" by many and were generally either ignored or discouraged by NASA.

The Alternate Space Shuttle Concepts (ASSC) study was initiated on 6 July 1970 in parallel with the Phase B studies of fully reusable shuttle configurations. The intent of the ASSC study was to define and investigate various alternatives to the fully reusable Phase B TSTO concepts so that meaningful comparisons of technical concerns and issues, operational approaches, and cost factors could be made. It also allowed NASA a fallback position if space shuttle funding was reduced, which was looking more and more likely.

Contracts were issued by MSFC to the Chrysler Space Division and the Lockheed Missiles and Space Company, and MSC subsequently issued an ASSC contract to a combined Grumman/Boeing team. The initial matrix of alternate concepts to be studied consisted of twenty-nine configurations in three general categories: "fractional" stages (stage-and-a-half), semireusable (expendable booster/reusable orbiter), and fully reusable TSTO alternatives similar to the Phase B vehicles. Chrysler, however, did not feel bound by the rules and pursued a SSTO vehicle powered by a unique aerospike engine. Their inventiveness was rewarded by being totally ignored by NASA and the rest of the aerospace community.[27]

Ultimately, ASSC concluded that the fully reusable TSTO vehicles being developed under Phase B were the "best" since they gave the greatest assurance of achieving low cost per flight into orbit, and would have the lowest "total" program[28] costs. However, if peak yearly funding and/or development risk was to be minimized, consideration should be given to a phased development approach of a reusable orbiter with an expendable booster with the option of developing a recoverable booster later.[29]

A New Tact

Midway through the Space Shuttle design process, reality set in. The Phase B studies had resulted in some truly ambitious two-stage vehicles designed to meet NASA's stated preference for a fully reusable Space Shuttle. In an era when the largest winged hypersonic vehicle was the 30,000-pound X-15, it would have been an enormous leap to the 1,000,000-pound hypersonic vehicles that most of the early shuttle designs used as a first stage. All of the concepts presented by the contractors would have been expensive and contained potentially large development risks, even if NASA and the contractors were unwilling to fully admit it. As more detailed design work was accomplished, the cost estimates for developing a completely reusable vehicle soared and quickly became prohibitive. The congressional attitude toward NASA cooled and budgets shrank, at least in real terms. A new tack would need to be taken.

In May 1971, the Office of Management and Budget told NASA that it could expect to get no budget increases in the next five years. This was a drastic blow because it meant the agency could not carry out the shuttle development program it had been planning for almost two years during the ILRV studies. Those plans called for a development cost of almost $10 billion, with a peak annual budget of some $2 billion. If limited to the $3.2 billion budget approved for fiscal year 1972, the most NASA could hope to put into shuttle development, and still maintain a balanced science and application program, was roughly $1 billion a year for five years. This brought a new sense of urgency to some of the ideas examined under the ASSC contracts.[30]

Most ideas for lowering the cost of the Space Shuttle had involved replacing the fly-back booster with some sort of expendable stage. This did not appear technically feasible, however, given the large size of the orbiter. The problem was resolved in June 1971 with a decision by NASA to endorse a variation the stage-and-a-half concept originally proposed by FDL's Draper and Cosenza in 1968 and initially investigated in detail by both Lockheed and McDonnell Douglas during the Air Force's RLV and ILRV studies as early as 1965. This concept moved the large LH2 tanks outside the orbiter's airframe, and made them expendable, allowing a smaller and lighter orbiter, with a significant reduction in development costs, but with a corollary increase in costs each time the orbiter was launched. Grumman had revisited this concept during the ASSC studies, marking the first time that the MSC had officially sanctioned a partially reusable concept. Once it had gained NASA's ear, Grumman vigorously pursued the external tank concept, and the Grumman proposal manager, Larry Mead, continued to argue that it was the economically preferred method. It would be economics, not technology, that eventually convinced NASA to listen more closely to what Grumman was telling it.[31]

Although NASA planners still preferred the fully reusable approach, economic considerations, supported by outside analysis, clearly dictated the program would use some variation of the stage-and-a-half concept. Actually, the program as a whole was already leaning toward what the ASSC study had termed a "fractional" stage or "semireusable" vehicle (expendable booster/

reusable orbiter), as opposed to a true stage-and-a-half configuration in which the orbiter constituted the only stage but used expendable external propellant tanks.

On 12 September 1971, both Phase B contractors along with two of the ASSC contractors (Grumman/Boeing and Lockheed) were told to reevaluate their proposals using the NASA-developed MSC-040 orbiter and "external tank" concept. The Chrysler ASSC proposal had been too far out of the mainline program, and Chrysler elected not to continue participating in the airframe competition.

The choice of boosters was left largely to the individual airframe contractors, but recommendations included modified Saturn IB and Saturn V stages, clusters of Titan-derived stages, as well as the development of new liquid and solid boosters. The choice to make the boosters expendable or reusable also rested with the contractors. This first extension to the Phase B contract was generally called Phase B Prime.

This represented a major change from the original approach of reducing the number of contractors as the phases progressed. The four contractors would now compete on a theoretically equal basis for the Phase C/D development and production contract. The desire for an increased payload capacity led to the development of the MSC-040C powered by three new high-chamber-pressure engines that would be used as the starting point by the contractors in their final Phase B and ASSC presentations, as well as the proposals for Phase C/D.

The Phase B Prime studies contained potential space transportation systems that could be designed, developed, and operated with significantly lower RDT&E program costs and reduced peak annual funding than those that had been defined during the initial Phase B studies. However, the costs for these vehicles still exceeded both the acceptable RDT&E costs, and the peak annual funding figures desired by NASA and the Congress. Phase B Prime and the associated ASSC extension were scheduled to terminate on 30 October 1971, concurrent with the release of the Phase C/D request for proposals.[32]

But since vehicle definition was still ongoing, NASA issued $2.8 million contracts to Grumman/Boeing, Lockheed, McDonnell Douglas/Martin Marietta, and North American Rockwell to continue reporting the results of various study efforts. These were collectively known as Phase B Double Prime, even though two of the contractors had not technically participated in Phase B. The primary "open technical issue" at this point involved the booster configuration, and the emphasis shifted to three alternate concepts: a fully recoverable booster using a new development pressure-fed liquid-fueled engine; a flyback pump-fed liquid-fueled booster based on the Saturn V first stage; and several configurations using various combinations of solid rocket boosters (SRB). The Phase B Double Prime efforts began on 1 November 1971 and would continue through 15 March 1972, the expected Phase C/D contract award date.

Several different configurations were advanced for the flyback pump-fed liquid-booster concept. The majority concentrated on the maximum use of components developed for the S-IC first stage of the Saturn V, including the basic tank structure. Attached to this were either a moderately swept wing with wing tip mounted vertical stabilizers or a delta wing with canard surfaces and a central vertical tail. These designs used the 1,522,000-lbf sea level (1,748,000-lbf vacuum) Rocketdyne F-1 engine originally used on the S-IC, although these engines were not considered reusable and would have needed replaced after every flight. The majority of the designs utilized the same five-engine layout as the original Saturn stage. Both manned and unmanned versions of this booster configuration were investigated, and it was expected that the booster alone would weigh in excess of 3.5 million pounds fully loaded. The orbiter and external tank were mounted on the nose of the booster. This all resulted in a truly enormous vehicle, rivaling some of the early Phase A/B designs in sheer size, although development costs were substantially less. The inability to reuse the relatively expensive F-1 engines was a serious economic drawback.[33]

The pressure-fed liquid booster was an MSFC-led attempt to further reduce costs by employing a relatively simple design approach with a minimum number of components. In the pressure-fed design, all of the engine turbo machinery was eliminated and replaced by a large pressurization system to provide the propellants at a usable pressure to the engines. This resulted in a lighter, and simpler, but much less efficient engine installation. Two different configurations were studied in detail, the first consisting of a 32.6-foot-diameter booster that was 159 feet long with the orbiter and ET mounted on the top (nose). This stage had seven 1,350,000-lbf engines and a gross liftoff weight of 5,791,000 pounds. The other design used three stages side by side with the orbiter and external tank mounted on the top of the middle booster. Each of the three stages had four 1,048,000-lbf engines, and the integrated vehicle weighed 5,279,000 pounds at liftoff.

A set of 120-inch-diameter solid rocket boosters had the highest cost of all the proposed designs, and was quickly eliminated from further evaluations. But a dual 156-inch-diameter solid rocket booster configuration was much more economical and held great promise, although the use of solids was disliked, and discouraged, by MSFC. Using a pair of 156-inch solids resulted in a gross liftoff weight of 4,898,000 pounds. The exhaust nozzles contained a noticeable precant to ensure the center of thrust went through the integrated vehicle's center of gravity and the nozzles could not be gimbaled in flight. Aft thrust termination ports were provided to reduce the thrust for abort planning. The SRBs contained 2,740,000 pounds of solid propellant and burned for 130 seconds, providing 2,939,000 lbf at sea level.

Further cost comparisons by the contractors between the pressure-fed liquid booster, the 156-inch solid, and the Saturn-based pump-fed liquid booster showed that the liquid systems had the lowest per flight costs. But the 156-

inch solids were by far the cheapest to develop. This cost analysis showed that
156-inch solid rocket boosters would cost $3.7 billion to develop, with a total
system cost of $10.8 billion, resulting in a cost of $228 per pound to orbit. The
pressure-fed liquid-booster system would cost $4.6 billion to develop, but the
total system would only cost $9.4 billion, with a pound to orbit costing $127.
The pump-fed liquid booster would put a pound into orbit for $125, after
costing $4.2 billion to develop and $8.9 billion to build. It should be noted
that the solids, although they had a higher total program cost, would require
less early-year funding than the liquid systems, but would yield higher per-
flight costs. Unfortunately, early-year funding issues were quickly becoming
the major driver in the selection process.[34]

It was felt that the liquid-propellant system was more flexible because of
the ability to tailor the thrust at almost any point in the ascent phase. However, the development risk appeared to be in favor of the solids because of
their greater simplicity and that recovery of the liquid boosters presented more
of a challenge, since they were more fragile. Ground handling of the solid
boosters was also considered easier.

Of course, life-cycle costs are greatly influenced by how often the vehicle
is used. Airlines routinely use jet aircraft for twelve to eighteen hours per day,
rapidly justifying their $100 million investment. NASA knew that to amortize
the development cost of the Space Shuttle it would have to fly a great deal
more frequently than it had ever done in the past. It also knew that the early
projections of 200–300 flights per year were completely baseless; not only was
it unlikely that the vehicle could physically accommodate such a rate, it was
hard to imagine that there were that many payloads that needed launched.

The final economic plan used to develop the operations scenario for the
shuttle was a five-orbiter fleet with a projected usage of forty flights per year
from the John F. Kennedy Space Center (KSC) in Florida and an additional
twenty flights per year from Vandenberg AFB in California. Early analyses
confirmed that forty flights from KSC could be achieved with a five orbiter
fleet if vehicle ground turnaround could be completed within 160 hours.

The fully reusable vehicle was history, a victim of its own development
costs and NASA's fiscal struggles. In its place, a reusable orbiter that would be
boosted into orbit by either liquid or solid rocket boosters and an expendable
external propellant tank. The thought was to keep the most expensive parts
(the orbiter and main engines) fully reusable and to discard the least expensive element (the ET). It was quickly becoming obvious that only the solid
rocket boosters had a development cost that could be afforded within the
budgets NASA expected to receive in the future.

Still, progress had been made toward defining a much more reasonable
space shuttle system. The two-stage fully reusable concepts were dead—and
at least Charles Donlan, an early Space Shuttle Program manager, would not
miss them: "It wasn't till the Phase B's came along and we had a hard look at

the reality of what we meant by fully reusable that we shook our heads saying, 'No way you're going to build this thing in this century.' As I say, 'Thank God for all the pressures that were brought to bear to not go that route.'"[35]

Program Approval

Monday, 3 January 1972, dawned like many before it. NASA was still unsure what sort of shuttle would be authorized for development, although they knew that the current Director of the OMB, George Shultz, privately supported the baseline NASA vehicle. In a meeting at 6:00 P.M. Shultz confirmed that he would agree to $200 million in startup funds in the fiscal year 1973 NASA budget. NASA was asked to prepare a draft of a presidential statement announcing the approval of the Space Shuttle Program.[36]

Fletcher and Low flew to the Western White House in California to meet with President Nixon before the announcement. Although the president had intended to spend only 15 minutes of a hectic day with the representatives from NASA, he ended up spending over half an hour with them. On 5 January Nixon released his formal statement:[37]

> I have decided today that the United States should proceed at once with the development of an entirely new type of space transportation system designed to help transform the space frontier of the 1970s into familiar territory, easily accessible for human endeavor in the 1980s and 90s. This system will center on a space vehicle that can shuttle repeatedly from earth to orbit and back. It will revolutionize transportation into near space, by routinizing it, It will take the astronomical cost out of astronautics. . . . This is why the commitment to the Space Shuttle program is the right next step for American to take, in moving out from our present beach-head in the sky to achieve a real working presence in space—because Space Shuttle will give us routine access to space by sharply reducing costs in dollars and preparation time.

Before the announcement, NASA had prepared a list of possible names for the new program, including Pegasus, Hermes, Astroplane, and Skylark. The White House favored Space Clipper, and this name appeared in several early drafts of the statement. But in the end, Nixon himself decided it would be better to refer to the vehicle simply as Space Shuttle.[38] This largely ended six months of maneuvering between the Office of Management and Budget and NASA over what a space shuttle should look like. Now NASA could get down to the business of actually building a reusable space vehicle.

Phase C/D

But as late as the end of November 1971, NASA was still undecided on whether the orbiter's engines should be ignited simultaneously with the booster's (the "parallel-burn" approach) or if a more traditional "staged" approach of letting the booster power the initial ascent then igniting the orbiter's engines for the final climb to orbit should be taken, although it was tending toward the parallel-burn design. Mathematica, the consulting firm conducting the shuttle economic analysis for NASA, reported that parallel-burn with an external propellant tank was the economically preferred concept, and that it could be developed for approximately $6 billion, with a cost per launch of around $6 million.[39]

In December 1971, NASA finally adopted the parallel-burn design but left open the decision of liquid or solid boosters. Total development costs were now estimated at $5.8 billion. The booster decision was finally announced on 15 March 1972, with the choice being 156-inch-diameter solids. NASA Administrator James Fletcher reported to the Office of Management and Budget that solid boosters could be developed faster and for $700 million less than an equivalent-thrust liquid booster, lowering estimated total development costs to $5.15 billion in 1971 dollars. It was a decision nobody liked, but it was the only one anybody could afford.

However, a solid booster would cost more to operate, and contained more unknowns, since NASA had little experience with this technology. This decision also placed a much larger share of the burden of paying for the Space Shuttle Program on its future users. The 1972 mission model for NASA, DoD, and other users called for some 580 flights over a twelve-year period (1979–90), an average of less than 50 flights per year. Thirty-eight percent of these flights were to be scientific research aboard Spacelab modules, 31 percent were to be in support of the DoD, and the remaining 31 percent were to be commercial satellite deliveries and space station support missions.[40]

Later (1974) models would call for over sixty flights per year, and as late as 1985, just prior to the *Challenger* accident, NASA was still predicting twenty flights per year. Using the 1972 model, launch and launch-related costs using existing expendable vehicles were estimated at $13.2 billion, while the total Space Shuttle launch costs were estimated at $8.1 billion, excluding all research and development costs. The total cost per shuttle flight was estimated at $10.5 million, or $175 per pound[41] for a full payload bay. It was expected the integrated vehicle could be brought to the pad and checked out, then placed on "standby" status for twenty-four hours at a time. An orbiter on standby would require only two hours' notice for cryogenic (LO2/LH2) loading and crew ingress prior to launch. This was a capability highly valued by the DoD, but one that would prove to be unattainable in actual operations.

Unlike the previous two phases of the Space Shuttle competition, Phase C/D would not involve government funding. An RFP was issued on 17 March 1972 to Grumman/Boeing, Lockheed, McDonnell Douglas/Martin Marietta, and North American Rockwell for their continued refinements of the MSC-040 design using an external propellant tank and solid rocket boosters. The technical proposals were due on 12 May, with the associated cost data due on 19 May.[42] See Image 9-3 for a summary of the design data developed during Phases A, B, and C.

Building and Testing

It had been a long, hard road, but finally, on 26 July 1972, NASA announced that North American Rockwell[43] had won the $2.6 billion contract to build the Space Shuttle Orbiter. The requirements included the ability to support sixty flights per year using a fleet of five Orbiters; each Orbiter was to be capable of one hundred missions.[44] The Rocketdyne Division of North American Rockwell had already been selected, on 13 July 1971, to design and build the Space Shuttle Main Engines (SSMEs). Subsequently, Martin Marietta was awarded the contract for the External Tank (ET), and USBI was selected to build the Solid Rocket Boosters, with Morton Thiokol providing the solid rocket motor segments.[45]

Things progressed quickly, and on 4 June 1974 the Orbiter *Enterprise*[46] (OV-101) was rolled out of the assembly plant in Palmdale, California. Ground testing consumed the better part of three years, but on 18 February 1977 *Enterprise* took to the air for the first time on top of a modified Boeing 747 Shuttle Carrier Aircraft (SCA). A total of eight "captive" flights would be conducted at Edwards AFB before the first of five free flights took place. On 12 August 1977, Fred W. Haise Jr. and Charles Gordon Fullerton took *Enterprise* on the first free flight, separating from the 747 at 310 mph at 24,100 feet. Just over 5 minutes later, OV-101 made its first solo landing, on the dry lake at Edwards where so much history had already been made.[47]

Now that NASA knew it was safe to fly the Orbiter on top of the 747, *Enterprise* ventured east to the Marshall Space Flight Center for additional ground tests. During early 1979 *Enterprise* would find itself at the Kennedy Space Center, providing practice to the launch team getting ready for the first orbital flight of the second Orbiter, *Columbia* (OV-102). A third airframe, STA-099, was being used as a static test article in Palmdale. Meanwhile, the other major parts of the shuttle—the ET, SRBs, and SSMEs—were undergoing development at their respective contractors.[48]

When the SSME design had been selected on 13 July 1971, it was still intended to use derivatives of the same engine on both the fly-back booster and the Orbiter. As later events played out, the fly-back boosters were abandoned in favor of the SRBs. The parallel-burn concept meant the Orbiter SSMEs would need to operate at sea level, something that had not been considered in their original design since they were meant to be ignited at altitude. This meant that Rocketdyne and NASA needed to redefine the SSME and limit the nozzle expansion area to 77.5 to 1 or less. This was held up because of a protest concerning the Rocketdyne contract and it was not until May

PHASE:	A	A	A	A	A	B	B	B
Date :	01 Nov 69	11 Nov 69	22 Dec 69	31 Oct 69	22 Dec 69	11 Dec 69	11 Dec 69	13 Nov 70
Contractor / Team :	MDC	MMC	Lockheed	Convair	NAR	MDC / MMC	MDC / MMC	NAR / GD
Design :	—	SpaceMaster	LS-112	FR-3A	—	LCR	HCR	NAR-134-C

ORBITER:

Length Overall (ft) :	107.00	181.20	163.00	179.00	202.00	147.60	171.00	192.30
Wing Span (ft) :	75.70	107.00	92.00	146.00	146.00	113.80	97.50	126.60
Height (ft) :	48.30	30.00	36.75	45.00	51.30	62.40	57.60	49.30
Wing Area (sq ft) :	—	1,812	—	—	2,830	1,375	—	—
Weight (empty – lbs) :	—	—	183,830	287,000	256,515	188,600	203,500	217,732
No. of Rocket Engines :	2	2	3	3	2	2	2	2
Thrust (Sea Level – lbs) :	415,000	415,000	415,000	415,000	510,000	415,000	415,000	415,000
No. of Air-Breathers :	4	none	4	2	4	4	4	2
Payload (pounds) :	25,000	25,000	50,000	50,000	50,000	15,000	15,000	20,000
Payload Bay Size (ft) :	15 by 30	22 by 30*	22 by 30*	15 by 60	15 by 60	15 by 30	15 by 30	15 by 60
Cross-Range (miles) :	—	1,700	1,725	1,500	1,240	230	1,726	1,726
External Tanks ? :	none	none	none	none	none	none	none	none

* 15 by 60 foot payloads could also be accommodated.

BOOSTER:

Type of Booster :	Flyback	Flyback	Flyback	Flyback	Flyback	Flyback	Flyback	Flyback
No. of Boosters :	1	1	1	1	1	1	1	1
Length Overall (ft) :	195.00	197.00	237.00	235.50	280.00	220.20	232.20	257.00
Wing Span (ft) :	151.00	148.00	200.50	203.10	244.00	151.00	151.00	142.00
Height (ft) :	69.50	58.00	56.90	47.50	—	56.90	56.90	50.00
Wing Area (sq ft) :	—	4,850	—	—	8,050	3,970	—	2,001
Weight (empty – lbs) :	—	448,369	322,550	517,000	587,522	452,800	483,900	—
No. of Rocket Engines :	10	14	13	15	10	13	14	12
Thrust (Sea Level – lbs) :	415,000	415,000	415,000	415,000	510,000	415,000	415,000	415,000
No. of Air-Breathers :	4	8	4	4	4	10	10	4

PHASE:	—	ASSC	ASSC	ASSC	Baseline	Award	Production	Production
Date :	27 Apr 70	06 Jul 71	04 Jun 71	30 Jun 71	30 Aug 71	18 Feb 72	08 Mar 79	07 May 91
Contractor / Team :	MSC	Grumman	Lockheed	Chrysler	MSC	MSC	RIC	RIC
Design :	DC-3	H-33	LS-200	SERV	MSC-040	MSC-040C	OV-102	OV-105

ORBITER:

Length Overall (ft) :	122.67	157.00	156.50	90.00 (dia)	121.88	123.75	122.17	122.17
Wing Span (ft) :	90.83	97.00	92.00	n/a	75.00	80.20	78.06	78.06
Height (ft) :	41.66	61.25	49.00	66.50	46.88	48.13	56.58	56.58
Wing Area (sq ft) :	1,175	1,060	1,229	n/a	—	2,500	2,690	2,690
Weight (empty – lbs) :	120,000	197,000	294,399	—	—	150,000	158,289	151,205
No. of Rocket Engines :	2	3	9	1	4	3	3	3
Thrust (Sea Level – lbs) :	297,000	415,000	550,000	5,800,000	325,000	375,000	375,000	375,000
No. of Air-Breathers :	6	4	6	28	0	0	0	0
Payload (pounds) :	15,000	60,000	65,000	116,439	60,000	60,000	60,000	55,250
Payload Bay Size (ft) :	8 by 30	15 by 60	15 by 60	23 by 60	15 by 60	15 by 60	15 by 60	15 by 60
Cross-Range (miles) :	200	1,265	1,500	—	1,500	1,500	975	1,085
External Tanks ? :	none	LH2	LO2/LH2	none	LO2/LH2	LO2/LH2	LO2/LH2	LO2/LH2

BOOSTER:

Type of Booster :	Flyback	Flyback	none	none	SRB	SRB	SRB	ASRM
No. of Booster :	1	1	n/a	n/a	2	2	2	2
Length Overall (ft) :	203.00	245.00	n/a	n/a	—	184.80	149.16	149.16
Wing Span (ft) :	141.00	177.50	n/a	n/a	n/a	n/a	n/a	n/a
Height (ft) :	74.00	88.30	n/a	n/a	13.00 (dia)	13.00 (dia)	12.17 (dia)	12.17 (dia)
Wing Area (sq ft) :	2,840	—	n/a	n/a	n/a	n/a	n/a	n/a
Weight (empty – lbs) :	290,000	220,135	n/a	n/a	—	150,000	154,023*	139,490
No. of Rocket Engines :	11	12	n/a	n/a	1	1	1	1
Thrust (Sea Level – lbs) :	297,000	415,000	n/a	n/a	2,500,000	2,750,000	2,940,000*	3,485,000
No. of Air-Breathers :	4	12	n/a	n/a	n/a	n/a	n/a	n/a

NOTES:
— = data not available
n/a = not applicable to this configuration

* These figures are for the original SRBs. The RSRMs weigh 150,023 pounds, and produce 3,310,000 pounds-thrust, each.

Image 9-3: Space Shuttle design criteria evolution between Phases A, B, ASSC, and C. (Dennis R. Jenkins. *Space Shuttle: The History of the National Space Transportation System—The First 100 Missions.*)

1972 that work began in earnest on the final SSME configuration. Once the Orbiter contractor was selected, definition progressed rapidly, and an interface control document was released on 9 February 1973.[49]

Main engine development testing was planned to be conducted at the NASA rocket test site in Mississippi beginning in late 1974. The Mississippi Test Facility (MTF) had been used for static testing various stages of the Saturn launch vehicle, and these facilities were modified during 1973 to accommodate SSME testing. While the modifications were underway, early SSME component testing was to be conducted at the Coca facility at Rocketdyne's Santa Susana Field Laboratory in Chatsworth, California. Various test stands at this facility needed to be modified to accommodate the turbopumps and combustion devices required for the SSME, and this work quickly fell behind, leading to a six-month schedule slip. A program realignment in the summer of 1974 led to the decision to abandon testing at the Santa Susana facility, and to conduct all tests at the National Space Transportation laboratory (NSTL, as the MTF has been renamed; now it is the Stennis Space Center). This decision would later be reversed again, and testing at Santa Susana resumed after the first engines were flight qualified.[50]

The SSME operated at much high chamber pressures than any previous liquid-fuel rocket engine—interestingly, it ran at the same chamber pressure that Eugen Sänger had predicted would be necessary for his Silverbird during the early 1930s. The new engine also incorporated "pre-burning" of the propellants to help generate a higher specific impulse. To begin obtaining test results before a complete SSME was available, Rocketdyne elected to build an integrated subsystem test bed (ISTB). This was essentially a complete engine assembly that was not built to flight-hardware specification, that is, it was larger and heavier but would provide for proof-of-concept and component testing. The first full thrust chamber test of the ISTB was conducted on 23 July 1975 on Test Stand A-1 at NSTL.[51]

A great many problems were identified and corrected during the component testing using the ISTB, and on 12 March 1976 the engine successfully demonstrated a 65 percent power level for 42.5 seconds. This test had been scheduled to last 50 seconds, but a failure in the high-pressure fuel turbopump caused it to be terminated early. On 24 March 1977, a failure of a high-pressure oxidizer turbopump caused a one-month delay in testing while the cause was investigated. No concise cause was identified, so several general modifications were made and additional instrumentation was added to the test cell. On 27 April 1977 testing was resumed, and twenty-five tests were run on two engines at the NSTL with no serious difficulties. However, problems with the new advanced-design turbopumps would continue to plague the engine program throughout the development process. Another critical failure involved the use of an incorrect welding wire, forcing an inspection of all welds. The next two years would see a series of failures and fires involving the SSMEs.

As part of the Orbiter contract Rockwell constructed MPTA-098 to allow Rocketdyne to test the SSMEs in a realistic structural environment. MPTA-098 consisted of an aft-fuselage, a truss arrangement which simulated the mid-fuselage, and a complete thrust structure including all main propulsion system plumbing and electrical systems. MPTA-098 was shipped from Palmdale to the NSTL on 24 June 1977 and mated with an External Tank (MPTA-ET) and three prototype SSMEs for propellant loading and static firing tests. The first static firing of a non-flight-rated engine took place on 21 April 1978 and lasted for 2.5 seconds. By the summer of 1979, flight hardware had been installed on MPTA-098, and testing resumed, but not without problems.[52]

On 2 July 1979, the main fuel valve on engine #2002 developed a major fracture that allowed hydrogen to leak into the enclosed aft compartment. The engine was commanded to shut down, but before this process could be completed the pressure in the aft compartment exceeded the structural capability of the heat shield supports, and MPTA-098 sustained major structural damage. Testing was resumed in September, but on 4 November 1979, a high-pressure oxidizer turbopump failed after 9.7 seconds of a scheduled 510-second test of a three-engine cluster. A completely successful static firing (the sixth of the series) finally took place on 17 December 1979 involving three non-flight-rated engines cycling between 70 and 100 percent power for 554 seconds.

The first of four flight configuration engines were assembled and acceptance tested during the first half of 1979. Engine acceptance testing included a 1.5-second start verification, a 100-second calibration firing, and a 520-second flight demonstration test. Engine #2004 was allocated to the preliminary flight certification program, and engines #2005, 2006, and 2007 were installed on *Columbia*. But perhaps the most important test milestone was established in terms of total accumulated test duration of the single-engine ground test program (excluding the MPTA). A goal of 65,000 seconds was established by Space Shuttle Program manager John Yardley as representing a sufficient level of confidence to consider the engine flight worthy, and this goal was soon established as a flight constraint. This requirement was achieved on 24 March 1980 during a test on engine #2004.

Then, on 16 April, temperature limits on a high-pressure fuel turbopump were exceeded, causing an engine to shut down 4.6 seconds into a 544-second test. Another success was followed by another failure. The tenth static firing, this time with flight-design nozzles, was shut down 105 seconds into the test due to a burn-through in a hydrogen preburner. During November 1980, a weak brazing section on a nozzle failed during a 581-second test, causing a hole in the nozzle several inches in diameter. Thanks to the automated engine control system, and the use of highly instrumented test stands, most of the SSME failures resulted in damage to components, not to the entire engine. This was because the failure could be detected and the engine shut down quickly enough to prevent the kind of catastrophic failures that had characterized the early space program, when even entire test stands were frequently destroyed. Nevertheless, the SSMEs were rapidly becoming the pacing item toward the first manned orbital flight.

After the engines were installed in *Columbia,* significant changes were made to the engine design as a result of problems discovered during the ongoing test program. Because of the number and complexity of the changes, it was decided to repeat the final engine acceptance tests after the flight engines were modified. During June 1980 the engines were removed from *Columbia* and returned to the NSTL where they all satisfactorily completed a 520-second flight mission demonstration test. The engines were shipped back to KSC and reinstalled in the Orbiter.

The preliminary flight certification (PFC) program began in early 1980 and was defined in terms of a unit of tests that were called "cycles." Each cycle consisted of thirteen tests and 5,000 seconds of test exposure which emulated normal and abort flight profiles. It was required to complete two PFC cycles on each of two engines of the flight configuration to certify that configuration for ten missions. All of the tests in each cycle had to be completely successful; if any test was less than perfect, the entire cycle did not count. Eventually, eight PFC cycles were completed prior to STS-1. On 17 January 1981, just three months prior to the first launch date, MPTA-098 successfully demonstrated a 625-second firing, complete with simulated abort profiles. All totaled, by the launch of STS-1 the SSME program had accumulated 110,253 seconds during 726 tests.[53]

Since the solid rocket boosters were relatively new technology for NASA, and none had ever been "man-rated"[54] before the shuttle program began, an extensive test program was planned. The first hydroburst test of an empty case was successfully completed on 30 September 1977 at the Morton Thiokol facility in Utah. This provided the data required to verify fracture mechanics and crack growth analysis, and to demonstrate the case's cyclic pressure load capability. A second hydroburst was successfully conducted on 19 September 1980 using the aft dome, two cylindrical segments, and the forward dome.[55]

Four development motor static firings were successfully conducted at Thiokol—DM-1 on 18 July 1977, DM-2 on 18 January 1978, DM-3 on 19 October 1978, and DM-4 on 17 February 1979. Three qualification motor static firings also were successfully completed—QM-1 on 13 June 1979, QM-2 on 27 September 1979, and QM-3 on 13 February 1980. These seven tests provided the data required to certify the solid rocket motor design. The data obtained included ballistic performance, ignition system performance, case structural integrity, nozzle structure integrity, internal insulation, thrust performance, thrust reproducibility, dynamic thrust vector alignment, nozzle performance and the flight readiness of the SRM. Interestingly, none of these tests attempted to duplicate the loads or conditions anticipated during ascent—an oversight that would gain particular importance after the *Challenger* accident in 1986. Compare these 7 tests with the 726 tests required to certify the main engines.[56]

The design of the External Tank contained some significant differences from the Saturn stages it so closely resembled in theory. The Saturn had its engines at the bottom of the stage, and the LH2 tank was on top of the LO2 tank, eliminating the need for the hydrogen tank to support the heavy mass of oxygen. This also reduced the moments of inertia, making the stage more responsive to steering inputs. But with the ET, the LO2 tank went on top. The reason was that the SSMEs did not thrust vertically through the centerline of the tank—they were mounted on the Orbiter, well off to one side. In order for the SSME thrust vector (after the SRBs had separated) to pass through the ET center of gravity, the LO2 tank had to be on top. This necessitated a much more robust LH2 tank and intertank than had been used on Saturn.[57]

The first "tanking" of an ET—loading it with LO2 and LH2—was conducted at the NSTL on 21 December 1977 using the ET contracted from the structural test components. The tanking test included flowing propellants as far the main engine inlet valves. A few days earlier the ET had been 40 percent filled and vibrated with three large shakers to provide information on its natural frequencies.[58]

Surprisingly, the thermal protection system on the ET is more complex than it appears. On Saturn, the insulation acted merely to reduce the boiloff rates to acceptable levels. If ice formed on the tank, it simply acted as additional insulation, and nobody much cared if the ice fell off during launch. But on the shuttle, engineers did care: ice could damage the Orbiter's fragile tiles. In addition, the ET needed more than simple insulation since the complex configuration of the shuttle stack ensured that the ET would suffer shock-impingement heating during ascent. The top of the tank would be safe—it sat out in front of everything else and would only have to deal with the high-speed airflow. But a little farther back the SRBs and Orbiter would be creating shock waves that would impinge on the sides of the ET, resulting in heating rates of over 40 BTUs per square foot per second. In addition, the attachment hardware for the Orbiter and SRBs would create complex flow fields, resulting in high local heating rates.

After extensive testing, the spray-on CPR-488 thermal protection system was deemed adequate for ascent protection, and the top coat of fire-retardant latex (FRL) was deleted, saving 595 pounds. This coating was also deleted from the last four standard-weight tanks, and only STS-1 and STS-2 used the distinctive white-colored tanks protected with the FRL.

Those Pesky Tiles

During the original studies of lifting reentry vehicles during the late 1950s and 1960s, there had been a great debate over the relative merits of active cooling systems versus passive systems for the vehicle structure. The active systems were attractive—on paper—but nobody could quite figure out how to make them work. Therefore the choices were largely narrowed to either a hot structure like that used on the X-15 or a more conventional structure protected by some sort of insulation. The hot-structure approach required the use of rare and expensive superalloys, and there was always a great deal of doubt if it would have worked on a vehicle as large as the shuttle. Generally most contractors seemed to prefer a fairly conventional structure made of titanium and protected by a series of metallic shingles with a thick layer of insulation in between the two. There was some investigation into ablative coatings, but the unhappy X-15A-2 experience made just about everybody shy away from this technology except as a last resort.

Things began to change as the Lockheed Missiles and Space Company made quick progress with the development of the ceramic reusable surface insulation (RSI) concept. This work had begun during the late 1950s, and by December 1960 Lockheed had applied for a patent for a reusable insulation material made of ceramic fibers. The first use for the material came in 1962 when Lockheed developed a 32-inch-diameter radome for the Apollo spacecraft made from a filament-wound shell and a lightweight layer of internal insulation cast from short silica fibers. But the Apollo design changed, and the radome never flew.

However, the experience led to the development of a fibrous mat that had a controlled porosity and microstructure called Lockheat®. The mat was impregnated with organic fillers such as methyl methacylate (Plexiglas) to achieve a structural quality. These composites were not ablative—they did not char to provide protection. Instead, Lockheat evaporated, producing an outward flow of cool gas. Lockheed investigated a number of fibers—silica, alumina, and boria—during the Lockheat development effort. By 1965 this had led to the development of LI-1500, the first of what became the shuttle tiles. This material was 89 percent porous, had a density of 15 pounds per cubic foot, appeared to be truly reusable, and was capable of surviving repeated cycles to 2,500 degrees Fahrenheit (F). A test sample was flown on an Air Force reentry test vehicle during 1968, reaching 2,300 degrees F with no apparent problems.

Lockheed decided to continue the development of the silica RSI, but decided to produce the material in two different densities to protect different heating regimes—9 pounds per cubic foot (designated LI-900) and 22 pounds per cubic foot (LI-2200). The ceramic consisted of silica fibers bound together and sintered with other silica fibers, and then glaze coated by a reaction-cured glass consisting of silica, boron oxide, and silicon tetraboride. Since this mixture was not waterproof, a silicon polymer was coated over the undersurface (i.e., nonglazed) side. This material was very brittle, with a low coefficient of linear thermal expansion, and therefore Lockheed could not cover an entire vehicle with it. Rather, the material would have to be installed in the form of small tiles, generally 6-by-6-inch squares. The tiles would have small gaps between them (averaging about 0.01 inch) to permit relative motion and allow for the deformation of the metal structure under them due to thermal effects. A second concern was the movement of the metal skin directly under an individual tile—since a tile would still crack under this loading, engineers decided to isolate the skin from the tile by bonding the tile to a felt pad then bonding the felt pad to the skin. Both of these bonds were done with a room-temperature vulcanizing (RTV) adhesive.

In their Phase C response, Rockwell had proposed using mullite tiles made from aluminum silicate instead of the Lockheed-developed tiles since the technology was better understood and more mature. But the mullite tiles were heavier, and potentially not as durable. Given the progress Lockheed had made subsequently, Rockwell and NASA asked the Battelle Memorial Institute to evaluate both candidate systems—an evaluation that the Lockheed product won. But the Lockheed material was not appropriate for all applications. Very high temperature areas of the Orbiter—the nose cap and wing leading edges—would use a reinforced carbon-carbon (RCC) material originally developed by LTV for the Dyna-Soar program. The RCC would provide protection above 2,700 degrees F yet keep the aluminum structure of the Orbiter comfortably below its 350 degrees F maximum. Tiles were used for the entire underside of the vehicle and for most of the fuselage sides and vertical stabilizer. Black tiles could protect up to 2,300 degrees F, while white tiles protected up to 1,200 degrees F. Flexible reusable surface insulation (FRSI) protected areas not expected to exceed 750 degrees F.[59]

Interestingly, NASA and Rockwell originally believed that the leeward side (top) of the vehicle would not require any thermal protection. But in March 1975 the Air Force Flight Dynamics Laboratory conducted a briefing for space shuttle engineers on the classified results of the ASSET, PRIME, and boost-glide reentry vehicle (BGRV) programs that indicated leeward side heating was a serious consideration. The thermal environment was not particularly severe but easily exceeded the 350 degrees F capability of the aluminum skin. FRSI blankets were subsequently baselined for this area.

But in the meantime another problem had developed—with the tiles themselves. As flight profiles were refined and aero-loads better understood, engineers began to question whether the tiles could survive the punishment. By mid-1979 it had become obvious that in certain areas the tiles "did not have sufficient strength to survive the tensile loads of a single mission." NASA immediately began a massive search for a solution that eventually involved outside blue-ribbon panels, government agencies, academia, and most of the aerospace industry. As LeRoy Day recalled, "There is a case [the tile crisis] where not enough engineering work, probably, was done early enough in the program to understand the detail—the mechanical properties—of this strange material that we were using."[60]

The final solution to the tile problem (at least this one) involved strengthening the bond between the tiles and the felt strain-isolation pads (SIP). Analysis indicated that while each individual component—a tile, the SIP under the tile, and the two layers of adhesives—each had satisfactory tensile strength, when combined as a system the components lost about 50 percent of their combined strength. This was largely attributed to stiff spots in the SIP (caused by needling) that allowed the system strength to decline as far as 6 psi instead of the baseline 13 psi. In October 1979 NASA decided on a "densification" process that involved filling voids between fibers at the inner moldline (the part next to the SIP pad) with a special slurry mixture consisting of Ludox (a colloidal silica made by DuPont) and a mixture of silica and water. Since the tiles had been waterproofed during manufacture, the process began by applying isopropyl alcohol to dissolve the waterproof coating, then painting the back of the tile with the Ludox. After air drying for twenty-four hours the tiles were baked in an oven at 150 degrees F for two hours. After a visual and weight check, each tile was re-waterproofed using Dow-Corning's standard Z-6070 product (methyltrimethoxysilane). The densified layer acted as a "plate" on the bottom of the tile, eliminating the effect of the local stiff spots in the SIP, bring the total system strength back up to 13 psi.[61]

But the installation presented its own problems. Rockwell quickly ran out of time to install tiles while *Columbia* was in Palmdale—NASA needed to present the appearance of maintaining a schedule, and *Columbia* moving to KSC was a very visible milestone. So in March 1979 *Columbia* was flown from Palmdale to KSC on the SCA and quickly moved into the Orbiter Processing Facility (OPF). Just over twenty-four thousand tiles had been installed in Palmdale, with six thousand left to go. But it now appeared that all of the tiles would need to be removed so that they could be densified.

The challenge became to salvage as many of the installed tiles as possible while ensuring sufficient structural margin for a safe flight. The approach developed to overcome this almost insurmountable challenge was called the tile proof test. This involved the application of a load to the installed tile so as to induce a stress over the entire footprint equal to 125 percent of the maximum flight stress experienced at the most critical point on the tile footprint. This approach could potentially salvage thousands of installed tiles.[62]

The device used for the proof test employed a vacuum chuck to attach to the tile, a pneumatic cylinder to apply the load, and six pads attached to surrounding tiles to react the load. Since any appreciable tile load might cause some internal fibers to break, acoustic sensors placed in contact with the tiles were used to monitor the acoustic emissions for any internal fiber breakage. The proof testing not only salvaged tens of thousands of installed tiles but also revealed those tiles (13 percent failed the proof test) with inadequate flight strengths. The tiles that failed would be replaced with densified tiles.

Two other techniques were developed to strengthen tiles while they were still on the vehicle. The first involved "thick" tiles—usually on the underside of the Orbiter—that were relatively small. As shock waves swept air over these tiles, they tended to rotate, inducing high stresses at the SIP bonds. The solution was to install a "gap filler" that prevented the tile from rotating. But this solution would not be very effective for small thin tiles, so a technique was developed where the filler bar surrounding the SIP was bonded to the tiles. This was done by inserting a crooked needle into the tile-to-tile gap and depositing RTV on top of the filler bar. This significantly increased the total bonded footprint and decreased the effects of a shock-imposed overturning moment.

For the next twenty months, technicians worked three shifts per day, six days per week testing and installing 30,759 tiles. By the time the tiles were installed, proof-tested, often removed and reinstalled, then re-proof-tested, the technicians averaged 1.3 tiles per man per week. Sometimes it seemed like the workers were making no progress at all. During June 1979, Rockwell estimated that 10,500 tiles needed to be replaced—by January 1980 over 9,000 of these had been installed, but the number remaining had ballooned to 13,100 as additional tiles failed their proof tests or were otherwise damaged. By September 1980 only 4,741 tiles remained to be installed, and by Thanksgiving the number was below 1,000. It finally appeared that the end was in sight.[63]

Between April 1978 and January 1979 a team from the AFFDL conducted a review of potential Orbiter heating concerns, and concluded that the OMS pod might have unanticipated problems. To support the conclusion, the Air Force ran a series of tests at the Arnold Engineering Development Center (AEDC) between May and November 1979, and further tests were run at the Naval Surface Weapons Center shock tunnel in May 1980. It was discovered that the OMS pod structure would deflect considerably more than originally anticipated—the thin 8-by-8-inch tiles were relatively weak under bending loads, and it was feared that the tiles might fracture and separate from the vehicle. Since the tiles had already been installed, a unique solution was developed where each tile was "diced" while still attached to the pod. This involved carefully cutting each of the 8-inch tiles into nine equal parts. Each cut was carefully monitored to ensure neither the tile nor the underlying structure was damaged. The technique proved so successful that it has subsequently been used on other areas of the Orbiter when needed.[64]

Getting Ready

On 8 March 1979 the first flight-rated Orbiter, *Columbia,* had been rolled out of the Rockwell International facility in Palmdale and transported to Edwards

AFB. In order to make the scheduled arrival date at KSC, many of the tiles were missing, as were several other items of equipment. The Orbiter was mated to the SCA and made a short test flight on 9 March, but several of the simulated tiles that had been glued and taped to the outside of the Orbiter fell off during take-off. Hasty repairs were made and a second test flight took place on 20 March, this time without incident. The following day the SCA ferried *Columbia* as far as Biggs Army Air Field near El Paso, Texas. A short hop to Kelly AFB in San Antonio was made on 22 March, and on 23 March the pair made it as far as Eglin AFB, Florida. The next day the Orbiter arrived at KSC and was placed inside OPF-1 to begin the painstaking process of getting the thermal protection system ready for the first flight.

The first flight ready ET had arrived at KSC on 29 June 1979, and on 3 November 1980 it was mated with the pair of SRBs destined for use on STS-1. Finally, after spending a record 613 days in the OPF, *Columbia* was rolled over to Vehicle Assembly Building (VAB) High Bay 3 on 24 November 1980 and mounted to the ET on 26 November. The entire stack was powered up in the VAB for the first time on 4 December. Mobile Launch Platform 1 (MLP-1), carrying the first complete flight-rated Space Shuttle stack, was rolled out to Launch Pad 39A on 29 December 1980, making the trip in just over ten hours. Propellant loading tests were performed on 22 January 1981 (LH2) and 24 January (LO2) to verify that the ground and flight system would work together correctly, and allowed ground personnel in the Firing Room to gain some real-life experience. However, during the propellant loading tests several pieces of the thermal insulation on the ET debonded. After investigating the problem, it was decided not to attempt repairs until after the Flight Readiness Firing (FRF), so a cargo net was draped over the affected area.

The space shuttle was finally ready for its first flight—two years behind the original schedule. This, however, is not a condemnation of the program. Almost every program that attempts to push the state-of-the-art to new levels runs into delays; the shuttle pushed harder than most, but the delay was not nearly as great as many other lesser programs. Although the final vehicle was not as robust or reusable as expected, it was still a magnificent achievement given the constraints (mainly financial) under which it was developed. Now it was time to use it.

First Launch

On 20 February 1981 *Columbia* participated in a 20-second FRF while the vehicle remained firmly secured to the launch pad at KSC as a final check of all systems. A launch attempt on 10 April was scrubbed due to a timing slew when the Backup Flight Software failed to synchronize with the Primary Avionics Software System. The trouble was quickly diagnosed, and a software patch was installed.

Image 9-4: Astronaut Bob Crippen at the controls of *Columbia* during the flight of STS-1 in April 1981. (NASA Photo)

STS-1 was launched 12 April 1981 on a two-day demonstration of *Columbia*'s ability to go into orbit and return safely. In addition, there were checkouts of the payload bay doors, attitude control systems, and the orbital maneuvering system. Its main payload was a Developmental Flight Instrumentation (DFI) pallet containing equipment for recording temperatures, pressures, and acceleration levels at various points on the vehicle. This marked the first use of solid rockets on a manned vehicle, and the first time astronauts had piloted a new type of spacecraft on its maiden flight. The flight landed at Edwards AFB on 14 April, and postflight inspection showed *Columbia* had suffered minor damage from an overpressure created by the SRBs at ignition—OV-102 had lost 16 thermal protection tiles, and 148 more were damaged. Otherwise, OV-102 was in good condition for a "used" spaceship and returned to KSC on 28 April 1981.[65]

Three more orbital flight tests would be conducted over the next fifteen months, culminating in the launch of STS-4 at its scheduled time on 27 June 1982. The OFT series had not uncovered any serious problems, and NASA declared the National Space Transportation System operational.

Space Shuttle Processing

The Kennedy Space Center has responsibility for all mating, prelaunch testing, and launch control ground activities until the SRB hold-down bolts are released. Responsibility is then turned over to the Mission Control Center (MCC) at the Johnson Space Center (JSC) in Houston, Texas. The Mission Control Center's responsibilities includes ascent, on-orbit operations, reentry, approach, and landing until landing run-out completion, at which time the Orbiter is handed over to the postlanding operations team at the landing site for turnaround and relaunch. The postlanding operations team is assigned to KSC and is deployed to wherever the Orbiter lands.[66]

A nominal flight follows basically the same routine every time. It starts with the Solid Rocket Boosters being stacked on the MLP in the VAB while the Orbiter is being serviced in one of the OPFs. The ET is then mated to the completely stacked SRBs. Then the Orbiter is towed from the OPF to the VAB (called "roll-over") and mated to the ET. Orbiter connections are made, the integrated vehicle is checked out, and the range safety system ordnance is installed while still in the VAB.

A crawler-transporter moves the MLP with the entire Space Shuttle stack to one of the launch pads approximately 3.5 miles away (called "roll-out"), where connections are made and servicing and checkout activities begin. If the payload was not installed in the OPF, it will be installed on the launch pad following prelaunch activities. The Orbiter and ET are filled with propellants, and additional checks are made until the SSMEs are cleared for ignition.

At launch, the three SSMEs are ignited and verified to be operating at the proper thrust level. A signal is then sent to ignite the SRBs, and the eight hold-down bolts are released explosively. The stack quickly accelerates over the Atlantic Ocean. Maximum dynamic pressure is reached early in the ascent, nominally 60 seconds after liftoff.

Approximately 120 seconds into the ascent, the two SRBs are separated from the ET. This is also accomplished explosively, and eight small rocket motors fire to ensure the boosters separate cleanly from the vehicle. At a predetermined altitude the SRBs deploy parachutes which lower them into the Atlantic approximately 141 miles down range from the launch site. The boosters are recovered by ships operating from the Cape Canaveral Air Force Station (CCAFS) adjacent to KSC. See Image 9-5 for a typical ascent profile and abort scenario.

Meanwhile, the Orbiter and External Tank continue to ascend using the thrust provided by the three SSMEs. Approximately 8 minutes after launch, and just short of orbital velocity, the main engines are shut down (main engine cutoff, or MECO), and the ET is jettisoned. The forward and aft reaction control system (RCS) thrusters provide the translation away from the Exter-

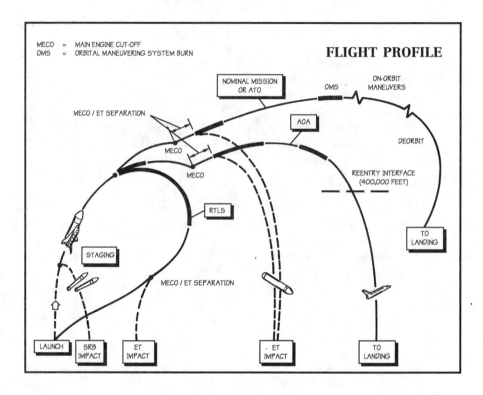

Image 9-5: A typical Space Shuttle flight profile. (NASA Photo)

nal Tank at separation and return the Orbiter to the proper attitude prior to the orbital maneuvering system (OMS) burn.

The External Tank continues on a ballistic trajectory and enters the atmosphere over either the Pacific or Indian Ocean (depending upon the ascent trajectory) where it disintegrates. This is assisted by a tumbling motion caused by unused liquid oxygen vaporizing and venting through the gaseous oxygen vent in the nose of the ET.

Normally, a single thrusting maneuver using the two OMS engines at the aft end of the Orbiter are used to complete insertion into Earth orbit and to circularize the spacecraft's orbit. On early flights, two OMS burns were made during ascent. The OMS engines are also used on-orbit for any major velocity changes. The orbital altitude of a mission is highly dependent upon the objectives of the mission. The nominal altitude can vary between 130 to 350 miles. The lowest orbit during the first 100 flights (not including classified DoD missions) was 138 miles during STS-68, while the highest was 385 miles during STS-82.[67]

The forward and aft reaction control system thrusters provide attitude (roll, pitch, and yaw) control of the Orbiter as well as any minor translation maneuvers along a given axis on-orbit. At the completion of orbital operations the Orbiter is oriented in a tail first attitude by the RCS and the two OMS engines are commanded to slow the Orbiter for deorbit.[68]

The RCS turns the Orbiter's nose forward for a reentry that occurs at 400,000 feet, slightly over 5,000 miles from the landing site. The Orbiter's velocity at reentry is approximately 17,000 mph with a 40-degree angle of attack. The forward RCS thrusters are inhibited by the GPCs immediately prior to reentry, and the aft RCS thrusters maneuver the vehicle until a dynamic pressure of 10 pounds per square foot is sensed, which is when the Orbiter's ailerons become effective. The aft RCS roll thrusters are then deactivated. At a dynamic pressure of 20 pounds per square foot the Orbiter's elevators become effective and the aft RCS pitch thrusters are deactivated. The Orbiter's speed brake is used below Mach 10.0 to induce a more positive downward elevator trim deflection. At approximately Mach 3.5 the aerodynamic rudder is activated, and the aft RCS yaw thrusters are deactivated at 45,000 feet.

Entry guidance must dissipate the tremendous amount of energy the Orbiter possesses when it enters the Earth's atmosphere to ensure the Orbiter does not either burn up (entry angle too steep) or skip back out of the atmosphere (entry angle too shallow). During reentry, excess energy is dissipated by the atmospheric drag on the Orbiter. A steep trajectory gives higher atmospheric drag levels which results in faster energy dissipation. Normally, the angle of attack and roll angle enables the atmospheric drag of any flight vehicle to be controlled. However, for the Orbiter, angle-of-attack variation was rejected because it creates exterior surface temperatures in excess of the insulation capabilities of the thermal protection system. The angle-of-attack schedule is loaded into the computers as a function of relative velocity, leaving only roll angle for energy control. Increasing the roll angle decreases the vertical component of lift, causing a higher sink rate and a higher energy dissipation rate. Increasing the roll angle does raise the surface temperature of the Orbiter somewhat, but not nearly as drastically as an equal angle-of-attack variation.

If the Orbiter is low on energy (current range-to-go greater than nominal at current velocity), reentry guidance will command lower than nominal drag levels. If the Orbiter has too much energy (current range-to-go less than nominal), reentry guidance will command higher than nominal levels to dissipate the extra energy, within the limits of the Orbiter to withstand the additional surface heating. The goal is to maintain a constant heating rate until the Orbiter is below 13,000 mph.

Roll angle is also used to control cross-range. Azimuth error is the angle between the plane containing the Orbiter's position vector and the heading alignment cylinder tangency point, and the plane containing the Orbiter's position vector and the Orbiter's velocity vector. When the azimuth error exceeds a predetermined number, the Orbiter's roll angle is reversed.

The equilibrium glide phase shifts the Orbiter from the rapidly increasing drag levels of the temperature control phase to the constant drag levels of the constant drag phase. The equilibrium glide flight is defined as flight in which the flight path angle, the angle between the local horizontal and the local velocity vector, remains constant. Equilibrium glide flight provides the maximum down-range capability and lasts until the drag acceleration reaches 33 fps squared.

At this point the constant-drag phase begins. The angle of attack is initially 40 degrees but begins to ramp down toward 36 degrees at the end of the phase. In the transition phase the angle of attack continues to ramp down, reaching approximately 14 degrees as the Orbiter reaches the terminal area energy management (TAEM) interface at approximately 83,000 feet and Mach 2.5.

The TAEM guidance steers the Orbiter to the nearest of two heading alignment circles (HAC), whose radii are approximately 18,000 feet and which are located tangent to, and on either side of, the runway centerline on the approach end. In TAEM guidance, excess energy is dissipated with S-turns and the speed brake is used to modify the drag coefficient as required. This increases the ground track range as the Orbiter turns away from the nearest HAC until sufficient energy is dissipated to allow a normal approach and landing. The spacecraft slows to subsonic velocity at approximately 49,000 feet about 25 miles from the runway.

The approach-and-landing phase begins at approximately 10,000 feet at an equivalent airspeed of 320 mph, roughly eight miles from touchdown. Autoland guidance is initiated at this point to guide the Orbiter to the minus 19-degree glideslope (about seven times as steep as a commercial airliner) aimed at a target one mile in front of the runway. The speed brake is used to control velocity, and at 1,750 feet above ground level a pre-flare maneuver is started to position the spacecraft for a 1.5-degree glideslope in preparation for landing. The final phase reduces the Orbiter's sink rate to less than 9 fps. Touchdown occurs approximately 2,500 feet past the runway threshold at a speed of roughly 220 mph.

A ground-support equipment air-conditioning purge unit is attached to the right-hand Orbiter T-0 umbilical so that cool air can be directed through the Orbiter's aft-fuselage, payload bay, forward fuselage, wings, vertical stabilizer, and OMS/RCS pods to dissipate the heat of reentry. A second ground cooling unit is connected to the left-hand T-0 umbilical Freon-21 coolant loops to provide cooling for the flight crew and avionics during the postlanding checks. The spacecraft fuel cells remain powered up at this time. The flight crew then exits the spacecraft, and a ground crew powers the Orbiter down.

If the landing was made at KSC, the Orbiter and its support convoy is moved to the Orbiter Processing Facility. If the landing was made at Edwards AFB the Orbiter is safed on the runway and then moved to the mate/demate device at Dryden. After a detailed inspection of the spacecraft, the tail cone is installed and the Orbiter mounted on one of the Boeing 747 Shuttle Carrier Aircraft for the flight back to KSC.

In the event of a return to an alternate landing site, a crew of about eight team members will deploy to the landing site to assist the crew in preparing the Orbiter for loading aboard the SCA for transport back to KSC. For landings outside the United States, personnel at the contingency landing site are provided minimum training on safe handling of the Orbiter with an emphasis on crash rescue training and how to safe the Orbiter and its propellants.

Upon its return to the Orbiter Processing Facility at KSC, the Orbiter is "safed" (ordnance devices removed), the payload (if any) is removed, and the Orbiter payload bay is reconfigured from the previous mission for the next mission. Any required maintenance, especially to the thermal protection system, is also performed in the OPF. A payload for the next mission may be installed in the Orbiter's payload bay in the OPF, or it may be installed when the Orbiter is at the launch pad.

Orbital Mechanics

Space Shuttle missions destined for equatorial orbits are launched from the Kennedy Space Center, and those requiring polar orbital planes were to be launched from the canceled Vandenberg Launch Site (VLS). Orbital mechanics, and the complexities of mission requirements, plus safety constraints and the possibility of infringing on foreign air space, prohibit polar orbit launches from the Kennedy Space Center. This leaves the United States without a human polar launch capability.[69]

Kennedy Space Center launches have an allowable path no less than 35 degrees northeast and no greater than 120 degrees southeast. These are azimuth degree readings based on due east from KSC being 90 degrees.

A 35-degree azimuth launch places the orbiter in an equatorial orbital inclination of 57 degrees. This means the spacecraft in Earth orbit will never exceed an Earth latitude higher or lower than 57 degrees north or south of the equator. A launch path from KSC at an azimuth of 120 degrees will place the spacecraft in an orbital inclination of 39 degrees (it will be above or below 39 degrees north or south of the equator).

These two azimuths—35 and 120 degrees—represent the normal upper and lower launch limits from the Kennedy Space Center. Any azimuth angles farther north or south would launch the Space Shuttle over a habitable land mass, adversely affect safety provisions for abort or vehicle separation conditions, or potentially present the politically undesirable possibility that the Solid Rocket Boosters or External Tank could land on foreign land or sea space. Steering the vehicle during the OMS burns ("dog-leg") after ET separation can attain slight variations in these azimuths.

Launches from the Vandenberg Launch Site would have had an allowable launch path suitable for polar insertions south, southwest and southeast. The launch limits at Vandenberg were 201 and 158 degrees. At a 201-degree launch azimuth, the spacecraft would be orbiting at a 104-degree inclination. Zero degrees would be due north of the launch site, and the orbital trajectory would be within 14 degrees east or west of the north-south pole meridian. With a launch azimuth of 158 degrees, the Orbiter would be at a 70-degree inclination, and the trajectory would be within 20 degrees east or west of the polar meridian. Like KSC, VLS had allowable launch azimuths that did not pass over habitable areas or involve safety, abort, separation, and political considerations.

Design problems at Vandenberg, as well as a desire within NASA to consolidate all Space Shuttle launches at the Kennedy Space Center, forced the Air Force to abandon plans for launching from VLS after the *Challenger* accident in 1986. Salvageable equipment was removed from SLC-6 during the early 1990s and brought to KSC to augment facilities at that launch site. There are no current plans to establish a launch site other than KSC.

Attempting to launch and place a spacecraft in polar orbit from the Kennedy Space Center and still avoid habitable land mass would be uneconomical because the Orbiter's payload would be reduced severely—down to approximately 17,000 pounds. A northerly launch into polar orbit of 8–20 degrees azimuth would necessitate a path over a land mass; and most safety, abort, and political constraints would have to be waived. This prohibits polar orbit launches from KSC. It should be noted, however, that NASA has accomplished the necessary technical analysis and planning to conduct such launches, and the flight software accommodates these unique flight profiles, so a polar launch from KSC could be accomplished if the national objective outweighed the economic and political considerations.

NASA's assessment of Orbiter ascent and landing weights incorporates all of the modifications accomplished during and since the *Challenger* stand-down and assumes a maximum Space Shuttle Main Engine throttle setting of 104 percent:[70]

> Kennedy Space Center Eastern Range (ER) satellite deployment missions. The basic payload-lift capability for a due east (28.5 degree inclination) launch is 55,250 pounds to a 126-mile orbit using OV-103, OV-104, or OV-105 to support a four-day mission. This capability is reduced approximately 115 pounds for each additional mile of altitude desired. The payload capability for the same satellite deployment mission with a 57-degree inclination is 41,000 pounds. The performance for intermediate inclinations can be estimated by subtracting 500 pounds per degree of plane change between 28.5 and 57 degrees. If OV-102 *(Columbia)* is used, the payload-lift weight capability must be decreased by approximately 8,050 pounds. This weight difference is attributed to an approximately 7,000 pound difference in inert weight.

Landing weight limits. After the 6.0 loads structural modifications, the Space Shuttle Orbiters are limited to a total vehicle landing weight of 250,000 pounds for abort landings and 240,000 pounds for nominal end-of-mission landings. It should be noted that each additional crew member beyond the five-person standard is chargeable to the payload weight allocation and reduces the payload capability by approximately 500 pounds, including life-support and crew-escape equipment.

What Was Launched Before 51-L?

Although it is hard to imagine in this day of widespread use of commercial satellites and such ambitious private systems as Iridium and Globalstar, the age of the commercial communication satellite did not really begin until the mid-1970s. Prior to 1975, essentially all satellites belonged to the government or to government-chartered corporations such as Comsat or Intelsat. But as the Space Shuttle was being developed, this began to change, and for the first time it was financially justifiable for individual corporations to purchase their own communication satellites for private use instead of buying services from Comsat or Intelsat. Large corporations and telephone companies around the world were beginning to order their own satellites in order to increase profits and provide a wider range of services. All of these satellites would need launched.[71]

The Space Shuttle was well positioned to provide launch services for this new generation of satellites. The most popular was the Hughes HS-376—and the Space Shuttle payload bay could, in theory, accommodate five of them on each flight. The shuttle only had to carry the satellites into low-Earth orbit since attached upper stages would boost them to their final orbit. Each HS-376 used essentially the same support equipment, allowing NASA to predict a quick turnaround between missions. It appeared that launching satellites would be a lucrative market.[72]

The Space Shuttle had, by legislation, acquired a monopoly in the domestic market, and NASA hoped to launch an Orbiter once per week. To ensure a full payload bay for each of these flights, NASA was marketing the unique capabilities of the Space Shuttle overseas and hoped to attract foreign satellites as well. As with any good marketing agency, NASA offered several introductory deals where customers could launch a number of satellites for a substantially discounted rate during the first five years of Shuttle operations. A similar deal had been made with the U.S. Department of Defense[73] to guarantee their continued support. In every case, NASA was actually losing money on each launch but was sure that once customers became accustomed to the services provided by the Shuttle, they would return at the full-price rate. Seven major telecommunications providers would take NASA up on this introductory offer—Aussat in Australia, Telesat in Canada, Perumtel in Indonesia, SCT in Mexico, and AT&T, RCA, and SBS in the United States. In addition, Hughes took advantage of the offer to manifest five of its HS-381 Syncom IV satellites that were being leased to the U.S. Navy as Leasats.

The Space Shuttle was declared operational after its fourth orbital flight—the first four flights had been used strictly to test the system to ensure it was safe and reliable. The fifth flight (STS-5) on 11 November 1982 carried the first commercial payload ever delivered to space by a manned spacecraft—the ANIK-C3 and SBS-C communication satellites. This was the second on-time launch of the National Space Transportation System and seemed to justify all of the hype and promise that had long been associated with the Space Shuttle.

The satellites were mounted vertically in separate cradles in the rear of the payload bay. Each HS-376 was essentially a drum covered with solar cells, and on station, the satellites would spin constantly to provide stability and to mitigate the thermal stresses encountered in space. But sitting in their launch cradles they would be baked on one side and frozen on the other as soon as the payload bay doors opened. To protect the satellites, clamshell covers were closed over them and electric heaters turned on until they were ready to be released. When it was time to release each satellite, *Columbia* faced its payload bay in the desired direction, and the satellite began spinning on a turntable in its cradle. Eight hours into the mission, SBS-3 became the first commercial satellite to be released by the Space Shuttle. The release was videotaped and relayed to the ground, showing the spinning satellite being ejected from its cradle and slowly drifting past the vertical stabilizer. The Shuttle crew advertised themselves as the "We Deliver" team.[74]

Over the next three years, twenty-four commercial satellites would be deployed from Shuttle, including four Leasats that were owned by Hughes and leased to the U.S. Navy. Each was boosted to geostationary orbit to serve as a communications relay. The Shuttle payload bay was technically capable of carrying up to five HS-376–class satellites, but in reality, three was the maximum number carried on any given mission. This was partially a factor of the marketplace—seldom were there more than three satellites ready at any given time—but also partially because NASA did not yet fully understand the effect of a forward center-of-gravity movement on the Orbiter's emergency landing characteristics.[75]

But things do not always go as planned. Although in each case *Challenger* performed its deployment mission perfectly, on 41-B (STS-11 was this mission's internal planning number) both satellites' PAMs failed to fire correctly, leaving Palapa-B2 and Westar-6 in useless low-Earth orbits. Despite the fact that the failure had nothing to do with the Shuttle, NASA felt its image had been tarnished. But the Shuttle offered a unique capability—one that can never be matched by an expendable vehicle. It can retrieve objects from space and return them to the earth. *Columbia* during STS-7 had first demonstrated this

Table 1 Pre-Challenger Shuttle Commercial Payloads[76]				
Shuttle Mission	Orbiter	Satellite	Type	Operator
STS-5	*Columbia*	Anik-C3	Hughes HS-376	Telesat (Canada)
STS-5	*Columbia*	SBS-3	Hughes HS-376	SBS (USA)
STS-7	*Challenger*	Anik-C2	Hughes HS-376	Telesat (Canada)
STS-7	*Challenger*	Palapa-B1	Hughes HS-376	Perumtel (Indonesia)
STS-8	*Challenger*	Insat-1B	Ford	ISRO (India)
41-B (STS-11)	*Challenger*	Palapa-B2	Hughes HS-376	Perumtel (Indonesia)
41-B (STS-11)	*Challenger*	Westar-6	Hughes HS-376	Western Union (USA)
41-D (STS-16)	*Discovery*	SBS-4	Hughes HS-376	SBS (USA)
41-D (STS-16)	*Discovery*	Telstar-3C	Hughes HS-376	AT&T (USA)
41-D (STS-16)	*Discovery*	Leasat-2	Hughes HS-381	Hughes (US Navy)
51-A (STS-19)	*Discovery*	Anik-D2	Hughes HS-376	Telesat (Canada)
51-A (STS-19)	*Discovery*	Leasat-1	Hughes HS-381	Hughes (US Navy)
51-D (STS-23)	*Discovery*	Leasat-3	Hughes HS-381	Hughes (US Navy)
51-D (STS-23)	*Discovery*	Anik-C3	Hughes HS-376	Telesat (Canada)
51-G (STS-25)	*Discovery*	Morelos-1	Hughes HS-376	SCT (Mexico)
51-G (STS-25)	*Discovery*	Arabsat-1B	Aerospatiale	ASCO (Arab League)
51-G (STS-25)	*Discovery*	Telstar-3D	Hughes HS-376	AT&T (USA)
51-I (STS-27)	*Discovery*	ASC-1	RCA S-3000	ASC (USA)
51-I (STS-27)	*Discovery*	Aussat-A1	Hughes HS-376	Aussat (Australia)
51-I (STS-27)	*Discovery*	Leasat-4	Hughes HS-381	Hughes (US Navy)
61-B (STS-31)	*Atlantis*	Morelos-2	Hughes HS-376	SCT (Mexico)
61-B (STS-31)	*Atlantis*	Satcom-K2	RCA S-4000	RCA (USA)
61-B (STS-31)	*Atlantis*	Aussat-A2	Hughes HS-376	Aussat (Australia)
61-C (STS-32)	*Columbia*	Satcom-K1	RCA S-4000	RCA (USA)

when the SPAS-01 (Shuttle Pallet Satellite) had been released on-orbit, retrieved, and returned to Earth. In November 1984, under contract to the insurance companies, *Discovery* retrieved Palapa-B2 and Westar-6 and returned them to Earth. The insurance companies had already paid Perumtel and Western Union for the loss of the satellites, so both satellites would be refurbished by Hughes and sold again to new customers, with the proceeds going to the insurance companies.

On 51-I (STS-27), NASA demonstrated another unique capability of the Space Shuttle. After deploying three communications satellites (which, incidentally, were of three different designs), *Discovery* maneuvered next to Leasat-3. This satel-

Table 2
Pre-Challenger TDRS and Major Science Payloads[77]

Shuttle Mission	Orbiter	Satellite	Type	Operator
STS-6	*Challenger*	TDRS-A	Tracking & Relay	NASA
41-C (STS-13)	*Challenger*	LDEF-1	Science	NASA
51-L (STS-33)	*Challenger*	TDRS-B	Tracking & Relay	NASA

lite had been deployed during 51-D (STS-23) but had failed to activate correctly after deployment. During extravehicular activities (EVAs) that totaled 11 hour and 27 minutes, astronauts successfully retrieved, repaired, and redeployed the satellite. All of this was televised live around the world; the publicity was priceless.

Government Payloads

Commercial satellites were not all that was deployed from the Space Shuttle before the *Challenger* accident. Immediately after the last Skylab mission, NASA had begun decommissioning many of its ground tracking and communications stations in order to save money. The plan for Shuttle had always included the use of space-based Tracking and Data Relay System (TDRS) satellites—ironically, the Space Shuttle would have to deploy the constellation before it could use them. The first TDRS satellite was deployed by *Challenger* on STS-6—*Challenger* was to be the workhorse for these satellites since *Columbia* was too heavy to effectively carry them. Unfortunately, the second satellite of the constellation was lost on *Challenger's* ill-fated 51-L (STS-33) mission.

Another government mission (STS-13) involved the Long-Duration Exposure Facility (LDEF), a large science payload designed to measure the effects of being in space for a prolonged period of time. On the same flight, astronauts conducted the first on-orbit repair of a satellite when they successfully (after several failed attempts) captured, retrieved, repaired, and redeployed the *Solar Maximum* science satellite. There were also a variety of missions that did not release payloads into space. Four Spacelab flights carried a large scientific module in the payload bay that allowed research into a variety of subjects—Spacelab modules were not released, and were brought back to Earth with the Orbiter landed. Two classified DoD flights were also conducted.

The publicity during some of these flights had been tremendous—around the world people watched in awe as astronauts maneuvered themselves and their Orbiter into position to launch or retrieve satellites. The Shuttle had successfully demonstrated that it was the carrier of choice for the current generation of communications satellites. This was significant since fully half the commercial[78] satellites deployed were for foreign customers—contracts which had been won against competition from Europe's Ariane. It appeared that NASA had successfully demonstrated that there was indeed a market for the Shuttle's services.[79]

But this apparent commercial success was deceiving. It was widely known that NASA had cut some good deals on multiple-satellite launch services, and this was somewhat expected for a new vehicle. What was not fully appreciated is that these "deals" were even better than they seemed. NASA's standard fee for deploying a satellite had been calculated from early estimates of flight rates, turn-around times, and operating costs. By the end of 1985, it was apparent that these estimates had been unrealistically low. The best flight rate the Shuttle had managed so far was nine missions in 1985—even if you looked at the best floating twelve-month window (late January 1985 to late January 1986), the best that had been managed was eleven missions, and one of those was the ill-fated 51-L (STS-33). It was a far cry from sixty flights per year.

At this lower flight rate, each mission was costing substantially more than had been envisioned. In fact, the cost of each mission was continuously increased as NASA invested more money in the Shuttle infrastructure, and the Orbiters demonstrated they required substantially more maintenance between flights than had been expected. But the fee being charged to most customers was fixed—and would be until 1988 at the earliest based on the introductory prices NASA had agreed with to attract customers. Unwittingly, the U.S. government was subsidizing the satellite launch business. A very dramatic event would cause all of this to change.

The Accident

The launch of 51-L (STS-33) was postponed three times and scrubbed once from the planned date of 22 January 1986. The first postponement was announced on 23 December 1985, adding an extra day (until 23 January) to accommodate the final integrated simulation that had slipped one day due to the late launch of 61-C (STS-32). On 22 January 1986, the date was slipped to 26 January, primarily because of KSC work requirements. The third postponement occurred on 25 January when forecasts indicated the launch site weather would be unacceptable. The new launch date was set for 27 January.[80]

The launch attempt on 27 January began with fueling the External Tank at 00:30 hours, Eastern Standard Time. The crew was awakened at 05:07, and events proceeded normally with the crew strapped into the Orbiter at 07:56. At 09:10, however, the countdown was halted when the ground crew reported a problem with the crew hatch exterior handle. By the time the hatch problem was solved at 10:30, the KSC recovery site designated for a return-to-launch-site abort had exceeded the allowable velocity for crosswinds. The launch attempt was scrubbed at 12:35, and rescheduled for the following morning.

The weather on 28 January was forecast to be clear and very cold, with temperatures dropping into the low twenties overnight. The management team directed engineers to assess the possible effects of temperatures on the launch. No critical issues were identified to NASA or contractor management officials, and while evaluation continued, it was decided to proceed with the countdown and loading of propellants into the ET.

A significant amount of ice had accumulated on Launch Pad 39B overnight, and it caused considerable concern for the launch team. In reaction, the ice inspection team was sent to the launch pad at 01:35, and returned to the launch control center at 03:00. After meeting to consider the team's report, the Space Shuttle program manager decided to continue the countdown and scheduled another ice inspection for T-3 hours.

At 08:44 the ice team completed its second inspection. After hearing the team's report, the program manager decided to allow additional time for the ice to melt. He also decided to send the ice team to perform one final inspection during the scheduled hold at T-20 minutes. At this point, the launch had been delayed two hours past its original scheduled time of 09:38.

At 11:15 the final ice inspection was completed, and during the scheduled hold at T-9 minutes, the flight crew, and all members of the launch team, gave a "go" for launch. The final flight of *Challenger* began at 11:38:00.010 Eastern Standard Time, 28 January 1986.

From liftoff until telemetry from the vehicle was lost, no flight controller observed any indications of a problem. The main engines throttled down to limit the maximum dynamic pressure, then throttled up to full thrust as expected. Voice communications were normal. The crew called to indicate the vehicle had begun its roll to head due east, and 55 seconds later, Mission Control informed the crew that the engines had successfully throttled up, and all other systems were satisfactory. Dick Scobee's acknowledgment of this call was the last voice communication from *Challenger.*

There were no alarms sounded in the cockpit, and the crew apparently had no indication of a problem before the rapid breakup of the vehicle. The first evidence of the accident came from live video coverage and when radar began tracking multiple targets. The flight dynamics officer in Houston confirmed to the flight director that "RSO [range safety officer] reports the vehicle has exploded," and 30 seconds later added that the Air Force range safety officer had sent the destruct signal to the Solid Rocket Boosters.

During the period of flight while the Solid Rocket Boosters are thrusting, there are no survivable abort options. There was nothing that either the crew, or the flight controllers, could have done to avert the catastrophe.

A combined Coast Guard/NASA/Air Force/Navy search team spent the next three months searching the Atlantic for the remains of *Challenger* and her crew. Approximately 30 percent of the Orbiter was recovered, including all three SSMEs, the forward fuselage with the crew module, the right inboard

| Table 3 Launch Vehicle Failures[81] | | | | | |
| Launch Vehicle | Launches | | Satellites [82] | | |
	Total	Failures	Total	Commercial	Lost
Shuttle	25	1	33	20	1
Ariane	10	1	16	13	2
Delta	12	0	13	6	0
Atlas	18	1	18	5	1
Titan III	13	1	Classified	0	Classified

and outboard elevons, a large portion of the right wing, the lower portion of the vertical stabilizer, three rudder/speed-brake panels, and portions of the midfuselage. The debris was evaluated by NASA and NTSB in typical accident investigation fashion.

Of course, the Shuttle was not unique in losing a vehicle. But it was unique in that it was a *piloted* vehicle—and the loss occurred live on national television. Of the five major launch vehicles in service, four of them experienced a launch failure during the first thirty-eight months the Space Shuttle was operational.[83]

During the time the Shuttle was actively launching commercial satellites, eight Arianes successfully deployed 13 commercial satellites while a single launch failure destroyed two satellites. In addition, an Ariane was used to launch the *Giotto* probe to observe Halley's Comet. Interestingly, only one HS-376 was launched on Ariane. Of the eighteen Atlas boosters launched, seventeen were successful, but only Intelsat used Atlas commercially—the rest of the payloads were for the Department of Defense, the National Oceanic and Atmospheric Administration, or NASA. All thirteen Titan III missions carried classified DoD payloads, and one was lost due to a launch failure. The Delta had long been the launch vehicle of choice for small communications satellites, and probably suffered most from the Shuttle entering the market. Only twelve launches were conducted—all successfully—delivering six commercial satellites and six government payloads.

Even discounting the Palapa-B2 and Westar-6 satellites that were left in useless orbits (not the Shuttle's fault), the Space Shuttle managed to deploy nearly as many commercial satellites as all of its competitors combined. This all would change.

The *Challenger* Investigation

President Reagan, seeking to ensure a thorough and unbiased investigation of the *Challenger* accident, announced the formation of a Presidential Commission on 3 February 1986. The commission was chaired by William P. Rogers, a former Secretary of State under President Nixon, and other members included astronauts Neil A. Armstrong and Sally K. Ride; Robert B. Hotz, former editor-in-chief of *Aviation Week & Space Technology;* Brig. Gen. Charles Yeager, USAF (Ret.); and numerous distinguished scientists and engineers.[84]

More than 160 individuals were interviewed and more than thirty-five formal panel investigative sessions were held, generating almost 12,000 pages of transcript. Nearly 6,300 documents, totaling more than 122,000 pages, and hundreds of photographs were examined and made a part of the commission's permanent record.

The report of the commission concluded, "The consensus of the Commission and participating investigative agencies is that the loss of the Space Shuttle Challenger was caused by a failure in the joint between the two lower segments of the right Solid Rocket Motor. The specific failure was the destruction of the seals that are intended to prevent hot gases from leaking through the joint during the propellant burn of the rocket motor. The evidence assembled by the Commission indicates that no other element of the Space Shuttle system contributed to this failure."

Although the commission found the 51-L (STS-33) accident was a direct result of the SRB O-ring failure, numerous other areas of concern were uncovered during the investigation. These areas included everything from the Orbiter braking system, to a decision-making process described as "flawed." It was found that in its efforts to produce an "operational" system, NASA had abandoned many of the procedures that had made it successful during its first twenty-five years. The commission provided NASA with nine major recommendations to help ensure a safe return to flight, with the primary one being to redesign the SRB joint and seal. But perhaps the most far-reaching recommendation was that "the nation's reliance on a single launch system should be avoided in the future."

This last recommendation resulted in a change to U.S. policy and law. No longer was the Space Shuttle to be the nation's only launch vehicle; in fact, its use to deliver commercial satellites was essentially forbidden. The new law read:

U.S. Code, Section 42—The Public Health and Welfare, Chapter 26—National Space Program, Section 2465a—Space Shuttle Use Policy.

(a) Use policy (1) It shall be the policy of the United States to use the Space Shuttle for purposes that (i) require the presence of man, (ii) require the unique capabilities of the Space Shuttle or (iii) when other compelling circumstances exist. (2) The term "compelling circumstances" includes, but is not limited to, occasions when the Administrator determines, in consultation with the Secretary of Defense and the Secretary of State, that important national security or foreign policy interests would be served by a Shuttle launch. (3) The policy stated in subsection (a)(1) of this section shall not preclude the use of available cargo space, on a Space Shuttle mission otherwise consistent with the policy described under subsection (a)(1) of this section, for the purpose of carrying secondary payloads (as defined by the Administrator) that do not require the presence of man if such payloads are consistent with the requirements of research, development, demonstration, scientific, commercial, and educational programs authorized by the Administrator.

Other sections of the law pertained to the NASA Administrator implementing the law and reporting on adherence to the law are not reproduced here.

Subsidies?

There were other reasons to remove commercial payloads from the Space Shuttle. First, it was decidedly non-cost-effective in a full-cost recovery environment to use the Shuttle to launch communication satellites. This was particularly pointed out when compared with Arianespace. Created in March 1980 as a private stock company by European aerospace firms, banks, and the French space agency, Arianespace soon took over operation of the multinational European Space Agency's (ESA) Ariane rocket. This included managing and financing of Ariane production, organizing worldwide marketing of launch services, and managing launch operations at Kourou, French Guiana.

Ariane launches began in December 1979, and the initial series of missions was conducted under ESA responsibility. The first full commercial mission under Arianespace control was the launcher's ninth flight in May 1984, when an Ariane I successfully lifted the GTE Spacenet 1 satellite into orbit. By the spring of 1985, Arianespace held firm orders for thirty satellites and had options for launching twelve more, representing a combined order book value of about $750 million. Of those orders, half were from satellite customers outside the European home market. Arianespace marketing combined the best of both worlds: the marketing freedom of a private company, plus the direct support of government agencies.

The Space Shuttle was not the only U.S. response to Arianespace competition. Transpace Carriers, a private U.S. firm created to provide McDonnell Douglas Delta launch services, attempted to halt trading by Arianespace in the United States on grounds of unfair subsidy pricing. Arianespace used a two-tier pricing policy, charging higher prices to the European Space Agency and its member states. The French space agency subsidized Arianespace launch and range services, as well as administrative and technical personnel, and ESA member states subsidized Arianespace insurance rates. Transpace Carriers filed a petition on 25 May 1984, with the Office of the United States Trade Representative (USTR) under Section 301 of the Trade Act of 1974. Transpace alleged that the European Space Agency was engaged in predatory pricing and other unfair trade practices in the sale of Ariane launch services. The USTR accepted the case on 9 July 1984, and meetings between European and U.S. government officials began in November 1984. Talks soon turned to pricing and subsidy comparisons of the Space Shuttle and Ariane. On 17 July 1985, President Reagan determined that the pricing and subsidy practices of Arianespace were neither unreasonable nor a restriction on U.S. commerce, because Arianespace practices were not sufficiently different from those of the United States Shuttle to warrant action under the Trade Act of 1974.Then came the *Challenger* accident. National Security Decision Directive 254, released shortly after the *Challenger* accident, took NASA and the Shuttle out of competition with U.S. commercial launch providers for commercial and foreign spacecraft payloads. This, in turn, was the impetus for the U.S.C 42 law cited above. NSDD 254 bore fruit in 1989, three years later, when the Office of Commercial Space Transportation (OCST) issued four commercial launch licenses, and had five more license applications pending. As the OCST reported to Congress in 1990, "Fiscal Year 1989 was a turning point for the Office of Commercial Space Transportation (OCST) and the U.S. commercial launch industry." The OCST had licensed the first U.S. commercial launches, thus marking "the beginning of a new era in the history of U.S. space endeavors."[85]

Post-*Challenger* Missions

Commercial satellites were essentially banned from the Space Shuttle when it returned to flight in 1988. In its place, the Reagan administration reverted to a "mixed fleet" strategy of using the Shuttle for missions that required the presence of humans in space, primarily for science, and of using expendable launch vehicles to place most satellites—commercial or government—in orbit. The DoD immediately bailed out of the Shuttle program, closing the mostly completed Shuttle launch site at Vandenberg AFB and ordering additional Atlas and Titan III boosters. In addition, the DoD began several programs to develop improved versions of the expendable boosters it would clearly rely upon for the foreseeable future.

The commercial users were left somewhat in a lurch. The few remaining Atlas and Titan boosters were all reserved for government use, and there were only five remaining Deltas in the inventory—two were reserved for the Strategic Defense Initiative Office, two for NOAA, and the last was used to compensate Indonesia for the embarrassing stranding of Palapa-B2. McDonnell Douglas had closed the Delta production line in the early 1980s when it became clear that the Shuttle would be the launch vehicle of choice. General Dynamics has scaled back Atlas production considerably, and Martin Marietta only had a few Titan IIIs rolling down the line as contingency vehicles. Only Ariane remained in series production in 1986 and was able to increase it operations immediately, and it quickly won contracts to launch three commercial satellites that had been scheduled for launch on the shuttle—Aussat-A3, Insat-1C, and SBS-5. The nine Ariane launches conducted during the thirty-

Table 4
Post-Challenger TDRS and Major Science Missions[86]

Shuttle Mission	Orbiter	Satellite	Type
STS-26R	*Discovery*	TDRS-C	Tracking & relay
STS-29R	*Discovery*	TDRS-D	Tracking & relay
STS-30R	*Atlantis*	*Magellan*	Venus probe
STS-34	*Atlantis*	*Galileo*	Jupiter probe
STS-31R	*Discovery*	*Hubble Space Telescope*	Great observatory
STS-41	*Discovery*	*Ulysses*	Formerly *Solar-Polar*
STS-37	*Atlantis*	GRO	Great observatory
STS-43	*Atlantis*	TDRS-E	Tracking & relay
STS-48	*Discovery*	UARS	Earth observatory
STS-54	*Endeavour*	TDRS-F	Tracking & relay
STS-61	*Endeavour*	HST Service Mission 1	HST service
STS-70	*Discovery*	TDRS-G	Tracking & relay
STS-82	*Discovery*	HST Service Mission 2	HST service
STS-93	*Columbia*	*Chandra* X-Ray Telescope	Great observatory
STS-103	*Discovery*	HST Service Mission 3	HST service

two months that the shuttle was grounded successfully orbited sixteen satellites and destroyed one other.[87]

Nevertheless, the Shuttle had a surprisingly busy launch schedule ahead of it when it returned to flight in 1988. There were still five classified DoD flights to be flown, and the TDRS constellation needed to be completed. The TDRS satellites had been designed specifically to be launched from the Shuttle and could not economically be reconfigured for another launch vehicle. Spacelab missions also continued to be flown, and NASA was looking forward to the construction of a space station in the not too distant future.[88] But perhaps more important, there were five major scientific payloads that required the unique capabilities of the Shuttle, and Shuttle missions were needed to service the Hubble Space Telescope.

Only one flight resembled the commercial satellite deployment flights that had characterized the early Shuttle manifest. In January 1990 *Columbia* deployed the Leasat-5 satellite, much like the four Leasats that had gone before. Again, the HS-381 bus had been designed for launch from the shuttle and could not easily be reconfigured for an expendable launch vehicle. What was different about this mission happened after the satellite was deployed—the Long Duration Exposure Facility was retrieved. LDEF had been left in orbit by 41-C (STS-13) in April 1984 for a one year mission; it ended up being almost six years.

During *Endeavour's* first flight, the Space Shuttle again proved its unique capabilities. The second launch of a Commercial Titan had left Intelsat VI (603) stranded in a uselessly low orbit in March 1990 when the attached kick motor failed to start. After much soul-searching, NASA agreed to try and rescue it. Fortunately, the satellite was a Hughes HS-393, which was a slightly larger version of the HS-381 that had been designed to be launched by the

Table 5 Spacelab and Spacehab Flights[89]			
Shuttle Mission	Orbiter	Spacelab Name	Purpose
STS-9	Columbia	Spacelab-1	General research demo
51-B (STS-24)	Challenger	Spacelab-3	Microgravity and life sciences
51-F (STS-26)	Challenger	Spacelab-2	Solar physics
61-A (STS-30)	Challenger	Spacelab-D1	Microgravity and life sciences
STS-35	Columbia	ASTRO-1	Astronomy
STS-40	Columbia	Spacelab SLS-01	Space life sciences
STS-42	Discovery	Spacelab IML-01	Microgravity
STS-45	Atlantis	ATLAS-1	Atmospheric studies
STS-50	Columbia	USML-1	Microgravity
STS-47	Endeavour	Spacelab-J1	Microgravity and life sciences
STS-56	Discovery	ATLAS-2	Atmospheric studies
STS-55	Columbia	Spacelab-D2	Microgravity
STS-57	Endeavour	Spacehab-1	Materials and life sciences
STS-58	Columbia	Spacelab SLS-02	Life sciences
STS-60	Discovery	Spacehab-2	Material sciences
STS-65	Columbia	Spacelab IML-02	Microgravity
STS-66	Atlantis	ATLAS-3	Atmospheric studies
STS-63	Discovery	Spacehab-3	Materials and life sciences
STS-67	Discovery	ASTRO-2	Astronomy
STS-71	Atlantis	Spacelab-Mir	Life Sciences
STS-73	Columbia	USML-2	Microgravity
STS-77	Endeavour	Spacehab-4	Materials and life sciences
STS-78	Columbia	LMS-1	Life and microgravity sciences
STS-83	Columbia	MSL-1	Materials sciences
STS-94	Columbia	MSL-1R	Materials sciences
STS-90	Columbia	Neurolab	Neurological life sciences
STS-95	Discovery	Spacehab-5	Life sciences

Table 6
Shuttle-Mir and International Space Station Flights[90]

Shuttle Mission	Orbiter	Purpose
STS-71	*Atlantis*	1st Shuttle - Mir Docking
STS-74	*Atlantis*	2nd Shuttle - Mir Docking
STS-76	*Atlantis*	3rd Shuttle - Mir Docking
STS-79	*Atlantis*	4th Shuttle - Mir Docking
STS-81	*Atlantis*	5th Shuttle - Mir Docking
STS-84	*Atlantis*	6th Shuttle - Mir Docking
STS-86	*Atlantis*	7th Shuttle - Mir Docking
STS-89	*Endeavour*	8th Shuttle - Mir Docking
STS-91	*Discovery*	9th and final Shuttle - Mir Docking Microgravity
STS-88	*Endeavour*	1st International Space Station Assembly Flight Microgravity and life sciences
STS-96	*Discovery*	2nd International Space Station Assembly Flight Atmospheric studies
STS-101	*Atlantis*	3rd International Space Station Assembly Flight
STS-102	*Atlantis*	4rd International Space Station Assembly Flight Microgravity

Shuttle. Only the grounding of the Shuttle fleet had forced Intelsat to use Ariane and Titan launchers instead. The April 1992 STS-49 mission of *Endeavour* successfully captured Intelsat VI, installed a new kick motor, and released the satellite, which later boosted itself into the proper orbit.

The Spacelab missions conducted by the Space Shuttle simply could not have been accomplished on any other launch vehicle, and at least partially vindicate NASA's choice of payload bay sizes. The Spacelab and privately funded Spacehab modules were sufficiently large to permit meaningful scientific work to be accomplished before a space station was completed. Ironically, as the assembly for the International Space Station kept getting delayed, NASA looked to Spacelab/Spacehab flights to keep the Shuttle launch rate at an economical level.

Space station–related flights began with a series of nine missions to the Russian *Mir* space station. In each case the Space Shuttle carried passengers to and from *Mir,* along with needed supplies. Finally, beginning with STS-88, crews from the Space Shuttle began the construction of the *International Space Station* on-orbit. The dream NASA had been chasing for over thirty years was becoming a reality.

Table 7

Fiscal Year Flight Rate Plan versus Actual											
	1992	1993	1994	1995	1996	1997	1998	1999	2000	2001	2002	2003
FY92 Plan	8	8	8	8	8	9						
FY93 Plan		8	7	7	8	7	8					
FY94 Plan			9	8	8	8	8	8	8			
FY95 Plan				7	7	7	7	8	8	8	8	
FY96 Plan					7	7	7	7	8	6	7	
FY97 Plan						7	7	8	8	7	7	7
FY98 Plan							6	8	9	9	8	8
FY99 Plan								7	9	9	8	8
Actual	7	7	8	6	8	7	4	4				

Flight Rate

One of the persistent problems of the Space Shuttle was its flight rate. Billed as a "space truck with the capability to fly to and from space on a routine basis," the shuttle carried with it the hopes and dreams of a generation of aerospace enthusiasts. But the shuttle proved to be neither as robust nor as cost effective as many had hoped. It was largely reusable, a truly major step forward, and it ended the costly and ridiculous practice of mobilizing the Navy to recovery spacecraft as had been done in Mercury, Gemini, and Apollo, but it never lent itself to routine operations on the model of the airline industry. The flight rate before the *Challenger* accident was never more than eight missions a year. After the *Challenger* accident, NASA intended to build to a flight-rate of twelve missions per year. However, the targeted planned annual flight rate was lowered to eight and subsequently seven plus or minus one as a cost-savings measure.[91]

From 1992 though 1997, the planned flight rate at the start of the year was typically achieved with a delta of plus or minus one mission. However, now that NASA has entered the International Space Station assembly era, there has thus far been a larger delta between planned versus actual flights. Whether or not payload delays will continue to cause a decrease in the planned flight rate in the future remains to be seen.

Figure 1

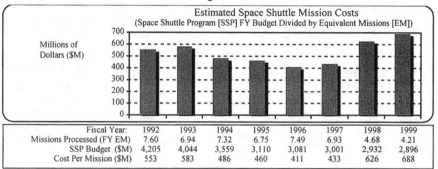

Estimated Space Shuttle Mission Costs
(Space Shuttle Program [SSP] FY Budget Divided by Equivalent Missions [EM])

Fiscal Year:	1992	1993	1994	1995	1996	1997	1998	1999
Missions Processed (FY EM)	7.60	6.94	7.32	6.75	7.49	6.93	4.68	4.21
SSP Budget ($M)	4,205	4,044	3,559	3,110	3,081	3,001	2,932	2,896
Cost Per Mission ($M)	553	583	486	460	411	433	626	688

The cost per mission may be figured in many ways, but the easiest is simply to divide the number of missions processed by the total yearly budget. This ignores amortizing the development cost, but give a reasonable recurring cost.

Cost Estimates

Next to the annual mission rate, probably the next most important measurements are the cost of the Shuttle program and the cost per mission. The Space Shuttle program budget from 1992 to 1995 was reduced a remarkable 26 percent. Since 1995, there has been a modest level of reduction, with the fiscal year 2000 budget representing a slight increase.[92]

Of more importance from a business perspective is the cost per mission. The cost per mission can be estimated by dividing the annual budget by the number of missions processed. Between 1993 and 1996, the cost per mission was reduced by nearly 30 percent—from $583 million to $411 million per mission. However, with the decreased flight rate in 1998 and 1999 along with a stable annual budget, costs ballooned to above $600 million for each mission.

Conclusion

It is popular to criticize the Space Shuttle for not achieving it stated goals of greatly reducing the cost of launching a pound of payload into orbit, and rightly so. However, it must be remembered that Shuttle was the first attempt to build a reusable spacecraft, and first attempts seldom accomplish all of the goals intended for them. As a first attempt, the Space Shuttle has succeeded relatively well, demonstrating that reusability can be made to work, and that a reusable spacecraft offers some unique possibilities that are not available with expendable vehicles (such as returning satellites from orbit).

At the same time, the Space Shuttle has always been hampered by fiscal constraints—first the development costs were capped in such a manner that less-than-optimum solutions (such as the SRBs) were forced upon the program. Then a series of events transpired to lower launch rates well below the projected rates (even the realistically projected rates—not the inflated 60 flight per year fantasies). The operational budget has never met the program's requirements, although in many instances the program wanted a great deal more money than was probably necessary—politics, rice-bowling, and other self-serving interests conspired to inflate the needs. However, the program has successfully demonstrated in the past several years that given money to fund improvements—such as the Block II SSMEs— that costs can come down and reliability can go up.

Despite this, it is wise to consider Shuttle an experimental vehicle, not an operational one. Hopefully we are learning some lessons that can be applied to the next vehicle. One thing we must be careful of, however, is to recognize the limitations of the hardware and technology we are using, and not to confuse those with the constraints that we impose upon ourselves because of management of political choices we have made. Many consider some of the latter to be more constrictive than the technological concerns. Only time will tell.

Notes

1. Richard P. Hallion and James O. Young, "Space Shuttle: Fulfillment of a Dream," Case VIII of *The Hypersonic Revolution: Case Studies in the History of Hypersonic Technology*, vol. 1, *From Max Valier to Project PRIME (1924–1967)*, Air Force Histories and Museums Program (Bolling AFB, Washington, D.C.: U.S. Air Force, 1998), p. 949.

2. "SR-89774 Reusable Space Launch Vehicle Report," prepared by The Boeing Company, Aero-Space Division, December 1959.

3. "Reusable Space Launch Vehicle Systems Study," prepared by the Convair Division of General Dynamics, undated; "Reusable Space Launch Vehicle Study," Convair report GD/C-DCJ-65-004, 18 May 1965; *This We Call Experience . . .* , a promotional brochure produced by Martin Marietta in support of their Phase A SpaceMaster concept, undated (but probably late 1969, early 1970).

4. It is, of course, much more complicated than this. For a more extensive look into early concepts, see Dennis R. Jenkins, *Space Shuttle: The History of the National Space Transportation System—The First 100 Missions* (Cape Canaveral, Fla.: Specialty Press, 2001).

5. C.W. Speith and W.T. Teegarden, *Astrorocket Progress Report*, report M-63-1 (Denver: Martin Company, December 1962).

6. The Manned Spacecraft Center in Houston was renamed the Lyndon B. Johnson Space Center (JSC) on 17 February 1973.

7. Report of the NASA Special Ad Hoc Panel on Hypersonic Lifting Vehicles with Propulsion (Houston: NASA, June 1964).

8. Hallion and Young, "Space Shuttle," pp. 957–62.

9. Request for Proposals, MSC-BG721-28-9-96C and MSFC-1-7-21-00020, 30 October 1968. Copies in both the JSC and MSFC History Office files.

10. The various NASA centers were responsible for technical monitoring only. All of the contracts were issued and administered by the MSC.

11. T.A. Heppenheimer, *The Space Shuttle Decision: NASA's Search for a Reusable Space Vehicle*, NASA SP-4221 (Washington, D.C.: NASA, 1999), pp. 117–18.

12. Day, who later became the Deputy Director for Space Shuttle development, was reluctantly reassigned from the historic Apollo program in January 1969 to head the SSTG.

13. Leroy E. Day, manager, *Space Shuttle Task Group Report*, 5 vols. (Washington, D.C.: NASA, 12 June 1969). Parts of volume 2 have been conveniently reprinted in *Exploring the Unknown, Volume IV, Accessing Space*, NASA SP-4407 (Washington, D.C.: NASA, 1999), pp. 206–10.

14. Spiro T. Agnew, manager, *The Post-Apollo Space Program: Directions for the Future* (Washington, D.C.: Space Task Group, September 1969), conveniently reprinted in *Exploring the Unknown, Volume I, Organizing for Exploration*, NASA SP-4221 (Washington, D.C.: NASA, 1999), pp. 270–74.

15. *The Next Decade in Space*, prepared by the Panel on Space Science and Technology (Washington, D.C.: PSAC, 1970).

16. John M. Logsdon, "The Space Shuttle Program: A Policy Failure?" *Science*, vol. 232, p. 1100; Heppenheimer, *Space Shuttle Decision*, pp. 270–74.

17. At some point, the term "National" was added to the original Space Transportation System, although most references continued to use the STS nomenclature.

18. Oskar Morgenstern and Klaus P. Heiss, *Economic Analysis of New Space Transportation Systems* (Princeton, N.J.: Mathematica, 31 May 1971); *Integrated Operations/Payloads/Fleet Analysis Final Report*, report ATR-72(7231)-1 (El Segundo, Calif.: Aerospace Corporation, 1 August 1971); *Economic Analysis of the Space Shuttle System* (Washington, D.C.: Mathematica, 31 January 1972), conveniently reprinted in *Exploring the Unknown, Volume IV, Accessing Space*, pp. 239–44; Logsdon, "Space Shuttle Program," p. 1100; Heppenheimer, *Space Shuttle Decision*, pp. 275–80.

19. See most of the individual Phase A and Phase B studies, in the files at KSC, JSC, and MSFC, and most are also at HQ. See also "Integrated Abort Analyses for a Fully Reusable Space Shuttle," MSC internal note 71-FM-339 (Houston: Contingency Analysis Section of the Flight Analysis Branch, 14 September 1971).

20. Maxine Faget and Milton Silveira, *Fundamental Design Considerations for an Earth-surface to Orbit Shuttle*, 70A-44618 (Washington, D.C.: CASI, October 1970).

21. Heppenheimer, *Space Shuttle Decision*, p. 210.

22. Ibid., pp. 211–12.

23. *Study Control Document: Space Shuttle System Program Definition (Phase B)* (Washington, D.C.: NASA OMSF, 15 June 1970).

24. See most of the individual contractor Phase A and Phase B studies, in the files at the JSC, MSFC, and KSC History Offices. See also "Integrated Abort Analyses for a Fully Reusable Space Shuttle."

25. Surprisingly, this is very similar to the current proposal to develop a pair of liquid fly-back boosters (LFBB—also known as the reusable first stage) for the existing Space Shuttle Orbiters.

26. An early classified Air Force concept, not to be confused with the much later X-30/S-30 National Aerospace Plane (NASP).

27. Heppenheimer, *Space Shuttle Decision*, p. 265.

28. Total program costs included research, development, test, production, and all recurring operational costs.

29. See the individual contractor ASSC studies, in the files at the JSC, MSFC, and KSC History Offices.

30. Logsdon, "Space Shuttle Program," p. 1101; Heppenheimer, *Space Shuttle Decision*, pp. 288–89.

31. Logsdon, "Space Shuttle Program," p. 1101.

32. "Minutes of the STS Committee," 27 October 1971, in the files of the JSC History Office.

33. See the individual contractor Phase B and ASSC Extension studies, in the files at the JSC, MSFC, and KSC History Offices.

34. Ibid.

35. John Mauer interview with Charles Donlan, Washington D.C., 19 October 1983, pp. 23–24, in the files of the NASA History Office.

36. Mauer interview with Donlan; Logsdon, "Space Shuttle Program," p. 1104; Heppenheimer, *Space Shuttle Decision*, pp. 408–11.

37. Memorandum for the Record, George M. Low, NASA, "Meeting with the President on 5 January 1972," 12 January 1972, in the files of the NASA History Office; Logsdon, "Space Shuttle Program," p. 1104. Richard M. Nixon, "Statement by the President," 5 January 1972; NASA Press Release 72-4, 6 January 1972.

38. Heppenheimer, *Space Shuttle Decision*, p. 413.

39. Logsdon, "Space Shuttle Program," p. 1101.

40. NASA news release 72–61, "Space Shuttle Decisions," 15 March 1972.

41. This was in fiscal year 1972 dollars—it would represent about $717 per pound in year 2000 dollars. In 1988, the Office of Management and Budget found that it cost $300 million per shuttle flight, or $5,000 per pound assuming a full payload bay—not exactly $700 per pound. Development had cost $6.651 billion in fiscal year 1971 dollars—$1.5 billion more than expected.

42. RFP 9-BC421-67-2-40P, "Space Shuttle Program" (Houston: NASA, 17 March 1972), cover letter; NASA Press Release 72-61, dated 15 March 1972 (but not released until 17 March).

43. North American Rockwell became Rockwell International on 16 February 1973.

44. Interestingly, this meant each Orbiter had a life expectancy of only eight years.

45. NASA news release, 26 July 1972.

46. OV-101 was to have been named *Constitution*, but a write-in campaign by fans of the TV series *Star Trek* influenced the final decision.

47. Jenkins, *Space Shuttle*, pp. 205–12.

48. Details on the SRB and SSME test programs, along with additional details on the other development and test efforts, may be found in T.A. Heppenheimer, *History of the Space Shuttle*, vol. 2, *Development of the Shuttle, 1972–1981* (Washington, D.C.: Smithsonian Institution Press, 2002); details may also be found in Jenkins, *Space Shuttle*.

49. Bob Biggs, "Space Shuttle Main Engine: The First Ten Years," a paper presented at the American Astronautical Society National Conference and Annual Meeting, 2 November 1989, Los Angeles.

50. Ibid.

51. Ibid.

52. Ibid.

53. Biggs, "Space Shuttle Main Engine"; *Space Transportation System Background Information* (Washington D.C.: NASA, September 1988), p. 563.

54. The Titan IIIM (or Titan 34D-7) version of the venerable Titan booster had been designed with man-rating in mind for the Manned Orbiting Laboratory (MOL) program, but according to most sources the test sequence was not completed before the program was canceled. However, UTC claims that both the UA-1205 (five-segment) and UA-1207 (seven-segment) motors were "man-rated designs." Also, no Soviet booster had used solids prior to 1993, and none had ever been man-rated.

55. *Space Shuttle Solid Rocket Booster*, SA44-80-2, (MSFC, Ala.: NASA, December 1980).

56. "Static Test Information," provided by Thiokol Propulsions communications group, 2 November 2000; *Space Shuttle Solid Rocket Booster*.

57. Heppenheimer, "History of the Space Shuttle," vol. 2, chap. 5 in the 27 April 1999 manuscript version.

58. Amos Crisp, "Tanking Test Conducted on Shuttle External Tank," MSFC news release 77-234, 23 December 1977.

59. Paul A. Cooper and Paul F. Holloway, "The Shuttle Tile Story," *Astronautics & Aeronautics* 19, no. 1 (January 1981): 24–34; *Technology Influence on the Space Shuttle Develop-*

ment, report 86-125C (Houston: Eagle Engineering, 8 June 1986), pp. 6–4 and 6–5; Hallion and Young, "Space Shuttle: Fulfillment of a Dream," pp. 1159–60.

60. First quote from Cooper and Holloway, "Shuttle Tile Story," p. 25; Day quote from an interview of LeRoy E. Day by John Mauer, 17 October 1983, pp. 5–6, in the files of the JSC History Office; Hallion and Young, "Space Shuttle: Fulfillment of a Dream," pp. 1161–66.

61. William C. Schneider and Glenn J. Miller, "The Challenging 'Scale of the Bird' (Shuttle Tile Structural Integrity)," a paper presented at the Space Shuttle Technical Conference (CR-2342) at JSC, 28–30 June 1983, pp. 403–13; Hallion and Young, "Space Shuttle: Fulfillment of a Dream," pp. 1165–66.

62. Schneider and Miller, "Challenging 'Scale of the Bird,'" pp. 403–13.

63. Hallion and Young, "Space Shuttle: Fulfillment of a Dream," p. 1166.

64. Cooper and Holloway, "Shuttle Tile Story," pp. 24–27; Schneider and Miller, "Challenging 'Scale of the Bird,'" pp. 409–10; Hallion and Young, "Space Shuttle: Fulfillment of a Dream," pp. 1160–63.

65. Contrary to popular belief, *Columbia* was not the first spacecraft to be "reused." The *Gemini 2* capsule had been refurbished and relaunched as a test of the Manned Orbiting Laboratory Gemini configuration. *Columbia* was, however, the first piloted spacecraft to be launched twice.

66. *National Space Transportation System Reference,* vol. 2, *Operations* (Washington, D.C.: NASA, September 1988), pp. 75–90.

67. *Shuttle Flight Data and In-Flight Anomaly List,* Revision U, "the green book" (Houston: JSC, June 1995).

68. *National Space Transportation System Reference,* vol. 2, pp. 75–90.

69. Jenkins, *Space Shuttle,* pp. 263–66.

70. Ibid., pp. 264–65.

71. David M. Harland, *The Space Shuttle: Roles, Missions and Accomplishments* (Chichester, England: Praxis Publishing, 1998), pp. 111–12

72. Harland, *Space Shuttle,* pp. 113–14.

73. During fiscal year 1982–83, the DoD had paid NASA a total of $268 million for nine military Space Shuttle launches.

74. Harland, *Space Shuttle,* pp. 115–16.

75. *Shuttle Flight Data and In-Flight Anomaly List;* Harland, *Space Shuttle,* p. 127.

76. *Shuttle Flight Data and In-Flight Anomaly List;* Harland, *Space Shuttle,* pp. 415–38; Jenkins, *Space Shuttle,* pp. 268–303.

77. Ibid.

78. Excluding the Leasats.

79. Harland, *Space Shuttle,* pp. 124–25.

80. William P. Rogers, Chairman, *Report of the Presidential Commission on the Space Shuttle Challenger Accident,* vol. 1 (Washington, D.C.: 6 June 1986).

81. Harland, *Space Shuttle,* p. 129.

82. Ibid., p. 128.

83. Satellites of all types; but in the case of the shuttle, not free-flyers returned on the same mission.

84. Rogers, *Report of the Presidential Commission.*

85. X-33 History Project, HTML version at http://www.hq.nasa.gov/office/pao/History/x-33/facts_1.htm.

86. *Shuttle Flight Data and In-Flight Anomaly List;* Harland, *Space Shuttle,* pp. 415–38; Jenkins, *Space Shuttle,* pp. 268–303.

87. Harland, *Space Shuttle,* pp. 128–29.

88. Or so it seemed at the time. A dozen years later we are still looking forward to the construction of a space station in the near future.

89. *Shuttle Flight Data and In-Flight Anomaly List;* Harland, *Space Shuttle,* pp. 415–38; Jenkins, *Space Shuttle,* pp. 268–303.

90. Ibid.

91. For fiscal year 1997, eight flights were launched; however, one of those flights (STS-94) was the reflight of STS-83.

92. Sources for budget data: fiscal years 1992 to 1994: written statement of Wayne Littles, Associate Administrator, Office of Space Flight, submitted as a part of his testimony to the House Subcommittee on Space and Aeronautics, *The Space Shuttle Program in Transition: Keeping Safety Paramount,* pt. 2, 9 November 1995; fiscal years 1955 to 1998: actual costs as reported in the OMB Budget Submittals for next years' budgets, that is, fiscal years 1997, 1998, 1999, and 2000, respectively; fiscal years 1999 and 2000: estimated costs in the fiscal year 2000 OMB Budget Submittal.

–10–

Eclipsed by Tragedy

The Fated Mating of the Shuttle and Centaur

Mark D. Bowles

Introduction

In August 1985 scientists and engineers at General Dynamics in San Diego, California, triumphantly unveiled the next generation launch vehicle of the space program. Such excitement had rarely been seen since the glory days of the Apollo program. The theme from the movie *Star Wars* accompanied the applause of more than three hundred officials from the company, the U.S. Air Force, and NASA. The focus of their attention was the new Centaur launch vehicle and the seminal role it would play for the American initiative in space. Craig Thompson, a General Dynamics operations representative, said, "I almost had tears in my eyes when they rolled that thing out."[1] Alan Lovelace, a former NASA Deputy Administrator, proclaimed it the "future of space probes in our generation." How had the "old-workhorse" been given new life twenty years after its first launch?

The Centaur rocket had its birth in the 1950s. With a daring new fuel system of liquid oxygen (LO2) and liquid hydrogen (LH2), it promised and delivered a radically new and powerful method for launching spacecraft. Management of the project moved from Marshall Space Flight Center to the Lewis Research Laboratory under the direction of Abe Silverstein in 1962. One year later the Centaur was mated as an upper stage for the Atlas booster. The successful November 1963 launch signaled the world's first in-flight ignition of a LO2/LH2 engine. Three years later the Centaur vehicle dramatically launched the Surveyor spacecraft, resulting in the first human artifact to land upon the lunar surface. Further successes in the 1970s included the Viking missions to Mars, the Mariner trips to Venus and Mercury, the *Helios* solar probe, the Pioneer flybys of Jupiter and Saturn, and the Voyager flights through the solar system and beyond. Demonstrating its broad capability, the Centaur was also the launch vehicle of choice for sending geostationary communication satellites into orbit. But by the late 1970s, under a new climate of disdain for all things expendable, the world's most successful expendable launch vehicle was to be redesigned to ride with the Space Shuttle.

This new, radical idea was considered the most significant and largest new program at NASA in the early 1980s. A new Centaur was necessary because while the Shuttle became the primary transport into space, it was limited to flying in low-Earth orbit. It was also unable to deploy unmanned missions, such as pushing a payload into a higher orbit or into the solar system. An upper stage like the Centaur promised to solve these problems. Once in Earth's orbit the astronauts could open the doors and launch the Centaur from inside the Shuttle's bay. The Centaur then provided the muscle to deliver its payload to the final destination. These payloads included interplanetary probes, communications satellites, and Earth-orbiting scientific spacecraft. The result was a unique, powerful, and reusable way to deliver payload to upper Earth orbit and into the far corners of the solar system.

The Shuttle/Centaur combination was heralded as the end of the era of the expendable launch vehicle and the convergence of manned and unmanned space flight. In the 1970s era of cost cutting and declining space budgets, reusable vehicles became a prime requisite for all future space initiatives. The combination of the next generation human program (the Shuttle) with the world's most powerful upper-stage rocket (Centaur) seemed the perfect fit. Because of these benefits, NASA engineer Jesse Moore believed that the new Centaur would become a "very integral, longtime part of the space shuttle program."[2] However, despite the hopes and dreams that surrounded the new rocket combination, ultimately the Shuttle never lifted a Centaur off the ground and NASA canceled the program in 1986, less than four months away from its first scheduled launch. The tears of pride shed by the scientists and engineers during the unveiling of the new Centaur turned to bitter disappointment in less than one year. After all the struggle and genius required to redesign the rocket, all that was left was the silence of a dream deferred.

Despite the failure, the story of Shuttle/Centaur is significant. The history of Shuttle/Centaur provides a lesson in the often-neglected topic of risk tolerance. Today most people attribute "safety failures" associated with liquid hydrogen as the central reason that Shuttle/Centaur never made the journey into space. But to the core group of engineers most responsible for designing and building the new Centaur, all of the safety precautions were addressed and all necessary emergency redundancy plans were in place. The more accurate explanation for why NASA canceled the launch of Shuttle/Centaur was that a devastating human tragedy forced the entire space program into an era of risk sensitivity—the catastrophic *Challenger* explosion. The lesson buried within this story is that quietly influencing political infighting, managerial skills, technical designs, and funding problems lies the subjective yet ever-changing tolerance for risk. The grand expectations of the end of the expendable launch vehicle, the years of engineering work, and the $1 billion price tag spent in the Centaur-to-Shuttle reconversion program were all fated to come to an end the day the first American astronauts died in space.

A Changing Vision of Space Travel

In 1979, Isaac Asimov, the great scientific popularizer and science fiction author, reflected upon the illustrious history of the space program. While impressed with the dramatic manned adventures to the Moon, he believed these were essentially political stunts, emblematic of technological bravado in a childhood stage. Asimov thought that human exploration of space would become mature only when spacecraft became *reusable*. Comparing space flight to oceanic voyages, he suggested that creating expendable launch vehicles was analogous to building an elaborate vessel like the *Queen Elizabeth II* only to destroy it after it crossed the Atlantic one time. To him this symbolized the economic wastefulness of the Apollo program. Reusability promised much more: a home for humans in space, a way to take advantage of other worldly resources, and the ability to extend the knowledge of the universe. Reusability heralded the coming of age of space flight and the emergence of what Asimov called "new pioneering horizons . . . stretching out as far as the eye can see or the mind conceive."[3] It was this type of enthusiasm, shared by many individuals in NASA, government, and industry, which was the driving force behind the Shuttle/Centaur revolution.

Despite the surge in enthusiasm over this "new" vision of space travel in the 1970s, the idea of reusability was an old one. It dated back to the 1930s, when a young Viennese engineer named Eugen Sänger developed plans for a liquid-fueled craft capable of reaching the stratosphere.[4] The idea of reusable rockets did not come to the forefront of space imagination in the United States until the early 1950s. German scientist Wernher von Braun was the most charismatic and effective promoter of this method of travel.[5] However, little funding was available to support von Braun's dreams because at the time, the United States government believed that "space" was an expensive and unnecessary place to explore. This belief changed quickly and dramatically with the Soviet's launch of *Sputnik I* in 1957. But reusable rockets remained in the background for another decade as the space competition between the United States and the USSR focused upon the race to the Moon. In an era of nearly limitless funds for cold war projects, reusability was not given a high priority.[6]

Yet once the Moon missions were successfully completed, a new direction was needed. NASA was at a crossroads. Faced with tightening budgets, its leaders decided that seemingly more economical ferrying systems and reusability were the most important goals for the future. While NASA engineers were initially disappointed that another high-profile mission was not going to replace Apollo, they soon learned to appreciate the challenge of reusability. As Apollo 11 astronaut Michael Collins recalled, reusability "implied a technological maturity, a feeling that space was here to stay, that launching ships into Kennedy's 'new ocean' would become routine." Collins concluded that for an aeronautical engineer, "the path to reusability was glorious."[7]

As the 1960s came to a close NASA began to officially investigate designs for a space shuttle.[8] Reusability became the primary goal, along with a manned orbiter capable of at least one hundred missions and a dry landing—very different from the oceanic splashdowns of the Apollo capsules.[9] In 1972, after a meeting with NASA Administrator James Fletcher and his Deputy George Low, President Richard Nixon officially announced the plan to dramatically change the American initiative in space.[10] He called the centerpiece of the new system a space vehicle that could "shuttle repeatedly from earth to orbit and back."[11] The Space Shuttle was born, and with it the dream to revolutionize space travel by making the voyage routine and the craft reusable.[12]

Despite the benefits associated with a reusable shuttle, the spacecraft had some significant limitations. Most important was that it was designed to only attain low-Earth orbits of about 300 miles. Scientists knew that a great number of missions would have to go beyond this orbit and the only way that the Shuttle might perform this task was through an additional propulsive stage. Geosynchronous orbit was 22,500 miles in space, nearly thirty-six times higher than the Shuttle could fly. The question was, What rocket was best suited to deliver the upper-stage payload to these orbits, or even farther into the solar system?[13] As one report concluded, one of the main "technology gaps" that needed to be filled was for a "propulsion system using liquid hydrogen and liquid oxygen."[14]

The Shuttle/Centaur Decision

The most logical technological choice to fill this upper-stage requirement was Centaur. But how difficult would it be to use the "old" expendable launch workhouse on a new piloted space vehicle? Engineers explicitly rejected Centaur for manned missions in the 1960s. In 1961 the Langley Research Center considered the possibilities of using the rocket for lunar and planetary missions but came to the conclusion that it was "out of the question." Centaur was not "man-rated" or considered safe to fly humans, and most predicted that it never would be.[15]

By the 1970s this opinion began to change. General Dynamics believed that a cryogenic stage was the answer to satisfy the new upper-stage requirements. It was clear which cryogenic rocket they had in mind—General Dynamics was the prime contractor of the Centaur that had already flown thirty-seven missions and had a flight backlog planned through 1979. In their 1973 "Reusable Centaur Study," they analyzed twelve key aspects of the program, including such important factors as safety, cost, schedule, weight, reliability, reusability, and flight complexity. Their final assessment was that the "Centaur programs are extremely low risk" and that no technology breakthroughs were required to achieve a reusable Centaur.[16]

Not everyone agreed that the Centaur was the answer for the Space Shuttle upper stage, and a heated controversy developed. Should NASA engineers redesign the Centaur as General Dynamics proposed, or should they develop an entirely new upper stage? As the journal *Science* reported, "Arcane though it sounds, the issue ignited a free-for-all between feuding congressional committees, the aerospace lobby, the Reagan White House, the Air Force, and NASA, with the latter caught in the middle."[17] The stakes were high because the solution would likely shape all American interactions with space for the foreseeable future. These interactions included the growing market for launching large new communication satellites, as well as the NASA's capability to send new probes farther into the solar system.

At first NASA opposed the General Dynamics plan and agreed with those who sought an alternative to the Centaur upper stage solution. One new plan, from the Air Force, developed an idea for a two-stage solid-rocket booster. The first stage was called the Inertial[18] Upper Stage (IUS), which was capable of launching medium-sized payloads, and NASA gave Boeing the lead as the prime contractor to develop the IUS. Boeing contended that their rocket offered a much greater chance for overall success. They concluded that while the Centaur required a major new development program, theirs was safer because it was a solid-fueled booster. It minimized risks for the crew, was mission flexible, and was Shuttle compatible. Boeing concluded, "The modified Centaur cannot satisfy the mission needs."[19]

NASA had numerous reasons for selecting the IUS as the best choice for the Shuttle. The consensus was that Centaur was too dangerous to fly in the Shuttle along with humans. Specifically, there were two main concerns with the rocket. The first was the common bulkhead. Centaur was never conceived as a man-rated machine, and as a result many of the design features of the rocket were at best marginally compatible with human flight. The common bulkhead was one of these features. Centaur had two tanks, one for the oxidizer at the bottom and the other for the fuel at the top. A vacuum bulkhead separated the tanks, but in reality it was just one large tank with a thin membrane keeping them apart. A leak in either tank would be dangerous, but a simultaneous leak in both tanks could be catastrophic. This feature of Centaur existed because as originally conceived it was not a man-rated machine; the worst that might happen was that the rocket would explode, resulting in a financial loss, but the performance rewards were worth the risk.

The move to a "man-rating" completely changed the equation. The most challenging of the safety related issues was what to do if the Shuttle mission was aborted before the Centaur was launched. What would happen if one of the crew became sick, or the bay doors would not open and the Centaur was not deployed? The crew could not land on a runway with the explosive liquid propellant in the payload bay. The only option was a complicated, but feasible, fuel dump. As soon as the possibility of letting it travel along with humans arose, the safety issues became a serious concern. These factors initially made the IUS the upper stage of choice.

But Boeing was unable to solve key problems with the IUS. The main limitation was that it was not powerful enough to launch a payload to a distant planet like Jupiter in a direct shot. To get to its destination the payload would have to perform "mission design tricks," which caused it to slingshot around several planets to gain a gravity-assisted speed boost. Most engineers considered this approach inelegant and too time-consuming. They didn't want to wait for additional years for their probes to reach their destinations. Also, by November 1980 new concerns emerged that the Inertial Upper Stage booster was too costly. Robert Frosch, upon leaving his post as NASA Administrator, argued that the possibility of readying the IUS to send an orbiter and probe to Jupiter for its scheduled launch was remote. Frosch decided that the only other alternative was Centaur. He argued that by allocating budget resources from 1981 and 1982, modifications of the Centaur could result in a "powerful combination." He concluded, "No other alternative upper stage is available on a reasonable schedule or with comparable costs. Shuttle/Centaur would offer both to commercial customers and to national security interests a highly capable launch vehicle with growth potential."[20]

In 1981 NASA withdrew support for the IUS and instead opted for the General Dynamics redesigned Centaur. Specifically, there were a number of technical factors that made the Centaur option more attractive. First and most important was that Centaur was more powerful. It had the capability of delivering the boost necessary to propel a payload directly to a planet in the deep solar system. Second, Centaur was a gentler rocket. Solid rockets had a harsh initial thrust that had the potential to damage their delicate payload. Liquid rockets generated their thrust more slowly, thus reducing this threat. Finally, liquid fuel had one other tremendous advantage over solid fuel: it could be turned off and on. Once ignited, a solid rocket burned until empty, while the liquid rocket was much more controllable.[21] The only remaining advantage of the IUS over the Centaur was safety. Liquid hydrogen was a dangerous fuel and presented a significant challenge. NASA decided to accept the risk and go with the Centaur.

Ultimately the responsibility for the Centaur redesign fell upon the shoulders of NASA Lewis Research Center in Cleveland, Ohio.[22] Before the decision was made, however, engineers at Lewis had to withstand a political play by the powerful Johnson and Kennedy Space Centers to keep the contract out of their hands.

Politics: The Push for Marshall

Lewis Research Center seemed the logical choice to manage the new Centaur program due to its decade of experience with the Centaur rocket as expendable launch vehicle. In 1979 John Yardley, Associate Administrator for Space Transportation Systems, instructed Lewis Research Center to determine the feasibility of integrating Centaur with the Shuttle. By contracting with General Dynamics Convair, Lewis engineers planned to develop a design that would minimize modifications to both vehicles. The main area of concern at this time was the "fulfillment of safety requirements imposed on all Shuttle cargo as a result of 'man-rating.'"[23] Lewis concluded that a cryogenic upper stage was feasible, it would provide increased mission flexibility for the Shuttle, and would meet the all-unique safety requirements. This positive finding was very important for Lewis. The Space Shuttle threatened the expendable launch program, including its own Centaur program. By proving the feasibility of Shuttle/Centaur, Lewis could maintain their Centaur work and significantly enhance its involvement in the space program.[24] Implicit in this preliminary work was the belief among Lewis engineers that if Shuttle/Centaur were approved, they would win the management of the Centaur conversion.

But the Lewis Research Center encountered strong opposition from within NASA. The leadership at Johnson Space Center (JSC), Marshall Space Flight Center (MSFC), and Kennedy Space Center (KSC) all opposed a Lewis-led Centaur program for the Shuttle, and they recommended that Marshall take the lead. In a dramatic display of unity and persuasion, in January 1981 the directors of these centers (Chris C. Kraft Jr., William R. Lucas, and Dick G. Smith, respectively) wrote a confidential joint letter to Alan M. Lovelace, then the acting NASA Administrator. They marked this letter "eyes only," meaning that they did not want their controversial position to leak out to the wider NASA community. Their argument was simple and straightforward: they did not want Lewis to take any leadership position with Centaur. They believed that Marshall was the best choice and wrote, "The Agency's decision to proceed with the development of the Wide Body Centaur as a planetary upper stage in lieu of the Inertial Upper Stage presents a most challenging assignment for the Agency. . . . Recognizing this complexity, it is essential that the Agency make the proper assignment of the management responsibilities for this program. . . . Therefore, we believe, when consideration is given to all relevant factors, it is in the best interest of the program and the Agency that the existing JSC/MSFC/KSC team be responsible for the Wide Body Centaur program."[25]

Within this team, the directors believed that Marshall should specifically shoulder the burden of managing Centaur for the following reasons. First, they argued that Marshall had a "unique capability" for developing cryogenic rocket engines and propulsion systems. Second, they argued that the Shuttle was a complicated manned system and only their centers had the "thorough understanding" of the technical and managerial relationships to integrate a new element like Centaur. As a result the three directors concluded that a Marshall-led team, working closely with Johnson and Kennedy, was the best choice.[26]

Despite the desire to keep this recommendation private and confidential, engineers at Lewis Research Center found out about it. In March 1981, Lewis director John F. McCarthy wrote his own persuasive letter to Lovelace, making the argument for his center to win the contract. McCarthy listed many reasons why Lewis was the right choice. First, Lewis had led the program to determine the feasibility of modifying Centaur for the Shuttle. This experience would be invaluable for building the new rocket. Second, Lewis engineers had the greatest Centaur experience among all the NASA centers. Specifically, they argued that their experience with the successful Titan/Centaur program would serve as a good model on which to base the Shuttle/Centaur program. The two were similar because the effort to reconfigure the Atlas/Centaur for the Titan included many of the same conversion issues that would arise with the Shuttle. Third, Lewis also had great experience with mission design and spacecraft integration. With Atlas/Centaur, Titan/Centaur, Earth-orbital missions, lunar missions with Surveyor, and Earth-escape missions like Viking, Voyager, and Pioneer, this experience meant, according to McCarthy, that "no other NASA center approaches the level of expertise in this area that Lewis has attained through the integration of such a wide variety of complex missions." In fact, the average number of years that each Lewis engineer had on the Centaur team was thirteen. McCarthy concluded, "Our staff uniquely possesses both the knowledge and experience and by virtue of its stability preserves an irreplaceable corporate memory. . . . I strongly believe that Lewis continues to be the best choice for Centaur management."[27]

One month later, in May 1981, Lovelace made his decision. In a letter to William R. Lucas, director at Marshall, he wrote that he had thoroughly reviewed his proposal for Marshall control of Centaur and acknowledged the support given him by both Johnson and Kennedy. He wrote, "Although I recognize each of the considerations which you mentioned as contributing to this recommendation, I have concluded that they are outweighed by another set of factors stemming from the long and continuing management of the Centaur program by the Lewis Research Center." Primarily, these factors included the Lewis experience and their excellent track record with the rocket. It was clear that Lewis was the right choice. As the authors of the history of the Marshall center argued, Lewis "had built the Centaur, and had staked out a role in advanced propulsion technology that Marshall could not expect to emulate."[28] Despite the significant competition, Lewis won the right to maintain the Centaur program and to integrate it with the Shuttle.[29]

Funding: Staking the Lewis Reputation on Centaur

The Lewis Research Center was ecstatic over their victory. But with victory came an awesome responsibility, and everyone at the laboratory knew that their reputation was at stake with the success or failure of Shuttle/Centaur. Larry Ross, Director of Space Flight Systems at Lewis (in 1990 he became the head of Lewis Research Center), said that there were two main reasons for this importance. First, the entire NASA agency had a strong political commitment to the Shuttle/Centaur program and any failure at Lewis would lead to "embarrassment *in extremis.*" Second, Lewis was going to play a major role in the project. Ross warned that if his center "failed in this vital Agency effort, I would despair of us ever again being relied on to do important work for NASA!"[30] As a result of the critical nature of this program, Lewis leadership devoted a great deal of personal attention to the project. This was, as Ross called it, the "world of Centaur. . . . A very exiting, albeit challenging, program area."[31]

With their spirits high after being awarded the contract and constantly aware of what this program would mean to Lewis, the center quickly began organizing contractors for the daunting task ahead of them. Lewis submitted to General Dynamics a proposal to provide materials, personnel, supplies, and services to create the new wide-body Centaur.[32] By June Lewis awarded four contracts totaling $7,483,000. Of the four contracts, Teledyne Industries developed a digital computer and several remote multiplexor units, Honeywell, Inc., developed part of the automatic navigation and guidance system, General Dynamic's Convair Division in San Diego developed two modified Centaur vehicles, and Pratt & Whitney developed four RL10A-3-3A rocket engines. All of these contracts were geared toward two key missions—the upcoming *Galileo* probe to Jupiter and the Ulysses mission to the Sun.[33] Despite the start of work and the flow of funds, uncertainties hovered over the project for over a year, constantly threatening to suspend the project. Would the funding and support for Shuttle/Centaur continue?

In 1982, President Ronald Reagan signed an Urgent Supplemental Appropriations bill that allocated $80 million for the design, development, and procurement of the Centaur upper stage. This was a dramatic increase over funding for this project in 1981, which amounted to only $20 million. Andy Stofan, the new center director, noted that the new funding "puts Lewis directly into the mainstream of the Shuttle Program."[34] This also meant that Lewis Research Center would have its first major involvement with the high-profile Space Shuttle program. Stofan reminded his employees of their new responsibility: "I consider this a major step forward for Lewis, and I think we should recognize that we have accepted responsibility for a highly visible and very important effort."[35]

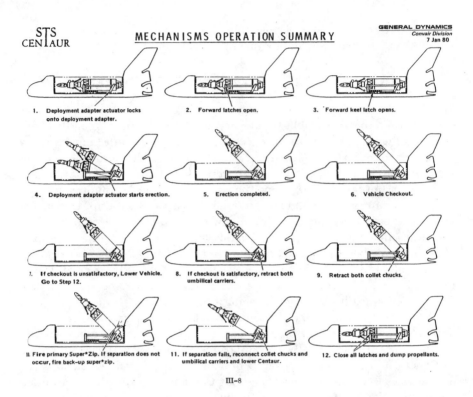

STS
CENTAUR

MECHANISMS OPERATION SUMMARY

GENERAL DYNAMICS
Convair Division
7 Jan 80

1. Deployment adapter actuator locks onto deployment adapter.

2. Forward latches open.

3. Forward keel latch opens.

4. Deployment adapter actuator starts erection.

5. Erection completed.

6. Vehicle Checkout.

7. If checkout is unsatisfactory, Lower Vehicle. Go to Step 12.

8. If checkout is satisfactory, retract both umbilical carriers.

9. Retract both collet chucks.

10. Fire primary Super*Zip. If separation does not occur, fire back-up super*zip.

11. If separation fails, reconnect collet chucks and umbilical carriers and lower Centaur.

12. Close all latches and dump propellants.

III–8

Image 10-1: Convair's sequence for launching the Centaur upper stage from the Space Shuttle, 1980. (Photo courtesy of General Dynamics)

Lewis formed the Shuttle/Centaur Project Office to serve as the central point of integration for the program. This office developed all main project objectives, as well as evaluating the progress of these goals. These goals included defining Centaur redesign requirements, creating the new vehicle design, determining how to best integrate it with the Shuttle, then finally developing and producing the new Centaur. This office also controlled the budget for the program as well as took responsibility for scheduling and coordinating all contractors and internal work related to the reconfiguration.[36] The project office also had extensive relationships with outside organizations, contractors, and other government agencies, including NASA Headquarters, the Department of Defense, Johnson Space Center, Kennedy Space Center, the Jet Propulsion Laboratory (JPL), and the European Space Agency.[37]

While this office relied upon the expertise of all of its workers, there were several unique individuals who helped to shape its future. William H. "Red" Robbins took over as head of the Shuttle/Centaur Project Office. There were two main divisions within this office: managerial and engineering. Robbins served as project manager and was responsible for the administrative functions, including budgets, schedules, and so on. Steven V. Szabo headed up the engineering division and in 1983 became the new chief of Lewis' Space Transportation Engineering Division. Another longtime Centaur engineer, Szabo began work in the original Centaur Project office in 1963. In his new position he led a one-hundred-member team responsible for not only keeping Atlas/Centaur operational but also implementing the new Shuttle/Centaur program.

Thus, Shuttle/Centaur was finally on its way to becoming a reality, and the "old" rocket was going to·be given new life. The funding was in place. The Lewis Research Center was designated as the integration point between NASA and the outside contractors. A young team of bright scientists and engineers was being put together to modify and build the new vehicle. Finally, the safety concerns were understood and plans were in place to reduce the risk to human life as much as possible and to achieve a "man-rating" for Centaur. With the infrastructure in place, NASA began to build the new horse.

Design: Building the New Horse

Many of the technical features of the new Centaur were similar to the Atlas/Centaur configuration, but despite the similarity there were some significant technical changes. NASA had to do everything in its power to keep the redesigns to a minimum. Every new adjustment increased the cost, decreased the reliability, and lengthened the development period. Working from a limited budget and time frame, NASA struggled to reduce the necessary modifications.

Specifically, there were two main redesigns required for the Shuttle to carry the Centaur. The first was a change to the rocket itself, which was necessary to fit the Centaur into the cargo bay of the Shuttle. Engineers developed two different configurations (called versions G and G-prime) of the Centaur. With roughly 60 feet of cargo space available in the Shuttle, both the Centaur and its payload had to fit within this length. The Centaur G was 20 feet long and capable of supporting a 40-foot payload. The larger payload was made possible in part by reducing propellant weight to 29,000 pounds. This G version was designed specifically for geosynchronous missions to place satellites into orbit. The U.S. Air Force funded half of the initial $269 million expense required to design and develop the Centaur G for use with the Shuttle.[38] The other configuration, Centaur G' (G-prime), was nearly 30 feet long and capable of holding payloads of up to 30 feet.[39] It was optimized for planetary missions, and the payload space was smaller due to the necessary increase in propellant weight (45,000 pounds). NASA funded the G-prime configuration itself.[40] Because both versions filled the entire cargo bay, each Shuttle mission to carry a Centaur would have to dedicate its entire mission to the launch. Overall, the two versions were very similar, having 80 percent of the design detail in common.[41]

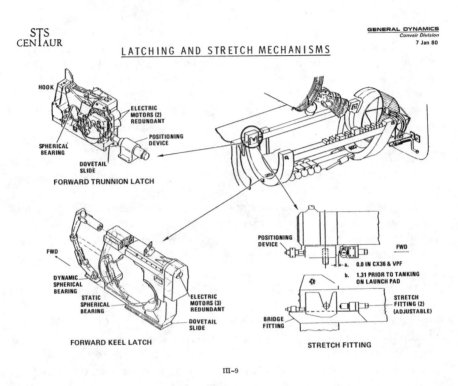

GENERAL DYNAMICS
Convair Division
7 Jan 80

STS CENTAUR

LATCHING AND STRETCH MECHANISMS

III-9

Image 10-2: The latching and stretching mechanisms for the Centaur upper stage in the Space Shuttle, 1980. (Photo courtesy of General Dynamics)

The second and most important new technical development was the launching mechanism.[42] The new deployment device was called the Centaur Integrated Support Structure (CISS), a 10-foot diameter aluminum structure that attached the rocket to the Shuttle. This device was of central importance · because it enabled the Centaur to fly in the Shuttle with a limited number of design modifications. The CISS was located within the payload bay on rotating support structures. It was responsible for all of the mechanical, electrical, and fluid interfaces between the Centaur and the Shuttle as well as most of the safety precautions.[43] The CISS was also fully reusable for ten additional flights.[44] The development of the CISS was considered one of the "more extensive and challenging tasks" of the entire Centaur modification project.[45]

The flight operation and the interactions between the Shuttle, Centaur, and CISS worked in the following way. The Shuttle launch was itself a standard flight. During the launch, engineers at Houston planned to monitor the Centaur through a Tracking and Data Relay Satellite via telemetry links to the ground. Once on-orbit, the Shuttle opened its payload bay doors and began a

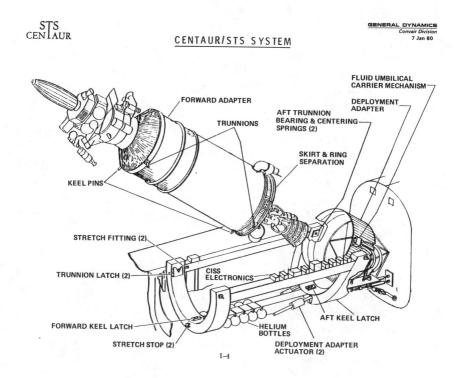

GENERAL DYNAMICS
Convair Division
7 Jan 80

Image 10-3: The deployment system for the Centaur upper stage in the Space Shuttle, 1980. (Photo courtesy of General Dynamics)

predeployment check of the Centaur. The upper stage then rotated 45 degrees out of the payload bay to a launch-ready position 20 minutes before it was to be released. Special spring thrusts on the CISS (twelve compressed coil springs built by Lockheed and called a Super*Zip separation ring) enabled the Shuttle crew to eject the Centaur from the deployment adapter into free flight at 1 foot per second.[46] The Centaur then coasted for 45 minutes prior to its main burn, allowing the astronauts to take the Shuttle a safe distance away. For planetary missions only a single burn was necessary. The final maneuver was to separate the spacecraft (communications satellite or interplanetary probe) from the Centaur. Once released the Centaur then was capable of maneuvering itself so that it would not interfere with the flight of the spacecraft, and would also prevent planetary impact.

Safety was always a key concern with the Centaur, and one of the major functions of the CISS was to address potential problems with the fluids and avionics systems. Increased safety meant "an ability to tolerate multiple failures before some unwanted consequences take place."[47] Special redundancy features were installed to comply with stringent safety considerations. The engineers believed that these would significantly reduce the dangers of using liquid hydrogen fuel. To minimize the risk, a unique propellant fill, drain, and venting system was installed on three shuttles (*Challenger, Discovery,* and *Atlantis*) designated to carry the Centaur.[48] This system enabled the Shuttle crew to dump the dangerous propellants in case of an emergency, allowing the Shuttle, Centaur, and its payload to safely return to the ground. The safety issues appeared worth the risk because this new launch system enabled larger spacecraft to engage in longer voyages into the solar system.[49] In short, the decision was made that this minimized risk was worth the scientific and commercial reward.

But one important question remained: Was NASA placing too much of its hopes for success on the Shuttle and the ideal of reusability? Launch systems like the Titan/Centaur were taken out of production because they were expendable vehicles. In the new era of reusability these old rockets did not satisfy the new vision of space flight. However, if the Shuttle failed to materialize or was unsuccessful, how would other payloads, such as unmanned planetary missions and satellites make the voyage into space? In 1981 one reporter for the *New York Times* speculated that a Shuttle problem would confirm the "worst fears of those who had criticized the agency for its failure to allow for backup systems to be kept in production while awaiting the shuttle."[50] Was the reusable vision myopic?

Management: To the Sun and Jupiter

The destinations for the first two Shuttle/Centaur launches were both extremely important for the scientific knowledge they would uncover about our solar system. The Sun and Jupiter, icons of mystery in the sky for as long as humanity has turned their gaze to the heavens, were targets for the first probes to be delivered by Centaur. Coupled with this importance was also a significantly high level of mission complexity. With only a six-day launch window between them, beginning 16 May 1986, there was no margin for error.[51]

In the 1980s, Project Galileo was NASA's only planned planetary mission in production. It had its genesis at JPL in the late 1970s. Named for the seventeenth-century Italian scientist who discovered Jupiter's moons, its objective was to explore the Jovian planetary system. The Galileo team was frustrated by the difficulty in securing a ride for their probe. Originally this mission was scheduled to be launched with a two-stage booster (IUS) from the Shuttle when the 1983 NASA budget included funding for Galileo but not for Centaur.[52] When Centaur was returned to funding status, NASA abandoned the IUS in favor of a faster ride for the *Galileo* probe. Initially the Galileo team was pleased with this decision, though they never suspected that in just four years they would again have to search for an alternate ride.

The second Shuttle/Centaur mission, also scheduled to launch in May 1986, was a joint venture with the European Space Agency. Originally called the International Solar Mission, the destination for this spacecraft was the Sun and the exploration of the solar environment. Renamed *Ulysses* in 1984, the mythological change provided allusions to Homer's hero and Dante's description of his urge to explore "an uninhabited world behind the Sun." Ironically, in journeying to the Sun the probe would first travel to Jupiter (the same destination as *Galileo*) and then use the gravitational mass of the planet to slingshot it over the solar poles. This would mark the first observations of these regions of the Sun. Scientists hoped to uncover knowledge about the solar wind, the heliosphere's magnetic field, the interplanetary magnetic field, cosmic rays, and cosmic dust.[53] But the question remained: Would Shuttle/Centaur be ready in time for the launches?

With the mission objectives established the main task became managing Centaur so it could become launch ready for May 1986. Even though the Lewis laboratory was thrilled to manage Centaur, they were concerned about how difficult the program would be to successfully execute. The significant funding for the project did not come until 1983, and the first planetary launches were scheduled for mid-1986.

The management relations between the various institutions involved in the project were complex. Although the project for Centaur G was jointly shared between NASA and the Air Force, the project management fell upon the shoulders of the Lewis lab while the Department of Defense and NASA Headquarters provided funding and overall direction for Shuttle/Centaur. Important interaction and communication was also maintained between other NASA centers, including Johnson Space Center for the Space Shuttle and Kennedy Space Center for the payload project office. Mission management users included the Jet Propulsion Laboratory for the Galileo project, the Air Force for Department of Defense projects, and the European Space Agency for the Ulysses project. The final area of interaction for Lewis was their contractors. These included General Dynamics for the Centaur vehicle and the CISS, Teledyne for the computer, Honeywell for the guidance system, and Pratt & Whitney for the engines. This complex management arrangement became a challenge in itself.[54]

To help cultivate teamwork Lewis project manager Larry Ross dug deep into what he called his "senior manager's bag of tricks." What he emerged with were trusted managerial and motivation techniques to concentrate the team's attention on May 1986, a date three years in the future. If this launch window was missed, the entire mission would have to be delayed by more than a year.[55] The cost resulting from such a delay was estimated at about $50 million. Ross relied on a motivational symbol to convey the urgency, focus, and sense of teamwork required. The image itself was an ancient Centaur, half man, half horse, emerging from the Shuttle, rotating backward and aiming his arrow into the sky. Ross insisted that his symbol become ubiquitous throughout the various laboratories and offices so that everyone who worked on the project was visually aware of their goal. Even contractors from industry agreed to display the symbol prominently in their work environment. Ross also emblazoned his symbol across a variety of project memorabilia. Campaign type buttons were printed, along with drink coasters and note pads. To reinforce the importance of meeting all deadlines the team members received a special Centaur pocket calendar. Unlike most twelve-month calendars, this one covered an unusual twenty-eight-month duration. It represented the months from January 1984 to May 1986, ending with the acceptable window of the Galileo launch.

Optimism remained strong in 1985 for both the Galileo and Ulysses missions. Fabrication for the G-prime version was well underway, and the structural testing was being completed at the Shuttle/Centaur test facility in Sycamore Canyon, California. Slightly behind schedule was the avionics box and fluid system testing, but this did not represent a major threat to the delay of the program.[56] But despite the technical and managerial success, safety concerns would not go away.

T-Minus One Year: The Countdown Begins

In May 1985, Kennedy Space Center was busy making plans for the Galileo and Ulysses launches. The two shuttles would be housed at Complex 39 on launch pads A and B. To complicate matters there were only six days scheduled in between the launches, with each having only a one-hour window. Officials at Kennedy felt pressure from both the media and the public to duplicate the success of the Viking and Voyager missions and to add to the dramatic images and scientific data these spacecraft returned. A Kennedy briefing warned, "Remember the public's response to . . . pictures of Saturn and Jupiter. They will be looking for more. They and the world will be watching."[57]

Lewis Research Center also felt the pressure for getting the two Centaurs launch ready. With ten months to go to before liftoff, they began their much-publicized "Centaur Countdown." They reported that while "much remains to be done, all activities support the crucial schedule."[58] In August 1985 General Dynamics had their roll-out of the first Shuttle/Centaur (SC-1) in San Diego. The vehicle then underwent three weeks of extensive checking to make sure that all specifications and requirements were met. After passing all of the tests, SC-1 was flown to Cape Canaveral, where the crucial CISS was waiting for its mate. The CISS had been flown to the Cape two months earlier. It was placed on a converted Atlas/Centaur launch pad, where it received its final assembly and checks. [59]

In October 1985, progress was still on track, and Larry Ross reported, "A lot of hard work remains, but all of the tasks required to launch the first Shuttle/

Centaur seven months from today are doable." A test rotation and separation of the Centaur was also completed successfully. Engineers hoped that this was the last time that this test would be performed before the real separation took place orbiting above Earth. SC-2 and CISS-2 were continuing their final inspections at General Dynamics in San Diego, and NASA anticipated transporting them across the United States in November. To reduce the difficulty in housing both the SC-1 and SC-2 at KSC at the same time, the Air Force made their Shuttle Payload Integration Facility available between November and December. This gave NASA the ability to simultaneously process both vehicles. Other good news included the report of a 109 percent engine thrust level for the Galileo mission. This thrust capability enabled a heavier payload to be launched. The extra weight would be used by completely filling the propellant tanks. More fuel enabled a greater launch window, because the additional propellant could be used for flight corrections once in space.[60]

With six months to go the SC-1 and the CISS-1 were still undergoing testing at the Kennedy Space Center. Special attention was given to leak checks of the hydraulic system. A few minor leaks were detected, but all were tracked down and repaired. The next key test was the cryogenic cold flow test, which was held in mid-November. Two successful tests were each performed for the liquid oxygen and liquid hydrogen systems. The first two phases of the Design Certification Review were also passed for SC-1 and CISS-1. The final stage was the presentation of the certification to NASA Associate Administrator Jesse Moore.[61]

In January 1986 the Centaur encountered two significant technical problems. During a test the mounts, which attached the propellant level indicating system to the support inside the liquid oxygen tank, failed. Engineers quickly redesigned the mounts that were then fabricated and installed on the SC-1. The new mounts passed all subsequent tests. The second main problem occurred during the liquid oxygen and liquid hydrogen cold flow tests. Specifically, the valves on the CISS-1 that controlled the fill, drain, dump, and engine-feed functions experienced erratic operations. Using a simulated system, engineers located this problem as well and developed a redesign. New tests were scheduled and 15 January became a "do or die" hurdle for the program.[62] Both propellant tanks were completely filled, pressurized at 110 percent of flight pressure, and all of the major system procedures were tested. Vernon Weyers, Deputy Manager of the Lewis Shuttle/Centaur Project Office, recounted, "Nobody close to the program dared to predict the success which was achieved. . . . The entire avionics system performed flawlessly."[63]

The elation experienced at this tanking was short-lived. Just thirteen days later NASA experienced its darkest moment. Ironically, the year that NASA believed would be the "most productive year in space science activities" turned out to be the most ill fated.[64] On 28 January, Shuttle *Challenger* lifted off from the Kennedy Space Center, and 73 seconds later it exploded over the Atlantic

Ocean, killing its crew. This was the greatest disaster in NASA history and the first time that any American astronaut was ever lost in flight.

Immediately the Centaur program was in jeopardy. Liquid hydrogen was carried in the Shuttle's disposable External Tank and was the reason the *Challenger* exploded so quickly when its O-ring failed. Centaur's first launch with the Shuttle was just four months away. With its reliance upon liquid hydrogen and technical problems associated with it that NASA experienced earlier in the month, the Shuttle/Centaur combination no longer appeared an acceptable risk. Despite years of work, the rocket quickly began to fall out of favor with both Johnson Space Center and NASA Headquarters.

Fighting For Survival

The Centaur team at Lewis was devastated. While the Shuttle/Centaur program was not immediately canceled, its future was in jeopardy, as were the Galileo and Ulysses missions. On 20 February 1986 Jesse Moore sent out the order to postpone the Galileo and Ulysses launches.[65] The very earliest that the two missions could be rescheduled was thirteen months away. This was the time when Jupiter would be on the direct opposite side of the Sun from the earth. In the meantime the *Galileo* probe was moved to the Vertical Processing Facility at KSC where it was mated with the Centaur. Engineers continued to perform compatibility and separation tests in space in the hopes that the missions might still get a green light.[66]

Acting NASA Administrator William R. Graham said that his decision to postpone did not mean that the next Space Shuttle launch would be delayed until after the Galileo and Ulysses launch opportunity the next May. The Shuttle might be back in operation earlier than thirteen months, but he said that two key factors forced his hand to delay these missions:

- That key personnel required to assure a safe and successful launch of either Galileo or Ulysses are preoccupied with the timely analysis of causes of the 51-L accident.
- The consequences of the accident have significantly eroded the schedule margins for launch-site processing required prior to the first flight of the Shuttle/Centaur upper stage.[67]

Despite the guarded optimism, the hopes for an eventual Shuttle/Centaur launch, even if an additional thirteen months of testing could be performed, were beginning to fade. The main concern was the safety factor. Not only was liquid hydrogen extremely dangerous, but the *Galileo* probe used a nuclear propulsion system. If the Shuttle malfunctioned, or if the Centaur failed, the result would not only be a loss of craft and crew, but a—remotely possible—nuclear release. In May 1986 Lewis held meetings on the Centaur safety program for all of those concerned about its performance in light of the Shuttle explosion. The attendees included representatives from General Dynamics, JSC, KSC, the Air Force, TRW, Boeing, Lockheed, Martin Marietta, and Analex. Their goal was to prove that despite the risk presented by liquid hydrogen, careful precautions ensured that the Centaur was safe.[68] But the effort was not enough because they could not agree that the Centaur presented an acceptable risk situation.

In late June, NASA Headquarters released a heartbreaking order to Lewis Research Center: "You are directed to terminate the Shuttle/Centaur upper stage program."[69] After spending nearly $1 billion transforming Centaur for the Shuttle, the program ended.[70] James C. Fletcher, NASA Administrator, gave the order to terminate. The official reason for ending the program was labeled "safety" concerns. One report stated, "The final decision was made on the basis that even following certain modifications identified by the ongoing reviews, the resultant stage would not meet safety criteria being applied to other cargo or elements of the Space Shuttle system." Fletcher stated that "although the Shuttle/Centaur decision was very difficult to make, it is the proper thing to do and this is the time to do it."[71]

The result was catastrophic for the space program. Suddenly the *Galileo* probe again lost its ride into space. Not only was it forced to wait until a solid-rocket upper stage could be built, but the reduced power capability of a solid rocket meant that the entire mission would take much longer. With a Centaur launch, *Galileo* would have arrived at Jupiter in two years. When *Galileo* finally did launch in October 1989 with the less powerful Boeing-built Inertial Upper Stage, it took six years to arrive in the Jovian atmosphere.[72] The loss of the Centaur also dramatically changed the navigational course that *Galileo* flew. With Centaur the *Galileo* flight path was to have been direct. With the solid IUS *Galileo* was required to take a much more circuitous route with the help of gravitational assists. *Galileo* actually flew by Venus once and Earth twice, each time using the planet's gravity to effectively "steal" some of its energy and direct it to the velocity of the craft. While this reduced the duration of the flight, the six years that it did take was three times as long as what Centaur was capable of delivering.

Not only did the Centaur cancellation result in nearly a decade delay of Galileo, it also significantly jeopardized the mission success. In April 1991, *Galileo* attempted to open the large high-gain antenna that was the primary mechanism by which the craft received data from Jupiter.[73] Soon after the attempt, telemetry revealed to the JPL engineers that something had gone wrong. The motors only partially opened the antenna and then they stalled. Over a period of weeks, more than one hundred experts at JPL analyzed the problem and concluded that the reason the antenna failed was because of excessive friction between antenna pins and sockets. Ironically, JPL engineers blamed this problem on the cancellation of Centaur. In 1985 the antenna was shipped from JPL in Pasadena to KSC in Florida. After the *Challenger* explosion, *Galileo* was shipped back across the country in 1986, and then back again in 1989, when it was finally launched. The vibration that the antenna experienced during the cross-country trips by truck loosened the pin lubricant and caused the main antenna to fail.

Like Galileo, the Ulysses mission also suffered from the cancellation of the Shuttle/Centaur program. Upon hearing news of the suspension of the Centaur program the Ulysses team reacted with "deep disappointment." But despite the wasted money and the delay to the Ulysses program, the program director said that "the overriding thought is the bitter blow this must be to all at Lewis and General Dynamics who worked with duration over intensely long hours to achieve an impossible schedule only to have the goal posts taken away and the game abandoned at the last minute."[74] Like *Galileo*, *Ulysses* also eventually made it into space. In May 1991 the spacecraft took its long journey to Jupiter for its eventual slingshot to the Sun.[75]

Ironically, the payload that the ill-fated *Challenger* was scheduled to deliver into space included a Tracking and Data Relay Satellite for placement into orbit. William Burrows later pointed out, "The fact of the matter was that seven people had died trying to get a satellite into orbit that could have been sent there on an expendable."[76] A Centaur, with an Atlas or Titan initial stage, could have delivered this payload without the risk to human life. This sentiment is shared by many of the core Centaur engineering contingent at NASA who never approved of the Shuttle idea. To them the expendable launch vehicle should have never been replaced. As Red Robbins argued, "I think they sold their soul to the devil to fund the Shuttle. . . . The expendable launch vehicle was a good animal and not nearly as expensive as the Shuttle program."[77]

Conclusion: A Changing Tolerance for Risk

President Nixon closed his 1972 announcement of the Shuttle program with a prophetic quote from Oliver Wendell Holmes: "We must sail sometimes with the wind and sometimes against it, but we must sail and not drift nor lie at anchor."[78] Few other statements could have so accurately predicted the future of the Space Shuttle and Centaur's relationship with it. Implicit within Holmes's words was a notion of risk. Sailing is a dangerous endeavor. When the wind blows at our back, that danger is reduced and our chance for safely completing the voyage is increased. However, when we set sail with the gusts directly in our face, churning the waters against the ship's bulkhead, the conditions for ultimate success are reduced. But, Holmes said, inherent in the nature of humanity is the desire to raise anchors and explore the unknown. We endure the dangers. We accept the risks of what we do for we believe that the rewards ultimately justify the sacrifice. The problem inherent in Holmes's eloquent argument for the adventurous spirit is that risk is not a constant factor, equally experienced by everyone. The tolerance for risk changes like the wind, deviating from one person, organization, or era to the next.

The official reason for the termination of the Shuttle/Centaur program is now routinely blamed upon safety reasons. Most recently, in 1999 one historian wrote, this Shuttle configuration was "eventually scuttled because of safety concerns stemming from the Centaur's common-bulkhead tank design and from the difficulties of dumping propellants in abort situations."[79] However, blaming the entire failure upon safety reasons obscures the engineering triumphs that made Shuttle/Centaur flight ready. These safety concerns were nothing new and did not emerge suddenly after the tragic *Challenger* loss. Instead, they were evident from the very start of the program and embedded in the Centaur design. Many engineers considered that the tank design and propellant dumping complexities were part of the acceptable risks of flight. They worked many years to reduce the risks even further to meet NASA standards. In this they were successful. Vernon Weyers concluded, despite the technical challenges and problems encountered, "At the time of the Challenger accident . . . there was little doubt that the Centaur would be ready on time to support the 1986 dual planetary launches."[80] So something more was at play in the decision to terminate beyond safety, and that component was a changing tolerance of risk.

Liquid hydrogen was such an alluring a fuel because of the capabilities that it gave to explore space. The tradeoff for greater power and control was heightened safety concerns. Liquid hydrogen was dangerous, and there was never any question that it was more volatile than the solid propellants. But during the 1970s and early 1980s these risks were considered acceptable. The *Challenger* explosion brought with it a new era of conservatism and a diminished capacity for risk. As Larry Ross argued, "After *Challenger* the equation changed." The design specifications that had been acceptable before *Challenger* were based upon theoretical and technical judgments. In the wake of the explosion new "emotional" judgments were added into the mix. Because of these strong emotional factors, Ross concluded, "there was no crew that would fly it because they were staggered by *Challenger*. It was not rational. It didn't have to be rational. It was human life."[81]

The astronauts were vocal about their concern over Shuttle/Centaur. The *Challenger* was the designated shuttle vehicle to launch the *Ulysses* probe. The commander was to be Frederick (Rick) H. Hauck, making this his third flight, after previously piloting a mission in 1983, and commanding another in 1984. Hauck was well aware of the concerns about putting the expendable upper stage in the Shuttle. He said, "It was considered to be probably one of the most hazardous projects that NASA had ever attempted to fly. . . . The chief of the Shuttle office at the time was John Young and he called Shuttle/Centaur, 'death star.'"[82] With the astronaut corps not behind the Shuttle/Centaur concept, it was doubtful it would ever win support again at Johnson Space Shuttle.

Those engineers who devoted years of their lives to making the Centaur rocket Shuttle-ready had a different perspective on this risk. They knew their rocket as well as they knew themselves, and they trusted it as no one else could have. Their pride was evident the day the first new Centaur rolled out of the General Dynamics assembly plant. The tears in their eyes represented hope for a new era of space exploration. Their enthusiasm did not blind them to the safety problems, but they believed they dealt with them and had reduced the risk to manageable levels for mission success and pilot safety.

There was also a risk-tolerance difference in 1986 between Johnson Space Center and Lewis Research Center that Rick Hauck called "cultural." He said that the Lewis engineers were "highly motivated professionals who believed they were exercising as much caution as needed to be exercised." But generally these engineers believed Centaur could fly in the Shuttle. As a whole, Lewis always believed that "the human spaceflight guys at Johnson were a bit too cautious." Hauck concluded, "There was a natural cultural difference between the groups. The issues about risk are tied up in the cultural issues." While Hauck was on the side that supported grounding the Shuttle/Centaur forever, he conceded that there was an irresolvable difference between the two. Eventually, the less risky solution won out.

Richard Kohrs was right in the middle of the debate between the engineers and the astronaut. Kohrs was the manager of the Technical Integration Office for the Space Shuttle from 1980 to 1985 at the Johnson Space Center. He was also the main point of contact between Lewis Research Center and Johnson, so he deeply understood both of the professional cultures. He conceded that Shuttle/Centaur seemed like a good idea, before they got to the "Now Let's Go Make It Work" phase. But when they began to go through all of the battles concerning weight issues, single point failures, interfaces, redundancy, etc., he and his colleagues at Johnson came to realize that the plan was less desirable. While initially the risks seemed worth taking, the risk tolerance changed as the program progressed. Kohrs recalled, "I think a lot of these risks were accepted before . . . the [technical] guys had to go make it happen. Then of course you got into the flight crew concern of this whole thing."[83] Because Johnson protected the astronauts, and because the astronaut crews became so against the Shuttle/Centaur plan, Kohrs and his colleagues also withdrew support.

In the wake of the Shuttle/Centaur cancellation NASA returned to the expendable launch vehicle program to lessen the backlog of science missions that were waiting for launch. One plan was to seek these services from the private sector. Fletcher said, "NASA's purpose in seeking expendable launch services is to lessen dependence on a single launch system, the Space Shuttle."[84] Besides looking to the private sector for this service, NASA also returned once again to the most reliable and powerful second-stage rocket—the old Centaur.

In 1990 a NASA advisory committee produced a report on the future of

Image 10-4: One of the most emotional experiences of spaceflight, the launch of a Space Shuttle from the Kennedy Space Center. (NASA Photo)

the United States space program and discussed the significance of risk for the agency. They wrote that risk has always been a central feature of all of the greatest human adventures. When Magellan first circumnavigated the earth in 1519 he started his voyage with five ships and a crew of 280. After the three-year voyage, only one ship and 34 crewmen returned. Early test pilots in the 1950s faced a similar risk factor, as one in four died while pursuing the limits of supersonic flight. The committee wrote, "Risk and sacrifice are seen to be constant features of the American experience. There is a national heritage of risk taking handed down from early explorers. . . . It is this element of our national character that is the wellspring of the U.S. space program."[85]

But despite this bold statement of our adventurous heritage, NASA worried that the "spark of adventure is flickering. As a nation, we are becoming risk averse." NASA engineers knew this from their own experience. They could see it in themselves with the risk aversion that kept them from launching Shuttle/Centaur. But risk is a subjective ever-present force that waxes and wanes throughout the minds of all that step into a NASA program. NASA and America would recover from the *Challenger* tragedy. They would never forget their fallen brethren, but they would move on and again accept the risk inherent in the exploration of space. Left behind, though, was Shuttle/Centaur. It would not survive and would forever remain an untested dream, eclipsed by a tragedy not of its own making. As Lewis trajectory engineer Joe Nieberding said, "To this day, the country is hurting for not having a liquid hydrogen upper stage coupled with the Shuttle."[86]

Notes

1. Quoted in Michael L. Norris, "New Booster Rolled Out in San Diego," *Los Angeles Times*, 14 August 1985.

2. Jesse W. Moore quoted in Robert Locke, "General Dynamics Shows Off New Probe-Launching Rocket," *San Diego Tribune*, 14 August 1985.

3. Isaac Asimov, "Coming of Age," foreword in Jerry Grey, *Enterprise* (New York: William Morrow, 1979), pp. 7–9.

4. Irene Sänger-Bredt, "The Silver Bird Story," *Spaceflight* 15 (May 1973): 166–81.

5. Roger D. Launius, "Prelude to the Space Age," in John M. Logsdon, gen. ed., *Exploring the Unknown: Selected Documents in the History of U.S. Civil Space Program*, vol. 1, *Organizing for Exploration*, NASA SP-4407 (Washington, D.C.: NASA, 1995), 1:17.

6. Stuart W. Leslie, *The Cold War and American Science: The Military-Industrial-Academic Complex at MIT and Stanford* (New York: Columbia University Press, 1993).

7. Michael Collins, *Liftoff: The Story of America's Adventure in Space* (New York: Grove Press, 1988), p. 202.

8. T.A. Heppenheimer, *The Space Shuttle Decision: NASA's Search for a Reusable Space Vehicle*, NASA SP-4221 (Washington, D.C.: NASA, 1999).

9. David Baker, "Evolution of the Space Shuttle, Part II," *Spaceflight* 15 (July 1973): 264–68.

10. Henry C. Dethloff, "The Space Shuttle's First Flight: STS-1," in Pamela E. Mack, ed., *From Engineering Science to Big Science: The NACA and NASA Collier Trophy Winners*, NASA SP-4219 (Washington, D.C.: NASA, 1998), p. 286.

11. White House Press Secretary, "The White House, Statement by the President," 5 January 1972, Richard M. Nixon Presidential Files, NASA Historical Reference Collection, NASA History Office, NASA Headquarters, Washington, D.C. (hereafter cited as NASA Historical Reference Collection), as found in "Nixon Approves the Space Shuttle," in Roger D. Launius, *NASA: A History of the U.S. Civil Space Program* (Malabar, Fla.: Krieger Publishing, 1994), p. 232.

12. For a comprehensive bibliography on the history of the Space Shuttle, see Roger D. Launius and Aaron K. Gillette, *Toward a History of the Space Shuttle: An Annotated Bibliography*, Studies in Aerospace History, no. 1 (Washington, D.C.: NASA, December 1992).

13. "Shuttle/Centaur Baseline IUS Study for SAMSO," 1975, NASA Glenn Research Center Archives, Glenn Research Center, Brook Park, Ohio (hereafter cited as GRC Archives), drawer H-D.

14. "Reusable Shuttle Planned for 70's," GRC Archives, *Lewis News*, 27 February 1970, pp. 3–4. After the United States committed to the idea of a space shuttle, one of the next important questions to answer was, What upper propulsive stage would it use to launch vehicles outside Earth's orbit? "Centaur/Shuttle Integration Study," General Dynamics, 1 August 1973, Contract NAS 3-16786, 1 August 1973, p. 2, GRC Archives, drawer 4-C.

15. James R. Hansen, *Spaceflight Revolution: NASA Langley Research Center from Sputnik to Apollo*, NASA SP-4308 (Washington, D.C.: NASA, 1995), p. 283.

16. "Reusable Centaur Study," General Dynamics, Contract NAS 8-30290, 26 September 1973, pp. 222, 233. NASA GRC Archives, drawer 4-C.

17. M. Mitchell Waldrop, "Centaur Wars," *Science* 217 (10 September 1982): 1012–14.

18. This was initially called the "interim" upper stage—the name was changed when it became apparent that the IUS was going to be around for a long time.

19. "Boeing rebuttal to Shuttle/Centaur," undated, NASA GRC Archives, drawer 4-C.

20. Robert A. Frosch, "Frosch on Centuar," GRC Archives, *Lewis News*, 30 January 1981, p. 1.

21. Waldrop, "Centaur Wars," p. 1012.

22. In 1999 this center was renamed the Glenn Research Center at Lewis Field.

23. Lawrence J. Ross, "Launch Vehicles," GRC Archives, *Lewis News*, 4 January 1980, p. 4.

24. Neil Steinberg, "Centaur Star Still Glows Brightly in Agency Plans," GRC Archives, *Lewis News*, 15 August 1980, p. 4.

25. Chris C. Kraft Jr., William R. Lucas, Dick G. Smith to Alan M. Lovelace, 19 January 1981, Larry Ross personal document collection.

26. Larry Ross interview with Mark D. Bowles, 1 March 2000.

27. John F. McCarthy Jr. to Alan M. Lovelace, 25 March 1981, Larry Ross personal document collection.

28. Andrew J. Dunar and Stephen P. Waring, *Power to Explore: A History of the Marshall Space Flight Center, 1960–1990*, NASA SP-4319 (Washington, D.C.: NASA, 1999), p. 138.

29. Alan Lovelace to William R. Lucas, 27 May 1981, Larry Ross personal document collection.

30. L.J. Ross to Chief Counsel, 7 December 1983, Box 2142, Division STED Project Management Office, Folder General Correspondence, NASA History Office, NASA Headquarters, Washington, D.C. (hereafter cited as Division STED Project Management Office).

31. L.J. Ross to Brigadier General Donald L. Cromer, 1 August 1984, NASA Historical Reference Collection, Box 2142, Division STED Project Management Office, Folder Correspondence with Air Force 1982–1984.

32. "Technical Evaluation of GDC Proposal, Shuttle/Centaur G Development and Production," 15 December 1983, NASA Archives, Box 2142, Division STED Project Management Office, Folder Centaur G Program.

33. Jim Kukowski, "Contracts Awarded for Shuttle Launched Upper Stage," 2 June 1981, NASA HQ Press Release, GRC Public Affairs Archives; Linda Peterson, "NASA Lewis Awards $7.5 Million in Contracts for Shuttle/Centaur Development," 2 June 1981, LeRC Press Release, GRC Public Affairs Archives; "Four Contracts Awarded for Modifying Centaur," GRC Archives, *Lewis News*, 2 July 1981, p. 3.

34. Paul T. Bohn, "NASA Lewis Gets Added Funding of $107.4 Million," 27 July 1982, LeRC Press Release, GRC Public Affairs Archives; "Center in Mainstream of Shuttle Program," GRC Archives, *Lewis News* (30 July 1982): 1; "Additional Funds Resurrect Important Programs," GRC Archives, *Lewis News*, 13 August 1982, pp. 1, 6.

35. Andrew J. Stofan to Lewis Employees, 17 September 1982, NASA Archives, Box 2142, Division STED Project Management Office, Folder General Correspondence.

36. Larry Ross, "Space Flight Systems," GRC Archives, *Lewis News*, 4 February 1983, p. 2.

37. Shuttle/Centaur Project Office, Functional Statements from the NASA Organizational Manuel, 1 October 1983, NASA Archives, Box 2143, Division STED Project Management Office, Folder Shuttle/Centaur Organization—LeRC.

38. NASA/DOD Agreement, NASA Archives, Box 2141.

39. Jeweline H. Richardson, "The Centaur G-Prime: Meeting Mission Needs Today for Tomorrow's Space Environment," May 1983, NASA Archives, Box 2142, Division STED Project Management Office, Folder Technical Paper Approvals by PM. "NASA Lewis Research Cen-

ter/USAF Space Division Agreement for the Management of the Shuttle/Centaur Program 10/25/82," NASA Historical Reference Collection, Box 2141.

40. W. F. Rector III and David Charhut, "Centaur for the Shuttle Era," May 1984, NASA Historical Reference Collection, Box 2142, Division STED Project Management Office, Folder Technical Paper Approvals by PM.

41. "Technical Evaluation of GDC Proposal, Shuttle/Centaur G Development and Production," 15 December 1983, NASA Historical Reference Collection, Box 2142, Division STED Project Management Office, Folder Centaur G Program.

42. Harold Hahn (General Dynamics), "A New Addition to the Space transportation System," July 1985, NASA Historical Reference Collection, Box 2142, Division STED Project Management Office, Folder Technical Paper Approvals by PM.

43. Omer F. Spurlock (aerospace engineer Lewis Research Center), "Shuttle/Centaur— More Capability for the 1980s," NASA Historical Reference Collection, Box 2142, Division STED Project Management Office, Folder Technical Paper Approvals by PM. This article is the best technical overview of the entire Shuttle/Centaur.

44. R. Wood and W. Tang, "Shuttle/Centaur Prestressed Composite Spherical Gas Storage Tank," July 1985, NASA Historical Reference Collection, Box 2142, Division STED Project Management Office, Folder Technical Paper Approvals by PM.

45. Larry J. Ross, "Space," GRC Archives, *Lewis News,* December 1981, p. 3.

46. R.E. Martin (from General Dynamics Space Systems Division), "Effects of transient Propellant Dynamics on Deployment of Large Liquid Stages in Zero-gravity with Application to Shuttle/Centaur," delivered at the 37th International Astronautical Congress, October 1986, Innsbruck, Austria, NASA Historical Reference Collection, Box 2142, Division STED Project Management Office, Folder Technical Paper Approvals by PM.

47. Larry Ross interview, 1 March 2000.

48. "Centaur for STS," 8 January 1986, NASA Historical Reference Collection, Box 2142, Division STED Project Management Office, Folder General Correspondence.

49. AP Press release, "General Dynamics," 14 August 1985, NASA Historical Reference Collection, Box 2141.

50. John Noble Wilford, "U.S. to Discontinue New Rocket System," *New York Times,* 18 January 1981.

51. John R. Casani to William H. Robbins, 9 April 1985, NASA Historical Reference Collection, Box 2143, Division STED Project Management Office, Folder Galileo Correspondence.

52. John R. Casani to Galileo Review Board, 6 January 1982, NASA Historical Reference Collection, Box 2143, Division STED Project Management Office, Folder Galileo Correspondence.

53. Jim Kukowski, "Ulysses New Name for International Solar Polar Mission," 10 September 1984, NASA HQ Press Release, GRC Public Affairs Archives.

54. William H. Robbins and Vernon Weyers, "Shuttle/Centaur Project Stages Challenge," GRC Archives, *Lewis News,* 22 March 1985, p. 1.

55. Vernon J. Weyers to Kevin L. Wohlevar, Cray Research, Inc., 16 August 1985, NASA Historical Reference Collection, Box 2141, Division STED Project Management Office, Folder General Correspondence.

56. Harold Robbins and Vernon Weyers, "Shuttle/Centaur Project Stages Challenge," GRC Archives, *Lewis News,* 22 March 1985, p. 1.

57. "Charter Planetary Panel Galileo and Ulysses Mission, Kennedy Space Center," NASA Archives, Box 2142, Division STED Project Management Office, Folder Kennedy Correspondence.

58. Vernon Weyers, "Centaur Countdown," GRC Archives, *Lewis News,* 9 August 1985, p. 2.

59. Vernon Weyers, "Centaur Countdown," GRC Archives, *Lewis News,* 6 September 1985, p. 3.

60. Vernon Weyers, "Centaur Countdown," GRC Archives, *Lewis News,* 1 November 1985, p. 5.

61. Vernon Weyers, "Centaur Countdown," GRC Archives, *Lewis News,* 29 November 1985, p. 4.

62. Vernon Weyers, "Centaur Countdown," GRC Archives, *Lewis News,* 10 January 1986, p. 3.

63. Vernon Weyers, "Centaur Countdown," GRC Archives, *Lewis News,* 7 February 1986, p. 3.

64. James F. Kukowski, "NASA Calls 1986 'A Year for Space Science,'" 27 November 1985, NASA HQ Press Release, GRC Public Affairs Archives; "1986: A Year for Space Science," GRC Archives, *Lewis News,* 10 January 1986, p. 3.

65. Jesse Moore, "Galileo and Ulysses Launch delay," 20 February 1986, NASA Historical Reference Collection, Box 2143, Division STED Project Management Office, Folder Status Reports.

66. "Galileo, Centaur Milestone Tests are Underway," GRC Archives, *Lewis News,* 16 May 1986, p. 2.

67. James F. Kukowski, "NASA Postpones Galileo, Ulysses, Astro-1 Launches," 10 February 1986, NASA HQ Press Release, GRC Public Affairs Archives; William R. Graham, "Galileo and Ulysses Missions Postponed for Year," GRC Archives, *Lewis News,* 21 February 1986, p. 6.

68. William E. Klein, "Minutes for Shuttle/Centaur G Safety Certification Process Briefing," 2 June 1986, NASA Historical Reference Collection, Box 2142, Division STED Project Management Office, Folder Technical Paper Approvals by PM.

69. "Shuttle/Centaur Termination Status" NASA Headquarters, 4 September 1986, NASA Historical Reference Collection, Box 2144.

70. John M. Logsdon, "Return to Flight: Richard H. Truly and the Recovery from the *Challenger* Accident," in Mack, *From Engineering Science to Space Science,* p. 360.

71. Sarah G. Keegan, "NASA Terminates Development of Shuttle/Centaur Upper Stage," 19 June 1986, NASA HQ Press Release, GRC Public Affairs Archives.

72. Barbara Selby, Leon Perry, and Terry Eddleman, "Upper Stage Selected for Planetary Missions," 26 November 1986, NASA HQ Press Release, GRC Public Affairs Archives.

73. Paula Cleggett-Haleim, "Galileo Antenna Deployment Studied by NASA," 19 April 1991, NASA HQ Press Release, GRC Public Affairs Archives.

74. D. Eaton to Larry Ross, 11 July 1986, NASA Archives, Box 2143, Division STED Project Management Office, Folder Galileo Correspondence.

75. Paula Cleggett-Haleim and Robert MacMillin, "Ulysses to Begin Jupiter Physics Investigations," 28 May 1991, NASA HQ Press Release, GRC Public Affairs Archives.

76. William E. Burrows, *This New Ocean: The Story of the First Space Age* (New York: Random House, 1998), p. 556.

77. Red Robbins, interview with Mark D. Bowles, 21 March 2000.

78. White House Press Secretary, "The White House, Statement by the President," 5 January 1972, as found in "Nixon Approves the Space Shuttle," Launius, *NASA,* p. 235.

79. Ivan Bekey, "Exploring Future Space Transportation Possibilities," in John M. Logsdon, gen. ed., *Exploring the Unknown: Selected Documents in the History of the U.S. Civil Space Program,* vol. 4, *Accessing Space,* NASA SP-4497 (Washington, D.C.: NASA, 1999), 4:509.

80. Vernon Weyers, "Development of the Shuttle/Centaur Upper Stage," October 1986, paper delivered at the IAF Congress in Innsbruck, Austria, NASA Historical Reference Collection, Box 2143, Division STED Project Management Office, Folder IAF Paper.

81. Larry Ross interview, 1 March 2000.

82. Rick Hauck, interview with Mark D. Bowles, 23 August 2001.

83. Richard Kohrs, interview with Mark D. Bowles, 13 July 2001.

84. Sarah Keegan, "NASA Plans Use of Expendable Launch Vehicles," 15 May 1987, NASA HQ Press Release, GRC Public Affairs Archives.

85. "Report of the Advisory Committee on the Future of the U.S. Space Program," December 1990, NASA publication, pp. 16–17.

86. Joe Nieberding, interview with Virginia P. Dawson and Mark D. Bowles, 15 April 1999.

–11–

The Quest for Reusability

Andrew J. Butrica

The quest for a reusable launcher has been long, yet it has met with only partial success. The single greatest touted advantage of the fully reusable rocket is that it reduces launch costs. Comparing reusable and expendable rockets is not simple, but a rather complicated task, not unlike the proverbial comparing of apples and oranges. In order to compare the costs of the two types of rockets, and therefore ascertain the cost advantage of reusable over expendable launchers, we must consider two types of costs, recurring and nonrecurring. Nonrecurring costs entail those funds spent on designing, developing, researching, and engineering the launcher (DDR&E costs). Recurring costs fall into two categories: expenses for building the launcher and the costs of operations and maintenance.

The costs of designing, developing, researching, and engineering reusable launchers are necessarily higher than those for expendable launchers, because creating reusable rockets is more challenging technologically. For example, a reusable launcher must have advanced heat shielding to allow it to reenter the atmosphere not once but many times. Throwaway rockets have no need for such heat shielding. In addition, we possess a profound knowledge of expendable rocket technologies thanks to our long experience (over half a century) with ICBMs and other throwaway rockets. Many of the technologies needed to build a fully reusable launcher, however, remain in the elusive future.

On the other hand, construction costs favor reusable launchers. For each launch, the cost of building a new expendable rocket is a recurring expense. For reusable launchers, construction cost are nonrecurring, but part of the up-front costs to be amortized over each launch. The result of this intrinsic difference between the two types of launchers leads to a trade-off between the lower DDR&E costs of expendable rockets and the lower recurring costs of reusable launchers. In making that trade-off, one must take into account a number of other realistic factors that favor expendable launchers.

Although one can amortize reusable rocket construction costs over many flights, reusable rockets are far more expensive to build than expendable rockets. For example, building a full-scale version of the VentureStar™ would cost (conservatively) more than the $1 billion NASA spent on the X-33 program. That same amount of money would buy ten expendable rockets at $100 million each. The knowledge gained in constructing a large number of a given disposable launcher actually helps to lower construction costs. Thus, in order to compete with the low DDR&E and construction costs of the established expendable industry, a reusable launcher would have to fly more than fifty times.

In light of the economies realizable with expendable launchers, and the technological difficulties and complexities of realizing a fully reusable rocket, the current launch market's appetite for disposable launchers is understandable. The gamble of the reusable launcher is that a small fleet of three to five craft could put payloads into orbit for less than the cost of an expendable rocket. A commercial builder and operator of reusable launchers, however, would be burdened by the need to amortize DDR&E and construction costs over each mission. Some obvious solutions would be to have the government pay for most or all of the DDR&E costs and for government to buy one or two reusable craft for its exclusive use.

The preceding discussion applies to a comparison of expendable rockets with fully reusable launchers. The economics of launching a reusable vehicle atop an expendable booster are rather different. Such hybrid systems are technologically more achievable than fully reusable single-stage or two-stage rockets. A variety of launchers that combine reusable and expendable stages are currently under development by companies and government, and they appear to promise reductions in the cost of placing payloads in orbit. Moreover, throughout the decades-long quest for reusability, the configuration of a reusable reentry vehicle atop a throwaway booster (a so-called boost-glide system) has dominated launcher thinking from as early as the 1920s and 1930s and continues even today. In these boost-glide systems, the upper-stage vehicle, once released from its booster rocket, climbs into orbit on its own power, then glides to a landing. Some reusable suborbital vehicles launch from a large jet, such as a B-52 or an L-1011. Until the advent of the Space Shuttle, no orbital boost-glide systems existed.

Aerospace Planes

Another early, but less prevalent, reusable vehicle concept was that of the aerospace plane. Aerospace planes essentially are airplanes that take off horizontally and fly into space powered by either rockets or special air-breathing jet (ramjet and scramjet) engines capable of operating in either the atmosphere or space. Devising a practical version of such a sophisticated engine has been and continues to be the major barrier to realizing air-breathing aerospace plane concepts. Like the boost-glide system, aerospace plane concepts appeared before World War II.

The American rocketeer Robert Goddard, for example, in a *Popular Science* article published in December 1931, described an aerospace plane ("stratosphere plane") with elliptically shaped wings and propelled by a combination air-breathing jet and rocket engine. The rocket engine drove the vehicle while it was outside the atmosphere, and two turbines moved into the rocket's thrust stream to drive two large propellers on either wing, thereby powering the vehicle while in the atmosphere.[1] German researcher Eugen Sänger, in his 1933 book on rocket flight, described a rocket-powered suborbital spaceplane known as the Silbervogel (Silver Bird), fueled by liquid oxygen and kerosene and capable of reaching a maximum altitude of 100 miles and a speed of Mach 10. Later, working with his future wife, the mathematician Irene Bredt, and a number of research assistants, Sänger designed the Rocket Spaceplane, launched from a sled at a speed of Mach 1.5. A rocket engine capable of developing 100 tons of thrust would boost the craft into orbit, where it could deploy payloads weighing up to one ton.[2]

Aerospace planes remained largely fictional concepts until 1957, the year the Soviet Union launched its first Sputnik satellite, when the U.S. Air Force initiated what became the Aerospaceplane program to develop a single-stage-to-orbit vehicle powered by an air-breathing engine. By 1959, the project had evolved into the Recoverable Orbital Launch System (ROLS), a single-stage-to-orbit design that would take off horizontally and fly into a 300-mile-high orbit. The ROLS propulsion system collected air from the atmosphere, then compressed, liquefied, and distilled it in order to make liquid oxygen, which mixed with liquid hydrogen before entering the engines. This complicated propulsion system, dubbed LACES (Liquid Air Collection Engine System), later renamed ACES (Air Collection and Enrichment System), as well as various scramjet engine concepts, underwent Air Force evaluation over time. Faced with the uncertainties of the single-stage design, the Air Force shifted the focus of the Aerospaceplane to two-stage-to-orbit concepts in 1962, and following the program's condemnation by the Scientific Advisory Board, the Aerospaceplane died in 1963. Congress cut fiscal 1964 funding, and the Pentagon declined to press for its restoration.[3]

Dyna-Soar

The more dominant reusable vehicle concept, however, has been the boost-glide system. The Peenemünde rocket group under Werner von Braun, which developed the A-4 missile (commonly known as the V-2), originally planned to develop a much larger missile, the A-10/A-9, capable of delivering a 1-ton bomb over 3,125 miles away. The A-10 first stage was a conventional booster rocket, while the A-9 upper stage was a winged vehicle that could glide at supersonic speeds before hitting its target. Additional Peenemünde work, but kept secret from the Nazis, included a manned version of the A-9 that would launch vertically and land horizontally, like the Space Shuttle. An even larger vehicle, the A-12, was a fanciful three-staged launcher whose top stage was a reusable winged reentry vehicle.[4] None of these concepts, however, were orbital vehicles.

At the end of World War II, as is widely known, Werner von Braun and much of the German rocket program became a vital part of the United States' own missile program and contributed to the development of boost-glide systems. Walter Dornberger, another key German rocketeer, became a consultant on guided missiles to the Air Materiel Command at Wright-Patterson Air Force Base. After leaving the Air Force in 1950, he became a consultant for Bell Aircraft and persuaded the firm to undertake a study of boost-glide technology. In 1952, that study led to the joint development by Bell and the Wright Air Development Center (WADC) of a manned bomber missile and reconnaissance vehicle called BoMi. A two-stage rocket would lift BoMi, which would operate at speeds over Mach 4. By 1956, the BoMi study work had evolved into a contract for Bell to develop Reconnaissance System 459L, commonly known as Brass Bell, a manned two-stage boost-glide reconnaissance system, while the manned bomber part of the BoMi work became RoBo, a manned hypersonic, rocket-powered craft for bombing and reconnaissance missions.[5]

The Air Force's ASSET (Aerothermodynamic/elastic Structural Systems Environmental Tests) boost-glide system originated in 1959. The program involved lofting small reusable hypersonic gliders from Cape Canaveral on top of expendable rockets. After two failed launches, the third ASSET vehicle flew in 1964. It was in excellent condition when McDonnell Aircraft disassembled it. Despite some water damage, the vehicle could have been refurbished and flown again. ASSET was the first practical experience the Air Force had with an actual lifting reentry vehicle returning from space at near-orbital velocities (in excess of Mach 15). However, potentially reusable the craft may have been, as none were ever flown more than once, it could not be turned around rapidly. At the end of the ASSET program, the Air Force concluded that a minimum of four months between flights was optimal for vehicle preparation and effectiveness.[6]

A major step in orbital boost-glide systems was the Dyna-Soar (for Dynamic Soaring) program. It was the final stage of a three-stage study of rocket-powered hypersonic flight initiated by the National Advisory Committee for Aeronautics (NACA) with Air Force participation. The study used a series of experimental aircraft ("X" vehicles) lifted into the sky by reusable aircraft. "Round One," to use the NACA nomenclature, consisted of the Bell X-1 series, the Bell X-2 series, and the Douglas D-588-II Skyrocket. Data derived from

Image 11-1: Dyna-Soar was one of the earliest attempts to fly to and from space with a winged, reusable vehicle. (NASA Photo)

Round One provided the basis for the next higher and faster stage. Round Two was the series known eventually as the X-15. Round Three called for testing winged orbital reentry vehicles.[7]

The Air Force's Dyna-Soar program emerged from a 1957 consolidation of NACA's Round Three and military hypersonic flight programs (including RoBo and Brass Bell). Eventually, NASA participated in the project as well. Dyna-Soar was a major milestone in the quest for reusability. Launched on an expendable booster, the Dyna-Soar X-20 would fly orbital or suborbital trajectories, perform reconnaissance at hypersonic speeds, and land horizontally like an aircraft at any one of a number of U.S. air bases. Although the Dyna-Soar vehicle was never built, a prototype was near completion when Secretary of Defense Robert McNamara terminated the program on 10 December 1963, only eight months before drop tests from a B-52. The first manned flight had been scheduled for 1964. Military space planners preferred instead a military version of NASA's Gemini program, known as Blue Gemini, as a preliminary step toward establishing a military space station, the Manned Orbiting Laboratory (MOL) (known also as the KH-10) capable of supporting a two-man crew.[8]

Dyna-Soar had a lot to offer the Air Force and the nation and might have changed history. The military might have benefited economically by possessing the world's first reusable orbital vehicle, and the Pentagon would not have become NASA's political ally in the space agency's political struggle to win funding for its Shuttle program. On the other hand, Dyna-Soar could have provided NASA a less expensive, but two-stage,Shuttle. The knowledge gained from the research program, which included over fourteen thousand hours of wind tunnel tests, could be applied to a number of applications from glide bombers to future spacecraft. Moreover, after termination of the program, Boeing carried out a small "X-20 continuation program" for several more years that involved testing various X-20 components and design features both in ground facilities and on flight research vehicles. The René 41 high-temperature nickel alloy developed for the X-20 reappeared in the 1970s as part of the airframe structure and heat shielding for Boeing's Reusable Aerodynamic Space Vehicle (RASV).[9]

Lifting Bodies

Also of note among these early boost-glide systems was a group of reusable suborbital vehicles known as lifting bodies. A lifting body is a wingless aerodynamic shape that develops lift, the force that makes winged craft fly, because of its peculiar body shape. Research on lifting bodies began in early 1957 at NACA's Ames Aeronautical Laboratory (now NASA's Ames Research Center). The NACA effort stimulated a small but growing nationwide interest in the potential of lifting bodies for future spacecraft. Following NASA's success with its wooden M2-F1, the Air Force joined NASA at Edwards AFB in the test flight program of the rocket-powered M2-F2, launched from a B-52 from 1966 until its crash in 1967.[10]

The most prominent of these lifting body craft was the Air Force's X-24B, built by Martin Marietta in 1972. A modified X-24B powered by aerospike engines was the Lockheed's shuttle concept, the StarClipper, while the X-24B's shape also inspired the design of what eventually became Lockheed Skunk Works' X-33 launch vehicle. Despite the apparent name similarity, the X-24B and X-24A had rather different shapes and distinct origins, though both were part of the Air Force's lifting body program.

The X-24A started in 1964 as a two-part Air Force program carried out in cooperation with NASA and dubbed START for Spacecraft Technology and Advanced Reentry Tests. One part of START, PRIME, consisted of hypersonic (boost-glide) flight tests, while PILOT (which became the X-24A) formed the low-speed test component. Three PRIME flights took place between September 1966 and May 1967, demonstrating precision accuracy in guiding the craft to a preselected point, while accomplishing the first cross-range maneuver of a spacecraft during flight. The third PRIME craft was capable of flying a second time, a milestone in demonstrating the reusability of lifting spacecraft, had the Air Force elected to undertake an additional flight test.[11]

The PILOT SV-5P was a manned, rocket-powered lifting-body glider released from a B-52. The craft came to be known as the X-24A in 1967, when it rolled out at the Martin Aircraft Company's Middle River, Maryland, plant. The first powered X-24A flight took place on 19 March 1970, at Edwards AFB, and the craft continued to fly until 4 June 1971. By approaching and landing from an altitude of 71,110 feet on one test flight, the X-24A achieved a major milestone: demonstration that an unpowered shuttlecraft could land on a conventional runway after returning from space.[12]

Like the X-24A, the X-24B was a rocket-propelled research lifting body aircraft dropped from a B-52. However, it had a rather different body shape, and was the product of research at the Air Flight Dynamics Laboratory (FDL) of Wright-Patterson AFB. Completed in the fall of 1972, the X-24B completed its first glide flight on 1 August 1973. The X-24B flight program ended with a series of flights in 1975 to demonstrate unpowered landing accuracy on concrete runways for NASA's Shuttle project, then underway.[13]

The RASV

Even as NASA and industry were building the Space Shuttle, the search for a reusable shuttle replacement was underway. As with lifting body research, NASA led the way. In 1972, the Langley Research Center, with the approval of NASA Headquarters, set up a small group to study the possibility of growing an aircraft known as the Continental/Semiglobal Transport (C/SGT) into a single-stage-to-orbit vehicle. The C/GST would take off, almost attain orbit, then land, delivering people or cargo to any place on Earth in less than two hours. Langley analysis of the vehicle suggested that with just a little bit more speed, the C/GST could achieve orbit.[14]

Using Space Shuttle technology as the starting point for their study of the structures, materials, and engines needed for a Shuttle replacement, the Langley analysis team evaluated the impact of improving structures and materials (such as composites) beyond the Shuttle on various configurations. The improved materials promised to reduce overall vehicle weight significantly, thereby bringing single-stage-to-orbit transport within the realm of the possible.[15] Next, in 1975, Langley funded two industry studies of single-stage-to-orbit rocket concepts carried out by teams from Martin Marietta–Denver and Boeing–Seattle. The stated purpose of the study was to determine the future technology development needed to build an operational rocket-powered, single-stage-to-orbit Space Shuttle replacement by 1995. Each team concluded that such a vehicle was feasible using technology available in the near term.[16]

Following this 1975 industry study, Langley continued internal studies of single-stage-to-orbit and other advanced launcher concepts into the 1980s, when it initiated a Shuttle replacement study known as Shuttle II.[17] NASA did not undertake the technology development outlined and solicited in those studies, because it was preoccupied with the existing Shuttle. Meanwhile, Boeing took their vehicle design from the Langley 1975 study, which the firm had been developing on its own since 1972, and tried to sell it to the Pentagon. The company's interest in the reusable single-stage-to-orbit vehicle was "based on the belief that the reusable airplane type operation of earth orbit transportation vehicles will allow considerable improvement in cost per flight and flexibility."[18] Those involved in the 1975 Langley study came to realize that the real key to lowering operational costs was to reduce the number of people doing operational and maintenance work using what they called (for want of a better name) "operations technology," that is, new electronic and computer hardware and software that would allow the operation and maintenance of reusable space vehicles with far fewer workers.[19]

The Boeing vehicle would have incorporated both proven and unproven technologies. The cylindrically-shaped, delta-winged reusable single-stage-to-orbit craft, powered by Space Shuttle Main Engines, would take off with the help of a sled and land horizontally on a conventional runway. It would have used a combination of aluminum-brazed titanium and René 41, a high-temperature nickel alloy developed for the Dyna-Soar X-20, for both its structure and heat shielding. The vehicle stored liquid hydrogen fuel in its body and liquid oxygen in its wings. The integration of the liquid hydrogen and liquid oxygen tanks into the load carrying structure (that is, the wings and the main body of the craft), combined with the metallic shell made of honeycomb panels, went far in reducing overall vehicle weight.[20]

Boeing soon interested the Air Force Space and Missiles System Organization (Los Angeles Air Force Station) in this vehicle concept. The Air Force dubbed it the Reusable Aerodynamic Space Vehicle and in 1976 provided funding for a seven-month preliminary feasibility study of the RASV concept. The study concluded (not surprisingly) that the RASV was feasible and that it would fulfill Air Force requirements. Among those requirements were flying 500 to 1,000 times "with low cost refurbishment and maintenance as a design goal" from a launch site in Grand Forks, North Dakota, into a polar orbit or once around the planet in a different orbit. The vehicle would have to reach "standby status within 24 hours from warning. Standby to launch shall be three minutes."[21]

An early Boeing appraisal estimated that going from concept development to testing of the first full-scale RASV would have taken ten years and would have cost over $3 billion.[22] Nonetheless, in talks with the Air Force, Boeing proposed a first flight in 1987. In all, the Air Force invested $3 million in the project for technology development. The Air Force had become convinced that the RASV potentially could provide a manned platform that could be placed above any point on the planet in less than an hour and which could perform a variety of missions, including reconnaissance, rapid satellite replacement, and general space defense. In December 1982, Boeing Chairman T.A. Wilson gave the RASV effort the go-ahead to propose a $1.4 billion prototype vehicle to the Air Force.[23] Boeing, however, would not build the RASV.

The problem was not the steep technological hurdles that the firm would have to leap, such as development of the sled to accelerate the RASV to a speed of 600 feet per second or achievement of fast turnaround time (twenty-four hours or perhaps as short as twelve hours) for the Strategic Air Command (SAC).[24] The Air Force ordered two classified studies of single-stage-to-orbit technologies, Science Dawn (1983–85) and Have Region (1986–89), conducted by industry partners Boeing, Lockheed, and McDonnell Douglas. They interpreted the study results as demonstrating the technological feasibility of the RASV for SAC.[25] Then, instead of proceeding with further RASV studies, the Air Force chose to develop a space vehicle that not only operated like an aircraft, as the RASV did, but also had air-breathing jet engines. That space vehicle would be known as the National Aero-Space Plane (NASP).

The National Aero-Space Plane

With NASP, the aerospace plane had returned. The conceptual resemblance of the aerospace plane to an airplane (complete with air-breathing engine) appealed to the Air Force and its pilot mentality. Jet engines are a technology that Air Force pilots and officers understand. Not surprisingly, then, in 1982, when the Air Force began studying shuttle replacement concepts in its TransAtmospheric Vehicle (TAV) program, air-breathing engines were a seri-

Image 11-2: The National Aero-Space Plane (NASP) served as a technology driver for the aerospace community throughout the mid-1980s, but the program never got to the flight stage. (NASA Photo)

ous, though not exclusive, consideration. The TAV program considered a variety of both single- and two-stage vehicle configurations, powered by either rocket or jet engines.[26]

The milestone moment for NASP was President Ronald Reagan's state of the union address, delivered on 4 February 1986, just days after the *Challenger* disaster. Reagan declared, "We are going forward with research on a new Orient Express that could, by the end of the decade, take off from Dulles Airport, accelerate up to 25 times the speed of sound attaining low-Earth orbit, or fly to Tokyo within two hours."[27] As portrayed by the president, the Orient Express would be both a high-speed aircraft and a single-stage-to-orbit vehicle, powered by air-breathing engines. Subsequently, the Air Force's TAV program merged with the classified three-phase Copper Canyon program of the Advanced Research Projects Agency (ARPA), which funded research on scramjet hypersonic vehicles. The new, larger program comprised the gamut of government agencies involved in hypersonic air-breathing engine studies at one time or another: NASA, ARPA, the Air Force, the Navy, and the Strategic Defense Initiative Organization (SDIO). On 1 December 1985, the title National Aero-Space Plane replaced all earlier designations.[28]

The NASP program initially proposed to design and build two research craft, the X-30, at least one of which was to achieve orbit by flying in a single stage through the atmosphere at speeds up to Mach 25. The X-30 would use a multicycle engine that shifted from jet to ramjet and to scramjet speeds as the vehicle ascended burning liquid hydrogen fuel with oxygen scooped and frozen from the atmosphere. The engine and vehicle designs had come from Tony du Pont, an aerospace designer who had developed a multicycle jet and rocket engine under contract with NASA then ARPA. Du Pont's vehicle design rested on a number of highly questionable assumptions, optimistic interpretations of results, and convenient omissions (such as landing gear).[29] In short, NASP would never have performed as du Pont alleged. Ivan Bekey, former Director of Advanced Programs at NASA Headquarters' Office of Space Flight, and a firm believer in single-stage-to-orbit transport, characterized NASP in stronger terms. He called it "the biggest swindle ever to be foisted on the country," because the program "was full of dubious aerodynamic claims and engine-performance claims and structural and thermal claims."[30]

NASP, like the aerospace plane program, fell victim to budget cuts, but this time as a result of the end of the cold war. Congress canceled NASP in 1992 during fiscal 1993 budget deliberations. Although the program never came near to building or flying hardware, NASP contributed significantly to the advance of materials capable of repeatedly withstanding high temperatures (on the vehicle's nose and body) or capable of tolerating repeated exposure to extremely low temperatures (the cryogenic fuel tanks). In order to keep the overall craft weight low, these materials also would have to be lightweight. Undertaking materials research and testing after March 1988 was in the hands of a consortium, the NASP Materials and Structures Augmentation Program (NASP MASAP), consisting of all major NASP contractors. By 1990, NASP MASAP had realized significant progress in titanium aluminides, titanium aluminide metal matrix composites, and coated carbon-carbon composites; government and contractor laboratories had fabricated and tested large titanium aluminide panels under approximate vehicle operating conditions; and NASP contractors had fabricated and tested titanium aluminide composite pieces.[31]

The Delta Clipper

The end of NASP was not the end of efforts to realize a fully reusable launch vehicle, however. In parallel with, but never in competition with, NASP was the Single Stage To Orbit (SSTO) Program of the SDIO. The program differed radically from its predecessors, which attempted to develop flight technology; instead, it tested the flight operations of a single-stage-to-orbit vehicle, the Delta Clipper Experimental (DC-X). Its intention was not to develop technology. The goal was to demonstrate "aircraft-like" operations, which included autonomous operations, minimal launch and operational crews, ease of maintenance, abort capability, and short turnaround time. The novelty of the SSTO Program also was to combine the goal of "aircraft-like" operations with the use of an "X" vehicle and a "lean" management approach by both government and industry in the hope of expediting the project and keeping costs low.

The SSTO Program began in the mind of Lockheed aerospace engineer Max Hunter. Shortly before retiring, Hunter came up with a design for a Shuttle replacement that he called alternately the X-Rocket and the XOP, for Xperimental Operational Program, to emphasize the importance of operations. Failing to find support within Lockheed, NASA, or the armed forces,[32] Hunter described his concept to a meeting of the Citizens' Advisory Council on National Space Policy in 1989. Fortuitously, Lt. Gen. Daniel O. Graham (Ret.) was among those in attendance. The former head of the Defense Intelligence Agency (September 1974 to December 1975), Graham had joined Ronald Reagan's staff during the 1976 and 1980 campaigns as an advisor on military affairs. Graham had many friends within the White House, and he initially figured among the small cabal of men meeting in the offices of Edwin Meese to formulate new policy on strategic defense in the fall of 1981. After Reagan's election, Graham founded High Frontier, a widely supported organization that advocated adoption of the Strategic Defense Initiative, of which Graham was one of the first champions.[33]

Graham used his political influence to open the door to the White House. On 15 February 1989, he, Max Hunter, and Jerry Pournelle, the head of the Citizens' Advisory Council on National Space Policy, briefed Vice President J. Danforth Quayle on Hunter's concept of developing a single-stage-to-orbit vehicle (now known as the SSX, SpaceShip Experimental) using "lean" program management and an "X" vehicle to demonstrate "aircraft-like" operations as the key to lowering launch costs. They urged that the Strategic Defense Initiative Organization be the agency in charge.[34] Quayle liked the idea, and, after a positive evaluation conducted by The Aerospace Corporation, a civilian analysis arm of the Air Force,[35] the SDIO initiated the Single Stage To Orbit Program in early 1990. The focus of the program would be on single-stage-to-orbit operations, not technology development, partly, at least, because Have Region and other classified studies already had investigated key technologies.

The ten-month-long Phase I consisted of design studies and the identification of critical technologies by Boeing, General Dynamics, McDonnell Douglas, and Rockwell International. Both McDonnell Douglas and General Dynamics settled on vertical-takeoff-and-landing configurations. Boeing proposed an improved version of its RASV, while Rockwell selected a vertical-takeoff-and-horizontal-landing vehicle.[36] In June 1991, following a review of Phase I concepts by NASA's Langley Research Center,[37] the SDIO solicited proposals for Phase II. The statement of work described the capabilities of the full-scale operational single-stage-to-orbit vehicle, which would loft SDI Brilliant Pebbles payloads into orbit, and the Phase II small suborbital X vehicle, its support infrastructures (such as the launch pad), and operational concepts. Of the three contractors competing, General Dynamics, McDonnell Douglas, and Rockwell International, the SDIO selected McDonnell Douglas in August 1991 to build its Delta Clipper Experimental (DC-X) in twenty-four months. The firm clearly understood the need to demonstrate operations rather than develop technology.[38]

McDonnell Douglas rolled out the 111-foot DC-X in record time, four months ahead of schedule, in April 1993. The company built the Delta Clipper out of modified existing hardware, some of which, like welding rods and hinges, they purchased literally from local hardware stores. Pressure regulators and cryogenic valves came from Thor missiles formerly positioned in Europe, and the constructor of the aluminum liquid oxygen and hydrogen tanks was not an aerospace firm, but Chicago Bridge and Iron of Birmingham, Alabama.[39] More important, McDonnell Douglas sought to achieve SSTO Program operational goals. The Flight Operations Control Center at the White Sands Missile Range, New Mexico, consisted of a compact, low-cost, 40-foot mobile trailer. Three people operated the ground support equipment and launched the DC-X, not the hundreds typically used for rocket launches. Former astronaut Pete Conrad was the "flight manager." McDonnell Douglas designed the DC-X so that they could fly it again after only three days.[40] Eventually, on 8 June 1996, the Clipper team demonstrated a one-day (twenty-six-hour) turnaround.

By the time the DC-X undertook its first short flight on 18 August 1993, the world had changed dramatically. The cold war was over, and Defense cuts were the order of the day. As DC-X flight trials took place, the future of funding for those flights, as well as for completion of the program, grew uncertain. Money for Phase III disappeared, and various bureaucratic maneuvers stymied White House and congressional approval of financing. The predicament grounded the Clipper after only three flights, until the NASA Administrator intervened financially in January 1994.[41]

NASA's New X Vehicles

By January 1994, the NASA Administrator had become interested in single-stage-to-orbit and other kinds of reusable space vehicles. His interest did not arise from any internal NASA studies, such as those conducted by the Langley Research Center as early as the 1970s, nor from the influence of high-level individuals at NASA Headquarters, such as Ivan Bekey, Director of Advanced

Image 11-3: This early artist's concept illustrated the proposed flyback booster and orbiter system. Various concepts of a reusable space shuttle system were identified under Phase B definition study contracts. (NASA Photo no. 9801791)

Programs in the Office of Space Flight,[42] although Bekey was to play a role. Rather, the Administrator was reacting to a September 1992 mandate from Congress to assess national space launch requirements, particularly in light of declining federal budgets.[43]

The NASA "Access to Space" study considered NASA, military, and commercial launch needs for the period between 1995 and 2030. It examined three different launcher alternatives ("options"): (1) upgrade the Shuttle and keep it flying until 2030, (2) develop a new expendable launcher, and (3) replace the Shuttle with a "next-generation, advanced technology system . . . a 'leapfrog' approach, designed to capitalize on advances made in the NASP and SDI programs to achieve order-of-magnitude improvements in the cost effectiveness of space transportation."[44] Ivan Bekey compared the results from the three "option" teams and wrote the final report. The Access to Space study strongly concluded in favor of pursuing development of a single-stage-to-orbit replacement for the Shuttle, especially because it would be the best way to reduce overall launch costs.[45]

Although Bekey long had supported single-stage-to-orbit concepts, what really determined the study's conclusion was the report of the Option 3 team that analyzed advanced space launchers. Crucial to their report was the influence of the SDIO's SSTO Program, that is, the Delta Clipper. The twenty-three-member Option 3 team had consisted of NASA personnel and three representatives from the Department of Defense, including the Ballistic Missile Defense Organization (BMDO), the successor to the SDIO. The BMDO representative, moreover, was Gary Payton, who had drawn up the original SSTO Program Phase I statement of work. Moreover, he wrote the section of the Option 3 report that discussed operations. Members of the Option 3 team visited the McDonnell Douglas hangar at Huntington Beach and the White Sands "launch complex." The BMDO Delta Clipper influence encompassed the agency's program management approach that emphasized short-term (faster) less costly (cheaper) projects run by small teams (smaller), in addition to the emphasis on "aircraft-like" operations.[46] Indeed, the Option 3 team became so captivated by the DC-X program that they proposed a NASA technology development program using an "X" vehicle, the X-2000 (for the program's final year of operation), to be built entirely by NASA with joint funding from the Pentagon. The X-2000, not by chance, closely resembled the SX-2, the next vehicle to be built in the Delta Clipper program.[47]

NASA, however, was not going to build the X-2000. National policy, instead, directed and shaped NASA's response to the "Access to Space." Shortly after Bill Clinton took office as president in January 1993, the White House Office of Science and Technology Policy (OSTP) held the first meeting of an interagency working group to decide what the administration's launch policy would be. The Pentagon, meanwhile, in response to the fiscal year 1994 Defense Authorization Act, undertook its own study of launch needs. The study results appeared as the Department of Defense Space Launch Modernization Plan, more commonly known as the Moorman Report, after the study's leader, Lt. Gen. Thomas S. Moorman Jr., Vice Commander of Air Force Space Command.[48]

Shortly before the release of the Moorman Report in May 1994, the White House released the draft National Space Transportation Strategy in April 1994. It determined that NASA would "be the lead agency for technology development and demonstration for advanced next generation reusable launch systems."[49] It also decreed, in Section III, paragraph 2(b), "Research shall be focussed on technologies to support a decision, no later than December 1996, to proceed with a sub-scale flight demonstration which would prove the concept of single-stage to orbit."[50] Thus, shortly after the release to Congress of the Access to Space results, national space policy committed NASA to the development of reusable and single-stage-to-orbit space vehicles.

In designating NASA as the lead agency for reusable launchers, and the Department of Defense as the lead agency for expendable systems,[51] the OSTP facilitated transfer of funds to the BMDO for the resumption of the DC-X test flights. It also paved the way for the transfer of the DC-X to NASA, where it formed the initial component of the agency's Reusable Launch Vehicle (RLV) Program. While NASA's DC-XA (the *A* stood for "advanced"), also known as the Clipper Graham, tested certain key operational concepts, such as a critical rotational maneuver and a seventy-two-hour turnaround time, the vehicle also became a technology demonstrator.[52]

In addition to the DC-XA, NASA's new RLV Program consisted of two additional X vehicles. One, the X-34, also known as the Reusable Small Booster Program, would demonstrate certain technologies and operations useful to smaller reusable vehicles launched from an aircraft. Among those were autonomous ascent, reentry, and landing, composite structures, reusable liquid oxygen tanks, rapid vehicle turnaround, and thermal protection materials.[53] The other, the X-33, known also as the Advanced Technology Demonstrator Program, was far more challenging technologically. Among the operations and technologies it would demonstrate were reusable composite cryogenic tanks, graphite composite primary structures, metallic thermal protection materials, reusable propulsion systems, autonomous flight control, and certain operating systems, such as electronics for monitoring vehicle hardware.[54]

As of 2002, the now defunct RLV Program had achieved only limited success. The DC-XA demonstrated an unprecedented twenty-six-hour turnaround time (a record that stands unchallenged four years later) and a crucial rotational maneuver, then after completing its fourth flight, the Clipper Graham tipped over. A series of three explosions followed by the explosion of the hydrogen tank destroyed the vehicle.[55] The X-34 also met serious obstacles. In January 1996, the joint venture between Orbital Sciences Corporation and Rockwell International ended over a dispute between the two companies. NASA bid the contract again and selected Orbital in June 1996.[56] Unfortunately, increasing costs brought about by changing requirements for the test vehicle (for instance, making many originally simple systems redundant to satisfy range safety concerns) resulted in the X-34 program being canceled in April 2001.

The X-33 program experienced even greater difficulties. After surmounting weight and control problems, the X-33 project encountered one delay after another because of complications and obstacles encountered in the design and construction of the aerospike engines and the construction and testing of the composite liquid hydrogen tanks. The vehicle's launch was postponed from the original March 1999 to sometime in 2003. In 2001, with costs totaling $1.4 billion and growing and after lengthy negotiations between NASA and Lockheed Martin, the vehicle's manufacturer, NASA terminated the X-33 pro-

Image 11-4: Pictured is an artist's concept of the experimental X-33 in flight. The X-33 program was designed to pave the way to a full-scale commercially developed reusable launch vehicle. The program was to put the United States on a path toward safe, affordable, reliable access to space by proving the latest technology is ready for space flight. Until canceled in April 2001, the X-33 was the flagship technology demonstrator for technologies that were supposed to dramatically lower the cost of access to space. (NASA Photo no. MSFC-9906386)

gram. Although additional NASA money could have been available for completion of the X-33 program through the Space Launch Initiative (SLI), NASA decided to pursue other options with SLI funds and officially withdrew from the X-33 agreement with Lockheed Martin in April 2001. Lockheed Martin then worked to interest the Air Force in completing the X-33 demonstrator, but this seems unlikely given the technical uncertainties facing the program.[57]

Shortly after the start of the RLV Program, NASA initiated the Pathfinder and Trailblazer programs to establish an ongoing effort to develop low-cost reusable space transport. Pathfinder involved technology experiments conducted on existing flight vehicles, such as the Space Shuttle. NASA can conduct these experiments quickly, at low cost, on a wide spectrum of technologies and applications. Trailblazer, on the other hand, entailed the construction of entirely new

X vehicles to demonstrate advanced space transport technologies and operations. In August 1998, NASA solicited proposals for Future-X, the first of the Trailblazer vehicles,[58] and in December, announced that it had entered into negotiations with Boeing to design and build the Advanced Technology Vehicle (ATV), the first X vehicle to fly in orbit and to reenter the atmosphere.[59]

Soon, the ATV became the X-37. The shuttle will carry the X-37 into space, then release it. The X-37 will orbit the planet, then return to Earth through the atmosphere, testing heat shielding and other advanced space materials and technologies. The X-37 shape derives from that of the X-40A, an unpowered Air Force craft designed and built by Boeing's Phantom Works. In August 1998, the Air Force drop tested the X-40A from an Army Black Hawk helicopter above Holloman Air Base, New Mexico, and the vehicle landed under remote control on a runway. The drop tests served to reduce risk prior to expanding tests with the larger X-37. The Air Force provided partial funding for the X-37 in the hope of realizing some of the objectives of its Space Maneuver Vehicle (SMV), a reusable winged craft capable of deploying satellites, weapons, and antisatellite devices, inspecting enemy satellites, and other military missions. The SMV would have remained in orbit for three months to a year, and would have been capable of a seventy-two-hour turnaround.[60]

No discussion of NASA's reusable X vehicles would be complete without at least a mention of the Crew Recovery Vehicle (CRV) that will serve as a lifeboat for the International Space Station. The X-38 is an experimental 80 percent scale version of the CRV. Drop tests at increasing altitudes from a B-52 began in 1999 and will continue into 2001. The basic idea for the CRV originated at NASA's Langley Research Center, although the X-38—as designed and built by the Johnson Space Center—uses the aerodynamic shape developed by Martin Marietta as part of a 1960s Air Force program that spawned the unmanned X-23A and manned X-24A test vehicles.

Commercial Launchers

NASA and the Air Force were not the only ones developing reusable launchers during the 1990s. The issuance of the first commercial launch licenses in 1989 to General Dynamics, McDonnell Douglas, and Martin Marietta by the Department of Transportation's Office of Commercial Space Transportation (OCTS) marked the beginning of the U.S. commercial launch industry.[61] Those launches were all on expendable rockets. Soon, as the global market for satellite launches grew throughout the decade, small start-up companies entered the field, but with plans for two-stage reusable vehicles.

Among those was Kelly Space & Technology, Inc., initially headed by Michael S. Kelly. Starting in 1993, with funding from NASA and the Air Force, the firm began developing the Astroliner, a reusable glider towed to launch altitude by a Boeing 747 aircraft using patented Eclipse towing technology.

An expendable stage launched from the Astroliner would place payloads in orbit. Recently, Kelly has received NASA funding to develop its reusable launcher.[62] A comparable two-stage system that combines a reusable first stage with an expendable second stage is Pioneer Rocketplane's Pathfinder. A two-seat aircraft powered by air-breathing and (RD-120) rocket engines, the Pathfinder will take off from Vandenberg AFB, take on additional liquid oxygen in midair from a Boeing 747 freighter, then climb outside the atmosphere. There, it will release the upper stage and its payload, then reenter the atmosphere and land like an aircraft.[63]

Pursuing development of a different two-stage launch system known as the K-1 is the Kistler Aerospace Corporation. The K-1 is an unpiloted vehicle powered by surplus Russian NK-33 and NK-43 engines. It will launch vertically and be capable of a turnaround of nine days. A system of parachutes and airbags (field tested in 1998) allows the company to recover and reuse both the booster and orbital stages. In 1998, the firm announced that it would hold test flights at Woomera, Australia, to avoid the regulatory difficulties of using Nellis AFB in Nevada. In January 1999, however, work on the K-1 stopped, because of a lack of cash arising from a tight bond market in the U.S. and the loss of Asian venture capital. Infusions of new capital came from Northrop Grumman, a K-1 subcontractor, and several Taiwanese banks. Like Kelly, Kistler has begun to receive NASA funding for its launcher.[64]

The only single-stage-to-orbit vehicle under commercial development is also the only one that has received no NASA funding. That is the Rotary Rocket Company's Roton. The firm's founder, Gary Hudson, has been pursuing single-stage-to-orbit concepts since the 1980s with funding from the private sector. Hudson, a staunch believer in private enterprise, received substantial backing for the Roton from author Tom Clancy. Like the Delta Clipper, the Roton would have taken off and landed vertically, but would have used rocket-powered rotors for the final descent and touchdown, much like a helicopter. Roton, like many start-up companies, eventually ran out of money before any significant progress had been made, although the company did successfully demonstrate low-speed flight of a prototype vehicle.[65]

METEOR

The trailblazing predecessor of these two-stage commercial and NASA reusable space vehicles was METEOR (Multiple Experiment Transporter to Earth Orbit and Return). METEOR was to be boosted into space atop a Conestoga rocket built by EER Systems of Vienna, Virginia. The vehicle pioneered the way for the X-33 and other reusable craft, because it was the first to seek a license from the OCST to fly over and set down on land and the first reusable reentry vehicle to seek such licensing. In order to minimize injury and property damage, launches occur at seacoast sites, and reentry vehicles touch down in the ocean. Licensing legislation applies only to launches, not to vehicle reentries or landings. This regulatory gap arose because legislators imagined that commercial launch companies would use only expendable rockets, and because rides on NASA's Shuttle are exempt from regulation. NASA intended METEOR to be a fully commercial affair from beginning to end and to jump-start the struggling microgravity industry in the United States.

METEOR began in 1990 as COMET, the COMmercial Experiment Transporter, at the University of Tennessee, with a grant from NASA. The COMET spacecraft would carry microgravity experiments into low-Earth orbit for a month, then parachute samples back to Earth in a reusable recovery module. Cost overruns and technical problems, however, caused delay after delay. The program was over budget and behind schedule. Eventually, Congress became involved. NASA finally took program management away from the University of Tennessee and signed a sole-source contract with one of the original contractors, EER Systems, in March 1995.[66]

Part of the understanding between NASA and EER Systems was the decision to change the recovery module's landing site from Utah to the Atlantic Ocean between Wallops Island and Bermuda. From NASA's perspective, the desert landing had raised the liability questions that, delays and cost overruns aside, had plagued the project. This decision also cleared the way for the OCST to issue EER a launch license for the Conestoga rocket. A license, however, was still needed to certify the safety of the payloads and the recovery module. In the end, the OCST licensed the METEOR reentry vehicle on the statutory basis that it constituted a payload on a licensed expendable rocket. The classification of METEOR as a payload, and therefore something the OCST could license, was possible because METEOR was part of a larger system placed in orbit. In this way, the OCST licensed the first landing of a reusable launch vehicle before passage of legislation that would empower it to license such landings.[67] The use of aircraft as part of a reusable launch system, such as Kelly's Astroliner and Pioneer's Rocketplane, raises additional complex administrative questions, because they require the coordination of FAA air and OCST space regulations.

In May 1995, NASA renamed COMET the METEOR. The first (and last) Conestoga launch with its METEOR reentry module took place on 23 October 1995. This also was the first launch of EER's Conestoga rocket. The flight began with a picture-perfect liftoff. Just 46 seconds later, though, in a scene reminiscent of the *Challenger* disaster, the Conestoga broke up 14 miles off the Virginia coast.[68]

Conclusion

The quest for reusability has suffered its losses (METEOR). There have been mistakes (NASP), overly ambitious projects (X-33), and seemingly fruitful

Image 11-5: This artist's conception depicts a future for space travel far into the twenty-first century. Pictured is a third-generation Reusable Launch Vehicle (RLV). Projected for the year 2025, this RLV is expected to introduce an era of space travel not unlike air travel today. (NASA Photo no. MSFC-9700451)

routes taken but abandoned (Dyna-Soar, RASV). Success has been partial for three major reasons: (1) full reusability and "aircraft-like" operations are major technological challenges, (2) the lack of an ongoing technology development program, and (3) the toll on the search for a new launch system taken by past space policy and political decisions. The Shuttle was the realization of the easier of two technological paths, a reusable assisted by an expendable booster versus a fully reusable two-stage or one-stage vehicle. Have we advanced beyond these boost-glide systems? Are we moving toward a fully reusable launcher?

NASA is pushing development of a so-called second-generation RLV to service the International Space Station through the Space Launch Initiative. The stated goal of this $14 billion effort[69] is to reduce the costs and technological risks involved in developing reusable launch vehicles. Specifically, NASA has stated that it seeks to improve crew safety over that of the Shuttle (to a crew risk of one in ten thousand missions) and to cut costs to about one thousand dollars per payload pound. According to the NASA research announcement NRA 8-30, issued 12 October 2000, the SLI will provide NASA "two viable commercial competitors" by 2005. The goal is for NASA to access the International Space Station on more than one launch vehicle in the year 2010.[70]

The SLI eventually may or may not assure NASA of two commercial reusable vehicles to service the International Space Station. The SLI is not immune from political changes that may occur in the near future, nor from any major shifts in space policy, although it has two years of approved funding. Budgetary priorities might thwart the SLI by diverting funds to other, more urgent areas, such as national defense or Social Security. Meanwhile, the future of the commercial launch industry also remains unclear. The declining number of commercial satellite payloads in the wake of the failure of the Iridium communications satellite constellation and the vicissitudes of the investment market raise questions about the survivability of that industry, as well as about the availability of speculative capital for investment in reusable vehicle technologies. Nonetheless, the perceived advantages of reusable launchers may continue to push technological development, whether by government or industry, in the direction of increased reusability of launch systems.

Notes

1. Russell J. Hannigan, *Spaceflight in the Era of Aero-Space Planes* (Malabar, Fla.: Krieger Publishing, 1994), p. 71. Materials in NASA Historical Reference Collection (hereafter cited as NHRC), NASA Headquarters, Washington, D.C., Folder 824 indicate that the article appeared in the December 1931 issue, pp. 148–49, and was titled, "A New Turbine Rocket Plane for the Upper Atmosphere."

2. Irene Sänger-Bredt, "The Silver Bird Story: A Memoir," Folder 7910, NHRC; Hannigan, pp. 71–73; Michael J. Neufeld, *The Rocket and the Reich: Peenemünde and the Coming of the Ballistic Missile Era* (New York: Free Press, 1995), pp. 7–10; Richard P. Hallion, "In the Beginning was the Dream . . . ," pp. xi–xv, in Hallion, ed., *The Hypersonic Revolution: Eight Case Studies in the History of Hypersonic Technology*, vol. 1, *From Max Valier to Project Prime, 1924–1967* (Dayton, Ohio: Special Staff Office, Aeronautical Systems Division, Wright-Patterson AFB, 1987) (hereafter cited as Hallion, *Hypersonic Revolution*, vol. 1).

3. Hannigan, pp. 77–78; T.A. Heppenheimer, *The Space Shuttle Decision: NASA's Search for a Reusable Space Vehicle*, NASA SP-4221 (Washington, D.C.: GPO, 1999), pp. 75–78; Richard P. Hallion and James O. Young, "Space Shuttle: Fulfillment of a Dream," pp. 949–51, in Hallion, ed., *The Hypersonic Revolution: Eight Case Studies in the History of Hypersonic Technology*, vol. 2, *From Scramjet to the National Aero-Space Plane* (Dayton, Ohio: Special Staff Office, Aeronautical Systems Division, Wright-Patterson AFB, 1987) (hereafter cited as Hallion, *Hypersonic Revolution*, vol. 2).

4. Hallion, "In the Beginning was the Dream," p. xviii; Neufeld, pp. 92–93, 121, 138–39, 156–57, 283; Hannigan, p. 73.

5. Clarence J. Geiger, "Strangled Infant: The Boeing X-20A Dyna-Soar," pp. 189 & 191–98, in Hallion, ed., *The Hypersonic Revolution*, vol. 1, a manuscript copy of which is in NHRC Folder 11,326, as Geiger, "History of the X-20A Dyna-Soar," October 1963, and items in

NHRC Folder 495; Hallion, "Editor's Introduction," p. II-xi, in Hallion, ed., *Hypersonic Revolution*, vol. 1; and materials in NHRC Folder 11,923.

6. Hallion, "ASSET: Pioneer of Lifting Reentry," pp. 449–50, 510, 512–13, 515–16, 518, & 523–24, in Hallion, *Hypersonic Revolution*, vol. 1.

7. Hallion, "In the Beginning was the Dream," p. xxi; Hallion, "Editor's Introduction," pp. I-iv-I-v & II-xi in Hallion, *Hypersonic Revolution*, vol. 1.

8. R&D Project Card Continuation Sheet, 23 August 1957, NHRC Folder 11,325; various items in NHRC Folder 11,340; Geiger, pp. 198–99, 201–4, 261, 263, 266, 276–78, 296–97, 299–301, 305, and 308–9.

9. Geiger, pp. 319–20 & 369; Andrew K. Hepler, interview by author, tape recording and transcript, Seattle, Wash., 11 July 2000, NHRC; Hepler and E.L. Bangsund, Boeing Aerospace Company, Seattle, Washington, *Technology Requirements for Advanced Earth Orbital Transportation Systems*, vol. 1, *Executive Summary*, NASA Contractor Report CR-2878 (Washington, D.C.: NASA, 1978).

10. R. Dale Reed, *Wingless Flight: The Lifting Body Story*, NASA SP-4220 (Washington, D.C.: NASA, 1997), pp. 9, 67, 69–72, 75, 87, 91, 96–98, 102, 106–9, & 116; John L. Vitelli and Hallion, "Project PRIME: Hypersonic Reentry from Space," p. 529, in Hallion, ed., *Hypersonic Revolution*, vol. 1.

11. Vitelli and Hallion, "Project PRIME," pp. 558, 566, 571, 577–96, 694–95, 699, 702–4, & 711.

12. Hallion and Vitelli, "The Piloted Lifting Body Demonstrators: Supersonic Predecessors to Hypersonic Lifting Reentry," pp. 901, 919 & 921–22, in Hallion, ed., *Hypersonic Revolution*, vol. 2.

13. Ibid., pp. 923, 925, 928–29.

14. Alan Wilhite, interview by author, tape recording and transcript, NASA Langley Research Center, Hampton, Va., 22 May 1997, NHRC.

15. Charles H. Eldred, interview with author, tape recording and transcript, NASA Langley Research Center, Hampton, Va., 20 May 1997, NHRC.

16. The two studies were Rudolph C. Haefeli, Earnest G. Littler, John B. Hurley, and Martin G. Winter, Martin Marietta Corporation, Denver Division, *Technology Requirements for Advanced Earth-Orbital Transportation Systems: Final Report*, NASA Contractor Report CR-2866 (Washington, D.C.: NASA, October 1977); and Andrew K. Hepler and E. L. Bangsund, Boeing Aerospace Company, Seattle, Washington, *Technology Requirements for Advanced Earth Orbital Transportation Systems*, 2 vols., *Executive Summary*, NASA Contractor Report CR-2878 (Washington, D.C.: NASA, 1978).

17. Eldred, "Evolution of NASA Rocket SSTO Concepts," AIAA 94-4673, paper read at AIAA Space Programs and Technologies Conference, Huntsville, Ala., 27–29 September 1994.

18. Hepler and Bangsund 1:13–14.

19. Eldred, interview.

20. Hepler and Bangsund 1:14–16; Hepler and Bangsund 2:191; Hepler, interview.

21. Boeing Aerospace Company, *Final Report on Feasibility Study of Reusable Aerodynamic Space Vehicle*, vol. 1, *Executive Summary* (Kent, Wash.: Boeing Aerospace, November 1976), pp. 5 & 35.

22. Hepler and Bangsund 1:24, estimated the total design, development, testing, and evaluation cost to run from $3.395 billion to $4.887 billion spread out over ten years.

23. Hallion, "Yesterday, Today, and Tomorrow: From Shuttle to the National Aero-Space Plane," p. 1334, in Hallion, ed., *Hypersonic Revolution*, vol. 2; P. Kenneth Pierpont, "Preliminary Study of Adaptation of SST Technology to a Reusable Aero-space Launch Vehicle System," NASA Langley Working Paper NASA-LWP-157, 3 November 1965; Jess Sponable, interview by author, tape recording and transcript, NASA Headquarters, Washington, D.C., 16 January 1998, NHRC; Gary Payton and Jess Sponable, "Designing the SSTO Rocket,"

Aerospace America, April 1991, p. 40; Boeing RASV proposal, December 1982, File 256, X-33 Archive. The X-33 Archive is a documentary collection created for a history of the X-33 program funded by NASA. It eventually will be integrated into the collections of the National Archives and Records Administration.

24. Hepler, interview.

25. Raymond L. Chase, "Science Dawn Overview," March 1990, File 235, X-33 Archive; Major Stephen Clift, "Have Region Program: Final Brief," September 1989, File 235, X-33 Archive; Jess Sponable, interview by author, tape recording and transcript, NASA Headquarters, Washington, D.C., 16 January 1998, NHRC.

26. Hallion, "Yesterday, Today, and Tomorrow," pp. 1337, 1340–41, & 1345.

27. Quoted in Scott Pace, "National Aero-space Plane Program: Principal Assumptions, Findings, and Policy Options," RAND publication P-7288–RGS, December 1986, p. 1.

28. T.A. Heppenheimer, *The National Aerospace Plane* (Arlington, Va.: Pasha Market Intelligence, 1987), p. 14; Hallion, "Yesterday, Today, and Tomorrow," pp. 1334 & 1362–64; Larry Schweikart, "The Quest for the Orbital Jet: The National Aerospace Plan Program, 1983–1995," manuscript, pp. I.30–I.31, NHRC; John V. Becker, "Confronting Scramjet: The NASA Hypersonic Ramjet Experiment," p. VI.xv, in Hallion, *Hypersonic Revolution*, vol. 2.

29. Schweikart, pp. I.11–I.12, I.19–I.20, I.23, I.28, & III.43.

30. Ivan Bekey, interview by author, tape recording and transcript, NASA Headquarters, Washington, D.C., 2 March 1999, NHRC.

31. Schweikart, pp. III.37–III.38 & III.41–III.42.

32. Max Hunter, interview by author, tape recording and transcript, San Carlos, Calif., 19 June 1998, NHRC; Memo, Gary C. Hudson to Thomas L. Kessler, "Comments on SSTO Briefing and a Short History of the Project," 17 December 1990, p. 2, File 242, X-33 Archive (hereafter cited as Hudson, "History"); Memo, Max Hunter to Pat Ladner, "Some History of the SSTO Program as of Sept 13 1990," 13 September 1990, n.p., File 242, X-33 Archive (hereafter cited as Hunter, "History"); Ivan Bekey, telephone interview notes, 5 May 1998; memo for the record, Steve Hoeser to Pat Ladner, "Past SDIO Director's Perspective and Decision Base Related to the SSTO," 7 December 1990, p. 1, File 242, X-33 Archive.

33. Graham, *Confessions of a Cold Warrior: An Autobiography* (Fairfax, Va.: Preview Press, 1995), pp. 101–4, 118–20; Donald R. Baucom, *The Origins of SDI, 1944–1983* (Lawrence: University Press of Kansas, 1992), pp. 145–46 & 150–51; Erik K. Pratt, *Selling Strategic Defense: Interests, Ideologies, and the Arms Race* (Boulder, Colo: Lynne Rienner Publishers, 1990), pp. 96–97 & 106.

34. Graham, pp. 205–6.

35. Jay P. Penn, C.L. Leonard, and C.A. Lindley, "Review of Pacific American Launch System SSX:Phoenix VTOL Concept," 19 July 1989, p. 2, File 255, X-33 Archive.

36. McDonnell Douglas Space Systems Company, "Single Stage to Orbit Program Phase I Concept Definition," 13 December 1990, File 267, X-33 Archive; General Dynamics Space Systems Division, "Concept Review Technical Briefing," 13 December 1990, File 265, X-33 Archive; Boeing Defense and Space Group, Space Transportation Systems, "Single Stage to Orbit Technology Demonstration Concept Review Technical Briefing," 12 December 1990, File 264, X-33 Archive; "Rockwell International, "SDIO Single Stage to Orbit Concept Review," 12 December 1990, File 259, X-33 Archive.

37. "NASA Evaluation of SDIO Phase I SSTO Concepts," n.d., File 294, X-33 Archive.

38. Sponable, interview.

39. Paul L. Klevatt, interview by author, tape recording and transcript, Tustin, Calif., 14 July 2000, NHRC; William Gaubatz, interview by author, tape recording and transcript, Huntington Beach, Calif., 25 October 1997, NHRC; Klevatt, "Design Engineering and Rapid Prototyping for the DC-X Single Stage Rocket Technology Vehicle," AIAA-95-1425, paper read at AIAA-ASME-ASCE-AHS-ASC Structures, Structural Dynamics, and Materials Conference, 10–12 April 1995, New Orleans.

40. Klevatt, interview; McDonnell Douglas Space Systems Company, "Single Stage To Orbit Program Phase I Concept Definition," 13 December 1990, File 267, X-33 Archive; Charles "Pete" Conrad, interview by author, tape recording and transcript, Rocket Development Company, Los Alamitos, Calif., 22 October 1997, NHRC; Luis Zea, "The Quicker Clipper," *Final Frontier*, October 1992, p. 4, File 267, X-33 Archive; Mark A. Gottschalk, "Delta Clipper: Taxi to the Heavens," *Design News*, September 1992, n.p. File 292, X-33 Archive; Leonard David, "Unorthodox New DC-X Rocket Ready for First Tests," *Space News*, 11–17 January 1993, p. 10.

41. Letter, George E. Brown Jr. to Les Aspin, 31 January 1994, File 293, X-33 Archive; Ben Iannotta, "DC-X Hangs by Thin Thread Despite Short-term Reprieve," *Space News*, 7–13 February 1994, p. 4; Iannotta, "Pentagon Frees Funds for More DC-X Flights," *Space News*, 9–15 May 1994, p. 4; Warren E. Leary, "Rocket: Program Faces Budget Ax," *New York Times*, 31 January 1994, p. 13A.

42. Bekey, interview.

43. U.S. House of Representatives, 102d Cong., 2d sess., *1993 NASA Authorization: Hearings before the Subcommittee on Space of the Committee on Science, Space, and Technology*, No. 137, vol. 2 (Washington, D.C.: GPO, 1992).

44. Memo, Arnold D. Aldrich and Michael D. Griffin to Daniel Goldin, "Implementation Plan for 'Access to Space' Review," 11 January 1993, File 197, X-33 Archive; NASA Office of Space Systems Development, "Access to Space Study Summary Report," January 1994, pp. 2–5, 8–58, File 100, X-33 Archive; Access to Space Study Advanced Technology Team, "Final Report, Volume 1: Executive Summary," July 1993, pp. iii & 38, File 85, X-33 Archive. According to Ivan Bekey, interview, the study initially was to compare Shuttle upgrades and a new expandable, or partially reusable, launcher. These alternatives ultimately became Option 1 and Option 2.

45. Bekey, interview.

46. NASA Office of Space Systems Development, "Access to Space Study Summary Report," January 1994, pp. 2–5, 8–58, File 100, X-33 Archive; Access to Space Study Advanced Technology Team, "Final Report, Volume 1: Executive Summary," July 1993, pp. iii & 38, File 85, X-33 Archive; Michael Griffin, interview by author, tape recording and transcript, Orbital Sciences Corporation, Dulles, Va., 18 August 1997, NHRC; Gary E. Payton, interview by author, tape recording and transcript, NASA Headquarters, Washington, D.C., 20 August 1997, NHRC.

47. Ben Iannotta, "Winged X-2000 Project Considered," *Space News*, 15–28 November 1993, p. 14; Single Stage to Orbit Advanced Technology Demonstrator (X-2000) briefing, August 1993, File 122, X-33 Archive; "Single Stage to Orbit: Advanced Technology Demonstrator," concept proposal briefing, SSTO Concept Proposal, X-2000, August 1993, File 162, X-33 Archive.

48. Letter, Gary L. Denman, Director, ARPA, to Sen. Charles S. Robb, 14 March 1994, File 293, X-33 Archive; "Space Launch Modernization Plan: Executive Summary," April 1994, File 142, X-33 Archive.

49. Draft, National Space Transportation Strategy, 26 April 1994, File 153, X-33 Archive.

50. Cited in NASA Press Release 95-1, 12 January 1995.

51. "Space Launch Modernization Plan: Executive Summary," p. 29; Iannotta, "Congress, NASA Dueling Over Reusable Rocket Management," *Space News*, 23–29 May 1994, p. 25.

52. After the death of Danny Graham, the DC-XA also took on the name Clipper Graham. The DC-XA differed from the DC-X in six areas: (1) a switch from an aluminum oxygen tank to a Russian-built aluminum-lithium alloy cryogenic oxygen tank with external insulation, (2) an exchange of the aluminum cryogenic hydrogen tank for a graphite-epoxy composite liquid hydrogen tank with a low-density reinforced internal insulation, (3) a graphite-epoxy composite intertank structure, (4) a graphite-epoxy composite feedline and valve assembly, (5) a gaseous hydrogen and oxygen auxiliary power unit to drive the hydrau-

lic systems, and (6) an auxiliary propulsion system for converting liquid hydrogen into gaseous hydrogen for use by the vehicle's reaction control system. Delma C. Freeman Jr., Theodore A. Talay, and R. Eugene Austin, "Reusable Launch Vehicle Technology Program," IAF 96-V.4.01, p. 3, paper read at the 47th International Astronautical Congress, 7–11 October 1996, at Beijing, China, File 92, X-33 Archive.

53. John W. Cole, "X-34 Program," in "X-33/X-34 Industry briefing, 19 October 1994," File 12, X-33 Archive, especially p. 1A-1216.

54. X-33 announcement in *Commerce Business Daily,* 29 September 1994, in File 276, X-33 Archive.

55. "NASA DC-XA Clipper Graham Mishap Investigation Report," 12 September 1996, "Executive Summary," p. 4-1, File 79, X-33 Archive. A more detailed account of the damage is in ibid, pp. 6-2 to 6-4.

56. Joseph C. Anselmo, "NASA Gives Orbital Second Shot at X-34," *Aviation Week & Space Technology,* Vol. 144, No. 25, 17 June 1996, Page 31, File 182; NASA Press Release 96-115, 10 June 1996, File 273; and NASA Press Release 96-177, 28 August 1996, File 273, X-33 Archive.

57. NASA Press Release 00-157, 29 September 2000; "Development Troubles Push First X-33 Flight Back to July 1999," *Aerospace Daily* [electronic version], 24 June 1997, article 34208, File 225, X-33 Archive; Brian Berger, "Activists Say Lockheed Should Not Compete for X-33 Funds," *Space News* 11, no. 39 (16 October 2000): 21.

58. NASA Press Release 98-141, 3 August 1998.

59. NASA Press Release c98-w, 8 December 1998.

60. NASA Press Release 99-139, 14 July 1999; Frank Sietzen Jr., "Air Force's Needs Shape Newest NASA X Rocket," 25 August 1999, http://www.space.com/business/aerospace/x37_briefing.html; "Space Maneuver Vehicle Drop Test Planned for Early August," *Aerospace Daily* (electronic edition), 21 July 1998, article 110718; "USAF Sets Aug. 4 Test of Space Maneuver Vehicle," *Aerospace Daily* (electronic edition), 30 July 1998, article 111407; "Competition Likely for Space Maneuver Vehicle Demonstrator," *Aerospace Daily* (electronic edition), 6 August 1998, article 111904.

61. "First Commercial Rocket Launch Successful," *Space News Roundup,* 31 March 1989, p. 4, Folder 10,784, NHRC.

62. Kelly press releases for 7 October 1996, 22 May 1997, and 2 February 1998, File 373, X-33 Archive.

63. "RLV Startups Have Enough Capital, but Worry About Regulation," *Aerospace Daily* (electronic edition), 13 February 1998, article 37503; "Rocketplane System," Pioneer Rocketplane website, http://www.rocketplane.com.

64. "RLV Startups Have Enough Capital, but Worry about Regulation," *Aerospace Daily* (electronic edition), 13 February 1998, article 37503; "Kistler May Shift Flight Tests to Australia," *Aerospace Daily* (electronic edition), 23 February 1998, article 37615; "Developments in the Field of Space Business Are Briefly Noted," *Aerospace Daily* (electronic edition), 7 July 1998, article 109711, and 27 July 1998, article 111101; Frank Morring Jr., "Tight Money Forces Slowdown at Kistler Aerospace," *Aerospace Daily* (electronic edition), 8 January 1999, article 122111; "Northrop Grumman Increases Stake in Kistler's K-1 Vehicle," *Aerospace Daily* (electronic edition), 22 March 1999, article 127002; "Kistler Has a Line on Remaining Financing, but Much Rests on Contingent Funds," *Aerospace Daily* (electronic edition), 2 June 1999, article 132104; "NASA Taps Kistler to Evaluate ISS Access Options," *Aerospace Daily* (electronic edition), 28 August 2000, article 163106; other materials in File 179, X-33 Archive.

65. Materials relating to Gary Hudson and the Roton rocket in File 348, X-33 Archive.

66. Jack Levine, interview by author, tape recording and transcript, Washington, D.C., 13 March 1998, NHRC; Jim Hengle, interview by author, tape recording and transcript, Futron

Corporation, Bethesda, MD, 25 February 1998, NHRC; Gregory Reck, interview by author, tape recording and transcript, NASA Headquarters, 17 February 1998, NHRC; Patrick Seitz, "Center Continues Working on Comet," *Space News,* 3 January 1994, p. 1; "This Week," *Space News,* 28 February 1994, p. 2; Seitz, "Goldin Halts Comet Funding," *Space News,* 9 May 1994, p. 1; "NASA Team to Discuss Congress' Comet Bid" in "This Week," *Space News,* 4 July 1994, p. 2; Seitz, "NASA Unfreezes Comet Funding," *Space News,* 25 July 1994, p. 1; Seitz, "Comet Talks Remove Logjam; Contractors, NASA Continue to Pursue Elusive First Launch," *Space News,* 8 August 1994, p. 4; Seitz, "Comet Budget Insufficient; Subcontractors Seek Additional Funding for First Launch," *Space News,* 17 October 1994, p. 4; "NASA Pulls Plug on COMET," *Defense Daily,* 5 May 1994, n.p., article reprint in File 10,783, NHRC; NASA News Release 95-44, 6 April 1995, File 10,783, NHRC.

67. Levine, interview; Jim Hengle, interview; documents from the Federal Register, the FAA web site, and other sources in File 341, X-33 Archive.

68. Levine, interview; Jim Hengle, interview; Warren Ferster, "Destruct Signal Fails in Conestoga Breakup," *Space News,* 30 October 1995, p. 6; "Faulty Internal Signal Cited in Rocket Explosion," *Washington Metropolitan Times,* 30 October 1995, p. C4, File 10,783, NHRC.

69. "Lawmakers Restore Space Launch Initiative," *Aerospace Daily* (electronic edition), 13 October 2000, article 166402.

70. Electronic version of NRA 8-30, 12 October 2000, posted on NASA Marshall Space Flight Center server, "Introductory Letter," http://nais.msfc.nasa.gov/EPS/EPS_DATA/083261-SOL-001–001.doc.

Epilogue: "To the Very Limit of Our Ability"

Reflections on Forty Years of Military-Civil Partnership in Space Launch

David N. Spires and Rick W. Sturdevant

Introduction

In an April 1960 memorandum, United States Air Force (USAF) Chief of Staff Gen. Thomas D. White informed his staff that the "Air Force must cooperate with NASA . . . to the very limit of our ability and even beyond it to the extent of some risk to our own programs."[1] Although cooperation has been a hallmark of civil-military relations, the partnership also has reflected competition and mutual dependence. Most recently, the partnership agreement between Air Force Space Command (AFSPC) and NASA represents the high-water mark of cooperation between the military and civilian space communities. While clearly motivated by budget and launch reliability imperatives, the current association also benefits from the absence of several contentious elements that in earlier decades had hindered full and genuine cooperation. This overview of the civil-military partnership that spans forty years, especially in the context of NASA and the Air Force, examines key elements of the relationship from an Air Force perspective. It focuses on space launch issues, especially those involving human space flight, and the challenge facing space enthusiasts to institutionalize space in an Air Force traditionally more committed to air than to space. Even today, the Air Force remains wedded more to an airplane culture than to space.

The Sputnik Crisis

An assessment of the USAF-NASA partnership must begin with the Air Force's quest to lead the nation's space program. Failure to realize their early, ambitious space agenda left Air Force space proponents frustrated, and this contributed to early friction between the service and NASA.

Prior to the launch of the first Sputnik on 4 October 1957, the military services dominated the country's space program. Civilian priorities remained secondary. Among the services, the Air Force believed that its responsibility for development of the Atlas and Titan ICBMs, the Thor IRBM, and the multifaceted military reconnaissance satellite system WS-117L, gave it pride of place as the lead service for space. At this time, the strongest support for Air Force space programs came from the Western Development Division in Englewood, California, where Maj. Gen. Bernard A. Schriever directed the ICBM and satellite programs. The Air Staff and civilian leaders in the Office of the Secretary of the Air Force had yet to accord space the level of support that Schriever and his team of "space cadets" thought it deserved.[2]

In the wake of the Sputnik flights, however, top Air Force leaders, like General White, embraced space with alacrity. On 24 January 1958, responding to a Defense Department request, the Air Staff submitted an imposing astronautics program—one they believed would give the Air Force leadership of a unified, DoD-oriented national space program. In the race among the services for the space mission, Generals White and Schriever led military spokesmen in defining space as a continuum of the atmosphere, a place potentially for military operations, and the logical arena for Air Force activities. Early in 1958, Air Force leaders coined a new term, "aerospace," to describe their service's legitimate role in space. Their claim also reflected an extensive biomedical research program that viewed human space flight as an extension in the chain of operational development from aviation medicine to space medicine.[3]

Of the twenty-one individual programs and projects included in the astronautics proposal, most striking were those involving human space flight, such as establishing a manned lunar base to be supplied by nuclear rockets and developing a single-seat hypersonic orbital bomber, designated Dyna-Soar, for dynamic soaring. By February 1958, Air Force Lieut. Gen. Donald L. Putt, the Air Staff's Deputy Chief of Staff for Development, had gained the support Dr. Hugh L. Dryden, Director of the National Advisory Committee for Aeronautics (NACA), for Dyna-Soar. Putt argued this project was a logical successor to the longstanding cooperative efforts between the NACA and the Air Force on X-series aircraft. While the NACA's interest in the space plane centered on reentry capabilities, the Air Force focused on its potential role as a reconnaissance and strategic bombardment platform. For many in the Air

Image 12-1: Five pioneers pose with scale models of their missiles they created in the 1950s. *Left to right:* Ernst Stuhlinger, a member of the original German rocket team who directed the Research Projects Office, Army Ballistic Missile Agency (ABMA); Maj. Gen. Holger Toftoy, who consolidated U.S. missile and rocketry development; Hermann Oberth, a rocket pioneer and Wernher von Braun's mentor; Wernher von Braun, director, Development Operation Division, ABMA; and Robert Lusser, who served as assistant director for reliability engineering for ABMA. This photograph was taken on 1 February 1956 by Hank Walker and appeared in the 27 February 1956 issue of *Life* magazine. (NASA Photo no. MSFC-9131100)

Force space community, Dyna-Soar represented the ideal system to transport Air Force personnel into space and, thereby, to gain support from the service's traditional flying element—the pilots.[4]

The Eisenhower administration quickly dashed Air Force leaders' high hopes. Under its policy of space for peaceful purposes, it had no interest in highlighting military initiatives that might provoke an unwanted Soviet response and endanger the highly classified reconnaissance satellite program. In early February, the administration, seeking to minimize interservice rivalry and avoid duplication of effort, created the Advanced Research Projects Agency (ARPA) to manage all DoD space programs. Lost by the Air Force were WS-117L and its human space-flight program. Moreover, that same month, President Eisenhower and his advisors removed a key element from the reconnaissance satellite program and awarded it to the CIA.[5] This would become Project Corona, the first intelligence satellite project of the future National Reconnaissance Office. Only Dyna-Soar, among major space programs, remained directly under Air Force management. Soon, it became apparent that other Air Force programs would be lost to the National Aeronautics and Space Administration (NASA), a new civilian organization that was based on, and absorbed, the NACA. Henceforward, Air Force leaders sought to escape ARPA's oversight; they wanted the service to become NASA's equivalent on the military side. Frustrated and disappointed, Air Force leaders now envisioned a significant role for the service in NASA operations.

Given Air Force pretensions to space leadership, friction between the service and NASA was to be expected. Inability to establish clear lines of responsibility between the military and civilian sectors worsened the problem. The Space Act of 29 July 1958 accorded the new civilian agency responsibility for all aeronautical and space activities, "except that activities peculiar to or primarily associated with the development of weapons systems, military operations, or the defense of the United States (including the research and development necessary to make effective provision for the defense of the United States) shall be the responsibility of, and shall be directed by, the Department of Defense."[6] The president was to decide which agency would be responsible for particular programs. Despite creating a Civilian-Military Liaison Committee (CMLC) to ensure coordination between the two agencies, neither the CMLC nor its successor, the Aeronautics and Astronautics Coordinating Board (AACB), nor formal agreements and various informal cooperative arrangements, proved capable of eliminating friction, duplication, and blurred lines of responsibility.[7]

The issue of duplication became especially troublesome when NASA, by late 1959, acquired the portion of the Army Ballistic Missile Agency that included Wernher von Braun's rocket team and its giant Saturn booster project.[8] Once NASA gained the Huntsville facilities, the nation had two separate rocket development operations that contributed to an increasingly fragmented space community. In effect, the Eisenhower administration's decision to support sepa-

Image 12-2: Launch of a Corona reconnaissance satellite, 1963. (USAF Photo)

rate military and civilian space efforts meant a tacit acceptance of duplication in the national space program.

NASA's acquisitions ended any major role in space for the Navy and Army, and boosted the Air Force to the forefront of military space. In achieving pride of place among the services, however, the Air Force had relinquished its manned space-flight mission to NASA.[9] Throughout the spring and summer of 1958 the Air Force's Air Research and Development Command had mounted an aggressive campaign to have ARPA convince administration officials to approve its Man-in-Space-Soonest development plan. But ARPA balked at the high cost, technical challenges, and uncertainties surrounding the future direction of the civilian space agency.[10]

In August, President Eisenhower assigned the human space-flight mission to NASA, and Man-in-Space-Soonest became part of the new agency's Project Mercury. With human space flight a NASA responsibility, it became increasingly difficult for Air Force space advocates to justify their expensive, technologically challenging, and single-seat strategic bombardment space plane. Although uniformed and civilian Air Force leaders still viewed human space flight as central in their effort to institutionalize space within the service, they now would pursue a role for military pilots in space largely in cooperation with NASA.

The civilian agency's absorption of Navy and Army space assets nevertheless left the Air Force as an indispensable ally of the fledgling agency. Despite NASA's acquisitions from the other services, it depended on the Air Force for biomedical and managerial personnel, space boosters for lunar probes, and range and tracking station support.[11] The Air Force expected NASA's initial dependency to continue, along the lines of the earlier relationship between the Air Force and the NACA, even though NASA, unlike the NACA, was an operating agency with substantial missions and facilities and an expansive space budget. Air force leaders quickly realized that genuine cooperation with, and support of, NASA was both in its own and the nation's best interest. The service might very well use its leverage as NASA's essential partner to convince Congress and, especially, DoD to sanction an Air Force–led military space program that included a manned military space presence and space-based weapons. Clearly, the Air Force and NASA needed each other from the outset.

Indeed, Gen. Thomas White and Administrator Keith Glennan worked together during the summer of 1959, when the issue of DoD support for Project Mercury became enmeshed in Joint Chiefs of Staff (JCF) discussions of Navy Chief of Operations Adm. Arleigh Burke's proposal to establish a joint military command for space operations. The Navy, with Army support, also favored supporting NASA through creation of a DoD Joint Task Force Mercury. General White opposed the plan, as did NASA Administrator Glennan, who did not want his agency answering to a task force commander. Instead, in September 1959, Defense Secretary Thomas S. Gates appointed the Air Force's Atlantic Missile Range commander as "DoD representative for Project Mercury support operations." This arrangement worked well until the advent of Project Apollo, when NASA's expansion and increased reliance on the Air Force's range structure at Cape Canaveral raised anew issues of cooperation and coordination. Secretary Gates also nixed the proposal for a joint military command for space, transferred specific space programs from ARPA back to the services, and assigned the Air Force responsibility for space boosters and systems integration. Payload assignments, however, would be made to the "appropriate military department."[12] While the defense secretary's memorandum confirmed the Air Force's predominance in the military space program, it did little to forestall duplication and redundancy among a growing military, civilian, and national security space community. This problem has continued to the present day, as reflected in widespread criticism during the past decade of separate "stovepipes" within the national space arena.

At the close of the Eisenhower administration, Air Force space enthusiasts continued to chafe under a "space for peace" policy that prohibited space-based weapons and a greater voice for the Air Force in the nation's space future. Despite General White's assurances of full, cooperative support to NASA, the Air Force quest for a larger role in space raised the specter among some congressmen and the general public of a possible Air Force takeover of NASA. These concerns increased when the incoming Kennedy administration indicated the Air Force might garner a larger role in the national space program.

Kennedy and USAF- NASA Cooperation in Space

On 12 January 1961, President-elect John F. Kennedy's Ad Hoc Committee on Space issued a report criticizing the "fractionated" condition of military space management and calling for "a single responsibility within the military establishment for managing the military portion of the space program." By implication, this organization would be the Air Force because, the report noted, that service had 90 percent of military space resources. Deputy Secretary of Defense Roswell Gilpatric offered the Air Force this responsibility if it would centralize its own space research and development effort. With Chief of Staff General White's promise to do so, Secretary Robert McNamara issued a directive on 6 March 1961 giving the Air Force control of all DoD space development. McNamara's directive stipulated, however, that the other services could perform preliminary research and that operational space systems still would be assigned to whatever service was deemed most appropriate. In response, the Air Force created an Air Force Systems Command on 17 March 1961 and assigned it responsibility for all Air Force research, development, and acquisition of space and missile systems. Its first commander was Gen. Bernard Schriever, the service's most aggressive spokesman for an expanded military space program.[13]

That same month, a committee chaired by former Assistant Secretary of the Air Force Trevor Gardner issued a report recommending a comprehensive, high-priority space development agenda similar to the crash ICBM program of the previous decade. The Gardner report called for a dynamic national military space effort that would include large booster development, a new

Space Development Force, military manned space flight involving a lunar landing, space research unrestricted by rigid performance criteria, military weapons in space, and an integrated national space program under Air Force
leadership.[14]

The Kennedy administration, however, refused to accept the more ambitious elements involving Air Force leadership and weapons in space. The crash
program it adopted was Project Apollo, but the Air Force received only a supporting role, under NASA's direction, in the lunar landing program. Even so,
Air Force leaders looked forward to a close, cooperative effort with NASA—
one that would enable it to achieve its own human space-flight objectives and
reenter the field of super-booster research, which had been a NASA preserve
since October 1959. Meanwhile, NASA grew phenomenally in terms of personnel, facilities, and funding for Project Apollo, and its growing prominence
strengthened its bargaining power with DoD and the Air Force.[15]

Air force expectations for a greater space role had also risen when the
Kennedy administration replaced Eisenhower's dual approach to space with
a single, integrated national space program requiring closer cooperation among
NASA, DoD, and the Air Force. Still, the challenge proved formidable. For
example, with both NASA and the Air Force developing launch vehicles, the
administration attempted to coordinate efforts through its newly designated
National Launch Vehicle Program. An agreement on 23 February 1961 between newly appointed NASA Administrator James E. Webb and Deputy Secretary of Defense Gilpatric stipulated that neither NASA nor DoD would initiate
a new space booster program unilaterally. Both would coordinate their efforts
through the Aeronautics and Astronautics Coordinating Board (AACB), which
in August 1960 had replaced the largely ineffective Civilian-Military Liaison
Committee.[16] In the summer of 1961 NASA and DoD formed the Large Launch
Vehicle Planning Group (LLVPG) to provide common specifications for developing large space boosters. Despite studying the issue for a year, the group
failed to resolve the different agencies' requirements and interests. NASA and
DoD continued to pursue separate booster programs focusing on the Saturn
C-1 and IVB and the Titan III, respectively. The nation's "integrated" launch
vehicle program still had two obvious facets.[17]

After the lunar landing project announcement in May 1961, the Air Force
quickly moved to create what Secretary of the Air Force Eugene M. Zuckert
termed an "equal partnership" with NASA. This meant playing a major role in
NASA's affairs and receiving formal designation from DoD as the executive
agent for military support to NASA. By 1962 the Air Force could cite a wide
range of support it provided the civilian agency. This included boosters for
Projects Mercury and Gemini and for unmanned space exploration; launch
and range facilities at Cape Canaveral; communication and tracking networks;
and funding and personnel assistance. Most notable among Air Force personnel seconded to NASA was Maj. Gen. Sam Phillips, who joined NASA in 1962

Image 12-3: A briefing being given by Maj. Rocco Petrone to President John F. Kennedy, Robert S. McNamara, Lyndon B. Johnson, James E. Webb, and others during a tour of Block-house 34 at the Cape Canaveral Missile Test Annex, 11 September 1962. (NASA Photo no. 62C-1443)

to manage Project Apollo. Stated policy in a 1962 Air Force report was to give NASA unqualified support, "even at the expense of USAF projects."[18]

To be sure, despite eighty-eight agreements between the DoD or the Air Force and NASA by the end of 1964 and creation of various coordinating mechanisms, both formal and informal, friction arose at nearly all interfaces between the Air Force and NASA. Particularly contentious were differences over reimbursement for support costs, as well as management and use arrangements for the Titan III and Saturn boosters at Cape Canaveral.[19] Nevertheless, cooperation remained the dominant theme, especially at the operational level. By February 1962, Secretary Zuckert had convinced Defense Secretary McNamara to formalize the Air Force role by assigning the service responsibility for "research, development, test, and engineering of satellites, boosters, space probes, and associated systems" involving NASA programs.[20] In all but name, the Air Force had achieved executive agent status for NASA support.

Air force leaders also solicited NASA's help to overcome a growing DoD tendency toward reliance on the civilian agency for satisfaction of military space needs. Furthermore, they sought to convince NASA that, despite the policy of an integrated national space program, the civilian agency alone could not meet all the military's requirements in the two vital areas of space exploration and human space flight. The record clearly indicates that the Air Force had more problems dealing with the McNamara Defense Department than with NASA. Secretary McNamara had stated that Air Force space programs must "mesh" with NASA's wherever possible. Air force leaders saw their support of NASA programs, and the civilian agency's support of service initiatives, as the wedge needed to maneuver a reluctant defense secretary into approving its programs for flying pilots in space: Dyna-Soar, Blue Gemini, and, later, the Manned Orbiting Laboratory (MOL).

In early 1962, Secretary McNamara approved a revised Dyna-Soar program that focused on orbital flight, as well as initial planning for a military space station that the Air Force wanted for inspection and neutralization of hostile satellites. Eventually termed Blue Gemini, the space station would consist of a permanent station test module, a Gemini spacecraft, and the Titan III "building block" launcher.[21] Although the revised spaceplane program's deemphasis of hypersonic flight lessened NASA's interest, the agency promised continued support.

As for Blue Gemini, the defense secretary had canceled it by December 1962 for reasons of cost-effectiveness, duplication, and the inability of Air Force leaders to provide convincing mission justification. Although continuing to express strong reservations about a military human space-flight role, he refused to close the door completely and, thereby, leave the field of manned space flight entirely to NASA. Attracted by NASA's Gemini program, with its objectives of space rendezvous and Earth observation, the defense secretary argued that the National Space Program represented a single program that should be jointly conducted. After first failing to convince Administrator James Webb that NASA's Project Gemini should be centralized under DoD's management, he proposed joint management. Webb objected and successfully upheld his position that the two agencies should coordinate their programs more effectively but should not jointly manage them. The agreement they concluded in September, which described the parameters of DoD-NASA cooperation on space station studies and development, reflected the NASA position.[22]

In 1963, as the Dyna-Soar program neared the production phase, Secretary McNamara became increasingly concerned about its high costs, its bombardment and satellite interception missions, and the fact that an Air Force space laboratory could better assess the military capabilities of humans in space. On 10 December 1963, he canceled Dyna-Soar and, with NASA's blessing, approved initial development of a Manned Orbiting Laboratory (MOL), which would use a Gemini capsule linked to a test module and launched by the Titan III. NASA Associate Administrator Robert Seamans immediately pledged the agency's full support for the Air Force MOL effort.[23]

The USAF-NASA partnership continued to improve during the latter half

of the decade. New Air Force leaders proved less assertive of Air Force pre-
rogatives for space and more interested in developing an effective working
relationship with DoD and NASA. They appreciated the wide range of support
NASA provided the MOL project, while NASA officials welcomed Air Force
support for the civilian agency's Apollo Applications Program.[24] More than
ever, both agencies understood their mutual dependency as they confronted
the challenges of reduced space budgets and declining public support for
manned space adventures in an era increasingly dominated by the Vietnam
war. Indeed, in the late 1960s critics argued that the costly MOL duplicated
the human space-flight and exploration missions already assigned to NASA
and the intelligence mission conducted by the National Reconnaissance Office. With the advent of the budget-conscious Nixon administration, MOL's
days were numbered. On 10 June 1969, Deputy Secretary of Defense David
Packard justified cancellation of the MOL on grounds of budget pressures
and the improved capabilities of robotic spacecraft.[25]

Significantly, former NASA Associate Administrator and now Air Force
Secretary Robert Seamans declared that "the cost of a manned [space] system
is too great to be borne at this time." He advised the Air Force to focus on
modernizing its tactical and strategic forces rather than on future exploitation of space.[26] Air Force manned space-flight enthusiasts could only look
back on the decade of the 1960s as a graveyard of false optimism; their dream
of achieving an expanded, "independent" Air Force space program lay dead, if
not buried. Space was far from institutionalized in the traditional Air Force.
Although the service would continue to closely cooperate with NASA, Air Force
leaders, especially the senior uniformed officers, would ever after be skeptical
about wholeheartedly embracing manned military space flight.

The Space Shuttle Challenge

Although Secretary Seamans and other civilian Air Force officials refused to
pursue an Air Force manned space program, they were not averse to cooperating with NASA's post-Apollo venture to develop a reusable national space
transportation system to provide routine space access for both civilian and
military agencies. Shuttle development and operation would significantly
broaden the scope of Air Force–NASA cooperation, but not without considerable friction. At the same time, the Shuttle would stimulate a major reorganization within the Air Force and somewhat restore its interest in manned
military space operations.

In June 1969, President Nixon's Space Task Group recommended approval
of a technically feasible, cost-effective, reliable "shuttle" to low-Earth orbit
(LEO). Apart from low-cost, routine access to space, the report claimed the
Space Transportation System (STS) could perform on-orbit repair of satellites, support a space station, launch high-energy missions, conduct short-

duration orbital missions and, consequently, contribute to the nation's international prestige.[27]

Civilian Air Force leaders saw in the Shuttle the promise of routine access to space, especially for future larger, heavier satellites, including the NRO's strategic reconnaissance spacecraft.[28] They also viewed the Shuttle as a means of reasserting Air Force influence in the national space program. The uniformed Air Force, however, proved far more reluctant to accept the multipurpose, reusable vehicle that NASA expected would become the sole means of launching all civilian and military satellites.

In February 1970, NASA Administrator Thomas Paine and Secretary Seamans established the joint NASA–Air Force Space Transportation System Committee to handle all aspects of the STS affecting NASA and the Air Force. The latter served as DoD executive agent for the STS.[29] Even before President Nixon formally approved the Shuttle in early January 1972, the STS Committee achieved considerable progress on design and performance specifications.

From the start, it was clear to NASA, DoD, and Air Force officials that NASA needed Air Force support to "sell" the project to a budget-conscious administration and Congress. In exchange for its support, the Air Force demanded a vehicle capable of launching its largest, heaviest payloads into polar orbit from Vandenberg Air Force Base and, under emergency conditions, returning to the California launch site. This meant developing a Shuttle with greater cross-range and payload capacity than NASA favored.[30] Rather than the NASA proposal of a straight-winged orbiter with a cargo bay measuring 12 feet in diameter by 40 feet in length, the approved version was delta-winged, with cargo bay dimensions of 15 feet by 60 feet. These requirements would lead to greater technical problems, increased development costs and, eventually, would jeopardize the initial projections of sixty flights per year at half the cost of expendable boosters.[31]

In late 1973, DoD created the Space Shuttle User Committee, chaired by the Air Staff's Director of Space, to focus on military responsibilities for Shuttle support. These included refurbishing the old Space Launch Complex-6 (referred to as "Slick Six") MOL launch site at Vandenberg Air Force Base, developing an upper-stage, orbit-to-orbit "space tug" and establishing a schedule for phasing out its fleet of Expendable Launch Vehicles (ELVs) during the 1980–85 time frame.[32]

By the mid-1970s, the attitude of both the Air Force's top brass and, especially, the middle-rank space enthusiasts had evolved from "resigned acceptance" to "cautious optimism." If the Shuttle lived up to predicted expectations, it could help achieve the long-sought institutional goal of normalizing space operations by means of standardized, reusable launch vehicles and, although hardly a priority at this point, perhaps preserve a military manned presence in space.[33]

NASA had assumed that all future DoD satellites would be launched on the Shuttle. To meet rising costs and shore up political support for the Shuttle, NASA officials insisted the Defense Department commit itself to a "Shuttle-only" policy and phase out its fleet of expendable launch vehicles. DoD and the Air Force, however, had never formally committed to the reusable space transportation vehicle as their *exclusive* launch system. DoD statements on Shuttle use and ELV phaseout always included a caveat that the Shuttle first needed to demonstrate its promised capabilities and low-cost operations before the military would discontinue ELV production completely. Despite their cautious optimism, DoD and Air Force leaders grew increasingly concerned about the Shuttle's high development costs, the growing pressure to phase out its ELV fleet, and the rising expense of supporting two launch programs. Also contributing to strained relations between the military and civilian agencies was the tendency of NASA officials to make program changes without notifying their Air Force counterparts.[34] In a 14 January 1977 memorandum of understanding, NASA, DoD, and the Air Force reaffirmed, but more clearly defined, their mutual responsibilities for Shuttle development and operations. Significantly for DoD, the STS would be the *primary*, rather than *exclusive*, launch vehicle for military payloads.[35]

In 1977, the Shuttle gained a strong ally in Dr. Hans Mark, who became Undersecretary of the Air Force and, consequently, Director of the NRO. He determined that all future NRO satellites should be configured for the Shuttle but not also for ELVs. He led the effort to designate the Shuttle the military's exclusive space launch vehicle. As one writer noted, the reusable Shuttle would make the expendable launcher truly expendable once and for all time.[36]

While the focus remained on USAF-NASA cooperative efforts to realize their Shuttle development commitments, the advent of the Shuttle also precipitated a major shift in Air Force thinking on space. Traditionally, the Air Force and DoD had assigned space systems on a functional basis to the command or agency with the greatest need. But in the late 1970s, systems possessing multiple capabilities and serving a variety of defense users, like the Defense Satellite Communications System (DSCS), the Shuttle, and the projected Global Positioning System (GPS), promised to blur the functional lines enormously. Moreover, the poorly defined line separating experimental from operational space systems meant that Air Force Systems Command (AFSC) had "ownership" of on-orbit spacecraft well beyond what many in the operational arena considered the legitimate responsibility of a research and development command. Would AFSC also serve as the military's "operational" organization for the Shuttle, or would Air Force space requirements be better served by creating a new, major command for space operations? The Shuttle generated an intense competition for operational responsibility among four major Air Force commands, each considering itself the logical choice to become the operational space command. The contest for control of the Shuttle would convince

Image 12-4: The Space Shuttle was a major step in the development of military-civil space access. It was to provide launch services for all government organizations. Here the 12 April 1981 launch at Pad 39A of STS-1, just seconds past 7:00 A.M., carries astronauts John Young and Robert Crippen into an Earth-orbital mission scheduled to last for fifty-four hours, ending with an unpowered landing at Edwards Air Force Base in California. (NASA Photo no. 81PC-0382)

Air Force leaders to deal with an increasingly fragmented military space community by creating, in 1982, the Space Command.[37]

By the early 1980s, Air Force and DoD concern approached alarm when faced with the prospect of the Shuttle's continued high costs, production delays, and reduced flight schedules. Air force critics increasingly faulted NASA's research and development mentality and called for more military involvement in Shuttle management. High-level policy reviews in March 1980 led to modification of the 1977 agreement, which resulted in assigning the military priority in Shuttle mission preparations and flight operations, and in integrating DoD personnel more directly into NASA's line functions.[38] Nevertheless, Air

Force leaders continued to worry about phasing out expendable launch vehicles even after the Shuttle became fully operational. By this time, the skeptics included new civilian Air Force leaders in the Reagan administration as well as the top brass. Following recommendations by the Air Force Scientific Advisory Board and the Defense Science Board, Air Force Chief of Staff Gen. Lew Allen, in October 1981, directed a major study of a "mixed fleet" strategy to complement the Shuttle with expendable launch vehicles. The following month, Undersecretary of the Air Force and NRO Director Edward C. "Pete" Aldridge, declared publicly that the Air Force "cannot continue to look to NASA as our country's Launch Service Organization in the Shuttle era." Aldridge, who would become a central figure in the space launch arena throughout the decade, went on to say that "it . . . seems illogical that our only 'truck' to deliver our goods to space be in the form of 3, or 4, or 5 highly complex launch vehicles."[39]

Even though President Reagan reaffirmed the Shuttle as the nation's primary launch vehicle in July 1982, the Air Force pursued its plans for a mixed fleet based on commercial production of the Titan III, the purchase of additional Titan 34Ds, and the refurbishment of Titan II ICBMs. At the end of 1983, Undersecretary Aldridge, proclaiming the need for "assured access to space," proposed the additional step of upgrading the Atlas, termed Atlas II, and procuring ten more powerful Titans. The latter, initially known as the Titan 34D7 because of its seven- rather than five-segment solid rocket motors, soon was designated the Complementary Expendable Launch Vehicle (CELV) and, later, the Titan IV.[40]

NASA officials found themselves on the defensive, pleased with neither the prospect of a competitive booster nor the mounting criticism for allowing "militarization" of the Shuttle program. In response, NASA vigorously defended its relationship with the military. Glynn Lunney, manager of the Space Shuttle program at the Johnson Space Center, for example, favored strengthening the already close ties between the military and civilian agencies. From his viewpoint, the military earlier had gained experience with unmanned space systems but had neglected manned space flight. "It is now time," he asserted, "for the DoD to fully embrace and exploit the manned space flight capabilities that NASA has developed for our nation." Doing so would put the military squarely behind the Shuttle. Not surprisingly, NASA officials fervently lobbied against the CELV because, they said, it would result in lower Shuttle flight rates and higher costs.[41]

Military officials, however, were not convinced. In early 1984, Secretary of Defense Caspar Weinberger, referring to the need for an assured launch capability, authorized the Air Force to acquire ten CELVs. At the same time, he rebuffed NASA's efforts to have DoD support its national space station proposal. Seeing no national security requirements that a space station could fulfill, he refused to commit DoD to a second major space project. The high cost of such a program, he argued, would affect the Strategic Defense Initiative (SDI) as well as other military and civilian space projects, while a space station would divert NASA from its primary focus, the Shuttle.[42] Despite NASA's best efforts, neither DoD nor the Air Force expressed significant interest in manned military space flight during the Reagan era.

By February 1985, "Pete" Aldridge, now Secretary of the Air Force, had convinced NASA Administrator James Beggs to agree to a limited number of expendable boosters. In exchange, the Air Force and DoD agreed to utilize at least one-third of all forthcoming Shuttle flights and to initiate joint studies leading to a second-generation STS. That same month, the National Security Council confirmed this agreement in its National Space Strategy directive, which authorized a "mixed fleet" approach to "assured access to space."[43] That summer Aldridge and top Air Force generals testified before Congress in support of a limited ELV program and against a fifth Shuttle Orbiter.[44]

On the eve of the *Challenger* disaster, the Shuttle remained the centerpiece of America's space launch program. Although the Air Force's commitment to the Shuttle as its primary launch vehicle had been tempered by diminishing expectations, it hoped that addition of a limited number of "mixed fleet" expendable boosters would help sustain the Shuttle without delaying military launch schedules.

In the wake of the *Challenger* shock wave, NASA conducted political damage control and turned to the military for outside assistance. It appointed Adm. Richard H. Truly as Associate Administrator for Space Flight, the Air Force's Space Division Commander, Lt. Gen. Forrest S. McCartney, as director of Kennedy Space Center, and also sought advice from former Apollo program manager Gen. Samuel C. Phillips.[45]

While NASA addressed the Shuttle issue, military space officials confronted additional difficulties. The Air Force had only begun to recover from the August 1985 failure of its Titan 34D rocket when, in April 1986, another Titan 34D exploded on liftoff at Vandenberg. The following month NASA lost a Delta rocket. After these launch disasters, space leaders effectively grounded the space program by prohibiting further flights of both the Shuttle and expendable launch vehicles until the problems could be solved. Concerned Air Force officials estimated that the launch delay, which lasted thirty-one months for the Shuttle, would affect twenty-five payloads and create a "ripple" effect that could not be overcome before 1992. By the time NASA resumed Shuttle operations on a conservative schedule, the military had reprogrammed eighteen of thirty-six previously manifested Shuttle payloads for expendable launchers. After 1992, the Defense Department would use the Shuttle only for SDI or research-and-development missions. In effect, the Air Force chose to abandon the standardized Shuttle, the "airliner to space," in favor of the diversification represented by expendable boosters.[46]

Looking back on the Shuttle experience, Air Force participation came at the behest of the service's civilian leadership. The "blue suit" Air Force, in contrast to earlier NASA initiatives, supported the Shuttle reluctantly. Air force leaders were no longer enamored with human space flight and only modestly enthusiastic about space in general. Ironically, the Shuttle helped provide support for an Air Force space focus by convincing Air Force leaders to centralize management responsibility for its increasingly effective unmanned space platforms in an operational space command. The central priority, however, became making space support essential to the warfighter, not flying pilots into space.

Looking ahead from the *Challenger* tragedy, Air Force generals who had come of age through the Shuttle years were not inclined to support major new cooperative ventures with NASA.[47] On the other hand, no one wanted to resort to business as usual, with its time-consuming practice of linking specific satellites to particular launch vehicles. The launch challenge for the 1990s would find NASA and the Air Force cooperating to develop an "assured launch strategy" based on lower costs and greater launch responsiveness.

The Rise of Commercialization

During the late 1980s and early 1990s, several trends or events converged to encourage a stronger cooperative spirit than ever before on the part of both the Air Force and NASA. Even before the loss of *Challenger*, the availability of less expensive, more responsive foreign launch capabilities had begun to erode customer demand for US launch services. That trend became even more visible after January 1986, when Shuttle flights ceased for nearly three years and subsequent policy changes placed significant restrictions on what would be launched from the Shuttle.[48] Operations Desert Shield and Desert Storm in the Persian Gulf region during 1990–91 spectacularly revealed how vital space systems had become to conducting successful military campaigns but, simultaneously, exposed the limitations of relying tactically on systems designed primarily for strategic, cold war purposes. The Air Force emerged from the Gulf War with a sense of urgency about developing a launch-on-demand capability to cope with tactical demand for space capabilities. That sense only increased with the collapse of the Soviet Union and the end of the cold war.[49] Not only was it in the best interest of the Air Force and NASA to push technology toward a cheaper, more responsive U.S. launch capability, but such a course was widely acknowledged as necessary for sustaining the nation's industrial base for launch vehicle production.

Growth in the commercial space sector and in dual-use aspects of space technology, including launch systems, also inspired closer coordination among the military services, NASA, other space-related government organizations, and industry. Anticipation of increasing demand for launch services—one estimate being that more than eleven hundred satellites would be launched between 1997 and 2006—especially from commercial interests intensified industry efforts to develop new lines of launch vehicles. Furthermore, as the 1990s drew to a close, statistics on space services revealed that roughly 80 percent of the total contribution came from commercial interests and only 20 percent from government. The burgeoning reliance of the military services on commercial space systems, especially communications satellites, and of civil and commercial users on military satellites, particularly the Global Positioning System, became unmistakable during the 1990s. It clearly made more sense for both the Air Force and NASA to encourage industry's pursuit of commercial-off-the-shelf products and services, rather than for those and other government organizations to continue paying astronomically higher prices for uniquely tailored systems.[50]

Yet another strong incentive for greater USAF-NASA cooperation in developing new, improved launch vehicles came from the fiscal environment of the 1990s. A combination of congressional pressure to reduce government spending and balance the budget, along with a shift in program priorities from the Republican administrations of Ronald Reagan and George Bush to the Democratic administration of Bill Clinton, resulted in demonstrably lower appropriations for both military and civil space programs. While NASA Administrator Daniel S. Goldin had loudly and vigorously attempted to implement a "faster, better, cheaper" approach to U.S. civil space programs since 1992, Air Force leaders had grudgingly stuck to traditional ways of doing business. Even as the latter spoke of an "air and space" force evolving toward a "space and air" force in the early twenty-first century, they resurrected the "aerospace" concepts and terminology that General White had used to defend the primacy of the Air Force's role in space during the late 1950s.[51] It was not long before ardent military space advocates, who perceived that the "fighter mafia" still maintained a too tight grip on the service's purse strings, began seeking innovative ways to circumvent the dilemma of dwindling in-house resources and to perpetuate their quest for an affordable, flexible space launch capability. One obvious solution lay in closer cooperation with other space-related government organizations.

One model for closer cooperation between the Air Force and NASA was the "joint program" arrangement. It had worked relatively well, at least from a test pilot's perspective, during the X-15 and lifting-body flight programs of the 1960s and 1970s.[52] But its reputation had been sullied somewhat by the more recent outcome of the technologically ambitious X-30A Hypersonic, TransAtmospheric Vehicle (TAV) program, which was specifically intended as the hardware testbed for a National Aerospace Plane (NASP). Beginning with the release of a classified request for proposals (RFP) in November 1985, two months before the *Challenger* disaster, the X-30 grew within two years to become the largest, most expensive research aircraft project ever undertaken, with seven teams overseeing technology development efforts and working on 125 critical tasks. Although the Defense Department officially sponsored the project with all funding consolidated under an Air Force budget line, several different organizations contributed money and other resources to it.

The Advanced Research Projects Agency, the Navy, the Air Force, and NASA occupied a joint program office (JPO) for NASP at Wright-Patterson Air Force Base, Ohio. Among other things, the JPO attempted to coordinate military and civilian users' requirements. Application of a "team" concept was supposed to insulate NASP from the vicissitudes of the budget cycle by providing the program with several "partners" who would form a stable constituency. Over the course of a decade, however, as some of the technological challenges began to seem insurmountable in the near term, interest and financial support of one after another of the original sponsors waned. In the long run, it became obvious that none of the various "partners" had been locked into a long-term commitment but, rather, that "ownership" had been diluted to the detriment of the overall NASP program. Congress drastically reduced funding for the program, and NASA withdrew its support. Consequently, the program died from fiscal starvation during 1993–95.[53]

Another USAF-NASA joint program attempt centered on efforts to produce a new National Launch System (NLS) to replace the existing fleet of expendable boosters as well as the Shuttle. A 10 December 1990 report from the Advisory Committee on the Future of the U.S. Space Program—better known as the Augustine Committee—called for ending reliance on the Shuttle and developing a new booster system as soon as possible to orbit larger, heavier payloads. On 2 January 1991, Vice President Dan Quayle, Chairman of the National Space Council, gave DoD and NASA only two months to prepare a joint program plan for space launch. This refocused attention on the Advanced Launch System (ALS) concept that had emerged in the immediate wake of the *Challenger* disaster but had foundered when DoD, NASA, and the Strategic Defense Initiative Organization (SDIO) could not reach a consensus on the program's direction on the nature of new launch vehicles.

Work on the joint program plan soon revealed general agreement between DoD and NASA in a number of areas. All levels of the national space community agreed that the launch program ought to be a joint DoD-NASA effort with each party covering half of the estimated $9 billion total cost. Similarly, all agreed that a "common core" vehicle should be developed to meet both military and NASA requirements. On 22 January 1991, the Office of the Secretary of the Air Force and NASA agreed that a single launcher capable of placing 70,000 to 80,000 pounds in low-Earth orbit could meet military needs as well as specifications associated with NASA's space station. The agreement, which delighted AFSPC's requirements staff because it contained lift parameters established in the command's ALS System Operational Requirements Document (SORD), represented a major concession by NASA to participate in a program not based on its own Shuttle-C concept—one which DoD had refused to accept. Things did not continue smoothly through February, however, because

NASA appeared less confident that the launch vehicle configuration would meet its space station requirements. Further meetings occurred among DoD, NASA, and administration officials, including a 16 April session of the National Space Council attended by OMB Director Richard Darman, Deputy Secretary of Defense Donald Atwood, NASA Administrator Richard Truly, and Assistant Secretary of the Air Force for Space Martin Faga. Still, NASA remained skeptical amid congressional efforts to pare its request for program funds from $175 million to $50 million.

Vice President Quayle's announcement of a new National Space Launch Strategy at Vandenberg Air Force Base, California, on 24 July 1991, did not alleviate the concern about NLS funding. The JPO, as well as the Directorate of Space and SDI Programs in the Office of the Secretary of the Air Force, interpreted congressional committee actions as meaning that fiscal year 1992 funding for NLS would amount to less than one-third of the requested $350 million. Insufficient funding meant program delays and, ultimately, inability to meet Quayle's target for a first flight in 1999.

Although funding remained the primary NLS issue throughout 1991, some senior Air Force officers also expressed reservations about a perceived managerial imbalance between the number of DoD people and the number of NASA people in the JPO. The NASA contingent clearly outnumbered the DoD group. Lt. Gen. Thomas S. Moorman Jr., commander of Air Force Space Command, feared the military might not be able to exercise sufficient control over the program to ensure its requirements were properly addressed. Uncertainty reigned over the type or types of launch vehicles the program ultimately would produce. In turn, that situation fueled parochialism, which prevented both development of a convincing set of NLS requirements and integration of program management. Consequently, with support for the NLS program unraveling, Congress directed termination of the program in the defense appropriations bill for fiscal year 1993.[54]

Planning for modernization of the nation's space launch capabilities limped along through 1993. Finally, in the National Defense Authorization Act for fiscal year 1994, Congress mandated that the Secretary of Defense develop a plan establishing and clearly defining "priorities, goals, and milestones regarding modernization of space launch capabilities for the Department of Defense or, if appropriate, for the Government as a whole." The task of initiating a study and preparing the plan went to John M. Deutch, Undersecretary of Defense for Acquisition and Technology, who approved the terms of reference for the study on 23 December 1993.[55]

The terms of reference established an Air Force–led study group under the chairmanship of Lieutenant General Moorman. "Spacelift systems are the enabling foundation for all military, intelligence, civil, and commercial activities in space," the document explained. "Over the past several years, a number of studies have examined and identified serious deficiencies in U.S. launch capabilities and competitiveness. To date, there is no consensus on the appropriate course of action to remedy these deficiencies." The study group included representatives from numerous organizations: Air Force headquarters, Air Force Space Command, and Materiel Command; the Advanced Research Projects Agency; the Ballistic Missile Defense Organization (BMDO); the Joint Staff; NASA; Naval Space Command; the Office of the Secretary of the Air Force; the Office of the Undersecretary of Defense for Acquisition and Technology; and United States Space Command. Representation on an ancillary steering group was even broader, including the Departments of Commerce and Transportation, as well as the NRO.

During the first three months of 1994, the interagency study group organized into five panels and performed its assigned task. The members listened to over 130 presentations from government agencies, industry, laboratories, and think tanks. They conducted interviews and roundtable discussions with congressmen and congressional staffers, industry executives, and current as well as past national space leaders. Of course, the group drew extensively from other recent launch studies: the "Report of the Advisory Committee on the Future of the U.S. Space Program" (the Augustine Report), December 1990; "The Future of the U.S. Space Launch Capability" (the Aldridge Study), November 1992; the "NASA Access to Space Study," 1993; and the "DoD Bottom-Up Review" (BUR), 1993.

Moorman sent the resulting "Space Launch Modernization Plan" to Deutch on 5 May 1994. It offered no specific programmatic approach but, rather, defined several "road-map" options to help reduce cost and improve operational effectiveness. The options included sustainment of existing systems, evolution of current expendable launch systems, development of a new expendable system, or development of a new reusable launch vehicle. Only "clean sheet" expendable and reusable designs would improve launch responsiveness. Since the estimated cost of pursuing those options ranged from $5 billion to more than $20 billion, Deutch labeled them "inaffordable" in the near term. For all options, however, the group recommended revitalizing the U.S. "core" space launch technology program.[56]

A new National Space Transportation Policy (PDD/NSTC-4) signed by President Clinton on 5 August 1994 fulfilled at least one recommendation from the Moorman report. It gave the Defense Department responsibility for developing an expendable launch vehicle line that would evolve from, and ultimately supersede, the so-called heritage systems—Atlas, Titan, and Delta. Since the Air Force was DoD executive agent for space launch, the task of ELV development devolved upon it.[57] Simultaneously, NASA received the nod to develop a Reusable Launch Vehicle (RLV) that ultimately would replace the Shuttle. Improvements in reliability and operability were major goals of this policy, but its primary objective was to reduce dramatically the cost of launching payloads into LEO. Although Evolved Expendable Launch Vehicle (EELV) and RLV managerial lines were clearly spelled out in President Clinton's state-ment, room remained for the Air Force and NASA to coordinate and cooper-ate, especially with regard to research and development of core technology. Furthermore, implementation of PDD/NSTC-4 compelled DoD, Commerce, Transportation, NASA, and the CIA to agree on a common set of requirements and a coordinated technology plan addressing the needs of the national secu-rity, civilian, and commercial space launch sectors.[58]

USAF-NASA Cooperation: Next-Generation Launchers

Even as the Air Force and NASA set out to develop acquisition strategies and other plans for the EELV and the RLV, respectively, renewed cooperation was in the wind. On 3 March 1995, Richard M. Mccormick, Air Force Deputy Assistant Secretary, Space Plans and Policy (SAF/SX), announced in a memorandum the establishment of an Air Force Space Launch Memorandum of Agreement (MOA) working group. This group would formulate "an agreed upon corporate Air Force policy and plan" that would include the role of the Air Force in NASA launch-related acquisition and technology efforts and, conversely, of NASA in Air Force activities. Once the Air Force had a clear position, McCormick in-tended to commence discussions with NASA leading to an MOA that would ensure all DoD and NASA space launch activities were "well coordinated and integrated" in compliance with President Clinton's policy. Although Air Force Space Command planners privately expressed "serious reservations about Mr. McCormick's office getting too involved in the details" of already "excellent tech-nology coordination between NASA Marshall and AF Phillips Lab," HQ (Head-quarters) AFSPC Director of Plans Brig. Gen. Roger G. DeKok told McCormick. "We wholeheartedly support this initiative." One result, by autumn 1995, was formation of several high-level integrated product teams (IPTs) chartered by the Aeronautics and Astronautics Coordinating Board.[59]

Cognizant of shortcomings in the many top-down DoD and Air Force coop-erative agreements with NASA that had come and gone over nearly four de-cades, AFSPC planners had begun focusing on a new bottom-up concept—partnership. While familiar to business and industry, this approach to expensive, risky undertakings was relatively foreign to military minds and experience. Nonetheless, as early as January 1995, midlevel officers in the AFSPC Headquarters Directorate of Requirements were referring to a "continuing part-nership with industry."[60] Convinced that a military spaceplane would have broad applications in the early twenty-first century, AFSPC Commander Gen. Joseph Ashy began calling later that year for DoD to support NASA's RLV technology lead. He discussed the fuller "synchronization" of RLV and EELV requirements with Administrator Dan Goldin and, in early 1996, enlisted the help of The Aerospace Corporation's CEO Pete Aldridge to facilitate that process.[61] A video teleconference sponsored by The Aerospace Corporation on 28 August 1996 kicked off a Future Spacelift Requirements Study involving AFSPC, Air Force Materiel Command (AFMC), and NASA. During this same period, an integrated concept team (ICT) chaired by HQ AFSPC/DR and involving over one hundred members from the Air Force and NASA began investigating the application of emerging technologies to the development of a military spaceplane.[62]

The breakthrough to a full-blown partnership between AFSPC and NASA for exploration and exploitation of future space launch technologies and activities came in early 1997. General Ashy's successor as AFSPC commander, Gen. Howell M. Estes III, with approval from the Air Force Chief of Staff, concluded an MOA with Dan Goldin that formed an AFSPC/NASA Partnership Council effective 28 February 1997. The council's explicit purpose was to improve the level of interaction between the organizations to achieve "harmonized long-range planning, more efficient resource allocation, expanded technology partnerships, and more compelling advocacy of programs." Co-chaired by the AFSPC Commander and the NASA Administrator, the partnership council embodied an expanded commitment on the part of the senior leaders of both organizations to promote "proactive coordination of activities in areas of mutual interest." The two leaders "envisioned a very fast track effort using a simple structure, focused on quick turn results." By midyear, the council had established study teams in eight areas, including one in support of NASA efforts to develop the Integrated Advanced Space Transportation Plan. The AFSPC and NASA members of each team had formulated mutually agreeable "terms of reference" to guide their deliberations and "deliverables."[63]

One of the earliest, most encouraging indications of the partnership's vitality was a 1 May 1997 memorandum of support from Keith R. Hall, Assistant Secretary of the Air Force (Space) and Director of the NRO. Not surprisingly, therefore, the 8 April 1998 meeting of the AFSPC/NASA Partnership Council included Hall as an attendee and resulted in expansion of the council to include the NRO as a full partner. Effective 23 November 1998, a new MOA signed by NASA Administrator Goldin, AFSPC Commander Gen. Richard B. Myers, and NRO Director Hall formalized the AFSPC/NASA/NRO Partnership Council.[64] All three organizations acknowledged a common interest in reducing costs and, simultaneously, improving capabilities within four essential areas: launch, infrastructure, command and control, and industrial base.

The Path Ahead

While it is far too early to assess historically the value of these partnering efforts, some preliminary, broad-brush analysis seems possible. On the positive side of the ledger, the decades-long tradition of sharing experienced management personnel continues. After twenty-three years in the Air Force, where he served as a launch controller, spacecraft design manager, astronaut, and Deputy for Technology in the Ballistic Missile Defense Organization, Gary E.

Payton became Director of the Space Transportation Division in NASA's Office of Space Access and Technology during the 1990s. Specifically, he managed the RLV program, including X-33 and X-34. Also, the deputy project manager for the X-37 at NASA's Marshall Space Flight Center (MSFC) was a lieutenant colonel from the Air Force Research Laboratory (AFRL). Both AFSPC and AFRL have liaison officers at NASA Headquarters, as does NASA at AFSPC Headquarters. Most recently, the Air Force actually has placed an Air Force colonel on the MSFC director's staff, and a senior NASA official sits on the AFSPC commander's staff. In 2002, AFSPC will establish a military detachment at MSFC. This lends credence to former Administrator Dan Goldin's opinion that the cooperative relationship between NASA and the Air Force is the best it has ever been.

On the other hand, one can find tension in the partnership. From the beginning, there has been a lot of friendly "back slapping" among senior leaders, and considerable "trench-level sharing" has prevailed among action officers in the various organizations. Trouble has sometimes surfaced, however, at the middle-management level—for example, lieutenant colonels and colonels in the Air Force. It is there, especially, that personalities, different cultural perspectives, and not necessarily pleasant experiences born of past relations have tended to breed wariness and suspicion about the others' motives. Division of a smaller "budget pie" among the same number of partners can increase tensions. Finally, the present arrangements are such that practically anybody can say no and only a selected few can say yes when the going gets tough.[65]

The fact remains that partnerships, in the sense that private industry thinks of them, are relatively unfamiliar territory for the Air Force. Increasingly, the Air Force, NASA, and NRO are being drawn toward closer relationships with the commercial sector. "Therein lies the need for change on the military side," one Air Force general officer recently stated. He concluded, "If we in the military are to be good partners with an industry driven by the pressure of business, then we must become better businessmen."[66] Is that really necessary in order to revitalize U.S. space launch capabilities for the 21st century? Time will certainly tell.

Notes

1. General Thomas D. White, Chief of Staff, United States Air Force, to General Landon, Air Force Personnel Deputy Commander, and General Wilson, Air Force Development Deputy Commander, 14 April 1960, reprinted in *Defense Space Interests, Hearings Before the Committee on Science and Astronautics*, U.S. House of Representatives, 87th Cong., 1st sess. (Washington, D.C.: GPO, 1961), p. 92.

2. Not only was WS-117 underfunded, but the administration prohibited public pronouncements by officials and military officers that might generate an international debate on "freedom of space" for military space flight. For example, after General Schriever raised the issue of offensive weapons in space in a speech in San Diego in February 1957, the Secretary of Defense ordered him to omit the word "space" in all future public discussion. David N. Spires,

Beyond Horizons: A Half Century of Air Force Space Leadership (Peterson Air Force Base, Colo.: Air Force Space Command, 1997), p. 47. For Schriever's speech, see Maj. Gen. Bernard A. Schriever, Commander, Western Development Division Headquarters, Air Research and Development Command, "ICBM—A Step Toward Space Conquest," address to the First Annual Air Force Office of Scientific Research Astronautics Symposium, San Diego, California, 19 February 1957, in Morton Alperin, Marvin Stern, and Harold Wooster, eds., *Vistas in Astronautics: First Annual Air Force Office Scientific Research Astronautics Symposium, International Series of Monographs on Aeronautical Sciences and Space Flight*, Theodore von Kármán and Hugh L. Dryden, eds., Division VII, Astronautics Division, vol. 1 (New York: Pergamon Press, 1958).

3. Spires, *Beyond Horizons*, pp. 53–55, 73–74. See note 51 for further analysis of the origin and use of the "aerospace" term.

4. Ibid., Appendix 2-1, p. 74; Lt. Gen. Donald L. Putt, Deputy Chief of Staff, Development, U.S. Air Force, to Dr. Hugh L. Dryden, Director, National Advisory Committee for Aeronautics, 31 January 1958. In April 1957, the Air Force had combined its previous hypersonic boost-glide studies into a single development plan that envisioned a highly maneuverable delta-wing vehicle boosted into a low-Earth orbital glide path before reentering the atmosphere and landing at a conventional airfield.

5. Spires, *Beyond Horizons*, pp. 58, 84–85; Dwayne A. Day, "Invitation to Struggle: The History of Civilian-Military Relations in Space," in John M. Logsdon, gen. ed., *Exploring the Unknown: Selected Documents in the History of the U.S. Civil Space Program*, vol. 2, *External Relationships*, NASA SP-4407 (Washington, D.C.: NASA, 1995), 2:250. In September 1958 ARPA reprogrammed WS-117L, also termed the Advanced Reconnaissance System, into separate component projects with revised designations: Sentry (soon changed to Samos) for the reconnaissance element; MIDAS for the infrared sensing system, and Discoverer, a cover for the covert Corona project. The latter, initially an interim measure, involved the Thor-Agena booster satellite combination, and the recoverable space capsule method of data retrieval. The more ambitious Samos, by contrast, called for an Atlas-Agena configuration and, initially, direct transmission of photographic and electromagnetic reconnaissance data to ground stations. When Samos continued to experience managerial and technical problems, President Eisenhower, in August 1960, created a separate management structure outside direct uniformed Air Force control to handle Corona and Samos. Years later, officials designated the structure the National Reconnaissance Office.

6. For the document, see John M. Logsdon, gen. ed., *Exploring the Unknown: Selected Documents in the History of the U.S. Civil Space Program*, vol. 1, *Organizing for Exploration*, NASA SP-4407 (Washington, D.C.: NASA, 1995), 1:334–35.

7. Significantly, in a December 1958 meeting involving NASA and the Air Force's Ballistic Missile Division, successor to the Western Development Division, Air Force representatives failed to inform their civilian counterparts of their Agena B upper stage development program. As a result, NASA pursed its own Vega program until a year later, when it became aware of the Agena. Duplication cost the government $16 million. Day, "Invitation to Struggle," p. 254.

8. Beginning with only NACA's aeronautical research facilities and personnel, NASA Administrator Keith Glennan turned to the Navy and Army for an infusion of space programs to achieve space capability quickly. While the Navy raised little objection to relinquishing Project Vanguard's personnel and facilities and over four hundred scientists from the Naval Research Laboratory, the Army strongly objected to Glennan's sweeping request to transfer Caltech's JPL and especially the von Braun team. ABMA director Maj. Gen. John Medaris could only achieve a compromise, whereby NASA acquired JPL by 3 December 1958, and the Huntsville complex at the end of the following year. Spires, *Beyond Horizons*, pp. 65–66.

9. NASA also received the Air Force's large rocket engine research project, and satellite tracking and satellite communications, meteorological, and navigation studies. Spires, *Beyond Horizons*, pp. 65–66.

10. Man-in-Space-Soonest called for a four-phase capsule orbital process, which would first use instruments, to be followed by primates, then a man, with the final objective landing men on the moon. The Air Force prepared seven development plans; appropriately, the final plan omitted the word "soonest." Spires, *Beyond Horizons*, p. 75. Also, see Lloyd S. Swenson Jr., James M. Grimwood, and Charles C. Alexander, *This New Ocean: A History of Project Mercury*, NASA SP-4201 (Washington, D.C.: NASA, 1966), pp. 33–97.

11. Specifically, the Air Force agreed to construct infrastructure facilities at Patrick Air Force Base, Florida, for NASA's space probes and provide Thor boosters for the Pioneer lunar probes and Tiros cloud cover satellite. The service also supported development of the hydrogen-and-oxygen-fueled Centaur high-energy upper stage, which it hoped to use in its own communications satellite program. Most important, the Air Force strongly supported Project Mercury by furnishing Atlas boosters and launching services, along with considerable technical, biomedical, and personnel assistance. Spires, *Beyond Horizons*, pp. 72–73.

12. Thomas S. Gates, Deputy Secretary of Defense, to the Secretaries of the Military Departments, et al., "Assignment of Responsibility for DoD Support of Project MERCURY," 10 August 1959, RG 218, Records of the U.S. Joint Chiefs of Staff, Chairman's File, General Twining, 1957–1960, Box 35, "471.94 (1959)," National Archives and Records Administration, Washington, D.C.; Neil McElroy, Secretary of Defense, to the Chairman, Joint Chiefs of Staff, "Coordination of Satellite and Space Vehicle Operations," 18 September 1959, Space and Missile Systems Center (SMC) Archive, SMC History Office, Los Angeles Air Force Base, California (hereafter cited as SMC Archive). In the spring of 1960, Admiral Burke once again unsuccessfully-proposed a joint command for military space operations. Spires, *Beyond Horizons*, pp. 76–77, 84.

13. "Report to the President-Elect of the Ad Hoc Committee on Space," 10 January 1961, SMC Archive; Robert S. McNamara, Secretary of Defense, to the Secretaries of the Military Departments, et al., "Development of Space Systems," 6 March 1961, with attached: Department of Defense Directive 5160.32, "Development of Space Systems," 6 March 1961, reprinted in *Defense Space Interests*, Hearings Before the Committee on Science and Astronautics, U.S. House of Representatives, 87th Cong., 1st sess. (Washington, D.C.: GPO, 1961), pp. 2–3. Expecting a major role in future military space ventures, one of the four new divisions of AFSC was Space Systems Division. Spires, *Beyond Horizons*, pp. 86–87, 89–90.

14. Headquarters Air Research and Development Command, "Report of the Air Force Space Study Committee," 20 March 1961, Personal Papers of General Bernard A. Schriever, United States Air Force, Air Force History Support Office (AFHSO) Archive, AFHSO, Naval Station Anacostia, Washington, D.C. With the Gardner report in hand, Secretary of the Air Force Eugene M. Zuckert directed General Schriever's Air Force Systems Command to prepare a detailed ten-year space proposal. The report from a committee chaired by Maj. Gen. Joseph R. Holzapple, AFSC's Assistant Deputy Commander for Aerospace Systems, described in detail necessary programs with appropriate costs and schedules. The Gardner report and Holzapple study provided the agenda for ambitious Air Force space planning over the next two years. Spires, *Beyond Horizons*, pp. 91–92.

15. By the spring of 1962 NASA had grown in one year from 57,500 to 115, 500 personnel, and a year later the figure was 218,000. By fiscal year 1963, the agency's budget totaled 66.7 percent of the total space budget, while DoD's portion declined from 45 percent in fiscal year 1961 to 28.5 percent in fiscal year 1963. Jane Van Nimmen and Leonard C. Bruno, with Robert L. Rosholt, *NASA Historical Data Book, 1958–1968*, vol. 1, *NASA Resources*, NASA SP-4012 (Washington, D.C.: NASA, 1976), 1:106. Spires, *Beyond Horizons*, pp. 119–20, Appendix 3-2.

16. Roswell L. Gilpatric, Deputy Secretary of Defense, to James E. Webb, Administrator, NASA, 23 February 1961, SMC Archive.

17. Nevertheless, the February 1961 agreement set the precedent for the cooperative agreements on Gemini and future space station projects. In the fall of 1961 agreements with NASA would authorize development of large liquid-propellant rockets (Nova engine) in tandem with the Air Force's work on large solid-propellant rockets until it became clear which would better support the lunar mission. Spires, *Beyond Horizons*, pp. 106–7; Mark C. Cleary, *The 6555th: Missile and Space Launches Through 1970* (Patrick Air Force Base, Fla.: 45th Space Wing History Office, 1991), pp. 193–94; Day, "Invitation to Struggle," pp. 258–59.

18. Report, "Air Force/NASA Space Program Management Panel 3 Report Ferguson Task Force) USAF Space Program FY 63–64," n.d. [December 1961–January 1962], SMC Archive.

19. By terms of a 24 August 1961 launch site agreement, NASA would acquire the large land parcel at Cape Canaveral needed for lunar operations and develop it according to existing range arrangements with the Air Force. The agreement also made NASA responsible for all costs associated with the lunar project and designated the Air Force range manager for the Apollo program. Despite the agreement, however, once NASA acquired 111,000 acres on Merritt Island north of Cape Canaveral, the Air Force and NASA became embroiled in a two year argument after the Air Force decided to locate the proposed Titan III launch site (LC-41) within NASA's area of operations at Cape Canaveral, to purchase an additional 11,000-acre buffer region to the north and to establish overflight procedures. Spires, *Beyond Horizons*, p. 111.

20. Robert S. McNamara, Secretary of Defense, Department of Defense Directive 5030.18, "Department of Defense Support of National Aeronautics and Space Administration (NASA)," 24 February 1962, SMC Archive.

21. Arguing that Dyna-Soar, in contrast to NASA's human space-flight program, offered the promise of routine space flight, launch readiness, and rendezvous in space, the revised Dyna Soar development plan stressed cost reduction by eliminating preliminary suborbital testing and using the Titan III to boost the glider to orbital velocity. Successful testing of rendezvous, docking, and transfer with Dyna-Soar, Air Force officials argued, would provide the necessary technical base for developing a space station that could inspect and neutralize hostile satellites. Spires, *Beyond Horizons*, p. 111; Day, "Invitation to Struggle," pp. 258–63.

22. James E. Webb, Administrator, NASA, and Robert S. McNamara, Secretary of Defense, "Agreement Between the National Aeronautics and Space Administration and the Department of Defense Concerning the Gemini Program," 21 January 1963, NASA Historical Reference Collection, NASA History Office, NASA Headquarters, Washington, D.C. (hereafter cited as NASA Historical Reference Collection); James E. Webb, Administrator, NASA, and Robert S. McNamara, Secretary of Defense, "Agreement Between the Department of Defense and the National Aeronautics and Space Administration Covering a Possible New Manned Earth Orbital Research and Development Project," 17 August 1963, with attached: "Procedure for Coordination of Advanced Exploratory Studies by the DOD and the NASA in the Area of Manned Earth Orbital Flight Under the Aegis of the Aeronautics and Astronautics Coordinating Board," Administrators Files, NASA Historical Reference Collection. The 21 January 1963 NASA-DoD agreement settled the conflict, largely in NASA's favor. NASA would continue to manage Gemini, while the joint Gemini Program Planning Board would determine the experimental program and DoD support. Most important, the agreement sanctioned new NASA and DoD human space-flight programs only by mutual agreement. The September agreement on space stations confirmed NASA's position of coordination rather than joint control of programs important to both agencies.

23. Both NASA and DoD officials argued that the MOL did not violate the September 1963 agreement on space stations, which was designed to avoid duplication, because the

MOL did not represent a future spacecraft "larger and more sophisticated than Gemini and Apollo." As such, it was not a national project needing presidential approval. Over the next twenty months, the Air Force conducted intense studies of organizational alternatives, system capabilities, and potential mission functions. The MOL's satellite inspection and reconnaissance missions served to convince DoD and President Johnson, on 25 August 1965, to approve full-scale development of the system. Following presidential approval, NASA expressed interest in participating in MOL experiments and provided a broad array of support for the military space station. Spires, *Beyond Horizons*, pp. 127–30; also see Robert C. Seamans Jr., Associate Administrator, NASA, to Honorable Harold Brown, Director of Defense Research and Engineering, DoD, 27 September 1965, NASA Historical Reference Collection.

24. Under auspices of the joint Manned Space Flight Policy Committee, which superseded the Gemini Policy Planning Board, NASA supported the MOL with three Gemini spacecraft, test capsules, a simulator, ground equipment, and subsystem hardware, as well as training aids, Apollo ships and tracking stations, and NASA engineers and technicians. The key feature of the Apollo Applications Program became the mini–space station, Skylab; Spires, *Beyond Horizons*, pp. 127–30.

25. David Packard, Deputy Secretary of Defense, Department of Defense, Office of Public Information, "Air Force to Develop Manned Orbiting Laboratory," News Release No. 491–69, 10 June 1969, SMC Archive; Spires, *Beyond Horizons*, pp. 132–33.

26. Robert Frank Futrell, *Ideas, Concepts, Doctrine: Basic Thinking in the United States Air Force*, vol. 2, *1961–1984* (Maxwell Air Force Base, Ala.: Air University Press, 1989), p. 683.

27. Department of Defense and National Aeronautics and Space Administration, "Joint DoD/NASA Study of Space Transportation Systems. Summary Report," 16 June 1969, Air Force Historical Research Agency Archives; Spires, *Beyond Horizons*, p. 180; Day, "Invitation to Struggle," pp. 263–64.

28. T.A. Heppenheimer, *The Space Shuttle Decision: NASA's Search for a Reusable Space Vehicle*, NASA SP-4221 (Washington, D.C.: NASA, 1999), p. 205; Spires, *Beyond Horizons*, pp. 180–81; Day, "Invitation to Struggle," p. 264.

29. Thomas O. Paine, NASA Administrator, and Robert C. Seamans Jr., Secretary of the Air Force, "Agreement Between the National Aeronautics and Space Administration and the Department of the Air Force Concerning the Space Transportation System," NMI 1052.130, Attachment A, 17 February 1970, NASA Historical Reference Collection.

30. Cross-range referred to the Shuttle's capability to maneuver to either side of its reentry track. Given the earth's rotation, a Shuttle launched into polar orbit from Vandenberg, unlike one lofted from Cape Canaveral, would need to make up 1100 nautical miles of crossrange in order to return to the California site following a single-orbit ninety-minute mission. Heppenheimer, *Space Shuttle Decision*, pp. 214–15; Day, "Invitation to Struggle," p. 264.

31. The Air Force wanted a Shuttle capable of launching a 65,000-pound payload into a low-inclination Earth orbit (28.5 degrees), and a 40,000-pound spacecraft into low-Earth (100 nautical miles) near-polar orbit (98 degrees). Spires, *Beyond Horizons*, pp. 181–82; Heppenheimer, *Space Shuttle Decision*, pp. 231–32; Day, "Invitation to Struggle," p. 264.

32. Air force planners also designed a future control facility, the Consolidated Space Operations Center (CSOC), which would combine a Shuttle Operations and Planning Complex (SOPC) and a Satellite Operations Center. Until the CSOC could be constructed at Falcon Air Force Base near Colorado Springs, Colorado, military missions would be controlled from the existing Sunnyvale, California, site and a special "classified" mode at NASA's Johnson Space Center. The SOPC was canceled following the *Challenger* tragedy. Spires, *Beyond Horizons*, p. 183, 197.

33. Ibid., p. 223; In addition to supporting the astronaut training program, the Air Force

supported a joint Air Force–Navy effort to fly payload specialists aboard the Shuttle. This Manned Space-flight Engineer (MSE) program ultimately included thirty-two officer scientists and engineers who underwent standard Shuttle astronaut training in preparation for conducting military scientific and medical experiments, as well as normal crew activities and satellite-Shuttle compatibility operations. The MSE program became one of the casualties of post-*Challenger* planning. Curtis Peebles, *High Frontier: The United States Air Force and the Military Space Program* (Washington, D.C.: Air Force History and Museums Program, 1997), p. 27.

34. Day, "Invitation to Struggle," p. 265; Spires, *Beyond Horizons,* pp. 223–24.

35. John F. Yardley, NASA Associate Administrator for Space Flight; John J. Martin, Assistant Secretary of the Air Force (Research and Development); James C. Fletcher, NASA Administrator; William P. Clements Jr., Deputy Secretary of Defense, "NASA/DoD Memorandum of Understanding on Management and Operation of the Space Transportation System," 14 January 1977, NASA Historical Reference Collection.

36. Given the Shuttle's special capabilities to launch, retrieve and service large on-orbit satellites, satellites designed for Shuttle use are seldom compatible with ELVs, too. A "Shuttle-only" policy led officials in the 1970s to reduce military satellite procurement until the Shuttle became operational, which would result in higher costs in following decade. Day, "Invitation to Struggle," pp. 265–66; Spires, *Beyond Horizons,* pp. 223–24.

37. Spires, *Beyond Horizons,* pp. 184, 202, 207, 215–19. Three years later, in September 1985, Air Force Space Command was joined at Peterson AFB, Colorado, by the unified operational command, United States Space Command.

38. Maj. Gen. James A. Abrahamson, for example, became NASA's Associate Administrator for Space Transportation Systems. Ibid., pp. 223–24.

39. Ibid., pp. 224–25.

40. Ibid., p. 225.

41. Secretary Aldridge's proposal for commercial production of ELVs also threatened NASA's Shuttle marketing operation. By 1984 the Reagan administration became sufficiently concerned about the projected shortfall in NASA's commercial operations to pass the 1984 Commercial Space Launch Act, which centralized all commercial launches under the Secretary of Transportation. The act effectively maneuvered NASA out of the private launch business and heightened interest in preserving and expanding ELV production for commercial purposes. Ibid., pp. 225–26.

42. Caspar Weinberger, Secretary of Defense, Memorandum for Secretaries of the Military Departments; Chairman of the Joint Chiefs of Staff; Under Secretaries of Defense; Assistant Secretaries of Defense; General Counsel, "Defense Space Launch Strategy," 7 February 1984, with attached: "Defense Space Launch Strategy," 23 January 1984, NASA Historical Reference Collection; Howard E. McCurdy, *The Space Station Decision: Incremental Politics and Technological Choice* (Baltimore: Johns Hopkins University Press, 1990), pp. 133–34; Caspar Weinberger, Secretary of Defense, to Honorable James M. Beggs, Administrator, NASA, 16 January 1984, NASA Historical Reference Collection. Beginning in 1985, key Air Force civilian leaders such as Undersecretary Aldridge expressed their interest in military manned space flight through a low-level program termed "Military Man-in-Space." But the Air Force continued to be hard pressed to justify specific roles for military officers in space. Nevertheless, Aldridge left open the door for a military human presence on the Shuttle—and potentially the space station. E.C. Aldridge Jr., Under Secretary of the Air Force, to General Larry D. Welch, Vice Chief of Staff, United States Air Force, "Air Force Policy on Military Man-in-Space—Information Memorandum," 1 August 1985, NASA Historical Reference Collection; Spires, *Beyond Horizons,* pp. 226–27.

43. Day, "Invitation to Struggle," p. 267; National Security Decision Directive 164, "National Security Launch Strategy," 25 February 1985, NASA Historical Reference Collection.

44. In the 1970s NASA had assumed—mistakenly—that the Air Force would be willing to fund a fifth Orbiter. Day, "Invitation to Struggle," p. 265; Spires, *Beyond Horizons*, p. 227.

45. Spires, *Beyond Horizons*, p. 228.

46. Ibid., pp. 228–29.

47. For an assessment of Air Force culture, see Day, "Invitation to Struggle," p. 269.

48. At least one authority, while acknowledging the decline in U.S. market share for commercial launch, has suggested that this did not signify a decline in the U.S. space industry's overall dominance of space-related markets. See Dr. Larry Schweikart, *The Quest for the Orbital Jet: The National Aero-Space Plane Program (1988–1995)*, vol. 3, in Richard P. Hallion, ed., *The Hypersonic Revolution: Case Studies in the History of Hypersonic Technology* (Bolling AFB, Washington, D.C.: Air Force History and Museums Program, 1998), pp. 303–7.

49. To address Air Force concerns about the role of space in the post–cold war era, Chief of Staff Gen. Merrill A. McPeak directed formation of a Blue Ribbon Panel on Space. The panel, chaired by Lt. Gen. Thomas S. Moorman Jr., vice commander of Air Force Space Command, included as part of its study issues external to the Air Force and reported its findings in February 1993. Among its findings, the panel determined that America's existing space launch capability was too expensive, relatively unresponsive to customers' schedules, and not competitive in the international market. History Office, Headquarters Air Force Space Command (HQ AFSPC/HO), *History of Air Force Space Command, January 1992–December 1993*, pp. 281–93.

50. At a Lockheed Martin Technology Symposium in Washington, D.C., 7 November 1995, Raymond S. Colladay, vice president for business development and advanced programs at Lockheed Martin Astronautics, observed, "The environment for today's launch business is evolving rapidly, spurred on in part by the global demand for commercial communication satellites. For now this increase has encouraged both national and international competition. . . . Now the Government can partner with industry, as the Air Force is doing with the EELV program, for example, to reap the benefits of corporate investment designed to remain competitive in the global commercial market." See Lockheed Martin press release, "Technology for the 21st Century: Lockheed Martin Launch Vehicle Strategy," National Press Club, Washington, D.C., 7 November 1995.

51. The origin, resurgent usage, and significance of the term "aerospace" await full, scholarly explication. It seems the word developed, primarily within U.S. Air Force senior leadership circles, in the wake of the first successful satellite launch on by the Union of Soviet Socialist Republics on 4 October 1957. Seeking to ward off Army, Navy and, later, NASA encroachment on what the Air Force perceived as its territory, the latter popularized the concept of an operational "air" and "space" continuum. The term "aerospace" first appeared in print in an Air Force news release on 8 July 1958. There was pressure, at the time, to establish a unified space command into which the individual military services would pool their resources. See Frank W. Jennings, "Genesis of the Aerospace Concept," *Air Power History*, 48, no. 7 (Spring 2001), pp. 46–55. In December 1981, congressman Ken Kramer from Colorado Springs, Colorado, and several others who wanted to see the Air Force place greater emphasis on space operations proposed changing the name of the service from U.S. Air Force to U.S. Aerospace Force. Secretary of the Air Force Verne Orr and senior service leaders, all of whom were pilots, opposed the name change. Air force senior leaders promoted a resurgence of the "aerospace" term during the late 1990s, however, in response to interservice sniping, intraservice challenges, congressional criticism, and public perceptions that the Air Force prioritized air ahead of space. Increasingly shrill calls for creation of a separate "space force" drove Air Force leaders to seriously contemplate changing the name of their service, as Kramer had suggested in 1981. See Futrell, *Ideas, Concepts, Doctrine*, vol. 1, pp. 10–11, 553–54, and vol. 2, pp. 166–67, 212–13; Rick W. Sturdevant, HQ AFSPC/HO, "Background Paper on Origin and Use of Term 'Aerospace,'" 16 September 1998.

52. Milton O. Thompson and Curtis Peebles, *Flying Without Wings: NASA Lifting Bodies and the Birth of the Space Shuttle* (Washington, D.C.: Smithsonian Institution Press, 1999), pp. 12–13.

53. Schweikart, *Quest for the Orbital Jet*, pp. 280–83, 312–17.

54. HQ AFSPC/HO, *History of the Air Force Space Command, January–December 1991*, pp. 146–51. Jointly funded, jointly managed USAF-NASA launch vehicle development efforts, though well intentioned, proved disappointing in terms of results. Differing goals and expectations among different constituencies, along with poor coordination, certainly contributed to lack of technological progress. One researcher recently concluded, however, that "Most of the management problems could be attributed to the funding uncertainties caused by the different congressional paths for obtaining budget authorizations and appropriations; there were a number of different committees and subcommittees for NASA and DOD, each subject to different priorities and pressures." See Ivan Bekey, "Exploring Future Space Transportation Possibilities," in John M. Logsdon, gen. ed., *Exploring the Unknown: Selected Documents in the History of the U.S. Civil Space Program*, vol. 4, *Accessing Space*, NASA SP-4407 (Washington, D.C.: NASA, 1999), 4:506.

55. Office of Assistant Secretary of Defense (Public Affairs) news release, "Space Launch Modernization Study Announced," 27 January 1994; HQ AFSPC/HO, *History of Air Force Space Command, January 1992–December 1993*, pp. 293–306.

56. Executive Summary, "Space Launch Modernization Plan," May 1994.

57. Explaining the need for EELV, one AFSPC officer observed, "Payload customers have always 'pushed the envelope' on the systems to maximize, not optimize, lift capability. Over the past seven years the commercial competitiveness of our US fleet of launch vehicles has declined. Our domestic launch industry has had trouble competing in the world market with Ariane and this trend may continue with the emergence of the Russian and Chinese launch vehicles." See Major Bachmann, HQ AFSPC/DOOL, "Why Evolved Expendable Launch Vehicle (EELV)," 2 January 1996.

58. The White House, Office of Science and Technology Policy (OSTP) fact sheet, "National Space Transportation Policy," 5 August 1994. By 2002, AFSPC planners agreed that the level of cooperation between the Air Force and NASA with respect to RLV programs was greater than ever before. Lt. Col. Michael Thome and Maj. Clive Paige, HQ AFSPC/XPXZ, interview with Rick W. Sturdevant, HQ AFSPC/HO, 4 March 2002.

59. Richard M. McCormick, SAF/SX, to SAF/AQ et al., "Air Force–NASA MOA Regarding Space Launch Activities," 3 March 1995; Lieutenant Colonel Randy Joslin, HQ AFSPC/XPXM, to Brigadier General Roger G. DeKok, HQ AFSPC/XP, "AF/NASA Space Launch Activities," 8 March 1995; Brigadier General Roger G. DeKok, HQ AFSPC/XP, to Mr. Richard M. McCormick, SAF/SX, 9 March 1995; HQ AFSPC/DRSV talking paper, "Synchronization of NASA RLV with EELV Requirements," ca. 29 April 1996.

60. HQ AFSPC/DRS staff summary sheet, "Draft Capstone Requirements Document for Spacelift," 12 January 1995.

61. Colonel Michael L. Heil, Phillips Laboratory(PL)/CC, to Brigadier General Lance Lord, HQ AFSPC/XP, "Release of FY96 Appropriated Funds for RLV Technology," 21 February 1996; HQ AFSPC/DR staff summary sheet, "Synchronization of NASA RLV and EELV Requirements," 29 April 1996.

62. Major Ed Bolton, HQ AFSPC/DRS, "Integrated Product Team (IPT) for Military Spaceplane," 6 August 1996; HQ AFSPC/DR staff summary sheet, "Aerospace," 29 August 1996, w/3 Tabs: List of Participants, Study Statement of Work, and Study Briefing.

63. General Howell M. Estes III, AFSPC Commander, and Daniel S. Goldin, NASA Administrator, Memorandum of Agreement (MOA), "Air Force Space Command–National Aeronautics and Space Administration Partnership Council," 28 February 1997.

64. General Richard B. Myers, AFSPC Commander; Daniel S. Goldin, NASA Adminis-

trator; and Keith R. Hall, NRO Director, Memorandum of Agreement (MOA), "Air Force Space Command, National Reconnaissance Office, National Aeronautics and Space Administration Partnership Council," 23 November 1998.

65. Status Report, "AFSPC/NASA/NRO Advanced Space Transportation Team," 19 March 1999; Rick W. Sturdevant, HQ AFSPC/HO, interview with Major Michael Dickey, HQ AFSPC/XPX, 15 October 1999; Thorne and Paige, interview. As early as 1987, an NRC publication observed, "A number of committees and liaison offices, such as the . . . Aeronautics and Astronautics Coordinating Board (AACB) . . . are designed to coordinate NASA and DOD activities in space technology research and development. Nevertheless, research coordination could be improved with benefits for both organizations. The committee agreed that the missions of both NASA and the DOD would be enhanced were cooperation between the two strengthened at top levels of management, at mid-levels, and at the working level." See Committee on Advanced Space Technology, Aeronautics and Space Engineering Board, Commission on Engineering and Technical Systems, National Research Council, *Space Technology to Meet Future Needs* (Washington, D.C.: National Academy Press, 1987), pp. 47–48. By late 1999, AFSPC and NASA officials from the highest levels downward agreed that the AACB was broken and, far from facilitating cooperation, actually hampered efforts to strengthen the AFSPC/NASA/NRO partnership.

66. Gen. George T. Babbitt Jr., Commander, Air Force Materiel Command (AFMC), quoted in Peter Grier, "Partners in Space," *Air Force Magazine*, February 1999, pp. 28–32.

Notes on Contributors

Matt Bille is an Associate with Booz Allen Hamilton in Colorado Springs. He holds a master's degree in space systems management. Matt has worked on small satellites and ICBM life extension programs for Air Force Space Command. He has published over twenty articles and papers on space policy, plus a book on the world's rarest animals.

Mark D. Bowles is a historian of technology working with History Enterprises, Inc. on a history of the Centaur upper stage.

Andrew J. Butrica, a graduate of the doctoral program in the history of science and technology at Iowa State University, is a research historian and author of numerous articles and papers on the history of electricity and electrical engineering in the United States and France and the history of science and technology in nineteenth-century France. He is the author of a corporate history, *Out of Thin Air: A History of Air Products and Chemicals, Inc., 1940–1990,* published by Praeger in 1990; *To Seen the Unseen: A History of Planetary Radar Astronomy,* published as SP-4218 in the NASA History Series in 1996; and editor of *Beyond the Ionosphere: Fifty Years of Satellite Communication* (NASA SP-4217, 1997).

Virginia P. Dawson is President of History Enterprises, Inc., in Cleveland, Ohio, working on a history of the Centaur rocket program. She is the author of *Engines and Innovation: Lewis Laboratory and American Propulsion Technology* (NASA SP-4306, 1991).

Kevin S. Forsyth is a wide- and local-area network consultant residing in Chicago, Illinois. His independent research led to the creation of *History of the Delta Launch Vehicle*, a web site that includes a log of all flights, current news updates, upcoming launch windows, and vehicle performance specifications; available on-line at http://kevin.forsyth.net/delta/. He holds a bachelor of arts degree in telecommunications from Michigan State University.

J.D. Hunley retired as historian at NASA's Dryden Flight Research Center in August 2001 and became the Ramsey Fellow at the Smithsonian National Air and Space Museum until September 2002. During that time, he continued research and began writing a history of the development of launch vehicle technology in the United States. He is the editor of numerous books, including *Toward Mach 2: The Douglas D-558 Program* (NASA SP-4222, 1999); *History of Rocketry and Astronautics*, vols. 19 and 20 (Univelt, Inc., AAS History Series, 1997–1998); *The Problem of Space Travel: The Rocket Motor* (NASA SP-4026, 1995); and *The Birth of NASA: The Diary of T. Keith Glennan* (NASA SP-4105, 1993). He has also published extensively on the development of rocketry technology in several scholarly journals, and has published *The Life and Thought of Friedrich Engels: A Reinterpretation* (Yale University Press, 1991).

Dennis R. Jenkins is a consulting engineer in Cape Canaveral, Florida, working on various aerospace projects including twenty years on the Space Shuttle Program. He is the author of *Space Shuttle: The History of the National Space Transportation System—The First 100 Flights* (Specialty Press, 2001) in addition to over thirty other works on aerospace history.

Pat Johnson is a Senior Systems Engineer for Analytic Services, Inc. (ANSER). She holds a Ph.D. in chemical engineering. Pat currently supports the U.S. Navy SM-3 development in the areas of plume effects and system testing. Previously, she developed ICBM engagement simulations for Air Force Space Command. She is a coauthor of ANSER's *Rocket Systems Launch Program (RSLP) Handbook* and forthcoming *ICBM Handbook*.

Robyn A. Kane of SenCom Corp is the Activity Based Costing Project Manager for the Electronic Systems Center (ESC) personnel resident at Peterson Air Force Base, Colorado. Her degrees include a master's in mathematics/statistics and a bachelor of science in mathematics/Spanish. She led the cost analysis for a variety of Analysis of Alternatives (AoAs) and estimated a twenty-year Life Cycle Cost for a Tactical Launch System using Minuteman II stages coupled with commercial boosters. She is a member of the Military Operations Research Society (MORS) and the Society of Cost Estimating and Analysis (SCEA).

Roger D. Launius is curator in the Division of Space History at the Smithsonian Institution, Washington, D.C. He has produced several books and articles on aerospace history, including *Reconsidering Sputnik: Forty Years Since the Soviet Satellite* (Harwood Academic, 2000); *Innovation and the Development of Flight* (Texas A&M University Press, 1999); *NASA & the Exploration of Space* (Stewart, Tabori, & Chang, 1998); *Frontiers of Space Exploration* (Greenwood Press, 1998); *Organizing for the Use of Space: Historical Perspectives on a Persistent Issue* (Univelt, Inc., AAS History Series, vol. 18, 1995); *NASA: A History of the U.S. Civil Space Program* (Krieger Publishing, 1994); and *History of Rocketry and Astronautics: Proceedings of the Fifteenth and Sixteenth History Symposia of the International Academy of Astronautics* (Univelt, Inc., AAS History Series, vol. 11, 1994).

Erika R. Lishock is a satellite engineer and writer in Colorado Springs. She holds a bachelor of science degree in mechanical engineering and is a member of the American Society of Mechanical Engineers. Her support to Air Force Space Command includes analysis and future planning for the Milstar system. She was a console analyst on the Milstar DFS-1 and DFS-2 launch team.

David N. Spires is a faculty member of the University of Colorado, Boulder. He is the author of *Beyond Horizons: A Half Century of Air Force Space Leadership* (Peterson Air Force Base, Colo.: Air Force Space Command, 1997).

Rick W. Sturdevant is Deputy Command Historian with the Air Force Space Command, Colorado Springs, Colorado. He is the author of numerous articles.

Ray A. Williamson is a Senior Research Scientist in the Space Policy Institute, focusing on the history, programs, and policy of Earth observations, space transportation, and space commercialization. He joined the institute in 1995. Previously he was a Senior Associate and Project Director in the Office of Technology Assessment of the U.S. Congress. He joined OTA in 1979. While at OTA, he was Project Director for more than a dozen reports on space policy, including *Russian Cooperation in Space* (1995), *Civilian Satellite Remote Sensing: A Strategic Approach* (1994), *Remotely Sensed Data: Technology, Management, and Markets* (1994), *Global Change Research and NASA's Earth Observing System* (1994), and *The Future of Remote Sensing from Space: Civilian Satellite Systems and Applications* (1993). Dr. Williamson has written extensively about the U.S. space program. He holds a bachelor's degree in physics from the Johns Hopkins University and a Ph.D. in astronomy from the University of Maryland. He spent two years on the faculty of the University of Hawaii studying diffuse emission nebulae and ten years on the faculty of St. John's College, Annapolis. He is a member of the faculty of the International Space University and is a member of the editorial board of *Space Policy*.

Index

A-2 rocket, 42–48, 235

A-4 (V-2) rocket, 42–48, 72, 73, 187–89, 235, 302, 445

Able, 54

Access to Space study, 456, 457

Adelman, Barnet R., 249, 250, 268, 272

Advanced Ballistic Reentry System (ABRES), 81, 83

Advanced Composition Explorer (ACE), 136

Advanced Research Projects Agency (ARPA), 60, 112, 266, 303, 340, 342, 452, 453, 473, 475, 476, 488, 490

Advanced Technology Demonstrator Program, 458

Advanced Technology Vehicle (ATV), 460

Advent Launch Services, 31

Advent satellite, 346

Advisory Committee on the Future of the U.S. Space Program, 488

AeroAstro, 220

Aerobee, 46, 49, 50, 188

Aerobee-Hi sounding rocket, 192

Aerojet Engineering Corporation, 189, 233

Aerojet General Corporation, 54, 57, 61, 97, 130, 151, 233, 241, 243–45, 251, 256, 257, 263, 264, 266–74, 276–78, 281, 315; Downey Division, 257, 258

Aeronautics and Astronautics Coordinating Board (AACB), 473, 477

Aerospaceplane, 359, 369, 370, 444, 445

Aerothermodynamic/elastic Structural Systems Environmental Tests (ASSET), 446

Agena, 26, 54, 59

Air Collection and Enrichment System (ACES), 445

Air Force–NASA partnership, 470–501

Air Force Space Launch Memorandum of Agreement (MOA), 491

Aldridge, Edward C. "Pete," 484, 485, 491

Aldridge Study, 490

Algol program, 209

Allegany Ballistics Laboratory (ABL), 107, 193, 232, 259, 261, 264

Allen, Gen. Lew, 484

Allen, H. Julian, 49

Alliant Aerospace Company, 172

Alliant Techsystems, 132, 214, 281

Altair, 110

Alternate Access to Station, 222

Alternate Space Shuttle Concepts (ASSC), 370–72

Altman, David, 268

American Association for the Advancement of Science, 328

American Bosch, 82, 153

American Chemical Society, 231

American Interplanetary Society (AIS), 38, 39

American Rocket Society (ARS), 38
American Telephone and Telegraph
(AT&T), 116, 117, 396
Ames Aeronautical Laboratory, 448
Analex, 433
Anik, 128, 129
Antares program, 209, 210, 213
antimatter propulsion system, 24
Apollo 7, 11
Apollo 11, 63
Apollo Applications Program, 325, 480
Apollo program, 10–12, 15, 59, 62, 164,
301, 304, 313, 315, 317, 324–26, 328,
329, 417, 418
Apollo Soyuz Test Project (ASTP), 12, 167
Arendale, Dr. W.F., 254
Ariane launch vehicle, 16, 170, 173, 399,
401, 403, 404, 407
Ariane program, 223
Arianespace, 20, 403, 404
Armour Institute, 236
Armstrong, Neil A., 301, 402
Arnold Engineering Development Center
(AEDC), 386
Arnold, Gen. Henry H. "Hap," 40
Arrowsmith Tooling, 261
Asimov, Isaac, 417
Astroliner, 460–61
astronauts, 58
Astroplane, 375
Astrorocket, 359, 360
ATK Thiokol Propulsion Company, 214
Atlantic Missile Range, 343
Atlantic Missile Range Impact Locator
System, 79, 80
Atlantic Research Corporation (ARC), 241,
243, 245, 259, 260, 263
Atlantis, 428
Atlas, recoverable, 92
Atlas I, 92
Atlas II, 30, 92
Atlas IIA, 93
Atlas IIAS, 19, 93
Atlas III, 93
Atlas IIIA, 95
Atlas IIIB, 96
Atlas V, 22, 96
Atlas A, 76, 77
Atlas B, 78, 79
Atlas C, 79, 80
Atlas D, 80, 81, 84, 86, 87, 90

Atlas E, 82, 86
Atlas F, 82–84, 86
Atlas G, 92
Atlas-Centaur program, 60, 138
Atlas ICBM, 7, 49, 53, 54, 56, 60, 75, 77,
105, 106, 147, 191, 198, 247, 334, 336,
338, 340, 350
Atlas launch vehicle, 2, 9, 12, 15, 18, 21, 42,
59, 70–102, 150, 153, 170, 194, 359, 404
Atlas Powder Company, 258
Atlas-SCORE, 78, 79
Atomic Energy Commission (AEC), 60–62,
75
Atwood, Lee, 317
Atwood, Sec. of Def. Donald, 489
Augustine Committee, 488
Augustine Report, 490
Aussat, 396, 404
AVCO, 82; Lycoming Division, 257
Ayers, Col. Langdon, 252

Baby Sergeant, 195, 197
Ballistic Missile Defense Organization
(BMDO), 211, 457, 458, 490
Ballistic Missile Division (BMD), 53, 56,
249, 251–53
Bantam small launch vehicle, 221
Bartley, Charles, 230–32, 234, 243
Battelle Memorial Institute, 274, 384
Battin, Richard H., 247
Beacon satellite, 197
Beal Aerospace, 31
Beggs, James, 485
Bekey, Ivan, 453, 456–57
Bell Aircraft, 88, 151, 339, 446
Bell Laboratories, 104, 110, 112, 153, 199
Bell X-1, 39, 73
Bell X-2, 37
Bendix, 258
BF Goodrich, 255
Big Stoop missile, 241
Bilstein, Roger, 305
Biosatellites, 123
Black, Sivalls and Bryson, 261, 266
Blue Gemini, 448, 479
Blue Goose, 367
Blue Scout, 212
Boeing Company, 21, 97, 137, 138, 140,
141, 252, 253, 257, 274, 282, 313, 328,
358–60, 419, 433, 448, 451, 454, 460
Boeing Phantom Works, 460

BoMi, 446
Bossart, Karel J. "Charlie," 53, 72, 74, 337–39, 345, 354
Boswell, Walt, 232
BPD Difesa e Spazio, 211
Brass Bell, 446, 448
Bredt, Irene, 445
Bristol Aerospace, 220
British Interplanetary Society, 39
British National Committee of Space Research, 115
Budd Company, 274
Bulova, 258
Bumper 8, 46, 47
Bumper Round 5, 187
Burke, Adm. Arleigh, 475
Burrows, William, 434
Bush (George W.) administration, 23
Bush (George) administration, 487
Bush, Pres. George, 62
Bush, Vannevar, 149, 237
Bussard, R.W., 61

Cal Doren Company, 279
Canright, Richard, 339
Case Institute of Technology, 274
Cassini mission, 282
Cassini spacecraft, 173, 178, 179
Centaur, 60, 90, 96, 108, 334, 334–56, 415–42
Centaur Project Development Team, 343
Central Intelligence Agency (CIA), 473, 491
Challenger, 18, 92, 164, 170, 171, 280–82, 365, 376, 381, 394, 396, 398–404, 408, 428, 431, 432, 434, 435, 437, 452, 485
chemical rocket technology, 40–42
Chicago Bridge and Iron, 455
Chrysler Corporation, 71
Chrysler Space Division, 370, 372
Citizen's Advisory Council on National Space Policy, 454
Civilian-Military Liaison Committee (CMLC), 473, 477
Clancey, Tom, 461
Clarke, Arthur C., 111, 357
Clementine, 165
Clinton administration, 487
Clinton, Pres. Bill, 457, 490, 491
Clipper Graham, 458
Cold War, 2–14, 70, 147, 187
Coleman Aerospace Company, 221, 222

Collins, Michael, 418
Colorado Center for Astrodynamics Research, 138
Columbia, 377, 380, 381, 385–88, 396, 398, 405
Commercial Experiment Transport (COMET), 462
Commercial Space Launch Act (CSLA), 16, 18, 19
commercialization, 486–91
Committee on Hypersonic Lifting Vehicles, 360
Common Booster Core (CBC), 139, 140
Common Core Booster (CCB), 96
Complementary Expendable Launch Vehicle (CELV), 170, 484
Comsat, 395
Conestoga, 219, 461
Conrad, Pete, 162, 455
Consolidated-Vultee, 72–75, 77, 99
construction, 41, 44, 49–51, 53, 59, 60, 72–74, 80, 81, 86, 88, 92–97, 105–7, 109, 110, 115, 126–28, 130, 132, 137, 139–41, 151, 152, 154, 157, 167, 168, 171, 173, 192, 193, 195–99, 202–6, 212, 214, 216, 235, 244–46, 248, 250, 255–58, 261, 262, 264, 270, 271, 273–80, 282, 283, 285, 304, 305, 314, 317, 336, 349, 371–74, 376, 377, 380–86, 388, 419, 420, 425–28, 430, 431, 450, 453, 455, 458, 481
Continental/Semiglobal Transport (C/SGT), 449–50
Contraves, 97
Convair, 60, 74, 75, 77, 78, 84, 147, 150, 304, 315, 359, 360, 369
Cooper, Gordon, 87
Corona imagery intelligence program, 107, 120, 123, 199
Corporal, 237, 238
Cosenza, Charles, 367, 370, 371
cost, 84, 93, 96, 110, 167, 170, 214, 216–18, 220–23, 225, 237, 250, 251, 282, 283, 305, 324, 328, 340, 343, 351, 360, 362, 363, 371, 374–76, 399, 403, 408, 409, 417, 423, 425, 429, 443, 444, 450–51, 458, 460, 463, 488, 489
Crew Recovery Vehicle (CRV), 460
Crosby, Joseph, 233
Crowley, John, 272
Crucible Steel Company of America, 274
Cuban Missile Crisis, 213

Cunningham, R. Walter, 11
Curate rocket, 232, 233
Curtiss-Wright, 256

Darman, Richard, 489
Davis, Deane, 336, 342, 344, 345
Day, LeRoy, 384
Deacon rocket, 259
Deep Space 1, 136
Defense Meteorological Satellite Program
    (DMSP), 165, 201
Defense Mobilization Science Advisory
    Committee, 239
Defense Advanced Research Projects
    Agency (DARPA), 214, 215, 220, 222,
    283
Defense Satellite Communication System
    (DSCS), 482
Defense Science Board, 484
Defense Support Program (DSP), 180
DeKok, Brig. Gen. Roger G., 491
Delco Electronics, 126
Delta Clipper, 453–55, 457
Delta II, 25
Delta IV, 21, 22
Delta Inertial Guidance Systems, 126
Delta launch vehicle, 2, 7, 10, 12, 15, 18, 21,
    42, 51, 54, 70, 103–46, 170
Delta Mariner, 140
Delta Redundant Inertial Measurement
    System, 130
Dempsey, James R., 343
Design Certification Review, 431
Deutch, John M., 489, 490
development, 33–44, 50, 54–57, 61, 71, 72,
    77, 80, 84, 86, 112, 117, 118, 120, 121,
    123, 128, 129, 132, 138–41, 151, 159,
    164, 166, 170, 187–91, 195, 197, 199,
    206, 214–17, 219–23, 225, 230–37, 241–
    52, 254–56, 259–67, 269–76, 278, 279,
    281–83, 286, 302–5, 308, 315, 316, 336–
    38, 340–48, 358–64, 366–68, 370–76,
    385, 386, 425–29, 445–51, 453–55, 457–
    58, 460–62
Diamant missile, 251
DIGS, 130
Directorate of Space, 489
Discoverer I, 199
Discovery, 397, 428
Donlan, Charles, 374
Dornberger, Walter R., 42, 44, 339, 446

Douglas Aircraft Company, 104, 105, 108,
    117, 141, 198, 274, 304, 315, 328, 360
Downhower, Dr. Walter, 196
Draper, Alfred, 367, 368, 370, 371
DRIMS, 132
Dryden, Dr. Hugh L., 471
Dunn, Louis G., 235, 238
Du Pont, Tony, 453
Dyna-Soar, 167, 303, 359, 370, 384, 445–48,
    471, 473, 479

E.I. duPont de Nemours & Company, 258
Early Bird, 120, 122, 123, 128
Earth Resources Technology Satellite
    (ERTS), 127
Earth Satellite Vehicle Program, 187
Eaton Canyon Project, 232
Echo, 111, 112, 119
EER Systems, 219, 461, 462
Egan, John, 176
Egan International, 176
Ehricke, Dr. Krafft A., 337–43, 346, 350, 354
Eisele, Donn F., 11
Eisenhower, Pres. Dwight, 50, 51, 78, 79,
    191, 239, 303, 473, 475, 477
Eisenhower administration, 7, 61, 149, 302,
    473, 476
Elias, Antonio, 215
Emergency Rocket Communication System
    (ERCS), 213
Emme, Eugene M., 25, 26, 32
Endeavour, 406, 407
Energetic Particle Explorers (EPE), 114
Energia, 21
engines and propulsion, 41, 48, 49, 57, 58,
    62, 72, 73, 78, 79, 81, 86, 90, 92, 93, 95–
    97, 99, 105, 106, 109, 120, 123, 126, 129,
    130, 132, 137–39, 152, 153, 166, 171,
    172, 190, 191, 198, 199, 201–3, 205, 216,
    219, 238, 251, 255, 256, 259, 269–74,
    276, 277, 280–82, 303, 304, 313, 340,
    342, 347, 348, 372, 373, 377, 379–82
Enterprise, 377
Environmental Science Services
    Administration, 121
Eros, 133
Estes, Gen. Howell M., III, 492
European Space Agency (ESA), 20, 178,
    209, 223, 403, 424, 429
European Space Research Organization
    (ESRO), 205, 209

Eutelsat, 96
Evans, Frances, 343
Evolved Expendable Launch Vehicle
    (EELV), 21, 139, 173, 181, 282, 490, 491
Exelco, 274
expendable launch vehicle (ELV), 14, 16,
    18–20, 22–25, 131, 168, 171, 357, 358,
    416, 417, 434, 436, 481, 482, 490
Explorer, 114, 123
Explorer I, 51, 52, 190, 196
Explorer 6, 107
Explorer VII, 52
Explorer VIII, 52
Explorer XI, 52
Explorer program, 223
Explorer satellites, 208
Extended Long Tank Thor, 127
Extra Extended Long Tank Thor, 132
extravehicular activity (EVA), 11, 164, 398
Extreme Ultraviolet Explorer (EUVE), 133

Faga, Martin, 489
Faget, Maxime A., 366–69
Falcon missile, 234, 235
Federal Communications Commission
    (FCC), 128, 129
Ferguson, H.L., 230
Fiedler, Willy, 246
Fischer, Rodney, 245
Fleeter, Rick, 220
Fletcher, James C., 171, 363, 375, 376, 418,
    433, 436
flight and orbit (Space Shuttle), 389–95
Flight Operations Control Center, 455
FLTSATCOM, 91
Ford Motor Company, Aeronautical
    Systems, 268
Frankford Arsenal, 274
Freedom 7, 10, 59
Frosch, Robert, 420
fuels and propellants, 34, 35, 41, 48, 55, 57,
    61, 62, 73, 86, 93, 96, 105, 106, 109, 126,
    137–39, 151, 154, 156, 171, 195, 196,
    201–5, 216, 219, 229–34, 236–42, 245,
    246, 254–64, 266, 269–73, 275–77, 280–
    82, 285, 286, 303, 304, 313–15, 319, 335,
    336, 339–41, 346, 350, 351, 354, 376,
    382, 415, 420, 432, 435, 445, 450
Fuller, John, 244
Fullerton, Charles Gordon, 377
Future Spacelift Requirements Study, 492

Gagarin, Yuri, 59
Gardner, Trevor, 75, 150, 338, 476
Gardner report, 476
Garlock Company, 257
Gates, Sec. of Def. Thomas S., 475
Geckler, Dick, 245
Gemini, 55, 59, 448
GenCorp Aerojet, 221
General Accounting Office (GAO), 278, 304
General Dynamics, 86, 92, 99, 274, 404,
    415, 418–20, 423, 429–31, 433, 436, 454,
    455, 460; Convair Division, 361, 362,
    421, 423; Convair/Astronautics Division,
    336, 339, 340, 342–48, 350, 351
General Electric Company (GE), 46, 88,
    104, 151, 162, 190, 235, 248, 256
General Motors, 274; Allison Division, 256,
    257
General Tire and Rubber, 245, 257
Genesis, 136
Geodetic Earth Orbiting Satellite (GEOS),
    121
German Rocket Society, 338
Gilpatrick, Roswell, 476
Gilruth, Robert, 159
Giotto probe, 401
Glasser, Col. Otto, 251
Glenn, John, 10, 59
Glennan, T. Keith, 304, 315, 336, 475
Glenn L. Martin Company, 46, 50, 54, 147,
    151–54, 159, 189, 192, 449
Global Positioning System (GPS), 131, 482
Globalstar, 395
Goddard Space Flight Center, 115, 117,
    285; Wallops Flight Facility, 205
Goddard, Robert H., 3, 33–37, 39, 60, 63,
    64, 73, 187, 303, 335, 445; estate, 37;
    patent infringement, 64
Gold, Marvin, 245
Goldin, Daniel S., 487, 491–93
Goodyear Tire and Rubber Company, 7,
    277
Government Polymer Laboratories, 255
Graham, Lt. Gen. Daniel O., 454
Graham, William R., 432
Grand Central Rocket Company, 193, 243,
    272
Graphite Epoxy Motors, 157
Grey, Maj. Rex, 253
Grissom, Capt. Virgil I. "Gus," 11, 59, 164,
    190

Grumman, 371
Grumman/Boeing, 372, 376
GTE Spacenet, 403
Guggenheim Aeronautical Laboratory at
    the California Technical Institute
    (GALCIT), 39, 40, 44, 230, 231, 233
Guggenheim Fund, 35
guidance systems, 54, 78–82, 90, 95, 106,
    110, 126, 132, 153, 195, 196, 206, 247,
    248, 252, 262, 265
Guided Missile Development Division, 71
Guided Missiles Study Group, 149

Hagen, John, 192
Haise, Fred W., Jr., 377
Hall, Keith R., 492
Hall, Maj. Edward N., 234, 242, 243, 248,
    250, 251
Hamilton Standard, 126, 130
Hansen, Grant, 343, 350
Hauck, Frederick (Rick), 435, 436
Have Region, 451
Heaton, Col. Donald H., 343
Helios missions, 416
Henderson, Charles B., 241, 242, 260
Henry, George, 244
Hercules Aerospace Company, 132, 214,
    281, 283
Hercules Powder Company, 233–44, 251,
    258, 259, 261, 262, 266
Hermes, 375
Hermes C1 missile, 71
Hermes program, 235
Heuter, Hans, 343, 344, 345
High Altitude Research Corporation
    (HARC), 222
Hines, William, 351
Holmes, Brainerd, 167, 315
Honest John, 259
Honeywell, Inc., 206, 343, 423, 429
Horizontal Integration Facility, 140
HOTROC, 197
Hotz, Robert B., 402
Hubble Space Telescope, 280, 405
Hudson, Gary, 219
Hughes Aircraft Company, 122, 123, 128,
    234, 395, 396, 406
Hunsaker, Jerome, 40
Hunter, W.E., 254
Huygens space probe, 173
Hyper Environmental Test System (HETS),
    212

ICBM. *See* intercontinental ballistic missile
Illinois Institute of Technology, 236
Imp, 127
Incorporated Thematic Mappers, 127
Inertial Upper Stage (IUS), 419–21, 428,
    433
Insat, 404
Integral Launch and Reentry Vehicle
    (ILRV) System, 360, 365, 368, 371
Integrated Advanced Space Transportation
    Plan, 492
Integrated Support Structure, 426
Intelsat, 124, 128, 395, 406, 407
Interagency Chemical Rocket Propulsion
    Group, 266
intercontinental ballistic missile (ICBM), 2,
    14, 39, 61, 147, 148, 150, 151, 244, 338;
    operational, 80–84; origins, 148
intermediate-range ballistic missile
    (IRBM), 7, 48, 49, 104, 198, 239, 244
International Business Machines (IBM),
    328
International Geophysical Year (IGY), 7,
    49, 50, 191, 192
International Launch Services (ILS), 21,
    95, 98
International Nickel Company, 274
International Satellite Communications
    Consortium, 120, 124
International Scientific Radio Union, 49
International Solar Mission, 429
International Space Station, 21, 25, 280,
    407, 408, 463, 464
International Union of Geodesy and
    Geophysics, 49
Interplanetary Monitoring Platforms, 124
Interstellar Magnetic Particle Explorers,
    121
Iridium, 395
Israeli Aircraft Industries (IAI), 221
Italian Space Agency, 209

Jet Propulsion Laboratory (JPL), 44, 46, 49,
    56, 133, 187, 191, 230–38, 249, 268, 339,
    350, 352, 354, 424, 429, 433, 434
jet-assisted takeoff (JATO), 36, 40, 56, 57,
    230, 233, 234, 259
jetavators, 247, 252
Johns Hopkins University, Applied Physics
    Laboratory, 189, 266
Johnson, Pres. Lyndon B., 318
Johnson, Vince, 351

Johnson Space Center (JSC), 330, 389, 421, 424, 429, 432, 433, 436, 460, 484
Joint Army/Navy/NASA/Air Force (JANNAF) Interagency Propulsion Committee, 266
Joint Chiefs of Staff, 54, 490
Joint Task Force Mercury, 475
Journal of the British Interplanetary Society, 39
JP Aerospace, 222
Juno, 49–52, 195–97, 217
Juno I, 51
Juno II, 52
Juno V launch vehicle, 303
Jupiter, 48, 51, 190, 195, 196, 198, 201, 213, 239, 241, 247
Jupiter A, 49
Jupiter C, 49, 51, 52
Jupiter Composite Re-entry Test Vehicle, 49
Jupiter IRBM, 52, 54, 302

Kármán, Theodore von, 40, 233, 268
Karth, Rep. Joseph E., 334, 351
Kelly, Michael S., 460
Kelly Space & Technology, Inc., 31, 460–61
Kennedy, Pres. John F., 58, 61, 301, 302, 308, 318, 324, 476
Kennedy administration, 12, 58, 477
Kennedy Space Center (KSC), 330, 359, 374, 377, 385, 387, 389, 392–94, 399, 421, 424, 429–34
Khrushchev, Nikita, 7, 21
Killian, James, 239
Killian Committee, 239
Kincaid, Dr. John F., 259
Kistler Aerospace Corporation, 31, 461
KIWI test engines, 61, 62
Klager, Dr. Karl, 245, 248, 263
Kloman, Erasmus, 352
Knemeyer, Frank, 198
Kohrs, Richard, 436
Korean War, 74, 189
Korolev, Sergei P., 8, 39
Kraft, Chris C., Jr., 421
Kubrick, Stanley, 357
Kvaerner, 21

Ladish Company, 278
Laird, Melvin R., 168
Landsat, 127
Langley Research Center, 159, 362, 418, 449–50, 455, 460

Laning, J. Holcomb, Jr., 247
Large [Solid] Rocket Feasibility Program (AFLRP), 243, 248
Large Launch Vehicle Planning Group (LLVPG), 166, 477
Large Segmented Rocket Motor Program, 267–78
Launch Operations Center, 318, 322
Launch Service Organization, 484
Lawrence Livermore National Laboratories, 61
Lear Siegler, 274
Leasat, 396, 397, 405
LeMay, Gen. Curtis, 250, 251
Lewis Research Center, 276, 346, 347, 350, 354, 415, 420–23, 425, 429, 430, 433, 436
Lewis Shuttle/Centaur Project Office, 431
Liberty launch vehicle, 219, 220
Lindbergh, Charles A., 35
Liquid Air Collection Engine System (LACES), 445
liquid hydrogen (LH2), 335, 336, 339–41, 346, 350, 351 354, 415
Lockheed, 56, 107, 274, 151, 360–62, 372, 376, 427, 433, 451, 454
Lockheed Martin, 21, 23, 93, 95–99, 139, 281, 220, 354, 459
Lockheed Martin Astronautics, 96, 98, 172, 173, 180, 181
Lockheed Martin Astronautics Corporation, 165
Lockheed Martin Commercial Launch Services, 98
Lockheed Missiles and Space Company, 364, 370, 383
Lockheed Missiles and Space Division, 61, 241, 246, 264
Lockheed Propulsion Company, 272–76, 278, 280, 281
Lockheed Skunk Works, 449
Loki rocket, 190, 191
Long-Duration Exposure Facility (LDEF), 398, 405
Long Tank, 123, 124
Low, George M., 14, 364, 375, 418
Lowry, Dr. Dean, 254
LTV Aerospace and Defense Company, 205
Lucas, William R., 421, 422
Lundin, Bruce, 351
Lunney, Glynn, 484
Lusser, Robert, 472

Malfunction Detection System (MDS), 162

Malina, Frank J., 39, 40, 44, 230, 233

Manhattan Project, 260, 301, 338

Mann, D., 260

Manned Orbiting Laboratory (MOL), 140, 167–69, 448, 479–81

Manned Spacecraft Center (MSC), 318, 322, 360, 366, 367, 369, 370

Manned Space Flight, 162

Mariner, 9, 88, 90, 415

Marion Power Shovel Company, 322

Mark, Dr. Hans, 482

Mars Global Surveyor, 133

Mars Pathfinder, 133, 136

Mars Surveyor program, 133, 136

Mars Viking landers, 60

Marshall Space Flight Center (MSFC), 305, 314–18, 322, 330, 335, 343, 345, 348, 352, 354, 360, 362, 365, 373, 377, 415, 422, 493; Engine Program Office, 314; Light and Medium Launch Vehicles Office, 343

Martin Marietta, 60, 72, 76, 99, 281, 358–62, 369, 377, 404, 433, 449, 450, 460

Martin Marietta Astronautics Group, 164, 166, 170

Martin Marietta Launch Systems Company, 173

Massachusetts Institute of Technology (MIT), 40, 153, 241, 247, 248

Mathematica, 364, 376

McCarthy, John F., 422

McCartney, Lt. Gen. Forrest S., 485

McCormick, Richard M., 491

McDonnell Aircraft Company, 360, 446

McDonnell Douglas, 125, 127, 129, 159, 361, 362, 369, 371, 404, 451, 454, 455, 457, 460

McDonnell Douglas Astronautics Corporation (MDAC), 130, 132, 139, 141

McDonnell Douglas Delta Launch Services, 403

McDonnell Douglas/Martin Marietta, 372, 376

McElroy, Sec. of Def. Neil H., 56, 195, 250, 251

McMillan, Dr. Frank M., 232

McNamara, Sec. of Def. Robert S., 167, 448, 476, 478, 479

Mead, Larry, 371

Medaris, Maj. Gen. John B., 192, 195

Medium Launch Vehicle (MLV) II program, 92

Mellon Institute, 274

Mercury-Atlas D, 9, 86–88

Mercury program, 190

Microcosm Sprite, 222

MicroSat Launch Systems, 220

MicroSat satellite, 22

MIDAS satellite programs, 88

Millikin, Robert A., 39

Milstar satellites, 171

Miniature Sensor Technology Integration (MSTI), 211

Minitrack system, 50

Minotaur, 221, 283–85

Minuteman ICBM, 56, 81, 84, 159, 201, 213, 219–21, 229, 248–300, 338

Mir space station, 407

Mission Control Center (MCC), 389

missions, 46, 52, 58–60, 88–91, 107, 110–12, 114, 116–18, 120–23, 127, 132–34, 136, 153, 154, 162, 164–68, 171, 172, 177–80, 190, 193, 196, 197, 199, 201–5, 208–13, 216, 282, 285, 310–12, 320–24, 352, 377, 379–87, 415, 416, 428–34

Mississippi Test Facility (MTF), 379

modern warfare, 52–58

Moore, Jesse, 416, 431, 432

Moorman, Lt. Gen. Thomas S., 457, 489, 490

Moorman Report, 457

Morton Thiokol, 377, 381

Moscow Group for the Study of Reactive Motion (MosGIRD), 39

MR-2, 58

Mrazek, Willie, 345

Mueller, George E., 162, 318

Multiple Experiment Transporter to Earth Orbit and Return (METEOR), 461, 462

Multi-Service Launch System (MSLS), 220

M.W. Kellogg Company, 260

MX-774, 72, 76

Myers, Dale D., 15

Myers, Gen. Richard B., 492

NASA (National Aeronautics and Space Administration), 9–12, 14, 15, 18, 20–23, 25, 31, 37, 54, 55, 57–61, 63, 88, 103, 108, 112, 114, 117, 121, 123, 128, 129, 131, 133, 136, 141, 159, 162, 166, 167, 170, 171, 196, 201, 205, 208, 211, 212,

215, 221, 238, 266, 268, 273, 276, 280, 285, 302–5, 308, 313, 315, 316, 318, 319, 322, 324–26, 328–30, 334–36, 340, 342–44, 347, 348, 350, 351, 354, 358–64, 366, 370, 371, 374–77, 379, 384, 385, 395–405, 407, 408, 415, 416, 418–20, 422–24, 428, 429, 431–34, 436, 437, 448, 449, 450, 453–57, 458–60, 462
NASA–Air Force Space Transportation System Committee, 481
NASA Great Observatories Program, 136
NASA Historical Reference Collection, 64
National Academy of Sciences, 40, 230
National Advisory Committee for Aeronautics (NACA), 49, 204, 205, 340, 446, 448, 471, 473, 475
National Aerospace Plane (NASP), 22, 23, 451–53, 462, 487, 488
National Aerospace Plane Materials and Structures Augmentation Program (NASP MASAP), 453
National Defense Authorization Act, 489
National Defense Research Committee, Explosives Research Laboratory, 259
National Defense Research Council, 149
National Launch System, 488
National Launch Vehicle Program, 477
National Oceanic and Atmospheric Administration (NOAA), 123, 164, 165, 401, 404
National Reconnaissance Office (NRO), 170, 283
National Security Council, 239, 485
National Space Council, 489
National Space Launch Strategy, 489
National Space Program, 479
National Space Transportation Laboratory (NSTL), 379–81
National Space Transportation Policy, 20, 220, 490
National Space Transportation Strategy, 457
National Space Transportation System (NSTS), 92, 396
National Space Vehicle Program, 108
National Synthetic Rubber Corporation, 255
National Traffic Safety Board (NTSB), 401
Navaho, 48, 53, 190, 313
Naval Ordnance Test Station (NOTS), 197, 198, 208, 214, 232, 233, 241, 246, 264

Naval Research Laboratory (NRL), 46, 50, 51
Naval Space Command, 490
Naval Weapons Laboratory, 274
NAVSTAR, 131, 132
Near Earth Asteroid Rendezvous (NEAR), 132
NEAR Shoemaker, 132, 133
Neumann, Prof. John von, 75, 150
Newell, Homer E., 334, 350
Newport News Shipbuilding Corporation, 276
New York Times, 35, 428
Nieberding, Joe, 438
Nike missile, 259
Nike Zeus antiballistic missile system, 81
Nitropolymer Program, 244
Nixon, Pres. Richard, 14, 30, 363, 375, 418, 434, 480, 481
Nixon administration, 480
nomenclature, 105, 125, 201, 208, 209, 375, 446
North American Aviation, 48, 60, 72, 73, 274, 315–17, 238; Autonetics Division, 252, 256, 262
North American Rockwell, 358–62, 369, 372, 376, 377
North Atlantic Treaty Organization (NATO), 49, 251
Northrop Grumman, 461
Northrop, 72
NOTSNIK, 197, 198, 208, 214
Nova booster, 108, 313
Nuclear Engine for Rocket Vehicle Application (NERVA) program, 61, 62
nuclear propulsion, 60–63

Oak Ridge National Laboratory, 61
Oberth, Hermann, 34, 38, 63, 303, 335–37, 472
Office of Commercial Space Transportation (OCST), 404, 460, 462
Office of Management and Budget (OMB), 364, 371, 375, 376
Office of Manned Space Flight, 167, 318
Office of Naval Research (ONR), 189, 190, 244, 245
Office of Scientific Research and Development, 237
Office of the United States Trade Representative (USTR), 403

Ohio State University, 244
Operation Desert Shield, 486
Operation Desert Storm, 486
Operation Dominic, 248
Operation Dominic-Fishbowl, 116
ORBCOMM communications satellites, 215
Orbital Express, 220
Orbital Sciences Corporation (OSC), 214, 220, 458
Orbital/Suborbital Program (OSP), 220, 221, 283
Orbiter Processing Facility (OPF), 385, 392, 393
Orbiting Solar Observatories (OSO), 114, 115, 118, 121, 123
Orient Express, 22

Pacific American Launch Systems, 219
Packard, David, 480
Paine, Thomas O., 364, 481
Palapa, 396, 404
Parette, R., 245
Parsons, John W., 230, 233
Pathfinder, 459, 461
Patrick, Joseph C., 230
Payload Assist Module (PAM), 131
Payton, Gary E., 457, 492, 493
Peacekeeper missile, 240
Pegasus launch vehicle, 186, 198, 214, 216, 220, 221, 222, 283, 285, 375
Percheron rocket, 218
Perry, Robert, 55
Perumtel, 396, 397
Phillips, Maj. Gen. Samuel C., 251, 253, 317, 318, 477, 485
Phillips Petroleum, 243; Rocket Fuel Division, 249, 263
Phoebus reactor, 62
PILOT, 449
Pilot Run Item Master Schedule Committee (PRIMISCO), 258
Pioneer, 88, 121, 123, 199, 416, 422
Pioneer 3, 52
Pioneer 4, 52
Pioneer Rocketplane, 31, 221, 461
Platform Internationals, 31
Pohl, Robert, 339
Polar, 136
Polaris missile, 56, 57, 238–50, 258, 260–62, 264
Polter, Tom, 272

Pournelle, Jerry, 454
Pratt & Whitney, 93, 138, 273, 340, 342, 347, 348, 349, 423, 429
Presidential Commission, 401, 402
President's Science Advisory Committee (PSAC), 363
PRIME, 449
Project Apollo, 39, 475, 477, 478
Project Argus, 197
Project Bumper, 187, 188
Project Galileo, 428–30, 432–34
Project Gemini, 10, 11, 159, 162, 164, 168, 301, 477
Project Hermes, 46, 48
Project Mercury, 9, 58, 80, 84, 86, 159, 162, 301, 477
Project MX-774, 53
Project Orbiter, 191
Project Paperclip, 339
Project Pilot, 197, 198, 214
Project Pluto, 61
Project RAND, 74, 150
Project Rover, 61, 62
Project Suntan, 340
Project Ulysses, 429, 432, 434, 435
Project Vanguard, 50, 51, 103
Proton launch vehicle, 21
Purdue University, 244
Putt, Lt. Gen. Donald L., 268, 471

Quayle, Vice Pres. J. Danforth, 454, 488, 489

R-7 ICBM, 8, 9
Ramo-Wooldridge Corporation, 75, 104, 249–51, 338; Space Technology Laboratories, 53
Ranger space probe, 9
Rapid Access, Small Cargo, Affordable Launch (RASCAL), 222
Rayborn, Adm. William F., 247, 249
RCA, 396
RCA Global Communications (Globcom), 128–30
RCA Satcom, 129, 130
Reaction Motors, Inc., 38, 72, 189
Reactor-in-Flight Tests (RIFT), 61, 62
Reagan, Pres. Ronald, 16, 18, 30, 401, 404, 423, 452, 484
Reagan administration, 16, 22, 170, 404, 419, 484, 487

Reconnaissance satellite program, 473
Recoverable Orbital Launch System
  (ROLS), 445
Redstone, 48, 49, 52, 58, 190, 195, 196, 213
Redstone Arsenal, 48, 51, 71, 104, 105, 189,
  233, 235, 237
redundant inertial flight control assembly
  (RIFCA), 132
Regulus II, 57
Relay, 117, 118, 119
reusability, 417, 418, 428, 443–64
Reusable Aerodynamic Space Vehicle
  (RASV), 448–51
Reusable Centaur Study, 419
reusable launch vehicle (RLV), 18, 21–25,
  99, 360, 371, 416, 458, 463, 490, 491
reusable space launch vehicle (RSLV), 358
Reynolds Aluminum, 255
Ride, Sally K., 402
Ritchey, Harold W., 272
Robbins, William H. "Red," 424, 425, 434
Robert Heller and Associates, 344
Roberts, Dr. Ernest R., 272
RoBo, 446, 448
Rocket Development Corporation, 221
Rocketdyne, 48, 71, 74, 104, 105, 128, 190,
  198, 243, 256, 263, 305, 313, 315, 328,
  340, 377, 379
Rocket Propellant Information Agency
  (RPIA), 266
Rocket Propulsion Engineering Company,
  222
Rocket Research and Development Center,
  233
Rocket Spaceplane, 445
Rocket Systems Launch Program, 283
Rockwell International, 384–86, 454, 455,
  458; Autonetics Strategic Systems
  Division, 262, 265
Rogers, William P., 402
Rosen, Milton W., 108, 167, 192, 193, 342,
  343
Ross, Larry, 423, 429, 430, 435
Rotary Rocket Company, 31, 461
Roton, 461
RTV-A-2 HIROC, 72
Rumbel, Keith, 241, 242, 260
Russian engines, 93, 99

Sänger, Eugen, 38, 39, 64, 359, 379, 417,
  445

Sapolski, Harvey, 249
satellite programs, 88
satellites, 51, 111, 112, 114, 116, 119, 120,
  122, 123, 128, 132, 138, 141, 180, 190,
  191, 193, 199, 209–12, 215, 280, 285,
  286, 395–99, 403–6, 482, 487
Saturn I, 10, 108
Saturn IB, 11, 12, 14
Saturn IV, 10
Saturn V, 10, 12, 13, 15, 19, 30, 61, 301–33
Saturn launch operations, 319–24
Saturn launch vehicles, 10, 302–13, 358,
  359
Saturn-N, 61, 62
Saturn origins, 302–5
Saturn program, 167, 304
Saturn S-IC, 61
Saturn S-II, 61
SBS, 396, 404
Schirra, Walter, 11
Schriever, Maj. Gen. Bernard A., 150, 243,
  249, 251, 338, 471
Schultz, George, 375
Science Dawn, 451
SCORE communications, 54
Scorpius launch vehicle, 222
Scout small launch vehicle, 51, 57
SCT, 396
Sea Launch, 21
Seaberg, Lt. Col. John, 342
Seaborg, Dr. Glen T., 62
Seamans, Robert, 315, 364, 365, 479–81
Sergeant, 191, 195, 201, 237, 238, 242
Sergeant Contractor Selection Committee,
  237
Sergeant test vehicle, 235, 236
Shafer, John I., 232
Shavit launch vehicle, 221
Shell Chemical Company, 232, 255
Shepard, Lt. Cmdr. Alan B., Jr., 59, 112, 190
Shoemaker, Eugene, 132
Shuey, Dr. Henry M., 259
Shuttle/Centaur, 415–42
Shuttle/Centaur Project Office, 424
Shuttle Payload Integration Facility, 431
Silbervogel (Silverbird), 359, 379, 445
Silverstein, Abe, 303, 304, 334, 340, 351,
  415
Silverstein Committee, 303, 305, 315
single-stage-to-orbit (SSTO), 22, 23, 369,
  453, 454

Skylab, 12, 324, 325, 398
Skylark, 375
Sloan, Dr. Arthur, 241, 260
Sloop, John, 336, 340
SLV-3 (standardized launch vehicle), 86
Small Explorer program, 211
small launch vehicles, 186–228
Smith, Capt. Levering, 241, 247, 259
Smith, Dick G., 421
Smithsonian Institution, 35
Sojourner, 133, 136
Solid Controlled Orbital Utility Test
    (Scout), 204–13
Solid Propellant Information Agency
    (SPIA), 266
Solid Rocket Booster (SRB) [shuttle], 278–
    81
solid-rocket launch technology, 229–300
Solid Rocket Motor (SRM) [Titan], 180
Solid Rocket Motor Upgrade (SRMU), 172,
    173, 281
Space Access, 31
Space and Information Systems Division,
    317
Space and Missile Systems Center, 98
Space and Missile Systems Division, 281
Space and Missiles System Organization,
    451
Space Clipper, 375
Spacecraft Technology and Advanced
    Reentry Tests (START), 449
Space Development Force, 477
Space Exploration Initiative (SEI), 63
Space Flight Development Programs, 334
Space Infrared Telescope Facility, 136
Spacelab, 376, 398, 407
Spacelab/Spacehab, 407
Space Launch Initiative (SLI), 459, 463–64
Space Launch Modernization Plan, 490
Space Maneuver Vehicle (SMV), 460
Space Nuclear Propulsion Office (SNPO),
    61
Space Services, Inc. (SSI), 218, 219
Space Shuttle Task Group (SSTG), 362
Space Shuttle User Committee, 481
Space Shuttle, 12, 15, 16, 18, 21–23, 25, 30,
    41, 57, 58, 62, 92, 95, 131, 132, 140, 168,
    171, 229, 237, 268, 278, 281, 282, 325,
    329, 330, 354, 357–442, 450, 459, 480–
    86
Space Task Group, 159, 480, 362

Space Technology Laboratories (STL)
    (Ramo-Wooldridge), 244, 251–53
Space Technology Program, 204
Space Transportation System (STS), 363,
    364, 421, 480
space walk. *See* extravehicular activity
Sparks, Brian, 351
Special Projects Office, 241, 247, 249, 260
Sperry Gyroscope Company, 237, 238
Sputnik I, 7, 9, 49, 51, 56, 78, 116, 193, 195,
    197, 340, 417, 471
Sputnik 2, 79
Sputnik 3, 78
Stanford Research Institute, 272
Starbuck, Brig, Gen. Randy, 180
StarClipper, 449
Starfish Prime event, 116
Starhunter Corporation, 222
Stennis Space Center, 379
Stewart Committee, 191, 192
Stewart, Homer, 191
Stofan, Andy, 423
Storms, Harrison, 317
Strategic Air Command (SAC), 54, 213,
    251, 254, 451
Strategic Arms Reduction Treaty (START),
    283
Strategic Defense Initiative Organization
    (SDIO), 220, 404, 453, 454, 455, 457,
    485, 488
Strategic Missile Squadron, 154
Straza Industries, 257
STS Committee, 364
Stuhlinger, Ernst, 472
Subcommittee on Space Sciences, 334
submarine launched ballistic missile
    (SLBM), 225
Summa Technology, 221
Sun Shipbuilding and Dry-dock, 277
Super Six, 123, 124
Surveyor, 90, 352, 415, 422
Syncom, 119, 120, 128
Szabo, Steven B., 425

tactical ballistic missile (TBM), 103, 104
Talbot, Harold E., 150, 151
Talos missile, 258
Teapot Committee (Air Force Strategic
    Missiles Evaluation Committee), 53, 75,
    149, 338
Technical Integration Office, 436

Technological Capabilities Panel, 239
Teledyne Industries, 423, 429
Telesat, 128, 396
Telestar, 116, 117
Terhune, Col. Charles H., Jr., 250
Terrier missile, 259
testing, 46, 48, 50, 51, 57, 58, 61, 73, 77,
    79–81, 95–97, 105, 106, 138, 152, 153,
    157, 159, 164, 168, 169, 172, 175, 193,
    196, 198, 205, 212, 213, 218, 235, 236,
    239, 240, 242, 244, 246, 248, 252–54,
    262, 267, 271, 274–77, 306, 317–19, 324,
    346–48, 377, 379–87, 430, 431, 437, 448,
    449, 451, 455, 458, 461, 462
Thackwell, H. Lawrence, Jr., 232
The Aerospace Corporation, 281, 362, 364,
    454, 491
Theil, Walter, 339
Thiel, Dr. Adolph K., 104, 105
Thiokol Chemical Corporation, 201, 204,
    230–37, 243, 244, 251, 253–55, 257, 258,
    263, 272–78, 280
Thompson, Craig, 415
Thompson, David, 214
Thompson, L.T.E., 247
Thompson Ramo Wooldridge (TRW), 251,
    264, 267, 276, 277, 433
Thor-Able, 108, 109
Thor IRBM, 2, 7, 10, 53, 54, 56, 70, 103,
    105, 128, 198–201, 239, 243, 247, 249
Thor launch vehicle, 2, 10, 194
Thor program, 103
Three Axis Reference System (TARS), 162
Thrust Augmented Improved Delta (TAID),
    120, 121
Thrust-Augmented Thor (TAT), 120
Thunderbird test vehicle, 232–34
Thurber Space Systems, 222
Tiny Tim, 187
Tiros (Television and InfraRed Observation
    Satellite), 107, 112, 114, 121, 199
Titan, 2, 12, 15, 18, 21, 42, 147, 153, 359,
    404, 406, 407
Titan I, 7, 10, 53, 54, 56, 70, 76, 84, 147,
    151, 152, 199
Titan II, 55, 59, 147, 148, 154–62, 164, 338,
    484
Titan III, 11, 14, 57, 59, 60, 70, 148, 167,
    229, 237, 268, 272, 273, 484
Titan IV, 19, 21, 30, 57, 148, 171, 172, 229,
    237, 268, 283

Titan solid rocket motors, 267–78
Toftoy, Maj. Gen. Holger, 472
Tracking and Data Relay Satellite (TDRS),
    398, 405, 425, 434
Trailblazer program, 459–60
TransAtmospheric Vehicle (TAV), 451–52,
    487
Transit navigation satellites, 209
Transpace Carriers, 403
Truax, Cmdr. Robert, 104, 105
Truly, Adm. Richard H., 485, 489
Truman administration, 73
Tsien, Hsue Shen, 39
Tsiolkovsky, Konstantin Eduardovich, 33,
    34, 60, 63, 303
Twining, Gen. Nathan F., 150
two-stage-to-orbit (TSTO), 359, 369

United Aircraft Corporation, 268
United Nations, 40
United Research Corporation, 251, 268
United Space Alliance (USA), 21
United Space Boosters Inc. (USBI), 278
United States: Air Force, 7, 11, 20, 22, 30,
    53–57, 59–61, 70–74, 76, 77, 84, 88, 92,
    93, 96, 98, 103, 104, 121, 131, 132, 138–
    40, 147, 150, 151, 154, 159, 166, 167,
    171, 173, 175, 178, 180, 191, 198, 205,
    211, 212, 215, 236, 243, 244, 246–50,
    256, 258, 263, 268, 270, 276, 283, 285,
    303, 322, 335, 337, 338, 340, 342, 343,
    348, 350, 358, 359, 365–67, 386, 400,
    415, 419, 429, 431, 433, 445–48, 451,
    452, 453, 454, 459, 460, 470, 471, 474–
    83, 485–88, 490, 491, 493; —, Space
    Division, 362; Air Force Air Research
    and Development Command (ARDC),
    151, 342, 475; Air Force Air Technical
    Services Command, 337; Air Force
    Ballistic Missile Division, 104; Air Force
    Flight Dynamics Laboratory (FDL), 360,
    362, 367, 370, 384, 386, 449; Air Force
    Materiel Command (AFMC), 446, 490,
    492; Air Force Propulsion Group, 242;
    Air Force Rocket Propulsion Laboratory
    (AFRPL), 273, 275, 276; Air Force
    Scientific Advisory Board (SAB), 238,
    484; Air Force Space Command, 20,
    470, 489–92; Air Force Strategic
    Missiles Evaluation Committee (Teapot
    Committee), 53, 75, 149, 338; Air Force

Systems Command (AFSC), 19, 20, 274, 482, 483; Air Force Wright Air Development Center (WADC), 446; Army, 38, 44, 46, 48, 51, 54, 71, 75, 191, 195, 196, 198, 233–35, 238, 241, 266, 335, 474, 475; Army Air Corps, 40, 230; Army Air Forces, 44, 48, 71, 149; Army Air Technical Services Command, 71; Army Ballistic Missile Agency (ABMA), 10, 50, 51, 189, 198, 205, 302, 304, 339, 473; Army Materials Research Agency, 274; Army Ordnance, 71, 232, 233, 235, 237; Army Signal Corps, 79; Coast Guard, 400; Congress, 16, 73, 93, 168, 170, 175, 210, 334, 346, 348, 351, 354, 456, 462, 475, 489; Department of Commerce, 490, 491; Department of Defense (DoD)), 16, 18–21, 31, 37, 49–51, 54, 58, 147, 149, 165–67, 170, 171, 175, 181, 187, 191, 192, 205, 210, 222, 223, 248, 250, 260, 266, 273, 276, 303, 363–65, 376, 395, 398, 401, 404, 405, 424, 429, 471, 473, 475, 477–85, 487–91; Department of Transportation (DoT), 460, 490, 491; House of Representatives, 31; Navy, 48, 54, 56, 75, 104, 187, 191, 192, 197–99, 201, 204, 205, 209, 210, 230, 232, 237, 238, 244, 246–49, 260, 264, 266, 285, 316, 322, 396, 400, 474, 475, 488, 490, 492, 493; Navy Bureau of Aeronautics, 233, 234, 241; Navy Bureau of Ordnance, 189; Space Command, 490; State Department, 221
United States Steel, 274
United Technologies Corporation (UTC), 121, 166, 251, 268, 272, 273, 278
United Technologies Corporation, Chemical Systems Division (CSD), 268
United Technology Center, 59, 268
Universal Space Lines, 221
University of California at Berkeley, 269
University of Tennessee, 462
Upper Atmosphere Rocket Research panel, 46
USAF-NASA Shuttle Coordination Board, 364
USS *Ethan Allen*, 248
USS *George Washington*, 57, 248

V-2 (Vengeance Weapon) (A-4), 42–48, 72, 73, 187–89, 235, 302, 445

Valier, Max, 38
Van Allen, James, 46
Van Allen radiation belts, 51, 107
Vanguard, 49–52, 192–95, 217, 223, 261
Vanguard satellite, 7, 191–94
Vega, 108
VentureStar, 23
Verein für Raumschiffarht (VfR), 38
Vertical Assembly Building (VAB), 322, 324, 387
Vertical Integration Facility (VIF), 96
Vicar rocket, 232
Viking, 46, 49, 50, 168, 170, 189, 192, 193, 197, 415, 422, 430
Viking 1, 46
Von Braun, Wernher, 10, 29, 34, 38, 42, 44, 45, 48, 50, 52, 53, 63, 64, 71, 187, 189, 190, 195, 302–4, 315, 325, 335, 339, 344, 348, 350, 351, 354, 359, 417, 446, 472, 473
Von Neumann, John, 53
Vought Astronautics, 205
Voyager, 168, 170, 416, 422, 430

WAC Corporal, 44, 46, 47, 187, 188
Washington Evening Star, 351
Waste King Corporation, 196
Webb, James E., 315, 326, 336, 477
Weinberger, Sec. of Def. Caspar, 484
Westar, 129, 396
Western Development Division (WDD), 53, 54, 150, 198, 243, 249, 338, 471
Western Electrochemical Company, 234
Western Union, 129, 397
Westinghouse Electric Corporation, 61
Weyers, Vernon, 431, 435
White, Edward H., II, 11, 164
White, Gen. Thomas D., 470, 471, 475, 476, 487
White House Office of Science and Technology Policy (OSTP), 457, 458
White Whizzer, 232, 233
Wickman & Propulsion, 222
Wide Body Centaur, 421, 423
Wilcox, Dr. Howard, 197
Wilson, Andrew, 208
Wilson, Charles E., 54, 149, 198
Wilson, T.A., 451
Wilson Memorandum, 198
WIND, 133
World War II, 148, 149

Wuerth, J.M., 265

X-33 program, 21
X-34 program, 21
X-37 program, 21
XLR43, 48
XLR83, 48
Xperimental Operational Program, 454

Yardley, John, 380, 421
Yeager, Brig. Gen. Charles, 402
Young, John W., 11, 164, 435
Young, Richard E., 260, 261, 266
Young Development Laboratories, 260
Yuzhnoye, 21

Zenit rocket, 21
Zuckert, Eugene M., 477